621.381 C532e FV
CHIANG
ELECTRICAL AND ELECTRONIC
INSTRUMENTATION
37.50

Electrical
and Electronic
Instrumentation

ELECTRICAL
AND ELECTRONIC
INSTRUMENTATION

HAI HUNG CHIANG
Biomedical Engineering Department
Chung Yuan Christian University
Taiwan, Republic of China

A WILEY-INTERSCIENCE PUBLICATION
JOHN WILEY & SONS
New York   Chichester   Brisbane   Toronto   Singapore

Library of Congress Cataloging in Publication Data:

Chiang, Hai Hung.
   Electrical and electronic instrumentation.

   "A Wiley-Interscience publication."
   Bibliography: p.
   Includes index.
   1. Electronic instruments.   2. Electric meters.
I. Title.

TK7878.4.C48   1984   681'.2   83-21742
ISBN 0-471-89624-1

Printed in the United States of America

10 9 8 7 6 5 4 3 2 1

Preface

The purpose of this book is to illustrate the basic theory and common-core concepts of electrical and electronic measuring instruments. An effort has been made to emphasize the complete circuits of electronic instruments so that the practical contemporary circuit-design technique will hopefully be more transparent.

This book is derived from my past 30-year working (as an engineer) and teaching (as a professor) experiences. It covers a systematic sequence of various fundamental topics and practical applications, and should provide sufficient background for those who intend to work in the electrical and electronic instrumentation area.

The first two chapters cover basic electrical instruments and meter movements. Chapters 3–5 discuss various bridges, potentiometers, and Q-meters. Chapters 6 and 7 outline the semiconductor devices, digital systems, and transducers. Chapter 8 gives a general description of various types of oscilloscopes. Chapter 9 analyzes the circuits of solid-state electronic voltmeters and describes the multimeter systems. Chapters 10 and 11 discuss oscillators, signal generators, comparators, and function and pulse generators (including the complete circuits of some typical pulse generators). Chapter 12 analyzes the circuits of the telemetering transmitters and receivers. Chapter 13 analyzes the circuit of a typical triggered-sweep dual-trace oscilloscope. Chapter 14 presents the digital multimeter design and Chapter 15 provides an introduction to a TV terminal using a microprocessor. Problems are included at the end of each chapter. A solution manual will be available upon request from the publisher.

Finally, I would like to express appreciation for the assistance given by Professors Hong-Shong Chang, Biing-Nan Hung, M. T. Chen, and J. K. Lin.

HAI HUNG CHIANG

Taiwan, Republic of China
March 1984

Contents

Electrical
and Electronic
Instrumentation

1

Basic
Electric Instruments

1.1 INTRODUCTION TO METER MOVEMENTS

Almost all meters used for measuring electric signals employ current detection as their basic indicating mode. An ideal ammeter, which is used to measure the current flowing in an electric circuit, presents a short circuit to the loop in which it is connected. An ideal voltmeter, which is used to measure the potential difference (voltage) between a pair of nodes of an electric circuit, presents an open circuit to the node pair. In either of these cases the circuit voltages and currents remain unchanged when the meter is connected.

A meter movement is the electromechanical device that produces a mechanical motion of an indicator in response to an applied electric signal; this motion is caused by the interaction of magnetic or electric fields generated at least in part by the current or the voltage to be measured. Four types of meter movements are in common use for electrical instrumentation and will be described briefly.

The most common meter movement is the *permanent magnet–moving coil* (*PM–MC*) type, often called the d'Arsonval or, more recently, the Weston movement. The PM–MC assembly is operated by the reaction between the current in a movable coil and the field of a fixed permanent magnet.

The *electrodynamometer movement* consists of a fixed coil and a movable coil. The current to be measured flows in both coils, and the interaction of the resulting magnetic fields produces a torque. Since the current flows in both coils, the torque is proportional to the square of the current, and thus the dynamometer is also a true root mean square (rms) movement.

The *moving iron vane movement* consists of a fixed coil and two iron

1

vanes inside the coil. One vane is rigidly attached to the coil frame and the other is connected to the instrument shaft, which can rotate freely. The torque in a moving iron vane movement is produced by the reaction of the magnetic field due to the current coil and the magnetic field due to the current induced in the iron.

The *electrostatic movement* is the only type that employs the forces due to electric fields and charges, and is the only movement that measures voltage directly rather than by the effect of a voltage-produced current. The electrostatic mechanism resembles a variable capacitor, where the attraction between fixed and movable plates is balanced by a restoring force.

1.2   TORQUE OF PM–MC METER MOVEMENTS

In the PM–MC meter movement, the pointer, attached to the coil, moves over a graduated scale and indicates the angular deflection of the coil and hence the current through the coil. The pointer deflection is accomplished by suspending the coil carrying the current to be measured in a steady magnetic field produced by the fixed permanent magnet. The torque (twist force) produced by the interaction of the current-produced field and that of the permanent magnet is opposed by the restoring spring. Thus, as the current in the moving coil changes, so does the steady-state angular position of the coil. The torque and motion relationships are shown in Fig. 1.2-1*a*. The equation for the developed torque, derived from the basic law for electromagnetic torque, is

$$\tau = (\text{force})(\text{arm}) = (NBIL)(W \cos \theta)$$

or

$$\tau = NBIA, \tag{1.2-1}$$

where τ = torque in units of newton-meters (N·m),
   N = number of turns on coil,
   B = flux density in the air gap in webers per square meter (Wb/m^2),
   I = current in moving coil in amperes (A),
   A = effective coil area in square meters (m^2).
If τ is in dyne·cm (1 dyne = 10^{-5} N, 1 cm = 10^{-2} m), B is in gauss (1 G = 1 line/cm^2 = 10^{-4} Wb/m^2, 1 Wb = 10^8 lines), I is in amperes, and A in square centimeters, then the torque expressed in the cgs system is

$$\tau = \frac{NBIA}{10}. \tag{1.2-2}$$

A typical panel PM–MC instrument, with a 9-cm case, a 1-mA range, and full-scale deflection of 100 degrees of arc, would have the following char-

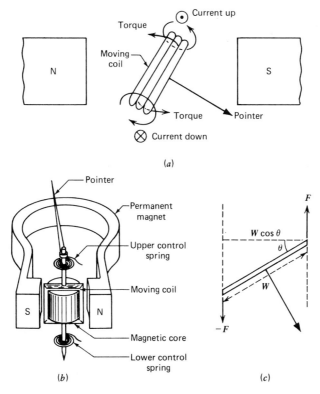

Figure 1.2-1 (a) Torque and motion relationships in a PM–MC meter movement; (b) magnet and coil of PM–MC movement; (c) force couple and arm.

acteristics:

$$\tau = 29.4 \text{ dyne} \cdot \text{cm},$$

$$N = 84 \text{ turns},$$

$$B = 2000 \text{ G},$$

$$A = 1.75 \text{ cm}^2,$$

$$\text{Coil resistance} = 88 \ \Omega,$$

$$\text{Power dissipation} = 88 \ \mu\text{W}.$$

If the permanent magnet field is uniform and the spring is linear, then the pointer deflection is also linear in coil current. Since an average torque can be produced only by a dc current, the PM–MC movement does not respond to ac current.

1.3 DC VOLTMETERS

A simple dc voltmeter can be constructed by placing a resistor or multiplier in series with a PM–MC movement and marking the dial scale to read the voltage across the resistor and movement. Figure 1.3-1 shows a practical arrangement of multiplier resistors with a range selector that switches the appropriate amount of resistance in series with the movement. Let R_s be the multiplier resistance and R_m be the moving coil resistance. Then the larger the sum of $R_s + R_m$, the larger the value of ohms/volt, or the higher the sensitivity of the voltmeter. The relationship between the sensitivity (S) and the full-scale current (I_{fs}) can be derived by referring to the circuit shown in Fig. 1.3-1b:

$$S = \frac{R_s + R_m}{V_{fs}} = \frac{V_{fs}/I_{fs}}{V_{fs}} = \frac{1}{I_{fs}},$$

where V_{fs} is the full-scale voltage, or

$$S \equiv \frac{1}{I_{fs}} \ \Omega/V. \tag{1.3-1}$$

Thus the sensitivity of a voltmeter is defined as the reciprocal of the full-scale current of the basic movement. This sensitivity has the dimensions of 1 divided by amperes or ohms per volt. The input resistance of a voltmeter is given by

$$R_i = R_s + R_m = \frac{V_{fs}}{I_{fs}} = SV_{fs}. \tag{1.3-2}$$

The voltmeter shown in Fig. 1.3-1a provides five dc voltage ranges. Since

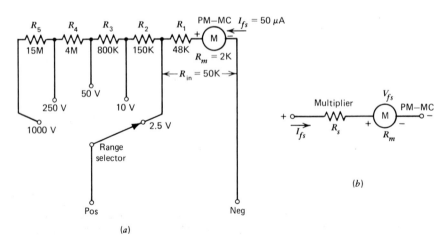

Figure 1.3-1 (a) A multirange voltmeter with a practical arrangement of multipliers; (b) a simple dc voltmeter.

the 50-μA movement has an internal resistance (R_m) of 2000 Ω, the voltmeter sensitivity is $1/(50 \times 10^{-6})$ or 20,000 Ω/V, and the total resistance ($=R_{in}$) of the instrument circuit is 50,000 Ω on its 2.5-V range. Thus the series resistor $R_1 = 50K - 2K = 48K$, and the voltmeter circuit draws 50 μA at full-scale indication:

$$I = \frac{2.5}{50,000} = 50\ \mu A.$$

Other series resistors on the remaining four ranges can be determined in the same way.

If the sensitivity of a voltmeter is high, the instrument circuit loading will be low, and the measurement error will be less.

1.4 DC AMMETERS

A dc ammeter consists of a PM–MC meter movement. The coil winding of a basic movement is small and light, and can carry only very small currents. It is thus necessary to extend the current range of dc ammeters. When high currents are to be measured, the main part of the current passes through a shunt, which is a resistor connected in parallel with the movement.

A multimeter or volt–ohm–milliammeter (VOM) provides several dc current ranges by means of a ring-shunt configuration, as shown in Fig. 1.4-1. The major advantage of a ring shunt is that the meter movement always remains shunted when the range selector is turned from one position to another. The movement is protected against accidental burnout when switching ranges with test leads connected into a "live" circuit.

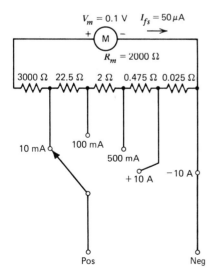

Figure 1.4-1 A multirange dc ammeter with a ring shunt.

Consider the operation of the milliammeter on its 10-mA range shown in Fig. 1.4-1. The input resistance of the circuit is:

$$R_T = (3000 + 2000) \parallel (22.5 + 2 + 0.475 + 0.025) \simeq 25 \ \Omega.$$

A 10-mA current produces a terminal voltage drop of

$$V_T \simeq 25 \times 0.01 = 0.25 \ \text{V}.$$

The voltage drop across the meter movement is

$$V_m \simeq \frac{0.25 \times 2000}{3000 + 2000} = 100 \ \text{mV}.$$

The current through the meter movement is

$$I_m \simeq \frac{0.25}{3000 + 2000} = 50 \ \mu\text{A} = I_{fs}.$$

Therefore, a current of 10 mA produces full-scale pointer deflection from a practical point of view.

1.5 OHMMETERS

The ohmmeter is based on the voltage divider network $R_1 - R_x$ shown in Fig. 1.5-1a. If R_1 is known, then measuring the voltage ratio V_1/V_0 will yield the unknown resistance R_x. In Fig. 1.5-1b, the voltage ratio is given by

$$\frac{V_1}{V_0} \simeq \frac{R_1}{R_1 + R_x} . \tag{1.5-1}$$

With the ohmmeter terminals short-circuited or $R_x = 0$, R_s is adjusted to give a reading of full scale (FS), which is 0 Ω. With R_x connected, the ohmmeter indicates the voltage ratio V_1/V_0 directly.

The ohmmeter is not accurate for higher values, which are crowded at the upper end of the scale. This can be seen in Fig. 1.5-2b where $R_x/R_1 = x$ and $V_1/V_0 = y$ are plotted.

In Fig. 1.5-1c, the actual input resistance is about 12 Ω, that is, the sum of 11.5 Ω and a small resistance in the test leads and cell circuit. If the test leads are connected across a 12-Ω resistor, the pointer will deflect to half scale, and the resistance indication will be 12 Ω. Or, if the test leads are connected across a 6-Ω resistor, the pointer will deflect to the 6-Ω point on the scale.

In order to use the same scale on all ranges of the ohmmeter, the circuit proportions must remain fixed on each range. For example, consider an ohmmeter operating on the $R \times 10K$ range, as depicted in Fig. 1.5-1d. Since

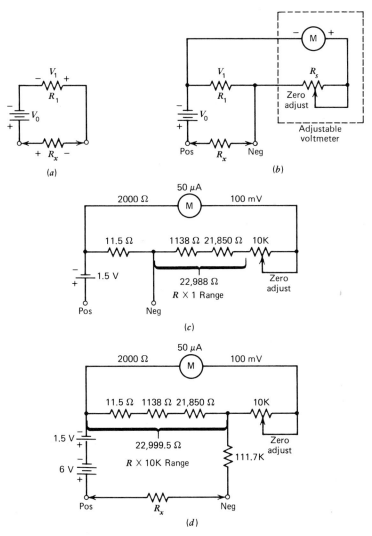

Figure 1.5-1 (a) Basic voltage-divider network; (b) simple ohmmeter circuit; (c) and (d) R ×
1 and R × 10K range ohmmeter sections of a Simpson 260 VOM. (Courtesy of Simpson Electric
Company.)

additional battery voltage is required on this high-resistance range, a 6-V
battery is automatically switched in series with the 1.5-V cell. Also, since
it is necessary to maintain a proportional resistance in the test lead and cell
circuit, a 117.7K resistance is included in series with one of the test leads.
If the test leads are connected across a 120K resistor, then half-scale deflection will result.

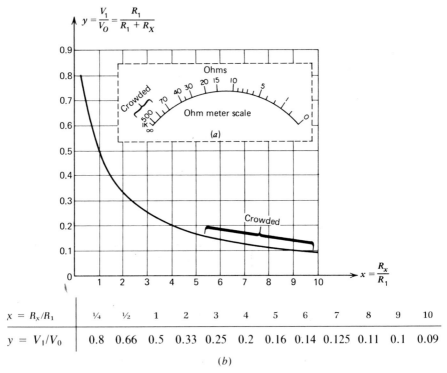

Figure 1.5-2 (a) Ohmmeter scale, showing the higher resistance values crowded at the upper end of the scale; (b) $x - y$ plot with $x = R_X/R_1$ and $y = V_1/V_0$.

1.6 PM–MC RECTIFIER-TYPE AC VOLTMETERS

Rectifier-type ac voltmeters generally use a PM–MC movement in combination with some chosen rectifier arrangement. A basic circuit of this type is shown in Fig. 1.6-1a. It must be emphasized that the reverse voltage across the rectifier D_1 be minimized whenever one chooses a value of R_s that is sufficient to cause the rectifier to operate within its reverse voltage rating. In practice, a protective diode D_2 is included in the circuit, as shown in Fig.

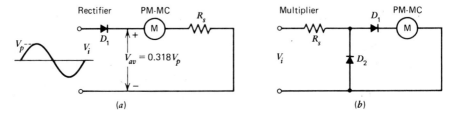

Figure 1.6-1 (a) Basic ac voltmeter of PM–MC rectifier type; (b) the rectifier-type ac voltmeter with a protective diode.

1.6-1*b*. Diode D_2 conducts on the reverse half-cycle of applied voltage and shunts a low value of forward resistance around diode D_1 and the meter.

The traditional instrumentation rectifier used to make dc meters read ac values is a copper oxide diode, which is suitable for use at ac frequencies well into the audio range; ac voltmeters that operate into the radio frequency (rf) range can be built from dc meter movements and silicon or germanium diodes. Typical types are 1N34 and 1N60 in Ge, and 1N914 and 1N4148 in Si. Instruments that are useful to over 500 MHz can be built with such diodes.

A typical VOM provides five ac voltage ranges, as shown in Fig. 1.6-2. This arrangement also includes two diodes, D_1 and D_2, for the purposes previously mentioned. The input resistance of this circuit on each of the ac voltage ranges is one-twentieth of the input resistance on a corresponding dc voltage range in the circuit of Fig. 1.3-1. Thus the sensitivity of the ac voltmeter is 20,000/20 or 1000 Ω/V. There are two basic reasons for this comparatively low sensitivity of the ac voltmeter:

1. The applied ac voltage is necessarily rectified, and the meter movement responds to 0.318 of peak value in this example. The ac voltmeter sensitivity is then reduced to 0.318 of its dc value, or 6360 (=0.318 × 20,000) Ω/V.

2. Rectification occurs on a highly nonlinear characteristic, and a swamping resistor R is employed to minimize "bunching" at the lower end of the ac scale. The result is further reduction in ac voltmeter sensitivity to 1000 Ω/V.

The PM–MC movement in Fig. 1.6-2 responds to 0.318 of peak value, whereas the ac voltage scales are calibrated to indicate 0.707 of peak value. If a dc voltage is applied in the forward direction to the ac voltmeter circuit, the pointer will indicate 0.707/0.318 or 2.22 times the value of the dc voltage.

The output-meter function of a VOM is similar to its ac voltage function except that a blocking capacitor is connected in series with the instrument circuit.

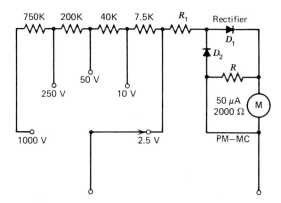

Figure 1.6-2   An ac voltmeter circuit with five ranges.

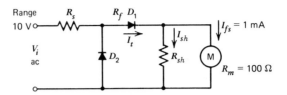

Figure 1.6-3 Example 1.6-1: the given ac voltmeter.

EXAMPLE 1.6-1 The ac voltmeter of Fig. 1.6-3 has a 10-V range and uses a 1-mA PM–MC movement with an internal resistance (R_m) of 100 Ω. The shunting resistance (R_{sh}) across the movement is 100 Ω. Diodes D_1 and D_2 have a forward resistance (R_f) of 400 Ω and are assumed to have infinite reverse resistance. Calculate (a) the value of multiplier R_s and (b) the sensitivity S of the ac voltmeter.

Solution. (a) The movement full-scale current $I_{fs} = 1$ mA and $R_m = R_{sh} = 100$ Ω; thus the current through R_{sh} is $I_{sh} = 1$ mA, and the total current is $I_t = I_{fs} + I_{sh} = 2$ mA. For H.W. rectification,

$$E_{dc} = 0.45E_{rms} = 0.45 \times 10 \text{ V}$$

$$= 4.5 \text{ V}.$$

Then, the total resistance of the current is

$$R_t = \frac{E_{dc}}{I_t} = \frac{4.5 \text{ V}}{2 \text{ mA}} = 2250 \text{ } \Omega.$$

Also,

$$R_t = R_s + R_f + \frac{R_m R_{sh}}{R_m + R_{sh}} \quad \text{or} \quad 2250 = R_s + 400 + 50.$$

Thus,

$$R_s = 2250 - 450 = 1800 \text{ } \Omega.$$

(b) On this 10-V ac range, $S = 2250 \text{ } \Omega/10 \text{ V} = 225 \text{ } \Omega/\text{V}$. Using the same movement, a dc voltmeter would have a sensitivity of $1/1$ mA or 1000 Ω/V.

1.7 THERMOCOUPLE INSTRUMENT

The thermocouple instrument shown in Fig. 1.7-1 can be used to measure ac and dc quantities. It consists of a PM–MC movement and a resistive-heating element AB coupled to a thermocouple. The thermocouple is a heat-to-voltage transducer which is made by joining together two dissimilar metals in a vee junction. When the junction C is heated, a voltage appears across terminals x and y. This phenomenon occurs because of a difference in the

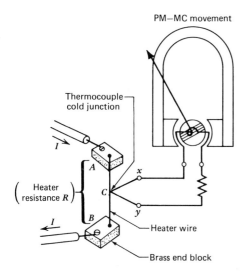

Figure 1.7-1   A basic thermocouple meter with heating element AB coupled to thermocouple xCy.

work function of the two dissimilar metals. The potential difference between x and y depends on the temperature of the so-called cold junction C. A rise in temperature causes an increase in the voltage, and this is used to advantage in the thermocouple. Heating element AB, which is in mechanical contact with the junction of the two metals at point C, forms part of the circuit in which the current is to be measured. ACB is called the hot junction. Heating energy produced by the current in the heating element raises the temperature of the cold junction C and causes an increase in the voltage developed across terminals x and y. The potential difference produces a dc current activating the PM–MC movement. Since the heating effect is proportional to I^2R, the resulting current which activates the PM–MC movement gives a true rms reading. (I is the current flowing in the heating element and R is the element's resistance.) Thus both the heat produced by the current and the temperature rise are all proportional to the square of the rms current.

Since the voltage between terminals x and y is linearly varied with the temperature at junction C only at the center portion of the voltage range, it is necessary to select a PM–MC movement which will deflect to the upper half of the voltage range for the current expected in the circuit.

The ac current being measured flows through heater resistance R and heats it, thereby causing the thermocouple to generate an output voltage. A voltmeter across the x-y terminals measures this millivolt-level potential, but is calibrated in amperes, or milliamperes; thus the measurement of this thermocouple ammeter represents the rms value of the current.

The chief advantage of a thermocouple meter is that its accuracy can be as high as 1%, up to frequencies of approximately 50 MHz. Current in the 0.5–20-A range and voltages up to 500 V can be measured with this instrument.

1.8 ACCURACY, ERRORS, AND PRECISION

Accuracy

The accuracy of a measurement denotes the extent to which we approach
a true or actual value. The actual value is not measurable, although we can
approximate this value by careful measurement. The accuracy is given by

$$\text{acc} = \pm \frac{\text{static error}}{\text{full scale}} \times 100\% = \pm \frac{|V_{\text{true}} - V_{\text{meas}}|}{V_{fs}} \times 100\%. \quad (1.8\text{-}1)$$

The static error is a difference of true value and measured value. The possible
percentage error is

$$\text{Percentage error} = \frac{|V_{\text{true}} - V_{\text{meas}}|}{V_{\text{true}}} \times 100\%. \quad (1.8\text{-}2)$$

Errors

Errors in measurement may be classified under the following headings: observation errors, estimation errors, systematic errors, and random errors
due to unknown causes. The observational error primarily results from the
fact that no two persons observe the same situation in exactly the same way
where small details are concerned. The cause of estimation error can be
understood from the simple example that various observers will disagree
concerning the exact position of the pointer over the dial of a meter movement. Errors of observation and errors of estimation are not arbitrary, but
tend to follow a law of normal (or gaussian) distribution, as shown by the
curve of Fig. 1.8-1. This law is formulated by

$$y = Ae^{-bx^2} \quad (1.8\text{-}3)$$

where y = number of times a certain value is observed,
      x = value,
   A,b = constants,
      e = 2.71828.

Notice that if a large number of observers read the scale indication as accurately as possible, the values they report will tend to cluster toward the
mean value depicted in Fig. 1.8-1.

Systematic error is explained as follows. An instrument has maximum
accuracy at standard temperature. If the room temperature changes substantially during the course of observing an indicated value, a systematic
error will occur. Observers who report at low ambient temperature will
disagree systematically with those who report at high ambient temperature.
In addition, various other sources of systematic error may be encounted.

Precision of Measurements

The precision of a measurement denotes its departure from the average of
a number of measured values. For example, suppose that we carefully meas-

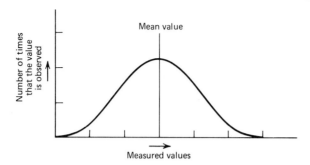

Figure 1.8-1   Normal distribution curve.

ure the terminal voltage of a dry cell six times and have six separate measured values: (1) 1.49; (2) 1.51; (3) 1.49; (4) 1.50; (5) 1.52; (6) 1.50. Then the sum of the measured value is 9.01 V, and the average is 9.01/6 or 1.50+ V. The first measurement has a precision of approximately 99.3%; this precision can also be stated as approximately −0.7% deviation from the mean. The fourth measurement has a precision of 100%, and so on.

Note that precision does not guarantee accuracy, although accuracy requires precision.

Accuracy Ratings of Instruments

The departure in indication of a particular voltmeter with respect to the established volt unit (the basic voltage standard) is stated as the accuracy of the voltmeter. Laboratory-type voltmeters have hand-drawn scales and thus provide comparatively high accuracy. Normally, this type has a rated accuracy of ±0.1% of full-scale reading at 25°C or an accuracy of ±0.25% at 15°C or 35°C. Most voltmeters have printed scales, and their rated accuracy is typically ±0.5% of full scale at 25°C. A utility-type voltmeter has an accuracy rating of ±3% of full scale at 25°C. The same accuracy considerations apply to ammeters and wattmeters.

EXAMPLE 1.7-1   A voltmeter with an accuracy of ±0.5% of full scale (at 25°C) is used on the 150-V scale to measure 100 V. (a) What is the possible error? (b) What is the range of the actual voltage readings if the scale indication is 100 V?

Solution.   (a) From Eq. (1.8-1), the possible error is

$$(\text{acc})_{V_{fs}} = \pm 0.005 \times 150 \text{ V} = \pm 0.75 \text{ V}.$$

(b) Also, from Eq. (1.8-1),

$$V_{\text{true}} = V_{\text{meas}} \pm (\text{acc})_{V_{fs}} = 100 \pm 0.75 \text{ V}.$$

Thus the scale indication of 100 V denotes the actual voltage value within the range from 99.25 to 100.75 V.

BIBLIOGRAPHY

Carr, J. J., *Elements of Electronic Instrumentation and Measurement*. Reston Publishing, Reston, Virginia, 1979, Chaps. 1 and 2.

Cooper, W. D., *Electronic Instrumentation and Measurement Techniques,* 2nd ed. Prentice-Hall, Englewood Cliffs, New Jersey, 1978, Chaps. 1 and 4.

Johnson, D. E. and J. R. Johnson, *Introductory Electric Circuit Analysis*. Prentice-Hall, Englewood Cliffs, New Jersey, 1981, Chap. 9.

Kantrowittz, P., G. Kousourau, and L. Zucker, *Electronic Measurements*. Prentice-Hall, Englewood Cliffs, New Jersey, 1979, Chap. 1.

McWane, J., *Introduction to Electronics and Instrumentation*. Breton Publishers (Division of Wadsworth, Inc.), North Scituate, Massachusetts, 1981, Chap. 1.

Wolf, S., *Guide to Electronic Measurements and Laboratory Practice*. Prentice-Hall, Englewood Cliffs, New Jersey, 1973, Chap. 2.

Bell, D. A., *Electronic Instrumentation and Measurements*. Reston Publishing, Reston, Virginia, 1983, Chap. 1.

Questions

1-1   A current meter is connected in _____ with the load.

1-2   A voltmeter is connected in _____ with the load.

1-3   A voltmeter can be made from a dc current meter movement if a _____ is connected in _____ with the meter movement.

1-4   What is the purpose of a multiplier resistor?

1-5   Which is more sensitive, a 10,000 Ω/V voltmeter or 100,000 Ω/V voltmeter? Explain.

1-6   How can a PM–MC movement be converted to measure ac voltage?

1-7   An ammeter should have a resistance that is _____ compared with resistances of the connected circuit.

1-8   A voltmeter should have a resistance that is _____ compared with resistance of the connected circuit.

1-9   The range of an ammeter, milliammeter, or microammeter can be increased if a _____ is used.

1-10   Can an ohmmeter be used in circuits with power turned on? Why?

1-11   On which part of an ohmmeter scale is the most accurate reading possible?

1-12   What quantities can a VOM generally be used to measure?

1-13   A thermocouple ac ammeter indicates the _____ value of the current waveform.

1-14   Thermocouple ac ammeters are used at frequencies up to _____ MHz.

1-15   Explain the term accuracy rating of a particular voltmeter.

1-16   Why do laboratory-type voltmeters have hand-drawn scales?

1-17   Explain the difference between the accuracy of a measurement and the precision of a measurement.

1-18   What is the law of gaussian distribution?

Problems

1.1-1   Find the current through the ideal ammeter (A) in the bridge of Fig. P1.1-1.

Answer.   25 μA.

1.2-1   Calculate the torque that is developed by the moving coil in a PM–MC meter movement when (a) N = 5000 turns, B = 4000 G, A = 2 cm², and I = 50 A; (b) N = 1000 turns, B = 5000 G, A = 2.5 cm², and I = 1 mA.

1.3-1   A 10-μA D'Arsonval movement with an internal resistance of 20K is used to construct a dc voltmeter having the ranges of 1.6 V, 8 V, 40 V, 160 V, 400 V, and 1600 V. Draw the instrument circuit and determine each series resistors for the different ranges.

Answer.   140K; 640K; 3.2M; 12M; 24M; 120M.

1.3-2   Derive the expression for the definition of the voltmeter sensitivity.

1.3-3   Check the values of the series resistors R_1 to R_5 shown in Fig. 1.3-1a for the different voltage ranges. Are all these resistances correct?

1.3-4   A battery is measured using a voltmeter that draws negligible current. (a) Sketch the measuring circuit diagram. (b) If the measured open-circuit voltage is E_{oc} and the load voltage is V_L (with load resistance R_L), what is the equation for expressing the internal resistance R_{int} of the battery?

1.3-5   The internal resistance (R_m) of a PM–MC meter movement is determined using a measuring circuit containing a battery of given voltage E and internal resistance R_{int} and a voltmeter that draws negligible current. Sketch the circuit diagram and write the expression for R_m.

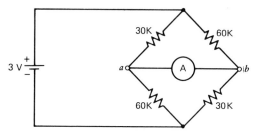

Figure P1.1-1

1.4-1   On the 10-mA range of the ammeter shown in Fig. 1.4-1, the input
resistance (R_T) of the circuit is not exactly 25 Ω. What is the error
assuming that $R_T = 25$ Ω?

Answer.   0.124 Ω.

1.4-2   (a) Calculate the current through the ammeter across the bridge output in Fig. P1.4-2. (b) What is the direction of the current?

Answer.   25 μA.

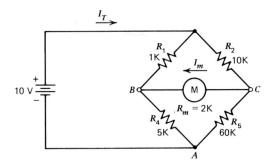

Figure P1.4-2

1.5-1   Prove that the ohmmeter scale of the instrument shown in Fig. 1.5-
1 is nonlinear.

1.6-1   An ac voltmeter having three ranges is constructed as shown in Fig.
P1.6-1. It uses a 1-mA meter movement with an internal resistance
(R_m) of 100 Ω. The shunting resistance (R_{sh}) across the movement
is 200 Ω. Diodes D_1 and D_2 have a forward resistance (R_f) of 200 Ω
and infinite reverse resistance. (a) Calculate the values of series resistors R_1, R_2, and R_3 if the required meter ranges are 10 V, 50 V,
and 100 V, respectively. (b) Determine the sensitivity of this ac voltmeter.

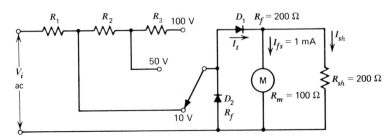

Figure P1.6-1

Answer. $R_1 = 2733.3\ \Omega;\ R_2 = 12K;\ R_3 = 15K;\ S = 300\ \Omega/V.$

1.6-2   In the circuit of Fig. P1.6-2, the voltage across the 500K resistor is exactly 10 V. If a voltmeter with a sensitivity of 20K Ω/V is used to measure the voltage between point *A* and *B*, what are the readings indicated on its 50 V, 15 V, or 5 V range ?

Answer.   8 V; 5.45 V; 2.86 V.

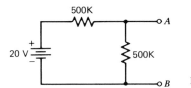

500K

20 V

500K

A

B   Figure P1.6-2

1.6-3   A 1-mA dc meter whose resistance is 10 Ω is calibrated to read rms volts when used in a bridge-rectifier circuit with semiconductor diodes. The effective resistance of each diode may be considered to be zero in the forward direction and infinite in the reverse direction. The sinusoidal input voltage is applied in series with a 5K resistance. Draw the circuit diagram and find the full-scale reading of the meter. Hint: $I_{dc} = 2I_{max}/\pi = \ ?$

Answer. $V_{rms} = 5.56$ V (full-scale).

1.7-1   A thermocouple ammeter is used to measure a 10-MHz sine wave, and it indicates a current of 2 A in a pure resistance. What is the peak current in this waveform?

Answer.   2.83 A.

1.8-1   A utility-type voltmeter with an accuracy of $\pm 3\%$ of full scale (at 25°C) is used on 300-V scale to measure 230 V. (a) What is the possible percentage error? (b) What range will the actual voltage fall within if the instrument reads 230 V?

Answer.   3.9%; 221–239 V.

1.8-2   Five students made the following readings on a very accurate voltmeter: 3.12 V, 3.15 V, 2.97 V, 3.10 V, and 2.99 V. (a) What is the most probable value of the voltage? (b) What are the precision and the deviation of each of the measurements?

Answer.   (a) 3.066 V;

(b) Measurement Number	Precision	Deviation from Mean
1	101.62%	$+1.62\%$
2	102.60%	$+2.6\%$
3	96.74%	-3.26%
4	100.98%	$+0.98\%$
5	97.39%	-2.61%

2
Various
Meter Movements

2.1 PM–MC TYPE VOLTMETERS AND AMMETERS

Basic Construction of a PM–MC System

As shown in Fig. 2.1-1, a moving coil (MC) of fine wire is suspended in a permanent magnet's (PM) field. Electric connections with the coil are made through the suspension wires, as illustrated, or through two helical springs, or through highly flexible bronze ribbons. When the current to be measured flows in the coil, it will rotate until the electromagnetic (EM) torque given by Eq. (1.2-1) is equal to the mechanical moment or restoring torque of the suspension wires. If the coil moves in a homogeneous field, the deflection of the coil or of a pointer or mirror rigidly connected to the coil is proportional to the current passing through the coil. The scale characteristic is linear.

The PM–MC system is damped by air resistance, by eddy currents arising during motion in the metallic frame on which the coil is wound (this can be avoided by using a frame with a nonconductive gap), and by current induced during motion in the windings of the coil (electrodynamic damping). None of these damping systems is very effective in meters that have a high restoring force.

The forerunner of the PM–MC system was a suspension galvanometer, basic to most dc-indicating movements currently used.

Voltmeters and Ammeters

The lowest level of input for a full-scale deflection of PM–MC voltmeters or ammeters is usually on the order of magnitude of 10 mV or 10 μA. Power requirements of these instruments (without series or parallel resistance) can

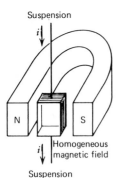

Suspension

i

N   S

i Homogeneous
     magnetic field

Suspension

Figure 2.1-1   Basic construction of a PM–MC system.

be as low as 0.1 μW, but are usually on the order of 1 mW. The coil resistance is in the range from 1 to several thousand ohms. A voltmeter is more sensitive the lower its coil resistance is. A current meter is more sensitive the higher its resistance is. However, a low-resistance voltmeter and a high-resistance current meter are likely to cause measurement errors.

Parallactic reading errors can be avoided by using pointers with flat ends and by having a mirror beneath the scale. Laboratory standard meters have an accuracy of about ±0.1% of full scale. Portable laboratory meters have an accuracy of 0.2–0.5% of full scale. Panel meters with a size larger than 3 in. and with a size up to 3 in. have accuracies of 1% and 2–5% of full-scale, respectively. Heavy magnet or iron surrounding the PM–MC system acts as a shield for external magnetic stray fields. Mirror galvanometers, which are the PM–MC systems with optical readout, are used primarily for recording on light-sensitive papers in motion at high speed or high frequencies.

2.2   LOW-INERTIA (OSCILLOGRAPHIC) PM–MC SYSTEMS

The concept of inertia is reviewed by referring to Fig. 2.2-1, which shows a rigid body rotating about the Z axis. A particle of mass m at a distance r from the Z axis moves in a circle of radius r with velocity ω about this axis and has a linear speed $v = \omega r$. Since

$$\tfrac{1}{2}mv^2 = \tfrac{1}{2}mr^2\omega^2, \tag{2.2-1}$$

the total kinetic energy is

$$K = \tfrac{1}{2}(m_1 r_1^2 + m_2 r_2^2 + \cdots)\omega^2,$$

or

$$K = \tfrac{1}{2}(\textstyle\sum m_i r_i^2)\omega^2 = \frac{I\omega^2}{2}, \tag{2.2-2}$$

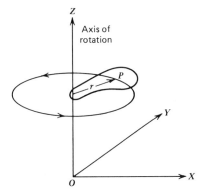

Figure 2.2-1 A rigid body rotating about the Z axis.

where $I = \sum m_i r_i^2$, is the moment of inertia of the body. I will be small if m_i and r_i are small.

In order to decrease the moment of inertia of the instrument, the coil is reduced to a narrow loop of fine wires under the tension of a spring, as shown in Fig. 2.2-2. A mirror cemented to the wires permits light-beam recording of the system movement. To ensure proper damping, the system is sealed in a container filled with a viscous fluid. Several systems may be combined in a common housing, along with a light source and the necessary optical and photographic equipment. Oscillographs of this type can reproduce input signals of a frequency up to 5000 Hz with a dynamic distortion of less than 1%. The frequency range can be extended by increasing the restoring force or by reducing the moment of inertia.

2.3   HIGH-INERTIA (INTEGRATING) PM–MC SYSTEMS

The high-inertia PM–MC system has a weight attached to the moving coil to increase the inertia; the ballistic meter and fluxmeter are two examples. The ballistic meter is usually a mirror galvanometer with a period of about 10 s or more. If a short current pulse i is applied, the system will respond with a deflection θ, and the electric charge is given by

$$Q = k_1\theta = k_2 \int i \, dt, \qquad (2.3\text{-}1)$$

where k_1 and k_2 are constants. Hence the charge is proportional to the deflection.

The fluxmeter contains a moving-coil system, which is greatly overdamped and has a negligible restoring force. If a dc signal (i) is applied, the system will move with a velocity ω that is directly proportional to current i. If the current ceases, then the movement stops, and the pointer indicates

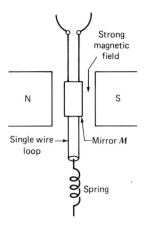

Figure 2.2-2 Low-inertia PM–MC system.

a deflection α:

$$\alpha \propto \int \omega \, dt \quad \text{and} \quad \alpha \propto \int i \, dt;$$

$$\int \omega \, dt = \int \frac{d\alpha}{dt} \, dt = \alpha = k \int i \, dt$$

and

$$iR = N \frac{d\phi}{dt},$$

thus, the flux is given by

$$\phi = \frac{R}{N} \int i \, dt = k'\alpha, \tag{2.3-2}$$

where N is the number of turns and k and k' are constants. The magnetic flux ϕ is proportional to the deflection α.

2.4 PM–MC SYSTEMS WITH TWO MOVING COILS

A PM–MC system may contain two moving coils. If the total torque produced is due to the two currents passing through the two separated coils, the instrument is a differential moving-coil system. If one of the two coils is used as a restoring spring instead, the system is a ratio meter.

Differential Moving-Coil System

This instrument is basically a PM–MC system with a spring-restoring moment but contains two coils wound on the same frame, as shown in Fig. 2.4-1. If input currents i_1 and i_2 are applied to both coils, the total torque (τ_t)

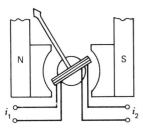

Figure 2.4-1 Differential galvanometer.

will be proportional to the sum of i_1 and i_2, or if one of the currents is reversed, the torque τ_t will be proportional to the difference of i_1 and i_2. It is an advantage of the instrument that the sum or difference of two currents can be formed in two electrically separated circuits. A disadvantage is that for difference measurements the torque in the vicinity of the zero position is proportional to the small difference of two currents, while the coils and springs are to be designed for the full current. As a galvanometer with optical readout, the system is well suited for very accurate resistance measurements.

Ratio Meters

The instrument shown in Fig. 2.4-2 contains two moving coils m_1 and m_2 on a common axis and has no mechanical restoring force. The coil m_2 moves in an nonhomogeneous magnetic field in such a way that the restoring torque exercised by m_2 increases with increased deflection. The restoring torque of coil m_2 is proportional to the current passing through it and to the angle of rotation. Coil m_1 moves in a homogeneous magnetic field; its torque is proportional to the current passing through it. When the currents i_1 in coil m_1 and i_2 in m_2 flow simultaneously, the entire system will rotate until both torques are equal. The final deflection is proportional to the current ratio $i_1 : i_2$.

2.5   PM–MC SYSTEMS WITH A NONHOMOGENEOUS MAGNETIC FIELD

The PM–MC systems may be constructed with nonhomogeneous fields for some special purposes, such as the zero-center scale or logarithmic scale indicated by the system.

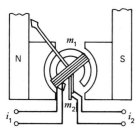

Figure 2.4-2 Ratio meter.

PM–MC System with Zero-Center Scale

A PM–MC meter with zero-center scale has a nonhomogeneous field, as shown in Fig. 2.5-1. The field is highest in the vicinity of the center (zero) position of the coil, so that in this position the meter has a high sensitivity, while the sensitivity decreases to both sides with increased meter deflection. Meters of this type are useful primarily as zero indicators, for example, in Wheatstone bridges.

PM–MC Logarithmic Meter

A PM–MC meter with an approximated logarithmic characteristic is shown in Fig. 2.5-2. Because of the nonhomogeneity of the magnetic field, the sensitivity of this meter is highest at the left-end position of the coil and diminishes gradually as the meter moves toward the right-end position. The range of such a logarithmic meter is usually from 0.1 to 1 V.

2.6   WIDE-ANGLE PM–MC SYSTEMS

For some applications it is desirable to use meters with deflections of the pointer over a range of almost 360°. A moving-coil system for such purposes is shown in Fig. 2.6-1. A hollow cylindrical core N forms one pole of a permanent magnet. It is surrounded by a cylindrical pole piece S, which forms the other magnetic pole. The magnetic field in the cylindrical air gap between N and S is radially directed and homogeneous. A rectangular coil C is mounted with one side in the magnetic field and its other side in the practically field-free center hole. The coil can turn around the central pivot point. Two spiral springs (not shown) provide the restoring force and act as connecting leads to the coil. The torque of such an instrument is only one-half the torque of an ordinary PM–MC system.

2.7   ELECTRODYNAMOMETERS

A moving-coil system in which the permanent magnet is replaced by an electromagnet is known as an electrodynamometer or dynamometer. It is

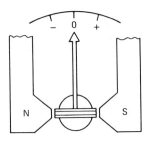

Figure 2.5-1   A PM–MC system with high sensitivity in the vicinity of the zero position for the Wheatstone-bridge detector.

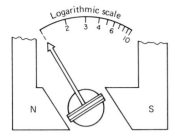

Figure 2.5-2 PM–MC-type logarithmic meter.

one of the most important ac movements and is often used in accurate ac voltmeters and ammeters in the range from the power-line frequency to lower audio frequencies. With some slight modifications, the dynamometer can be used as a wattmeter, a power-factor meter, a frequency meter, etc.

As shown in Fig. 2.7-1, the dynamometer contains a fixed coil, split into two equal halves, connected in series with a moving coil. The fixed coils carrying the current under measurement provide the magnetic field in which the movable coil rotates. The movable coil carries a pointer, which is balanced by counterweights. Its rotation is controlled by springs. The complete assembly is surrounded by a laminated shield to protect the movement from stray magnetic fields that may affect its operation. Damping is provided by aluminum air vanes, moving in sector-shaped chambers.

The electromagnetic torque developed by a coil suspended in a magnetic field is equal to the product of the flux density (B), the current (I), and the coil constants $(A$ and $N)$, as expressed by Eq. (1.2-1). In the dynamometer the flux density (B) depends on the current (I) through the fixed coil and is therefore proportional to the deflection current (I). Since the moving-coil dimensions and the number of turns on the coil frame are fixed quantities for any given meter, the developed torque becomes a function of the squared current (I^2). For ac use, the meter deflection is therefore a function of the mean of the squared current. The scale of the dynamometer is usually calibrated in terms of the square root of the average current squared, and the meter therefore reads the rms or effective value of the ac.

The dynamometer may serve as a calibration or transfer instrument because it can be calibrated on dc and then used directly on ac, establishing

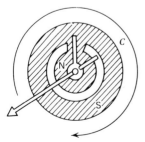

Figure 2.6-1 Wide-angle PM–MC system.

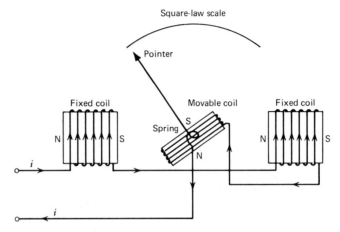

Figure 2.7-1   Schematic diagram of an electrodynamometer movement.

a direct means of equating ac and dc measurements of voltage and current. It is very accurate at the power-line frequencies and is therefore often used as a secondary standard. The addition of a series resistor converts the dynamometer (current meter) into a voltmeter, which again can be used to measure dc and ac voltages. The sensitivity of the dynamometer voltmeter is approximately 10–30 Ω/V (compare this to the 20K Ω/V of a PM–MC meter). The dynamometer has high power consumption, a direct result of its construction. Its flux density is typically 60 G (compare this to the 1000–1400 G of a PM–MC movement).

2.8   DYNAMOMETER-TYPE FREQUENCY METER

Frequency measurements can be made with the dynamometer-type frequency meter depicted in Fig. 2.8-1. In this instrument the field coils form part of two separate tuned circuits. Field coil F_1 is in series with inductor L_1 and capacitor C_1, and forms a tuned circuit that resonates just below the low end of the scale. Field coil F_2 is in series with inductor L_2 and capacitor C_2, and forms a tuned circuit that reasonates just above the high end of the scale. The center-scale frequency indication is 60 Hz.

At frequencies below 60 Hz, coil F_1 has the stronger effect on deflection, and the pointer moves counterclockwise. At frequencies above 60 Hz, coil F_2 has the stronger effect, and the pointer moves clockwise. The torque on the movable element is proportional to the sum of currents i_1 and i_2 through the moving coil. The restoring force is primarily provided by a small iron vane (not shown) mounted on the moving coil.

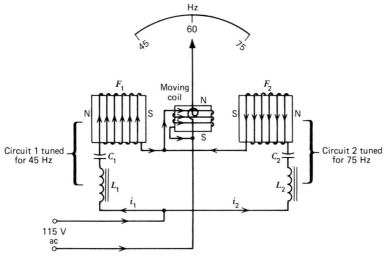

Figure 2.8-1 Schematic diagram of a tuned-circuit frequency meter used for monitoring the frequency of a power system.

2.9 THE DYNAMOMETER WATTMETER USED AS A SINGLE-PHASE POWER METER

When a dynamometer is used as single-phase power meter, its coils are connected in a modified arrangement, depicted in Fig. 2.9-1a. The two fixed coils (current coils) are connected in series and carry the total line current (i_c). The moving coil (potential coil), located in the magnetic field of the fixed coils, is connected in series with a current-limiting resistor (R) across

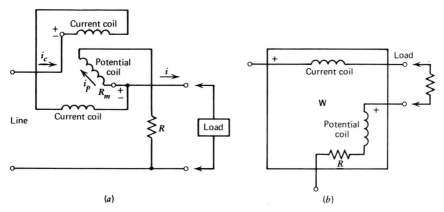

Figure 2.9-1 (a) A dynamometer wattmeter connected to measure the power of a single-phase load. (b) Simplified schematic representation of the dynamometer wattmeter.

the power line and carries a small current (i_p). The instantaneous current i_p equals v/R_p, where v is the instantaneous voltage across the power line and R_p is the sum of R and R_m (and is the moving-coil resistance). The deflection of the moving coil is proportional to the product of the currents i_c and i_p, or

$$\theta_{av} = K_1 \left(\frac{1}{T}\right) \int_0^T i_c i_p \, dt, \tag{2.9-1}$$

where θ_{av} = average angular deflection of the potential coil,
$\quad\quad K_1$ = instrument constant,
$\quad\quad\; T$ = period of the sine wave,
$\quad\quad\; i_c$ = instantaneous current in the current coils,
$\quad\quad\; i_p$ = instantaneous current in the potential coil.

The instantaneous value of the load current $i \gg i_p$, and, therefore, $i_c = i + i_p \simeq i$. Using the values for $i_c \simeq i$ and $i_p = v/R_p$, Eq. (2.9-1) reduces to

$$\theta_{av} = K_1 \left(\frac{1}{T}\right) \int_0^T i \frac{v}{R_p} \, dt = K_2 \left(\frac{1}{T}\right) \int_0^T vi \, dt. \tag{2.9-2}$$

Since the average power in a circuit is

$$P_{av} = \frac{1}{T} \int_0^T vi \, dt, \tag{2.9-3}$$

Eq. (2.9-2) indicates that the dynamometer–wattmeter configuration in Fig. 2.9-1 has an average deflection proportional to the average power. If $v = V_m \sin \omega t$, and $i = I_m \sin(\omega t \pm \theta)$, then, referring to the formula

$$\sin x \sin y = \tfrac{1}{2}[\cos(x - y) - \cos(x + y)],$$

Eq. (2.9-2) becomes

$$\theta_{av} = K_2 V_m I_m \left(\frac{1}{T}\right) \int_0^T [\tfrac{1}{2} \cos(\pm\theta) - \tfrac{1}{2} \cos(2 \omega t \pm \theta)] \, dt$$

$$= K_2 V_m I_m \left(\frac{1}{T}\right) \left(\frac{T}{2} \cos(\pm\theta) - \frac{1}{4\omega} \int_0^{T=1/f} \cos(2\omega t \pm \theta) d(2\omega t \pm \theta)\right)$$

$$= K_2 V_m I_m \left(\frac{1}{T}\right) \left(\frac{T}{2} \cos(\pm\theta) - \frac{1}{4\omega} [\sin(4\pi \pm \theta) - \sin(\pm\theta)]\right)$$

$$= \tfrac{1}{2} K_2 V_m I_m \cos(\pm\theta),$$

or

$$\theta_{av} = K_3 V_m I_m \cos\theta = K_4 VI \cos\theta, \tag{2.9-4}$$

where V and I represent the rms values of the voltage and current and θ represents the phase angle between V and I.

2.10   THE DYNAMOMETER WATTMETERS USED IN POLYPHASE POWER MEASUREMENTS

Polyphase power measurements require the use of two or more wattmeters. The total real power is the algebraic sum of the readings of the individual wattmeters. Real power can be measured by one less wattmeter element than the number of wires in any polyphase system, provided that one wire can be made common to all the potential circuits. Figure 2.10-1a shows the configuration of two wattmeters to measure the total power consumed by a balanced three-wire delta-connected three-phase load.

The phasor diagram in Fig. 2.10-1b shows the phase voltages V_{ab}, V_{bc}, and V_{ca}, and the phase currents $I_{a'b'}$, $I_{b'c'}$, and $I_{c'a'}$. The voltage V_{ab} is taken as a reference phasor. The phase sequence is assumed to be of abc. Thus, V_{bc} lags V_{ab} by 120° and V_{ca} lags V_{bc} by 120°. The vector sum of V_{ab}, V_{bc}, and V_{ca} is zero. The delta-connected load is assumed to be inductive, and so the phase currents lag the phase voltages by an angle θ. The vector sum of the phase currents $I_{a'b'}$ and $I_{a'c'}$ is the line current $I_{aa'}$, which is carried by the current coil of wattmeter 1, while the voltage across its potential coil is the line voltage V_{ab}. The vector sum of the phase currents $I_{c'a'}$ and $I_{c'b'}$ is the line current $I_{cc'}$, which is carried by the current coil of wattmeter 2, while the voltage across its potential coil is the line voltage V_{bc}.

First, taking $I_{a'b'}$ as a reference current, we find

$$I_{aa'} = I_{a'b'} - I_{c'a'} = \frac{I_m}{\sqrt{2}} \angle 0° - \frac{I_m}{\sqrt{2}} \angle 120°$$

$$= I(1 + j0) - I(-0.5 + j0.866)$$

$$= I(1.5 - j0.866) = \sqrt{3}\, I\ \angle{-30°}.$$

Then, using V_{ab} as a reference instead of $I_{a'b'}$, we can write

$$I_{aa'} = \sqrt{3}\, I\ \angle{-(\theta + 30°)}.$$

Since the delta-connected load is balanced, we can write

$$|\,V_{ab}\,| = |\,V_{bc}\,| = |\,V_{ca}\,| = V$$

and

$$|\,I_{aa'}\,| = |\,I_{bb'}\,| = |\,I_{cc'}\,| = I.$$

Then, the power indicated by each wattmeter is

$$P_1 = |\,V_{ab}I_{aa'}\,|\cos(30° + \theta) = VI\cos(30° + \theta), \qquad (2.10\text{-}1)$$

$$P_2 = |\,V_{cb}I_{cc'}\,|\cos(30° - \theta) = VI\cos(30° - \theta), \qquad (2.10\text{-}2)$$

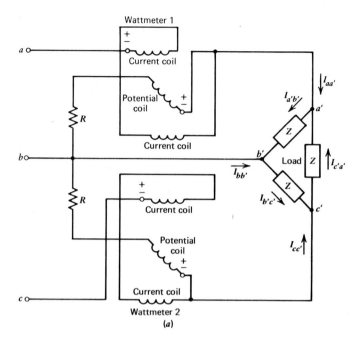

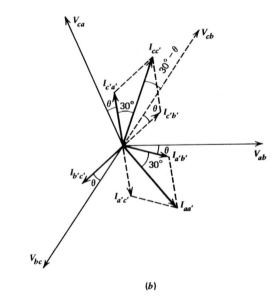

(b)

Figure 2.10-1 (a) Configuration of two wattmeters to measure the total power consumption of a balanced three-wire delta-connected three-phase load. (b) Phasor diagram of voltages and currents in the three-phase three-wire system. The phase sequence of the supply system is assumed to be "abc." The load is assumed to be inductive and so the phase currents lag the phase voltages by an angle θ.

and

$$P_1 + P_2 = VI[(\cos 30° \cos \theta - \sin 30° \sin \theta)$$
$$+ (\cos 30° \cos \theta + \sin 30° \sin \theta)$$
$$= \sqrt{3} \, VI \cos \theta.$$

Thus, the total real power is

$$P_T = P_1 + P_2 = \sqrt{3} \, VI \cos \theta. \tag{2.10-3}$$

The configuration shown in Fig. 2.10-1 is called the two-wattmeter method, not only used for delta-connected loads, but also suitable for Y-connected loads.

The two-wattmeter method can also be used to measure the power factor $\cos \theta$ and the reactive power Q_r (expressed in volt-ampere reactive or VAR) for a balanced three-phase load, as shown in Fig. 2.10-1. The tangent of a power triangle is $\tan \theta = Q_r/P_T$, where the real power $P_T = P_1 + P_2$ and the reactive power Q_r is given by

$$Q_r = \sqrt{3}(P_2 - P_1) = \sqrt{3}[VI \cos(\theta - 30°) - VI \cos(\theta + 30°)]$$
$$= \sqrt{3} \, VI[(\cos \theta \cos 30° + \sin \theta \sin 30°)$$
$$-(\cos \theta \cos 30° - \sin \theta \sin 30°)]$$
$$= \sqrt{3} \, VI \sin \theta.$$

Thus

$$\tan \theta = \frac{\sqrt{3}(P_2 - P_1)}{P_2 + P_1} = \frac{\sqrt{3}(1 - P_1/P_2)}{1 + P_1/P_2}. \tag{2.10-4}$$

From Eq. (2.10-4) we may conclude as follows: When the readings P_1 and P_2 are positive and equal, $\tan \theta = 0$ and $\cos \theta = 1$, indicating that the load is purely resistive. When one of the two wattmeter readings is zero, $\tan | 60° | = | \sqrt{3} |$ and $\cos \theta = 0.5$. When both readings are equal in magnitude but opposite in sign, $\tan 90° = \infty$ and $\cos 90° = 0$, indicating that the load is purely reactive. The sign of $\tan \theta$ in Eq. (2.10-4) depends on the phase sequence and which wattmeter reading is assigned as P_1 or P_2.

2.11   MOVING-IRON METER MOVEMENTS

The moving-iron instruments can be classified into attraction and repulsion types. The repulsion moving-iron systems are more commonly used. The basic repulsion-type instrument is shown in Fig. 2.11-1.

The movement contains a stationary coil of many turns, which carries the current to be measured. A fixed iron vane (FIV) and a moving iron vane

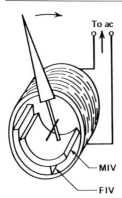

Figure 2.11-1   Basic repulsion-type iron-vane movement.

(MIV) are placed inside the coil. The moving vane is connected to the instrument shaft, which can rotate freely. The coil current magnetizes both vanes with the same polarity. The two magnetized vanes experience a repelling force, and the displacement of the moving vane is an analog of the magnitude of coil current. The repelling force is proportional to the current squared, but the pointer deflection is not of a perfect square law owing to the effects of frequency and hysteresis. A spring attached to the moving vane opposes the electromagnetic torque and permits the scale to be calibrated in terms of a current or a voltage value. When used as a voltmeter, a suitable multiplier resistance should be connected in series with the coil.

2.12   ELECTROSTATIC VOLTMETERS

Figure 2.12-1 shows a basic electrostatic meter movement, which is basically a variable capacitor. Its capacitance increases as the pointer moves to the right. Opposite charges on the plates force the pointer to move to the right. The pointer will come to rest when the torque caused by the electrical at-

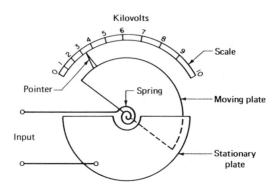

Figure 2.12-1   An electrostatic voltmeter.

traction equals the opposing torque of the coil spring. The deflecting torque is proportional to the square of the applied voltage, whatever its waveform, and the deflection of the instrument may be calibrated in rms volts. Its accuracy is $\pm 1\%$ to $\pm 0.5\%$ of full scale.

The electrostatic voltmeters responding to dc and ac are normally used for voltages in the 10-V to 100-kV range, and require infinitesimal power at dc. The instruments used on ac have a range from 10 pF in a 100-kV movement to about 200 pF in a 100-V movement. Since the movement capacitance represents an infinite impedance at dc, the electrostatic voltmeter is sometimes called the electrometer, which is a high-input-impedance voltmeter.

As mentioned above, the deflecting torque (τ_{av}) is proportional to the square of the input voltage (V_{rms}), or

$$\tau_{av} = KV_{rms}^2, \tag{2.12-1}$$

where K is a constant. The deflecting torque is based on the moment applied to the rigid body, as shown in Fig. 2.12-2. In time dt, point $P(t)$ moves a distance ds along the arc and rotates through an angle $d\theta$. The work dW done is

$$dW = F_s ds = (F \cos \phi)ds = (F \cos \phi)rd\theta. \tag{2.12-2}$$

The term $(F \cos \phi)r$ is the instantaneous torque (τ) exerted by force F_s on the rigid body about the axis perpendicular to the page through origin 0, so that

$$dW = \tau d\theta \quad \text{or} \quad \tau = dW/d\theta. \tag{2.12-13}$$

The instantaneous voltage across the capacitor is $v = q/c$, and the instantaneous energy stored in the electric field between the capacitor plates is $dW = cv^2/2$. Then the instantaneous torque of the electrostatic voltmeter will be

$$\tau = \frac{\partial W}{\partial \theta} = \frac{\partial}{\partial \theta} (\tfrac{1}{2}cv^2) = \tfrac{1}{2}v^2 \frac{\partial c}{\partial \theta} . \tag{2.12-4}$$

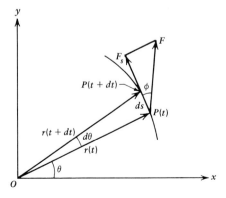

Figure 2.12-2 A rigid body rotating about the axis perpendicular to the page through origin O.

The average torque over an entire period T of the ac voltage is

$$\tau_{av} = \frac{1}{T} \int_0^T \tau \, dt = \frac{1}{T} \int_0^T \tfrac{1}{2} v^2 \frac{\partial c}{\partial \theta} \, dt = K V_{rms}^2.$$

Thus, the average torque is proportional to the square of the input voltage, whatever its waveform.

BIBLIOGRAPHY

Carr, J. J., *Elements of Electronic Instrumentation and Measurement*. Reston Publishing Company, Reston, Virginia, 1979, Chaps. 1 and 2.

Cooper, W. D., *Electronic Instrumentation and Measurement Techniques*, 2nd ed. Prentice-Hall, Englewood Cliffs, New Jersey, 1978, Chapt. 5.

Herrick, C. N., *Instruments and Measurements for Electronics*. McGraw-Hill, New York, 1972, Chap. 3.

Johnson, D. E. and J. R. Johnson, *Introductory Electric Circuit Analysis*. Prentice-Hall, Englewood Cliffs, New Jersey, 1981, Chap. 19.

Lion, K. S., *Elements of Electrical and Electronic Instrumentation*. McGraw-Hill, New York, 1975, Chap. 3.

Wolf, S., *Guide to Electronic Measurements and Laboratory Practice*. Prentice-Hall, Englewood Cliffs, New Jersey, 1973, Chap. 4.

Questions

2-1   What is the purpose of the two helical springs used with the moving coil of a D'Arsonval meter movement?

2-2   Name two types of PM–MC systems each of which has two moving coils.

2-3   Name three types of voltmeters that will indicate either dc or ac volts.

2-4   Why are ac voltmeter scales commonly calibrated in rms value?

2-5   A _____ is an ac ammeter constructed from a movable coil in the magnetic fields of two stationary coils. All three coils are connected in _____ with each other.

2-6   A _____ is used to extend the current range of an electrodynamometer.

2-7   Describe the principle of operation of the electrodynamometer movement.

2-8   Explain the principle of operation of the dynamometer-type frequency meter.

2-9   Discuss the connection of an electrodynamic mechanism as a single-phase power meter.

2-10   Iron-vane meters read the _____ value of the ac current waveform.

2-11   Why are most iron-vane voltmeters limited to the frequency range 25–135 Hz?

2-12   Describe the principle of operation of the electrostatic voltmeter.

Problems

2.1-1 (a) Determine the full-scale voltage indicated by a 500-μA meter movement with internal resistance of 250 Ω if no multiplier is used. (b) What is the sensitivity if this instrument is used to measure dc voltages?

2.1-2 (a) What is the relation between the terminal current I and the meter movement current I_m for the circuit in Fig. P2.1-2? (b) When the shunt resistor R is disconnected from the circuit, the basic meter movement has a full-scale deflection of 100 μA and a resistance of 900 Ω. When R is switched into the circuit, the ammeter has a full-scale deflection when I equals 1 mA. What is the resistance value of R?

Answer. $I = (R + R_m)I_m/R = 100$ Ω.

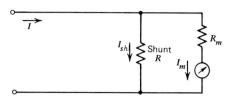

Figure P2.1-2 Ammeter with shunt.

2.1-3 A dc voltmeter has a 100,000 Ω/V rating. Its PM–MC movement has an internal resistance of 20,000 Ω. (a) What series resistance (multiplier) does it present when the range selector is on a 5-V dc scale? On a 25-V dc scale? (b) What is the meter-movement current for full-scale deflection?

2.7-1 (a) What is meant by transfer instrument? (b) Explain why the dynamometer can be used as a transfer instrument. (c) Prove that the torque of the dynamometer is a function of the current squared.

2.10-1 Assume that the readings of the two wattmeters in Fig. 2.10-1 are $P_1 = 1360$ W and $P_2 = -740$ W. What is the total power consumed by the load?

2.10-2 A three-phase, three-wire supply system has a phase sequence of *abc* and three balanced line-to-line voltages of $V_L = 110$ V at 60 Hz. This source feeds the three impedances of value $Z = 5 \underline{/45°}$ Ω connected in delta, as shown in Fig. P2.10-2. Find the line currents (I_L) and draw the phasor diagram (with V_{bc} as a reference voltage).

Answer. $I_{aa'} = 38.1 \underline{/45°}$ A; $I_{bb'} = 38.1 \underline{/-75°}$ A; $I_{cc'} = 38.1 \underline{/165°}$ A.

2.10-3 The three-phase emf of $V_\phi = 240$ V at 60 Hz feeds a balanced three-wire delta-connected three-phase load of $Z_\phi = 12 \underline{/70°}$ Ω. If the

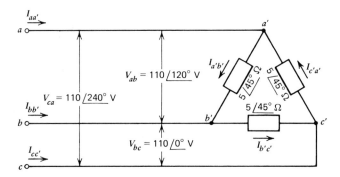

Figure P2.10-2

two-wattmeter method shown in Fig. 2.10-1 is used to measure the power consumed by the load, what will be the reading of each wattmeter? (*Hint:* The phasor diagram may be drawn referring to Fig. 2.10-1*b*.)

Answer. $P_T = 6.368 - 1.444 = 4.924$ kW.

2.10-4   A delta-connected three-phase inductive load uses the two-wattmeter method to measure its total power. One wattmeter reads 12 kW and the other reads 18 kW. After the load is changed, the wattmeters read 5 kW and 8 kW, respectively. (a) What is the power factor of the first load? (b) What is the power factor of the second load? (c) Is the second load inductive or capacitive?

Answer. $\cos \theta = 94.49\%$; $\cos \theta = 92.8\%$; the second load is still inductive.

2.10.5   A three-phase source of 440 V at 60 Hz supplies a power to a balanced inductive load. The two-wattmeter method is used to measure the power. Assume that the readings indicated are $P_1 = 4.9$ W and $P_2 = -1.37$ kW. Find the line current. [*Hint:* Use Eqs. (2.10-4) and (2.10-3).]

Answer. 15 A.

2.10-6   The two-wattmeter method is used to measure a known three-phase inductive load and results in the readings $P_1 = 12$ kW and $P_2 = 7$ kW. Assume that an unknown load is in use instead of the known load and the wattmeters read $P_1 = -18$ W and $P_2 = 24$ W. Calculate: (1) the two power factors of unknown and known loads; (2) the two load currents with a line voltage of 600 V applied.

Answer. (1) 0.08; 0.91. (2) 72 mA; 20.1 A.

3

Potentiometers and Resistance Bridges

3.1 BASIC POTENTIOMETER CIRCUITS FOR DIFFERENT USES

Introduction

Variable resistors usually have three leads, two fixed and one movable. If contacts are made to only two leads of the resistor (stationary lead and moving lead), the variable resistor is being used as a rheostat. If all three contacts are used in a circuit, it is termed a potentiometer or pot.

A basic potentiometer circuit contains a constant-voltage source V_i connected to a variable resistor R (potentiometer), as shown in Fig. 3.1-1a. If r is the resistance between points C and B, the voltage between C and B is a product of current I and resistance r, or

$$V_0 = Ir = V_i \frac{r}{R}. \qquad (3.1\text{-}1)$$

Equation (3.1-1) indicates that the output voltage from the potentiometer depends only on the input voltage and the contact setting, that is, on the ratio r/R when the load resistance $R_L \gg r$. The potentiometric circuit is commonly used as an attenuator in analog circuits and as an instrument for the comparison and measurement of voltages without drawing current from the source to be measured.

Variable resistors are usually made from a resistance wire wrapped around a mandrel or from carbon. By the appropriate shaping of the mandrel the resistance can be made to increase linearly with the rotation of the potentiometer or in any other function (square, sinusoidal, or logarithmic). Frequently, the resistance material is arranged in a helical fashion. This

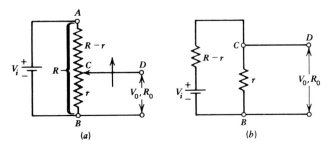

Figure 3.1-1   Basic potentiometer circuit: (a) potentiometer voltage divider; (b) equivalent potentiometer circuit.

construction permits a fine adjustment of the resistance and the use of a long resistance wire.

Any device that converts energy from one form to another is referred to as a transducer. Potentiometers are frequently used as transducers for converting displacements or forces into electric signals. The resistance varies linearly with the displacement; the deviation from linearity in precision potentiometers is sometimes not more than $\pm 0.05\%$.

Potentiometer Voltage Dividers

The basic potentiometer shown in Fig. 3.1-1a is also called the voltage divider. The internal resistance of the voltage source V_i is assumed to be zero. The output voltage V_0 is zero when r is zero, that is, when the variable contact C is at B. V_0 increases linearly with r and reaches a maximum when the variable contact C is at A. The output resistance R_0, that is, the resistance looking back into the terminals B and C, can be found from the equivalent circuit shown in Fig. 3.1-1b. The partial resistances r and $R - r$ appear to be connected in parallel; thus the output resistance with V_i short circuited is equal to the parallel equivalent resistance, that is,

$$R_0 = \frac{(R - r)r}{R - r + r} = \frac{(R - r)r}{R} . \tag{3.1-2}$$

If the variable contact C is at B, r is zero, and so R_0 is zero. If the variable contact C is at A, $r = R$, and so R_0 is again zero. The maximum value of R_0 is $R/4$ at $r = R/2$. In order to avoid the effect of variable output resistance, this basic potentiometer circuit may be replaced by an L-pad or T-pad attenuator, as shown in Figs. 3.1-2a and 3.1-2b. The L-pad attenuator is a system of two mechanically coupled resistors; the mechanical coupling (dotted line) is such that the resistance R_2 decreases as the resistance r increases, thereby the output resistance between terminals D and B remains constant when the setting of contact C is varied. In the T-pad voltage divider, the resistance between output terminals D and B and the resistance between

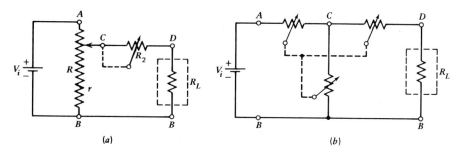

Figure 3.1-2 (a) L-pad voltage divider; (b) T-pad voltage divider.

input terminals A and B remain constant while the ratio of output voltage to input voltage is varied.

3.2 INTRODUCTION TO POTENTIOMETRIC INSTRUMENTS (POTENTIOMETERS)

A potentiometric instrument is usually referred to as a potentiometer. It is designed to measure an unknown voltage by comparing it with a known voltage standard. Measurements using the comparison method are capable of a very high accuracy since the result obtained depends only on the accuracy of the known voltage standard to which the comparison is made. The voltage determination by the potentiometer is quite independent of the resistance of the unknown emf since no power is consumed from the circuit containing the source to be measured when the instrument is balanced. As a result, even portable manual potentiometers are capable of measuring voltages to within ±0.05%. Much higher accuracies are available in the more-elaborate self-balancing types of potentiometers (±0.01% or less). Such accuracy makes potentiometers ideal for the task of calibrating other voltmeters.

The PM–MC meter movements have much lower accuracies (±0.1% or greater). Their error sources include bearing friction, nonlinearity in the restraining springs, and human reading errors, which are all eliminated by potentiometers.

The manual and self-balancing instruments are the two basic classes of potentiometers. The manually operated types require an observer to adjust the dials until the equality between the known and unknown voltages is located. The self-balancing types are widely used in industry since they are automatic devices which seek out the condition of equality themselves. These automatic models are usually equipped with a marking device and a moving chart system, which records their readings permanently and automatically on a graph.

3.3   BASIC MANUAL POTENTIOMETERS

Operation Principle

The principle of operation of all manual potentiometers is based on the circuit of the precision slide-wire potentiometer shown in Fig. 3.3-1. The resistance R of the portion ac of the slide wire is proportional to the length L of this portion. With key K open, the battery V_w initially supplies working current I through the rheostat R_h and the slide wire ab. With switch S in the "measure" position and key K closed, the slider contact c is adjusted so that the deflection of galvanometer G is zero. The unknown emf, E_x, then, is equal to the voltage drop V across portion ac of the slide wire. Hence,

$$E_x = V = IR, \tag{3.3-1}$$

where R is the resistance of portion ac of the slide wire. Determination of E_x now becomes a matter of evaluating $V = IR$. The working current I is calibrated or standardized by reference to a known-reference source or standard cell E_s.

Calibration Procedure

The following example illustrates the procedure for calibrating the potentiometer of Fig. 3.3-1.

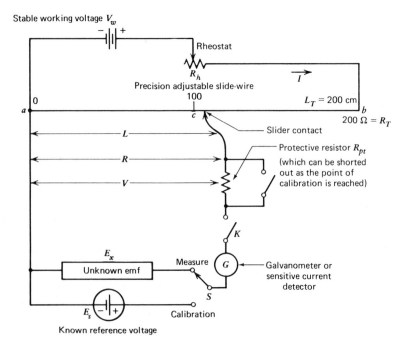

Figure 3.3-1 Circuit diagram of a basic manual potentiometer. The long slide wire is usually replaced by a combination of precision decade resistors and a circular slide wire.

The slide wire has $I_T = 200$ cm and $R_T = 200$ Ω. The reference voltage E_s is of 1.019 V. Switch S is in "calibrate" position, and slider contact c is at 101.9-cm mark. Rheostat R_h is adjusted to provide a value of current I such that the galvanometer shows no deflection when key K is depressed. Under this condition of null or balance, the IR drop along the 101.9-cm portion of the slide wire is $V = E_s = 1.019$ V. Then $(101.9 \ \Omega)I = 1.019$ V. Hence $I = 1.019$ V/101.9 $\Omega = 10$ mA. The voltage at any point along the slide wire is proportional to the length of the slide wire and is obtained by converting the calibrated length into the corresponding voltage, simply by placing the decimal point in the proper position, for example, 134.5 cm = 1.345 V. Once calibrated, I is never varied.

After calibration, E_x may be measured when switch S is in the "measure" position. At the null condition, the slide-wire scale reading is converted into its corresponding voltage value.

Main Features

The main features of manual potentiometers are as follows:

1.   When a null condition is reached, the unknown voltage is being measured without any current being drawn by the potentiometer. This means that the impedance of the potentiometer at balance is essentially infinite. An infinite impedance corresponds to the condition existing in an ideal voltmeter.

2.   The mechanical aspects of the galvanometer operation (such as bearing friction and component nonlinearities) are bypassed because there is no deflection required of the detector at the time the final measurement is made.

3.   The accuracy to which the resistance value along any point of the precision slide wire is known is typically $\pm 0.02\%$ or less.

4.   The galvanometers used as the null detectors have sensitivities that range from 2 mA/mm (portable type) to 0.0001 mA/mm (light beam, laboratory type).

5.   The working voltage source is usually a mercury battery, which provides a constant-voltage value over long periods of time and is thus well suited for its specified task.

6.   The standard cell is a very accurate reference-voltage source. If used and maintained properly, its output value drifts less than 30 μV per year. To prevent more than 100 μA from ever flowing through the standard cell during calibration procedures, the galvanometer branch also contains a large protective resistor. As the point of calibration is reached, this resistor can be removed from the branch to increase the accuracy of the calibration. Some potentiometers use Zener diodes as their voltage-calibration standards instead of standard cells.

7.   The slide-wire potentiometer is a rather impractical form of construction. Most commercial potentiometers use a group of precision decade re-

sistors and a circular slide wire, thus reducing the size of the instrument. The measured voltage is obtained from the dial settings by adding the values of the decade resistors to the value of the slide-wire resistance.

Operating Procedure

The operating procedure for manual potentiometers is as follows:

1. Set the decade resistors and the slider on the precision resistor to the value of the standard cell voltage E_s.

2. Set the function switch S to the "calibrate" position. Then, by tapping on the key K, open and close the circuit containing the galvanometer. While tapping the key, watch the galvanometer scale to observe the extent of the pointer deflection. Adjust the rheostat connected to the working battery until the galvanometer shows zero deflection or null (balance) condition.

3. When the null condition is approached, short out the protective resistor in the galvanometer branch. Make a final adjustment of the rheostat so that the most-sensitive null condition is achieved. This indicates that the potentiometer is calibrated to the value of the standard cell.

4. Turn the switch S from the "calibrate" to the "measure" position.

5. Connect the unknown voltage to the proper terminals of potentiometer.

6. Leave the protective resistor in the circuit with the galvanometer. Vary the decade resistor and the position of the slider on the precision resistor until a null reading is obtained.

7. When the null condition is reached, short out the protective resistor so that the most-sensitive null setting can be located. Proceed to approach this null by adjusting the value of the slide-wire resistor.

8. As a final null is reached, the dial settings of the potentiometer will be equal to the unknown voltage.

9. Working current is checked by returning to the "calibrate" position. If a null condition still exists and the dial settings are exactly the same as in the original calibration procedure, a valid measurement has been made.

3.4 SELF-BALANCING POTENTIOMETERS

Operation Principle

In a self-balancing potentiometer, a chopper-type converter is used instead of the galvanometer in the manual instrument. The unbalanced voltage is applied to an amplifier via the converter. The output of the amplifier drives a two-phase induction motor which moves the potentiometer slider to balance. The converter, inserted between the potentiometer output and amplifier input, converts the dc unbalanced voltage into an ac unbalanced voltage that can be amplified easily to the desired value by an ac amplifier.

Figure 3.4-1 shows a self-balancing potentiometer used for measuring temperature by a thermocouple (TC). The vibrating reed (1) of the converter is driven synchronously by the 60-Hz line voltage and operates as a switch which reverses the current through the split winding of the input transformer primary for each vibration of the reed. As a result, the transformer output is a 60-Hz voltage, which, proportional to the dc input of the converter, is applied to the amplifier (1). The control winding and exciting winding of the two-phase induction motor are supplied, respectively, by the amplifier output voltage and the ac line voltage which has been shifted approximately 90° in phase by the capacitor in series with the exciting winding. Depending on the polarity of the unbalanced dc voltage applied to the converter input terminals, the phase of the amplifier output voltage will either lead or lag by approximately 90° the fixed voltage applied to the exciting winding. The direction of rotation of the motor is determined by the phase relationship

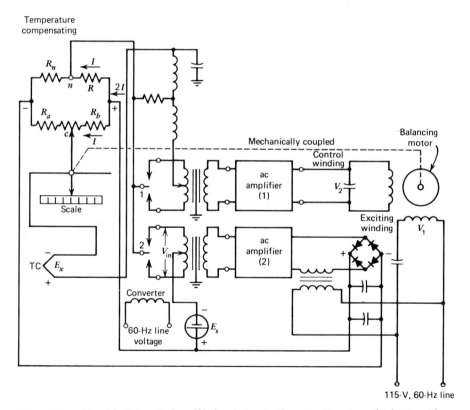

Figure 3.4-1 Simplified circuit of a self-balancing potentiometer. (Courtesy of Yokogawa Electric Works, LTD.)

between the two voltages at the two windings. If E_x, the emf being measured, is greater than the balancing voltage produced by the potentiometer, the motor will rotate in one direction. If E_x is smaller than the balancing voltage, then the amplifier output will be shifted by 180° and the motor will rotate in the other direction. The shaft of the motor is connected mechanically to the slider in such a way, that the rotation of the motor decreases the imbalance in the potentiometer circuit. Rotation will stop when a balance condition is reached or the control current is zero. When the potentiometer current is kept constant (at balance), the position of the slider C indicates the temperature. If the motor is mechanically coupled to a pen mechanism, any movement of the slider will be followed by a simultaneous movement of the pen on a strip chart.

Temperature-Compensating Potentiometer

In a basic potentiometer circuit, the balance error may have resulted from ΔT_{am}, the change in the ambient temperature. This error is avoided by using the temperature-compensating circuit, as shown in Fig. 3.4-1, where R_a, R_b, and R are the manganin resistors with zero temperature coefficient, and R_n is a nickel or copper resistor with a large temperature coefficient. Let ΔR_n be a change in R_n due to ΔT_{am}, ΔV_{nc} be a change in V_{nc} due to ΔR_n, and ΔE_x be a change in E_x due to ΔT_{am}. If ΔV_{nc} equals ΔE_x, then the variation of ambient temperature will be compensated.

Self-Regulating Reference Current

It is required to always keep the current in the potentiometer side and in the compensating resistance R_n constant. For the constant-current source, an ac current is rectified and smoothed as a dc current. To keep the current as a steady dc of constant value, it should be automatically regulated by a reference-current-adjusting circuit, an example of which includes a standard cell E_s, a section of the converter, an input transformer, an ac amplifier (2), a rectifier, and a filter, as shown in Fig. 3.4-1.

Now, resistance R can be used to check whether the current value is correct or not. Assume that R makes the current (I) in it equal to that in the compensating resistance R_n. If the current in R_n is kept at a constant value, then that in R also becomes a constant value. The value of R is determined so that when a standard current I flows in it, the voltage drop across the resistor is equal to the voltage of the standard cell, or

$$v_R = E_s = RI. \tag{3.4-1}$$

When the current I becomes I' due to a variation of source voltage, the input of the amplifier (2) becomes

$$v_{in} = E_s - V'_R = R(I - I'), \tag{3.4-2}$$

which is then amplified and rectified, and also $I - I'$ is canceled out of the circuit current. The circuit current is within $I \pm 0.1\%$ when the ac source voltage is within 100 V $\pm$ 10 V.

ac Control Motor

In low-power-level control systems, where the maximum outputs required from the motor range are a fraction of a watt up to a few hundred watts, two-phase induction motors are often used. The two-phase induction motor consists of a squirrel-cage rotor or the equivalent and a stator with an exciting winding and a control winding displaced 90 electric degrees from each other in space. The ac voltages applied to the two windings are generally phase displaced from each other 90° in time. When the voltage magnitudes are equal, the equivalent of balanced two-phase voltages is applied to the stator. The resultant stator flux is then similar to that in a three-phase induction motor. The voltage applied to the exciting winding is a fixed voltage obtained from a constant-voltage source. The voltage applied to the control winding is usually supplied from an amplifier at the controller output. The two voltages must be synchronous, which means that they must be derived from the same ultimate ac source. They must also be made to be approximately in time quadrature by introducing a 90° phase shift either in the amplifier or in the source of exciting voltage V_1. If the control voltage V_2 has a nonzero value leading V_1 by approximately 90°, rotation in one direction is obtained; if V_2 has a nonzero value lagging V_1, rotation in the opposite direction results.

Improved Circuit of Self-Balancing Potentiometer

The improved self-balancing potentiometer shown in Fig. 3.4-2 is of the model ER180 recording pyrometer, the newest type produced by Yokogawa Electric Works. In Fig. 3.4-2a, the range card amplifies input voltage E_s and yields an output E_0. The amplifier card converts dc voltage E_0 to an ac voltage e_0 with the same frequency as the power line and applies e_0 to the balancing (servo) motor control coils. Since the phase of voltage e_0 is different by 90° from that of voltage e_1 (the phase is shifted by capacitor C_1) the balancing motor (BM) rotates and moves slidewire SL and brush S. When E_s equals reference voltage E_r, that is, $E_0 = 0$ and $e_0 = 0$, the balancing motor stops. The BM rotates either clockwise or counterclockwise, depending on the polarity of E_0. Therefore, input voltage E_s is proportional to a voltage E_f on the slidewire brush S (i.e., proportional to the printer assembly position).

The thermoelectromotive force (thermo-emf) of a thermocouple changes depending on the difference between temperature to be measured and reference-junction temperature (temperature at the recorder input terminals). A compensating circuit shown in Fig. 3.4-2b is provided to compensate for the effect of the input terminal temperature changes on the thermo-emf, that is, to keep the reference-junction voltage equivalent to the voltage at 0°C. A fraction of voltage V_{BE} from transistor Q_{251} is applied to the (+) terminal

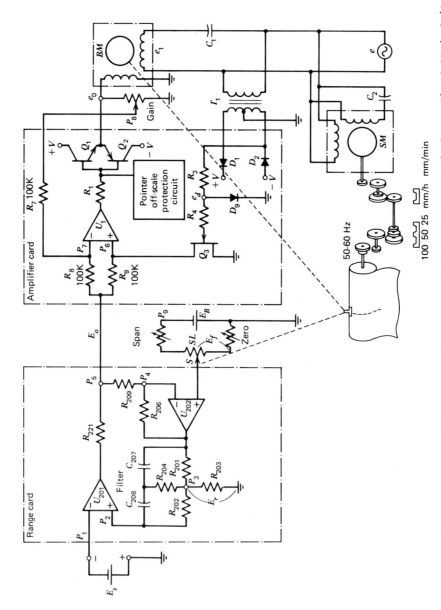

Figure 3.4-2a The circuit of model ER180 recorder with dc potentiometer input. (Courtesy of Yokogawa Electric Works, Ltd.)

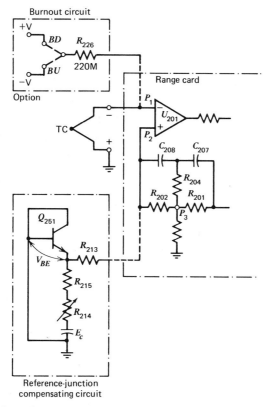

Figure 3.4-2*b* Reference-junction-compensating circuit for thermocouple input.

of U_{201} at point P_2. V_{BE} decreases as the BE junction temperature increases or vice versa, thereby compensating for changes in the reference-junction voltage. A burnout circuit option can be provided to detect thermocouple burnout quickly. In the burnout circuit, the $(+)$ or $(-)$ range-card power supply voltage is applied to $(-)$ input terminal of U_{201} through a high-value resistor. When a thermocouple burns out, the pointer moves off-scale to the 100% side with upscale burnout protection and to the 0% side with downscale burnout protection. The nonlinear thermocouple temperature-versus-emf curve is linearized by using a nonlinear slidewire potentiometer.

The resistance-temperature-detector (RTD) input circuit is shown in Fig. 3.4-2*c*. Resistance (temperature) changes in the RTD are converted into voltage signals by means of a constant-current source (I), which provides the current to the RTD. This is the same principle that is used in potentiometric recorders. Current from the constant-current source (I) flows through R_t, and the voltage drop caused by R_t is fed as an input to op-amp U_{201}. A voltage equal to the voltage drop in the leadwire resistance is fed to the U_{201} $(+)$ input through a compensating network to compensate for

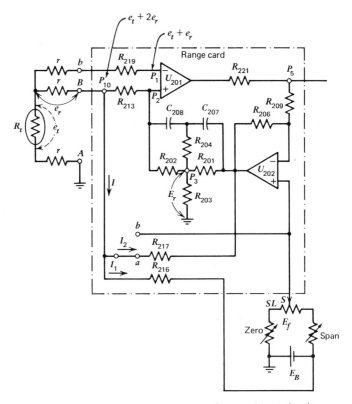

Figure 3.4-2c Resistance-temperature-detector input circuit.

the leadwire resistance. A linearizing circuit feeds back output voltage from U_{201} to the $(-)$ input of U_{202} to linearize the output of U_{201}.

3.5 WHEATSTONE BRIDGES

Operation Principle

The Wheatstone bridge is the most well-known and widely used resistance bridge. It is used for measuring resistance from 1 Ω to the low megohm range. It is simply a circuit of two potentiometers connected to the same source and a galvanometer (G) connected between the two variable contacts, as shown in Fig. 3.5-1a. When both the potentiometers are adjusted so that the galvanometer indicates no current ($I_G = 0$), the bridge balance or null condition is obtained. The balanced-bridge circuit is drawn as in Fig. 3.5-1b, where

$$I_G = 0, \qquad I_1 R_1 = I_2 R_2,$$

$$I_1 R_3 = I_2 R_4.$$

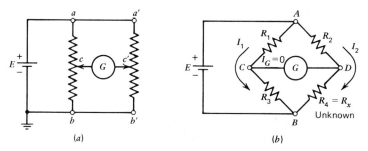

Figure 3.5-1 Wheatstone bridge: (a) two potentiometers connected to the same source; (b) balanced-bridge circuit.

Thus

$$\frac{R_1}{R_3} = \frac{R_2}{R_4} \quad \text{or} \quad R_x = R_4 = R_3 \left(\frac{R_2}{R_1}\right). \tag{3.5-1}$$

The unknown resistor R_x can be found from reading the resistance values R_3, R_2, and R_1 at balance. These resistors are the bridge arms. R_3 is called the standard arm, and R_2 and R_1 are called the ratio arms.

In practical balanced bridges, the resistor R_3 is made continuously variable, while the ratio R_2/R_1 is controlled by a switch that changes this ratio in steps of 0.01, 0.1, 1, 10, 100, ... When a balance condition is obtained, the resistance can be read directly from the dials, since these dial settings correspond to the variables of Eq. (3.5-1).

For measuring the resistors employed in high-frequency applications, the Wheatstone bridge should use an appropriate ac source instead of a battery and an oscilloscope or an electron-ray-tube detector instead of a galvanometer.

Errors of the Balanced Bridge

The possible errors which arise from using the bridge include the following:

1. Discrepancies between the true and rated values in the three known resistors. This error can be estimated from the resistor tolerances. (See Problems 3.5-1 and 3.5-2).

2. Changes in the known resistance values due to the heating effect of the current through the resistors.

3. Thermal emfs in the bridge of galvanometer circuits caused by different materials in contact and at slightly different temperature.

4. Balance-point error due to insufficient sensitivity of the galvanometer.

5. Lead and contact resistances introduced when measuring low-resistance.

Most commercial Wheatstone bridges are accurate to approximately 0.1%, and thus far more accurate than the ohmmeters.

The Bridge Used in the Deflection Method

In addition to the balanced-bridge method described above, the Wheatstone bridge can also be used in the following deflection method: The unknown resistor R_x is inserted in arm 4, and the bridge is first balanced by adjusting R_2, R_3, and R_1 so that the galvanometer current is zero. Any change of R_x by the amount ΔR_x causes a bridge imbalance so that a galvanometer current I_G flows, and a deflection of the galvanometer G can be noted. The calibration characteristic $I_G = f(\Delta R_x)$ is experimentally established, and the value of R_x is found from the galvanometer deflection and the calibration characteristic. The deflection method is frequently employed so that the bridge is calibrated to indicate in percent the variation $(\Delta R_x/R_x) \times 100$.

3.6 KELVIN DOUBLE BRIDGE

Effects of Connecting Leads

Low resistances are difficult to measure precisely with a Wheatstone bridge, since lead and contact resistances are introduced in the connections of the unknown to the bridge terminals. Therefore, a Kelvin double bridge is used to measure resistance values less than 1 Ω.

Consider the bridge circuit shown in Fig. 3.6-1, where the connecting-lead resistance $r_c = r_{mo} + r_{no}$. Obviously, the galvanometer G should not be connected to point m or n. Now, G is connected to a point o, in between the two points m and n, in such a way that

$$\frac{r_{mo}}{r_{no}} = \frac{R_1}{R_3} . \tag{3.6-1}$$

Then

$$1 + \frac{r_{mo}}{r_{no}} = \frac{R_1}{R_3} + 1,$$

$$\frac{r_{mo} + r_{no}}{r_{no}} = \frac{R_1 + R_3}{R_3},$$

$$r_{no} = \left(\frac{R_3}{R_1 + R_3}\right) r_c. \tag{3.6-2}$$

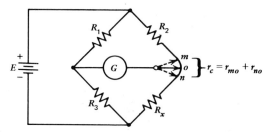

Figure 3.6-1 A Wheatstone bridge circuit, showing the resistance of the connecting lead from point m to point n.

Also

$$1 + \frac{r_{no}}{r_{mo}} = \frac{R_3}{R_1} + 1,$$

$$\frac{r_{mo} + r_{no}}{r_{mo}} = \frac{R_1 + R_3}{R_1},$$

$$r_{mo} = \left(\frac{R_1}{R_1 + R_3}\right) r_c. \tag{3.6-3}$$

The usual balanced bridge has the relationship

$$\frac{R_2 + r_{mo}}{R_x + r_{no}} = \frac{R_1}{R_3} \quad \text{or} \quad R_2 + r_{mo} = \frac{R_1}{R_3}(R_x + r_{no}). \tag{3.6-4}$$

Substituting Eqs. (3.6-2) and (3.6-3) into Eq. (3.6-4), we obtain

$$R_2 + \left(\frac{R_1}{R_1 + R_3}\right) r_c = \frac{R_1}{R_3}\left[R_x + \left(\frac{R_3}{R_1 + R_3}\right) r_c\right], \tag{3.6-5}$$

which reduces to

$$R_x = R_3 \left(\frac{R_2}{R_1}\right). \tag{3.6-6}$$

Equation (3.6-6) is the balance equation (3.5-1) for the Wheatstone bridge, and it indicates that the effect of the resistance of the connecting lead from point m to point n has been eliminated by connecting the galvanometer to the intermediate point o.

Kelvin Bridge Circuit

A second set of ratio arms is employed in Kelvin double bridge, as shown in Fig. 3.6-2a. The second set of ratio arms r_a and r_b is used to connect the galvanometer to a point 0 at the appropriate potential between m and n,

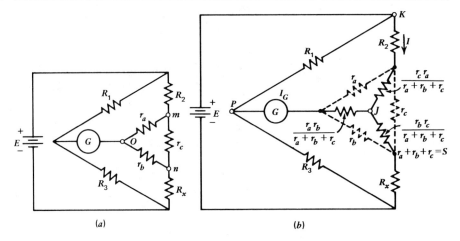

Figure 3.6-2 (a) Kelvin bridge and (b) its equivalent circuit.

eliminating the effect of r_c on the m-to-n connection. In this case, we set $r_a/r_b = R_1/R_3$. When this is done, we can then calculate the unknown value from the equation

$$R_x = R_2 \left(\frac{R_3}{R_1} \right), \qquad (3.6\text{-}7)$$

where R_2 is usually a variable standard resistor. Equation (3.6-7) is derived referring to Fig. 3.6-2b: At balance, $V_{PK} = V_{JK}$, $I_G = 0$,

$$V_{PK} = \frac{R_1 E}{R_1 + R_3} = \frac{R_1}{R_1 + R_3} I \left(R_2 + \frac{(r_a + r_b)r_c}{s} + R_x \right);$$

$$V_{JK} = I \left(R_2 + \frac{r_c r_a}{s} \right);$$

$$\frac{R_1}{R_1 + R_3} \left(R_2 + \frac{(r_a + r_b)r_c}{s} + R_x \right) = R_2 + \frac{r_c r_a}{s};$$

$$R_x = \frac{r_1 + R_3}{R_1} \left(R_2 + \frac{r_c r_a}{s} \right) - R_2 - \frac{(r_a + r_b)r_c}{s}$$

$$= \frac{R_3}{R_1} R_2 + \frac{r_c r_a}{s} \left(\frac{R_3}{R_1} - \frac{r_b}{r_a} \right),$$

where $R_3/R_1 - r_b/r_a = 0$, thus, obtaining Eq. (3.6-7).

The advantage of the Kelvin double bridge is that the value of r_c is eliminated, as seen from Eq. (3.6-7).

3.7 STRAIN-GAGE BRIDGE CONFIGURATION

Basic Strain Gages

The strain gage is one of the most commonly used passive transducers; it converts a mechanical force into a proportional change of resistance. Hence the gage can detect the force. Many other quantities such as displacement, torque, pressure, weight, and tension involve mechanical-force effects, and thus can also be measured by strain gages. A basic strain gage is constructed from a special wire, and its strain (defined to be a fractional change in linear dimension caused by an applied force) takes the form of a lengthening of the gage wire. If the wire is bonded to a thin paper or plastic base, the gage is called a bonded strain gage and is arranged as shown in Fig. 3.7-1. This type of strain gage is used to detect displacements caused by large forces. The bonded strain gage is cemented with a special adhesive to the structure or mechanical surface that is being measured. When extensive or compressive forces are applied to the structure to which the gage is attached, the resistance of the wire bonded to the base changes, more or less. However, the magnitude of the resistance change of bonded gages is only on the order of 0.1% of the unstrained resistance value. Thus, the output signal from a strain gage must be monitored by a well-designed Wheatstone bridge circuit. Bonded strain gages are manufactured in sizes from about $\frac{1}{8}$ in. $\times$ $\frac{1}{8}$ in. to a maximum of 1 in. $\times$ 1 in. The initial resistance of metal-wire strain gages usually lies between 120 and 400 Ω.

Strain-Gage Sensitivity or Gage Factor

When a gage wire as fine as 25 μm is strained within its elastic limit, the resistance of the wire changes because of changes in the diameter, length, and resistivity. The resulting strain gages may be used to measure extremely small displacements, on the order of nanometers. The basic equation for the resistance R of a wire with resistivity ρ (ohms-meter) length L (meters), and cross-sectional area A (meters squared) is given by

$$R = \frac{\rho L}{A}.$$ (3.7-1)

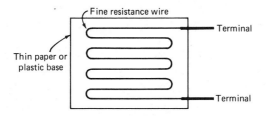

Figure 3.7-1 Bonded strain gage (as one active arm of Wheatstone bridge).

The differential change in R is expressed as

$$dR = \frac{\rho}{A} dL - \frac{\rho L}{A^2} dA + \frac{L}{A} d\rho. \tag{3.7-2}$$

Dividing this equation by Eq. (3.7-1) and assuming that the differentials can be replaced by the corresponding small, but finite, change Δ, we find that the fractional resistance change is

$$\frac{\Delta R}{R} = \frac{\Delta L}{L} - \frac{\Delta A}{A} + \frac{\Delta \rho}{\rho}. \tag{3.7-3}$$

Using the definition of Poisson's ratio μ, the fractional change in diameter is given by

$$\frac{\Delta D}{D} = -\mu \frac{\Delta L}{L}, \tag{3.7-4}$$

where $\Delta D/D$ is the strain in the lateral direction and $\Delta L/L$ is the strain in the axial direction. Since $A = \pi D^2/4$, $dA = (\pi/2)D dD$, or $\Delta A = (\pi/2)D \Delta D$; thus

$$\frac{\Delta A}{A} = 2 \frac{\Delta D}{D}. \tag{3.7-5}$$

Substituting Eq. (3.7-4) into Eq. (3.7-5) we find

$$\frac{\Delta A}{A} = -2\mu \frac{\Delta L}{L}. \tag{3.7-6}$$

Substituting Eq. (3.7-6) into Eq. (3.7-3) we obtain

$$\frac{\Delta R}{R} = \underbrace{(1 + 2\mu) \frac{\Delta L}{L}}_{\text{Dimensional effect}} + \underbrace{\frac{\Delta \rho}{\rho}}_{\text{Piezoresistive effect}}. \tag{3.7-7}$$

Therefore, the fractional resistance change is the sum of the term due to dimensional changes and the term caused by the change in resistivity with strain, referred to as the piezoresistive effect.

The gage factor or sensitivity of a strain gage is defined by

$$G \equiv \frac{\Delta R/R}{\Delta L/L} = 1 + 2\mu + \frac{\Delta \rho/\rho}{\Delta L/L}; \tag{3.7-8}$$

this equation is useful in comparing the performance of various gage materials. For most metals, $\mu \simeq 0.3$, then Eq. (3.7-8) becomes

$$G = 1.6 + \frac{\Delta \rho/\rho}{\Delta L/L}. \tag{3.7-9}$$

The gage factor for metals is primarily a function of dimensional effects; whereas, for semiconductors, the piezoresistive effect is dominant. The metal materials, constantan (Ni_{45}, Cu_{55}, advanced) and alloy 479 (Pt_{92}, W_8), have $G = 2.1$ and 3.6–4.4, and ρ_{TC} (temp. coeff.)($°C^{-1} \times 10^{-5}$) $= 2$ and 24, respectively. The semiconductors, p-type Si and p-type Ge, have $G = 100$–170 and 102, and $\rho_{TC} = 70$–700 and xx, respectively. The gage factor for semiconductors is approximately 50 or more times that of the metals. The desirable feature of higher gage factors for semiconductor devices is offset by their higher resistivity temperature coefficient. Thus, semiconductor strain gages must be used in systems with effective temperature-compensation designs.

Unbonded Strain Gages

Since bonded strain gages require large forces to measurably change their dimensions, smaller forces must be measured with more-sensitive devices of a different design, such as the commonly used unbonded strain gages. As shown in Fig. 3.7-2a, the unbonded gage typically consists of a wire 0.002 cm or less in diameter wound between a fixed frame and a movable member. This design requires a smaller force to change the wire's length. There are four sets of identical wires which have been wound under tension so that the movable member can be displaced by a small amount in either direction without reducing the tension in any of the wires to zero. When an applied force causes a motion of the movable member to the right, elements A and D increase in length causing the resistance of R_1 and R_4 to increase, whereas elements B and C decrease in length causing the resistance of R_2 and R_3 to decrease. To measure the change in resistance, the four elements are incorporated into a Wheatstone bridge in the order indicated in Fig. 3.7-2b. The unbalance current, indicated by meter M, is calibrated to read the magnitude of the displacement of the movable member. The balance arrangement provides its self-compensating action for any environment effects.

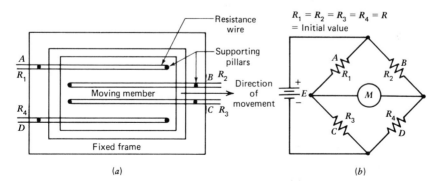

Figure 3.7-2 (a) Unbonded strain gage and (b) bridge.

The maximum fractional change in length that can be permitted with wire gages is typically 5000×10^{-6}, giving a 1% change in resistance when the gage factor is 2.0. Typically, the linearity is better than 0.2%, the full-scale output voltage is 40 mV (5 V bridge excitation), and the full-scale displacement is 0.005 cm, corresponding to a 30-g force. Resistance values are usually from 100 to 1000 Ω.

Null-Balance and Deflection-Balance Bridge Circuits

The bridge shown in Fig. 3.7-3 is balanced when $R_1/R_2 = R_3/R_4$. Strain gages or other resistance-type transducers may be connected as one or more active arms of a bridge circuit. The change in resistance can be detected by either null-balance or deflection-balance bridge circuits. The null-balance bridge results when the resistance change of the transducer is balanced out by a variable resistance of an adjacent arm in the bridge. The calibrated adjustment required for the null indicates the change in resistance of the transducer. On the other hand, the deflection-balance arrangement utilizes the amount of bridge imbalance to determine the change in transducer resistance.

Assume that all the active arms R_1 to R_4 have initial resistance R, which is much less than R_m, the internal resistance of the readout meter. Then an increase in resistance, ΔR, of all resistances still results in a balanced bridge. However, if $R_1 = R_4 = R + \Delta R$ and $R_2 = R_3 = R - \Delta R$, as shown in Fig. 3.7-2a, then the output voltage

$$\Delta v_0 = \left(\frac{R_4 + \Delta R}{(R_2 - \Delta R) + (R_4 + \Delta R)} - \frac{R_3 - \Delta R}{(R_1 + \Delta R) + (R_3 - \Delta R)} \right) E$$

$$= \frac{\Delta R}{R} E. \tag{3.7-10}$$

Equation (3.7-10) shows that Δv_0 is linearly related to ΔR for the four-active-arm bridge.

In Fig. 3.7-3, potentiometer R_x and resistor R_y bring the bridge into balance so that zero voltage output results from "zero" input of the measured parameter. To minimize loading effects, R_x is approximately 10 times the resistance of the bridge leg and R_y limits the maximum adjustment.

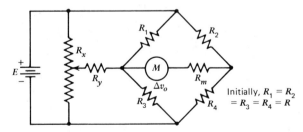

Figure 3.7-3 Wheatstone bridge including four active arms.

BIBLIOGRAPHY

Bell, D. A., *Electronic Instrumentation and Measurements*. Reston Publishing Company, Reston, Virginia, 1983, Chaps. 5 and 6.

Cobbold, R. S. C., *Transducers for Biomedical Measurements*. Wiley, New York, 1974, Chap. 6.

Cooper, W. D., *Electronic Instrumentation and Measurement Techniques*, 2nd ed. Prentice-Hall, Englewood Cliffs, New Jersey, 1978, Chaps. 6 and 7.

Lion, K. S., *Elements of Electrical and Electronic Instrumentation*. McGraw-Hill, New York, 1975, Chap. 2.

Webster, J. G., *Medical Instrumentation Application and Design*. Houghton Mifflin, Boston, 1978, Chap. 2.

Wolf, S., *Guide to Electronic Measurements and Laboratory Practice*. Prentice-Hall, Englewood Cliffs, New Jersey, 1973, Chap. 11.

Questions

3-1 Why are potentiometers more accurate than conventional voltmeters?

3-2 Describe the differences between manual and self-balancing potentiometers.

3-3 Describe the construction of a slide-wire potentiometer.

3-4 Describe in your own words how to measure an unknown potential E_x by using a standard cell E_s and a slide-wire potentiometer.

3-5 Describe the operating principle of the self-balancing potentiometer.

3-6 Describe the temperature-compensating potentiometer.

3-7 Describe the possible errors that arise from using the Wheatstone bridge.

3-8 Why are resistances with values less than 0.1 Ω difficult to measure with a simple bridge?

3-9 What is the purpose of the Kelvin double-bridge circuit?

3-10 Discuss the difference between the Kelvin and Wheatstone bridges.

3-11 Explain how stress or strain on a strain gage is measured with a Wheatstone bridge.

Problems

3.1-1 Prove that the maximum value of R_0 given by Eq. (3.1-2) is $R/4$ at $r = R/2$. (*Hint:* Start from setting $dR_0/dr = 0$.)

3.3-1 The voltage of a standard cell is measured with a manual potentiometer, which indicates a reading of 1.01892 V. When a 2-MΩ resistor is connected across the standard cell terminals, the potentiometer reading drops to 1.01887 V. Calculate the internal resistance R_{int} of the standard cell.

Answer. 98 Ω.

3.3-2   The slide-wire potentiometer of Fig. 3.3-1 has a working battery of 3 V and negligible internal resistance. The length of the slide wire is 200 cm and its resistance is 400 Ω. A 200-cm scale, placed along the slide wire, has 1-mm scale divisions and interpolation can be made to one-quarter of a division. The instrument is calibrated against a voltage reference source of 1.0180 V with the slider tap set at the 101.8-cm mark on the scale. Calculate (a) the working current I; (b) the resistance of the rheostat R_h; (c) the measurement range; (d) the resolution of the instrument, expressed in mV.

   Answer.   (a) I = 1.018 V/203.6 Ω = 5 mA; (b) R_h = 1 V/5 mA = 200 Ω; (c) IR_T = (5 mA)(400 Ω) = 2 V; (d) Resolution = (0.25 mm/2000 mm) $\times$ 2 V = 0.25 mV.

3.3-3   A manual potentiometer indicates a reading of 1.0 V when used to measure the voltage between two points in a dc circuit. A 20,000 Ω/V dc voltmeter reads only 0.5 V on its 2.5 V scale when connected to the same two points in the circuit. What is the circuit resistance between the two measured points?

   Answer.   50K Ω.

3.3-4   A manual potentiometer contains the following components: a working battery of 3.0 V and negligible internal resistance; a standard cell having an emf of 1.0191 V and an internal resistance of 200 Ω; a 200-cm adjustable precision slide wire having a total resistance of 200 Ω; a galvanometer having an internal resistance of 50 Ω in series with a protective resistance. The rheostat of the instrument is set so that the potentiometer is calibrated at the 101.91-cm mark of the precision adjustable resistance. (a) Calculate the current (I) flowing in the rheostat as well as its resistance value (R_h). (b) Calculate the value of the protective resistance (R_{pt}) necessary to limit the current in the galvanometer to 10 μA.

   Answer.   (a) I = 10 mA; R_h = 100 Ω. (b) R_{pt} = 101.76K Ω.

3.5-1   A resistor is measured using a Wheatstone bridge (Fig. 3.5-1*b*) and a balance is reached for the fixed resistors R_2 = 100 Ω ± 0.02%, R_1 = 1000 Ω ± 0.02%, and the variable resistor R_3 = 120.3 Ω ± 0.04%. In the worst case, the errors of R_3 and R_2 will be in the plus direction, and the errors of R_1 will be in the minus direction. What will be the unknown resistance in the worst case?

   Answer. R_x = (12.03)(1 ± 0.0008) = 12.03 ± 0.01 Ω.

3.5-2   A Wheatstone bridge (Fig. 3.5-1*b*) has R_1 = 250 Ω ± 0.02%, R_2 = 750 Ω ± 0.02%, and R_3 is a decade box with steps from 100 to 0.1 Ω. The resistors of the decade box are known to within ±0.05%. If R_3 is set to 153.7 Ω when a balance is obtained, find (a) R_x and (b) the percentage error which exists in the calculated value of R_x.

Answer. (a) R_x = 461.1 (1 ± 0.0009) = 461.1 ± 0.415 Ω. (b) Percentage error = 0.09%.

3.5-3 A Wheatstone bridge (Fig. 3.5-1*b*) has R_1 = 100 Ω, R_2 = 1000 Ω, R_3 = 200 Ω, and R_4 = 2002 Ω. The battery voltage is 5 V and its internal resistance is negligible. The galvanometer *G* has a current sensitivity of 10 mm/μA and an internal resistance R_G = 100 Ω. Determine the deflection of *G* caused by the 2-Ω imbalance in R_4. (*Hint:* Convert the bridge circuit into its Thevenin equivalent circuit, which is determined with respect to the galvanometer terminals. E_{TH} ≃ 1.11 mV; R_{TH} ≃ 730 Ω; I_G = 1.34 μA.)

Answer: d = 1.34 μA × 10 mm/μA = 13.4 mm.

3.5-4 A Wheatstone bridge (Fig. 3.5-1*b*) is used to measure high resistances and has R_2 = 10K and R_1 = 10 Ω. The variable resistor R_3 may be adjusted from 0 to 10K. The battery voltage is 10 V and its internal resistance is negligible. The galvanometer has an internal resistance of 50 Ω and a current sensitivity of 200 mm/μA. Calculate (a) the maximum measurable resistance and (b) the imbalance in R_x needed to produce a galvanometer deflection of 1 mm when the maximum resistance of part (a) is connected to the R_x terminals.

Answer: (a) 10 M Ω; (b) imbalance in R_x ≃ 50K.

3.6-1 The ratio arms (R_1, R_3, r_a, and r_b) of the Kelvin bridge in Fig. 3.6-2*a* are 1000 Ω each. The galvanometer has an internal resistance of 100 Ω and a current sensitivity of 500 mm/μA. A dc current of 10 A is passed through the standard arm (R_2) from a 2.2-V battery in series with a rheostat. The standard resistor is set at 0.1000 Ω and the galvanometer deflection is 20 mm. The connecting-lead resistance (r_c) may be neglected. Find R_x.

Answer. 0.1000088 Ω.

3.7-10 The wire in a strain gage is 10 cm in length and has an initial resistance of 150 Ω. An applied force causes a change in its length of 0.1 mm and a change in resistance of 0.3 Ω. Assume that the increase in resistivity with length is linear. Calculate (a) the strain caused by this force and (b) the gage factor of the device.

Answer. (a) σ = 0.001; (b) *G* = 2.

3.7-2 A resistance strain gage with a gage factor of 2 is fastened to a steel member subjected to a stress of 1050 kg/cm². The modulus of elasticity of steel is approximately 2.1 × 10⁶ kg/cm². Find the fractional resistance change of the device due to the applied stress. [*Hint:* Use Hooke's law, σ = *s*/*E*, where σ = strain = $\Delta L/L$, *s* = stress (kg/cm²), *E* = Young's modulus (kg/cm²).]

Answer. $\Delta R/R$ = 0.1%.

4
Capacitance Bridges and Their Applications

4.1 CAPACITANCE COMPARISON BRIDGES

Series-Capacitance Comparison Bridge

If the dissipation factor (D) of a capacitor is small ($0.001 < D < 0.1$), the series-capacitance comparison bridge of Fig. 4.1-1 is commonly used. As in the case of the Wheatstone bridge for dc measurements, the balance condition in this ac bridge is reached when the detector response is zero. The balance equation of the ac bridge is given by

$$\frac{Z_3}{Z_1} = \frac{Z_4}{Z_2},$$

or

$$\frac{R_3 - jX_{c3}}{R_1} = \frac{R_x - jX_{cx}}{R_2}. \qquad (4.1\text{-}1)$$

Two complex numbers are equal when both their real terms and their imaginary terms are equal. Equating the real terms of Eq. (4.1-1), we find $R_3/R_1 = R_x/R_2$, or

$$R_x = \left(\frac{R_2}{R_1}\right) R_3. \qquad (4.1\text{-}2)$$

Equating the imaginary terms of Eq. (4.1-1), we find $1/\omega C_3 R_1 = 1/\omega C_x R_2$,

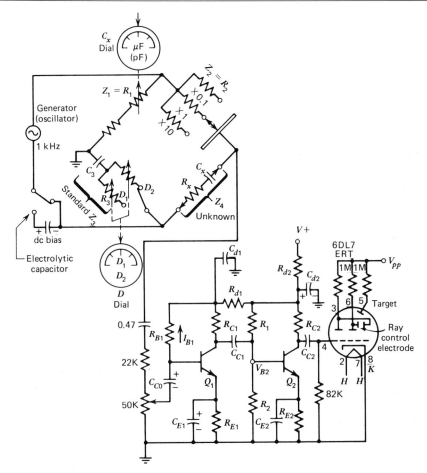

Figure 4.1-1 Series-capacitance comparison bridge with the electron-ray indicating device as the detector.

or

$$C_x = \left(\frac{R_1}{R_2}\right) C_3, \tag{4.1-3}$$

where R_1 and R_2 are the ratio arms, and C_3 is the standard capacitance. Equations (4.1-2) and (4.1-3) describe the two balance conditions that must be met simultaneously and they also indicate the two unknowns C_x and R_x expressed in terms of the known bridge components.

The dissipation factor (D) of a series RC circuit is defined as the cotangent of the phase angle, or

$$D \equiv \cot \theta = \frac{R_x}{1/\omega C_x} = \omega R_x C_x. \tag{4.1-4}$$

Note that D is the reciprocal of the quality factor Q, which is defined by

$$Q \equiv \tan \theta = \frac{1}{D} = \frac{1}{\omega R_x C_x}. \tag{4.1-5}$$

Substituting Eqs. (4.1-2) and (4.1-3) into (4.1-4) we find

$$D = \omega R_3 C_3. \tag{4.1-6}$$

This equation shows the dissipation factor expressed in terms of the known bridge components. The dissipation factor tells us something about the quality of a capacitor; that is, how close the phase angle of the capacitor is to the ideal value of 90°.

To satisfy both the balance conditions, R_1 and R_3 are selected as the two variable elements. Balance adjustment to obtain a null response is made by varying R_1 and R_3. The dials of R_1 and R_3 are calibrated, respectively, in C_x and D, and, thus, indicate the measured readings at balance. Measured ranges are controlled by the selector switch. In the electrolytic capacitor bridge the variable resistor R_3 is usually calibrated in units of power factor ($F_p = \cos \theta$).

Electron-Ray Indicating Circuit

The detector across the output of an ac bridge may be a meter, a set of earphones, or an electron-ray indicating device. As shown in Fig. 4.1-1, the bridge output is connected to an amplifier, which precedes the electron-ray tube (ERT). Note that in the ERT, each plate (pin 3, pin 6) connects a ray-control electrode (RCE), which will control the electron rays striking the target. When the bridge is being adjusted, the unbalanced signal is sufficiently amplified so that the ERT is operated. If the signal is larger, the control grid (pin 4) is more positive, so that both plate currents increase and both plate voltages decrease, resulting in a decrease of the RCE voltages; thus more electron rays strike the target, resulting in a smaller shadow area on the target, similar to the "magic eye" being closed. On the other hand, when the bridge becomes balanced, no signal is applied to the grid, so that the plates, and hence the RCEs, are at higher potentials, resulting in a larger shadow area on the target, similar to the eye being open.

Transistors Q_1 and Q_2 are connected in common-emitter configurations. They operate in active region with their BE junction forward biased and their BC junction reverse-biased. $V_{BE(\text{active})} \approx 0.3$ V for Ge npn, or $V_{BE(\text{active})} \approx 0.7$ V for Si npn. The voltage drop produced by the base current I_{B1} through resistor R_{B1} provides the reverse-bias for the BC junction of Q_1. The voltage divider $R_1 - R_2$ provides the appropriate dc bias voltages to Q_2 so that the bias circuit for Q_2 is independent of its beta. The values of coupling capacitors C_{c0}, C_{c1}, and C_{c2} must be large enough (usually, several microfarads each) so that the signal can be passed. The values of by-pass capacitors C_{E1} and C_{E2} are determined by using the relation $X_{C_E} \leq \frac{1}{10} R_E$. The $R_{d1} - C_{d1}$ and $R_{d2} - C_{d2}$ decoupling networks are used to suppress the oscillation due to the regenerative feedback via the internal resistance of the common collector source. Since the output voltage is 180° out of phase with the input voltage in a common-emitter amplifier stage, the phase of the output signal voltage from Q_2 is the same as that of the input voltage to Q_1.

In the bias circuit of the Q_1 stage, the emitter feedback resistor (R_{E1}) provides a means of bias stabilization so that the change in collector current due to the reverse-saturation current (I_{co}) would not cause a large shift in the operating point. Assuming $\beta = 100$, $V_{BE} = 0.7$ V, $V_{cc} = 20$ V, $V_{CEQ} = 10$ V, and $I_c = 2$ mA (operating point), the design calculation is carried out as follows:

$$V_{E1} = V_E \simeq \frac{1}{10} V_{cc} = \frac{20 \text{ V}}{10} = 2 \text{ V},$$

$$R_{E1} = R_E = \frac{V_E}{I_E} \simeq \frac{V_E}{I_C} = \frac{2 \text{ V}}{2 \text{ mA}} = 1\text{K},$$

$$R_{c1} = R_c = \frac{V_{cc} - V_{CEQ} - V_E}{I_c} = \frac{(20 - 10 - 2) \text{ V}}{2 \text{ mA}} = 4\text{K},$$

$$I_{B1} = I_B = \frac{I_c}{\beta} = \frac{2 \text{ mA}}{100} = 20 \text{ } \mu\text{A},$$

$$R_{B1} = R_B = \frac{V_{cc} - V_{BE} - V_E}{I_B} = \frac{(20 - 0.7 - 2) \text{ V}}{20 \text{ } \mu\text{A}} = 0.87 \text{ M}\Omega.$$

The bias circuit of the Q_2 stage in Fig. 4.1-1 provides stabilization both for leakage current and current gain changes. The values of R_{E2}, R_{C2}, R_1, and R_2 must be obtained for a specified operating point. Assuming $I_C = 10$ mA, $V_{CEQ} = 8$ V (operating point), $V_{cc} = 20$ V, $V_{BE} = 0.7$ V, and $h_{FE} = 80$, the design calculation is carried out as follows:

$$V_{E2} = V_E \simeq \frac{1}{10} V_{cc} = \frac{1}{10} (20 \text{ V}) = 2\text{V},$$

$$R_{E2} = R_E \simeq \frac{V_E}{I_c} = \frac{2 \text{ V}}{10 \text{ mA}} = 200 \text{ }\Omega.$$

$$R_{C2} = R_C = \frac{V_{CC} - V_{CEQ} - V_E}{I_C} = \frac{(20 - 8 - 2) \text{ V}}{10 \text{ mA}} = 1\text{K},$$

$$V_{B2} = V_B = V_E + V_{BE} = 2 \text{ V} + 0.7 \text{ V} = 2.7 \text{ V}.$$

R_1 and R_2 are determined using the two equations

$$V_{B2} = V_B = \frac{R_2}{R_1 + R_2} (V_{cc})$$

and

$$R_2 \leq \frac{1}{10} (\beta R_E),$$

where βR_E is the approximate resistance seen looking back into the base. Solving these equations results in $R_1 \simeq 10\text{K}$ and $R_2 \simeq 1.6\text{K}$.

Parallel-Capacitance Comparison Bridge

If the dissipation factor of a capacitor is large ($0.05 < D < 50$), the parallel-capacitance comparison bridge of Fig. 4.1-2 is usually employed. The balance equation of this bridge is given by

$$Z_1 Y_3 = Z_2 Y_4$$

or

$$R_1 \left(\frac{1}{R_p} + j\omega C_3 \right) = R_2 \left(\frac{1}{R_x} + j\omega C_x \right). \qquad (4.1\text{-}7)$$

Separation of the real and j terms yields $R_1/R_p = R_2/R_x$ or

$$R_x = \left(\frac{R_2}{R_1} \right) R_p \qquad (4.1\text{-}8)$$

and $R_1 C_3 = R_2 C_x$ or

$$C_x = \left(\frac{C_3}{R_2} \right) R_1. \qquad (4.1\text{-}9)$$

The quality factor of a capacitor is the tangent of the phase angle, or

$$Q = \tan \theta = \frac{1/X_{C_x}}{1/R_x} = \omega C_x R_x. \qquad (4.1\text{-}10)$$

Substituting Eqs. (4.1-8) and (4.1-9) into (4.1-10), we find

$$Q = \omega C_3 R_p. \qquad (4.1\text{-}11)$$

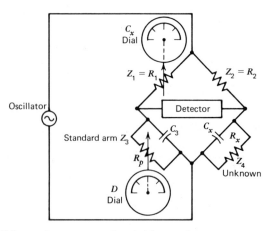

Figure 4.1-2 Parallel-capacitance comparison bridge used to measure capacitors of large dissipation factors.

Thus the dissipation factor is

$$D = \frac{1}{Q} = \frac{1}{\omega C_3 R_p}.$$ (4.1-12)

In this bridge R_2 and C_3 are fixed, and R_1 and R_p are variable. At the condition of balance, both C_x and D are read directly from the R_1 and R_p settings, respectively.

4.2 SCHERING BRIDGE

For measuring insulating properties of capacitors with extremely low dissipation factors, the Schering bridge shown in Fig. 4.2-1 offers more accurate readings than either of the capacitance-comparison circuits. The balance equation of this bridge is

$$Z_x = Z_2 Z_3 Y_1,$$ (4.2-1)

or

$$R_x - \frac{j}{\omega C_x} = R_2 \left(\frac{-j}{\omega C_3} \right) \left(\frac{1}{R_1} + j\omega C_1 \right).$$ (4.2-2)

Expanding,

$$R_x - \frac{j}{\omega C_x} = \frac{R_2 C_1}{C_3} - \frac{jR_2}{\omega C_3 R_1}.$$ (4.2-3)

Equating the real terms and the j terms, we obtain

$$R_x = \frac{R_2 C_1}{C_3}$$ (4.2-4)

and

$$C_x = \frac{C_3 R_1}{R_2},$$ (4.2-5)

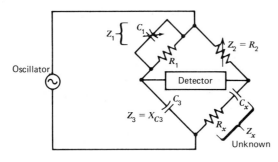

Figure 4.2-1 Schering bridge for measuring capacitance.

where C_1 and R_2 are chosen as two variables for balance adjustment. The dial of R_2 is calibrated in terms of C_x.

The power factor (F_p) of the unknown is the cosine of the phase angle θ. For $\theta \simeq 90°$, $X_x \simeq Z_x$. Thus

$$F_p = \cos \theta = \frac{R_x}{Z_x} \simeq \frac{R_x}{X_x} \simeq \omega C_x R_x. \qquad (4.2\text{-}6)$$

The dissipation factor of the unknown is given by

$$D = \frac{1}{Q} = \cot \theta = \frac{R_x}{X_x} = \omega C_x R_x = F_p. \qquad (4.2\text{-}7)$$

Substituting Eqs. (4.2-4) and (4.2-5) into (4.2-7), we find

$$D = \omega R_1 C_1 = F_p. \qquad (4.2\text{-}8)$$

If the value of R_1 is fixed, the dial of C_1 may be calibrated in terms of D.

4.3 A TYPICAL AMPLIFIER CIRCUIT FOR AC BRIDGE NULL DETECTION

Figure 4.3-1 shows a typical detector–amplifier circuit contained in Hewlett-Packard Model 4265B universal bridge. The amplifier circuit can be used in capacitance bridge or other ac bridges for null detection. The circuit consists of a preamplifier, a filter amplifier, and a meter amplifier. The circuit amplifies a minimized unbalancing-signal to provide a sufficient output so that the meter or crystal earphone can be driven. A 1-kHz band-pass filter is provided to reject other unwanted signals. An external tuned null detector may be used instead of the earphone.

The preamplifier is composed of input limiter R_1, D_1, and D_2, source follower Q_1, direct-coupled amplifier $Q_2 - Q_4$, and feedback dynamic resistors D_3 and D_4. An unbalancing signal from the ac bridge is fed into the input limiter, which limits the input level of Q_1 to 1.2 V peak-to-peak. Source follower Q_1 is an FET impedance converter. The input signal is amplified by the direct-coupled amplifier whose gain in the small-signal domain is approximately 200, as determined by the ratio of $(R_{10} + R_7)/R_7$. In the large signal domain, feedback dynamic resistors D_3 and D_4 work to decrease the gain gradually since they are in parallel with R_{10}.

The filter amplifier is selectable so as to apply the flat-gain amplifier or the 1-kHz filter amplifier with 1 kHz/flat switch S_1. With the flat-gain amplifier, the gain is about 10 from 50 Hz to 10 kHz. With an active 1-kHz filter amplifier, the gain is about 100 at 1 kHz. The selectivity is better than 26 dB. The feedback dynamic resistors D_5 and D_6 are similar to D_3 and D_4 and limit the maximum output to 1.2 V peak-to-peak. This output is fed to both the meter amplifier and the earphone jack J_2 provided for the earphone or external tuned null detector. The output impedance of J_2 is about 40 kΩ.

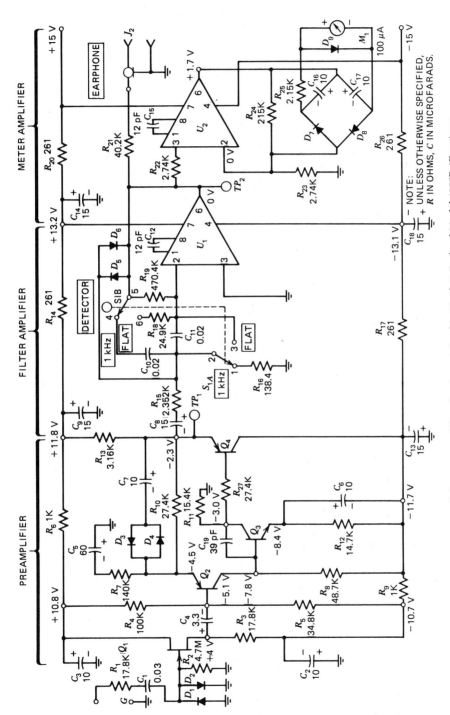

Figure 4.3-1 Typical detector amplifier circuit contained in Hewlett-Packard Model 4265B. (Courtesy of Yokogawa-Hewlett-Packard LTD.)

The meter amplifier is a negative feedback amplifier, and the meter circuit is inserted into the negative feedback network. The meter is driven by a constant-current driver whose current is determined by the ratio of the voltage of $U_2(3)$ and R_{23}. Thus, the meter is driven adequately without suffering from low sensitivity even at null conditions.

4.4 PHASE DETECTORS

The capacitance and inductance bridges require a phase detector to indicate the output condition. This detector is only needed to indicate relative phase—that is, whether or not a signal is in phase or out of phase with a reference signal. A typical phase detector circuit is shown in Fig. 4.4-1. The

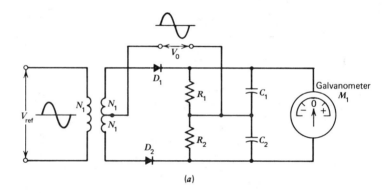

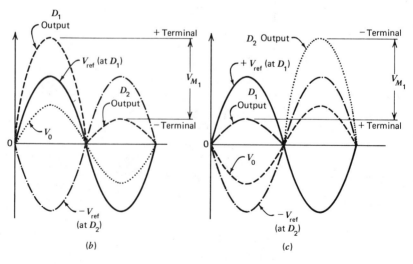

Figure 4.4-1 (a) Phase detector circuit. Signal relationships in phase detector: (b) when V_0 and V_{ref} are in phase; (c) when V_0 and V_{ref} are out of phase.

reference signal V_{ref} is a sine wave, and is usually the same signal that is used for bridge excitation.

When the bridge output signal V_0 is zero, diode D_1 conducts on one-half of the reference signal cycle, while D_2 conducts on the alternate half-cycle. A net potential across the meter movement M_1 is zero, since these alternations over the entire cycle are averaged by the effect of C_1, C_2, and the natural mechanical damping of M_1.

Since signal V_0 is not zero, it aids V_{ref} when in phase and opposes V_{ref} when out of phase. V_0 is applied to diodes D_1 and D_2 in parallel, while V_{ref} is applied in push–pull. On one-half of its cycle V_0 will forward bias both diodes simultaneously, whereas on the opposite half-cycle it will reverse bias both diodes. When V_0 and V_{ref} are in phase and on their positive half cycles, diode D_1 is forward biased by the sum of V_0 and V_{ref}, while the anode of D_2 sees a lower potential created by the difference of $V_0 - V_{ref}$. This situation causes the output of D_1 to be higher than that of D_2, thereby meter M_1 swings positive. On the second half-cycle the polarities are reversed, but the output relationships seen by M_1 are the same. When signal V_0 and V_{ref} are out of phase, exactly the opposite occurs, hence the output of D_2 is more positive than that of D_1, and so the meter M_1 swings negative.

The amplitude of V_{ref} is constant, but the amplitude of V_0 changes with bridge balance conditions. This detector circuit is sensitive to both phase and magnitude.

4.5 APPLICATIONS OF THE CAPACITANCE BRIDGE

Measurement of Small Capacitance

In order to accurately measure a small value of unknown capacitance C_x, the following substitution method is often used. The resistance of the two ratio arms are made equal, and a small capacitor with a known value on the general order of C_x is connected across the unknown terminals. The bridge is balanced and the capacitance dial reading C_a is noted. Then, the unknown capacitor is connected in paralled with the known capacitor, and when the bridge is balanced, the reading C_b is noted. The reading difference $C_b - C_a$ gives the value C_x of the unknown capacitor with high accuracy. The chief advantage of the substitution method is that the experimental error due to stray capacitances is minimized.

Measurement of the Winding Capacitance of a Transformer

Figure 4.5-1 shows a capacitance bridge used to measure the winding capacitance of the primary or secondary of a transformer to core (ground). Since the winding capacitance is usually a small value, the bridge frequency must be sufficiently low so that inductive reactance and resonance effects can be neglected. A 1-kHz (or 60 Hz) capacitance bridge may be used when the winding inductance is comparatively small (or large).

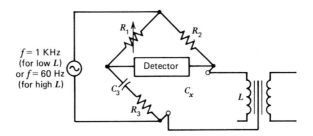

Figure 4.5-1   A capacitance bridge used to measure the winding capacitance.

Other Applications

The capacitance bridge can also be used for other measurements. Two examples are described as follows.

When the length of a coaxial cable is determined by using a capacitance bridge, its length is measured in terms of capacitance per foot. For example, a coaxiable cable of type RG-58/U (for general laboratory use) has a capacitance of about 30 pF/ft; if its measured capacitance is 0.009 μF, then the length of the cable is 300 ft.

A capacitance ratio of a two-section ganged variable capacitor can be measured by using a capacitance bridge which provides a ratio function.

4.6   CAPACITANCE BRIDGE EQUIVALENT CIRCUITS

The bridge of Fig. 4.1-1 has a standard arm consisting of capacitor C_3 in series with resistor R_3 and is commonly used for measuring the capacitors of small dissipation factors. The bridge of Fig. 4.1-2 has a standard arm consisting of capacitor C_3 in parallel with resistor R_3 and is commonly used for measuring the capacitors of large dissipation factors. When a capacitor to be measured is connected across the unknown terminals, the bridge is adjusted to reach a balance. Whether the balance is obtained on the basis of series R and C, or on the basis of parallel R and C, depends on how the standards are connected in the bridge circuit. The bridge will indicate series components if the standards are connected in series, or the bridge will indicate parallel components if the standards are connected in parallel.

A capacitor can be represented by the equivalent series $R_s C_s$ circuit or parallel $R_p C_p$ circuit to account for the dissipation. In order to change a bridge measurement to the most useful form for the problem at hand, the series $R_s C_s$ circuit may be converted into the equivalent parallel $R_p C_p$ circuit, or vice versa. In the series $R_s C_s$ circuit, its impedance and admittance are

$$Z_s = R_s - jX_{cs} \quad \text{and} \quad Y_s = \frac{1}{Z_s} = \frac{1}{R_s - jX_{cs}} = \frac{R_s + jX_{cs}}{R_s^2 + X_{cs}^2}.$$

Thus

$$Y_s = \frac{R_s}{R_s^2 + X_{cs}^2} + j\frac{X_{cs}}{R_s^2 + X_{cs}^2} = \frac{1}{R_p} + j\frac{1}{X_{cp}} = \frac{1}{Z_p}, \qquad (4.6\text{-}1)$$

where Z_p is the impedance of the equivalent parallel circuit,

$$R_p = \frac{R_s^2 + X_{cs}^2}{R_s}, \qquad (4.6\text{-}2)$$

and

$$X_{cp} = \frac{R_s^2 + X_{cs}^2}{X_{cs}}. \qquad (4.6\text{-}3)$$

Equations (4.6-2) and (4.6-3) give the necessary conditions for the conversion of the series $R_s C_s$ circuit to its equivalent parallel $R_p C_p$ circuit.

The quality factor of a series $R_s C_s$ circuit is $Q = X_{cs}/R_s$, thus

$$X_{cs} = QR_s. \qquad (4.6\text{-}4)$$

Substitution of Eq. (4.6-4) into Eq. (4.6-2) yields

$$R_p = \frac{R_s^2 + Q^2 R_s^2}{R_s} = (1 + Q^2)R_s. \qquad (4.6\text{-}5)$$

Substituting $R_s = X_{cs}/Q$ into Eq. (4.5-3), we obtain

$$X_{cp} = \frac{(X_{cs}/Q)^2 + X_{cs}^2}{X_{cs}} = \left(1 + \frac{1}{Q^2}\right) X_{cs}. \qquad (4.6\text{-}6)$$

The parallel $R_p C_p$ circuit will be converted into the equivalent series $R_s C_s$ circuit if

$$R_s = \frac{R_p}{1 + Q^2} \qquad (4.6\text{-}7)$$

and

$$X_{cs} = \frac{X_{cp}}{1 + 1/Q^2}, \qquad (4.6\text{-}8)$$

where Q is the quality factor of the parallel $R_p C_p$ circuit and is given by

$$Q = \frac{R_p}{X_{cp}}. \qquad (4.6\text{-}9)$$

Substituting $Q = 1/D$ into Eqs. (4.5-5), (4.5-6), (4.5-7), and (4.5-8), we obtain

$$R_p = \left(1 + \frac{1}{D^2}\right) R_s = \left(\frac{1 + D^2}{D^2}\right) R_s, \qquad (4.6\text{-}10)$$

$$X_{cp} = (1 + D^2)X_{cs} \quad \text{or} \quad C_p = \frac{C_s}{1 + D^2}, \qquad (4.6.11)$$

$$R_s = \left(\frac{D^2}{1 + D^2}\right) R_p, \qquad (4.6\text{-}12)$$

and

$$C_s = (1 + D^2)C_p, \qquad (4.6\text{-}13)$$

where

$$D = \omega R_s C_s = \frac{1}{\omega R_p C_p}. \qquad (4.6\text{-}14)$$

For $D < 0.1$, $C_s \simeq C_p$.

4.7 WIEN BRIDGE AND ITS APPLICATIONS

Wien Bridge for Frequency Measurement

The Wien bridge of Fig. 4.7-1 may be used to measure frequency, to determine the frequency in oscillators, or to be used for other applications. The description of its operation starts from its balance condition, that is,

$$\frac{Z_1}{Z_3} = \frac{Z_2}{Z_4} \quad \text{or} \quad \frac{Z_2}{Z_4} = Z_1 Y_3. \qquad (4.7\text{-}1)$$

Substituting the corresponding impedance and admittance values in this general equation, we obtain

$$\frac{R_2}{R_4} = \left(R_1 - j\frac{1}{\omega C_1} \right) \left(\frac{1}{R_3} + j\omega C_3 \right).$$

Expanding,

$$\frac{R_2}{R_4} = \frac{R_1}{R_3} - j\frac{1}{\omega C_1 R_3} + j\omega C_3 R_1 + \frac{C_3}{C_1}. \qquad (4.7\text{-}2)$$

Equating the j terms, we obtain

$$\frac{1}{\omega C_1 R_3} = \omega C_3 R_1; \qquad (4.7\text{-}3)$$

hence

$$\omega = 2\pi f = \frac{1}{\sqrt{R_1 R_3 C_1 C_3}},$$

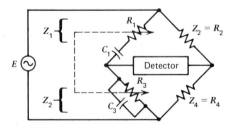

Figure 4.7-1   Wien bridge. Usually $R_1 = R_3$ $= R$, $C_1 = C_3 = C$. At balance, $R_2 = 2R_4$, $f = \frac{1}{2}\pi RC$.

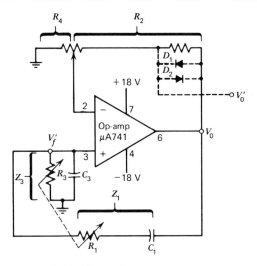

Figure 4.7-2 A Wien bridge oscillator. $R_1 = R_3 = R$, $C_1 = C_3 = C$.

or

$$f = \frac{1}{2\pi\sqrt{R_1 R_3 C_1 C_3}}. \qquad (4.7\text{-}4)$$

Equating the real terms, we obtain

$$\frac{R_2}{R_4} = \frac{R_1}{R_3} + \frac{C_3}{C_1}. \qquad (4.7\text{-}5)$$

The components are usually chosen such that $R_1 = R_3 = R$ and $C_1 = C_3 = C$. This reduces Eq. (4.7-5) to $R_2/R_4 = 2$ and Eq. (4.7-4) to

$$f = f_0 = \frac{1}{2\pi RC}, \qquad (4.7\text{-}6)$$

which is the general equation for the frequency of the Wien bridge. Hence, the Wien bridge is balanced or at resonance when $R_2/R_4 = 2$ and $f = 1/2\pi RC$. Provided that $R_2 = 2R_4$, the bridge may be used as a frequency-determining device balanced by a single control. This control may be calibrated in terms of frequency.

Wien Bridge Oscillator

The Wien bridge oscillator of Fig. 4.7-2 consists of an operational amplifier (op-amp) (see Section 6.7) and a Wien bridge circuit. The output terminals of the bridge are connected between the inverting input and noninverting input of the op-amp.

There are two feedback paths in the Wien bridge oscillator of Fig. 4.7-2: positive (regenerative) feedback through Z_1 and Z_3, which determine the frequency of oscillation, and negative (degenerative) feedback through R_2 and R_4, which affect the amplitude of oscillation. Oscillation can take place only at the frequency ($f_0 = 1/2\pi RC$) that permits the signal voltage (V'_f) across R_3 to be in phase with the output voltage (V_0), and also provided that the positive feedback voltage exceeds the negative feedback voltage and cancels out the circuit losses. The loop gain of the positive feedback circuit is given by $-\beta A$, where

$$\beta = \frac{V'_f}{V_0} = -\frac{Z_3}{Z_1 + Z_3} \quad \text{and} \quad A = 1 + \frac{R_2}{R_4};$$

then

$$-\beta A = \frac{\dfrac{-jR/\omega C}{R - j\omega C}}{\left(R - \dfrac{j}{\omega C}\right) + \left(\dfrac{-jR/\omega C}{R - j/\omega C}\right)} \left(1 + \frac{R_2}{R_4}\right)$$

$$= \frac{-jR/\omega C}{(R - j/\omega C)^2 - jR/\omega C} \left(1 + \frac{R_2}{R_4}\right)$$

$$= \frac{-j\omega RC}{(\omega RC)^2 - 1 - j3\omega RC} \left(1 + \frac{R_2}{R_4}\right).$$

Let $\alpha = \omega RC$, then

$$-A\beta = \frac{\alpha}{3\alpha - j(1 - \alpha^2)} \left(1 + \frac{R_2}{R_4}\right)$$

$$= \frac{\alpha[3\alpha + j(1 - \alpha^2)]}{(3\alpha)^2 + (1 - \alpha^2)^2} \left(1 + \frac{R_2}{R_4}\right). \quad (4.7\text{-}7)$$

In order to cause a feedback system to oscillate, the Barkhausen criterion ($-A\beta = 1$) must be satisfied (see Section 10.1). This condition implies both that $|A\beta| = 1$ and that the phase of $-A\beta$ is zero. This criterion is consistent with the feedback formula $A_f = A/(1 + \beta A)$. For if $-\beta A = 1$, then $A_f \rightarrow \infty$, which may be interpreted to mean that there exists an output voltage even in the absence of an externally applied signal voltage. The Barkhausen criterion ($-A\beta = 1$) applied to Eq. (4.7-7) requires that $\alpha = 2\pi f_0 RC = 1$ and $\frac{1}{3}(1 + R_2/R_4) = 1$. Thus,

$$f_0 = \frac{1}{2\pi RC} \quad \text{and} \quad R_2 = 2R_4. \quad (4.7\text{-}8)$$

In this case, $A = 1 + R_2/R_4 = 3$.

Continuous variation of frequency f_0 is accomplished by changing simultaneously the two variable resistors. The maximum frequency of oscillation is limited by the slew rate of the amplifier. The slew rate is the time rate of change of the closed-loop amplifier output voltage under large-signal conditions.

For the improvement of amplitude stabilization, the element R_2 in Fig. 4.7-2 may be shunted by two diodes D_1 and D_2 with D_1 (D_2) anode connected to D_2 (D_1) cathode. If output V_0 increases (for any reason), the current in the forward-biased diode will increase, causing the value of R_2 to decrease. Then the decreased R_2/R_4 ratio will result in a reduction of the gain A.

4.8   BRIDGE CIRCUITS FOR DIFFERENTIAL CAPACITANCE TRANSDUCERS

The capacitance (C) depends on plate area (A), plate separation (d), and absolute permittivity (ϵ) of the medium between the two parallel plates of a capacitor. It is possible to monitor the plate displacement by changing any of the three parameters A, d, or ϵ. However, the method that is easiest to implement and that is most commonly used is to change the separation between the plates. If the effects of electric field fringing at the edges can be ignored, the capacitance is given by

$$C = \epsilon_0 \epsilon_r \frac{A}{d} = \epsilon \frac{A}{d}, \qquad (4.8\text{-}1)$$

where ϵ_0 is the dielectric constant of free space, ϵ_r is the relative dielectric constant of the insulator (1.0 for air), and the absolute permittivity $\epsilon = \epsilon_0 \epsilon_r$ = $8.85 \times 10^{-14} \epsilon_r$ F/cm. The sensitivity S of a capacitive transducer to changes in plate separation Δd is found by differentiating Eq. (4.8-1):

$$S = \frac{\Delta C}{\Delta d} = -\frac{\epsilon A}{d^2}, \qquad (4.8\text{-}2)$$

which indicates that the sensitivity increases as the plate separation decreases.

Differential-capacitor systems such as shown in Fig. 4.8-1 are commonly used when accurate measurements are required. The parallel-plate differential capacitor has the advantage of linearly relating displacement to the fractional difference $(C_1 - C_2)/(C_1 + C_2)$. Let d be the equilibrium plate separation and x be the displacement in the direction of C_1. Then

$$C_1 = \frac{\epsilon A}{d - x} \quad \text{and} \quad C_2 = \frac{\epsilon A}{d + x}. \qquad (4.8\text{-}3)$$

Hence,

$$\frac{C_1 - C_2}{C_1 + C_2} = \frac{x}{d}. \qquad (4.8\text{-}4)$$

The bridge of Fig. 4.8-1a contains the differential-capacitor values C_1 and

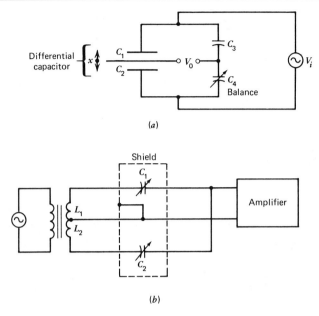

(a)

(b)

Figure 4.8-1 Bridge circuits for differential three terminal capacitors. (a) Capacitance bridge with output proportional to fractional difference in capacitance; (b) bridge circuit consisting of differential capacitor and transformer ratio arms.

C_2, and hence provides an output voltage proportional to the fractional difference in capacitance given by Eq. (4.8-4). Since C_1 and C_2 are equal at the equilibrium position and C_4 is balanced to equal C_3, the output voltage is given by

$$V_0 = \frac{V_i}{2d} x = \frac{V_i}{2} \frac{C_1 - C_2}{C_1 + C_2}. \tag{4.8-5}$$

A transformer ratio-arm bridge is shown in Fig. 4.8-1b. At balance, $C_1 = C_2$ and no current flows into the amplifier. The amplifier current is directly related to the degree of bridge imbalance $(C_1 - C_2)$ that may be found from Eq. (4.8-3):

$$C_1 - C_2 = \frac{2\epsilon A x}{d^2 - x^2}. \tag{4.8-6}$$

Note that the capacitance difference $(C_1 - C_2)$ is linear with the displacement x under the normal operating condition $(d \gg x)$. This type of bridge circuit has a sensitivity in the order of 10^{-5} pF. It is capable of high accuracy, and the balance is independent of the third terminal shield.

4.9   BRIDGE-TYPE ATTENUATORS

A series $R_1 - R_2$ combination may be used as a resistance attenuator, which is frequently connected across the input circuit of a stage of amplification.

The Thevenin equivalent resistance ($= R_1/R_2$) and the input capacitance of the stage make up an RC low-pass network, which ordinarily introduces an unacceptable rise time of the waveform applied to the amplifier. The attenuator may be compensated, so that its attenuation is independent of the frequency, by shunting R_1 by a capacitance C_1, as shown in Fig. 4.9-1a. The circuit is a simple capacitance bridge. At balance, $I_{xy} = 0$. Then

$$\frac{1/\omega C_2}{1/\omega C_1} = \frac{R_2}{R_1},$$

or

$$C_1 = \frac{R_2 C_2}{R_1}. \tag{4.9-1}$$

Thus for $C_1 = R_2 C_2/R_1 \equiv C_p$, the compensation is perfect, and the output voltage is given by

$$v_0 = av_i = \frac{R_2}{R_1 + R_2} v_i, \tag{4.9-2}$$

where $a = R_2/(R_1 + R_2)$. Note the output independent of the frequency.

Consider the appearance of the output voltage for a step input of magnitude V when the compensation is incorrect. An infinite current through C_1 and C_2 exists at $t = 0$ for an infinitesimal time, so that a finite charge $q = \int_{0-}^{0+} i\, dt$ is delivered to each capacitor. At $t = 0+$ (immediately after $t = 0$),

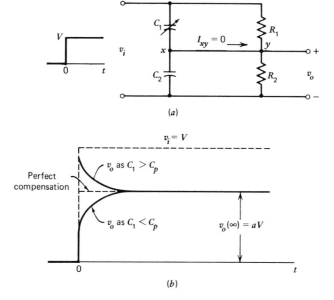

(a)

(b)

Figure 4.9-1 (a) Compensated attenuator operated as a bridge. (b) Response of the attenuator to a step-voltage input.

$$V = \frac{q}{C_1} + \frac{q}{C_2} = \frac{(C_2 + C_1)q}{C_1 C_2},$$

thus the output voltage at $t = 0+$ is

$$v_0(0+) = \frac{q}{C_2} = \frac{C_1}{C_1 + C_2} V. \tag{4.9-3}$$

Since a capacitor acts as an open circuit under steady-stage conditions for an applied dc voltage, the final output voltage at $t = \infty$ is determined by the resistors. Thus,

$$v_0(\infty) = \frac{R_2}{R_1 + R_2} V. \tag{4.9-4}$$

Because of the Thevenin resistance $R = R_1 R_2/(R_1 + R_2)$ in parallel with $C = C_1 + C_2$, the decay of the output from initial to final value takes place exponentially with a time constant $\tau = RC$. The responses of the attenuator for $C_1 = C_p (\equiv R_2 C_2/R_1)$, $C_1 > C_p$, and $C_1 < C_p$ are indicated in Fig. 4.9-1b. Perfect compensation is obtained if $v_0(0+) = v_0(\infty)$, equivalent to $C_1 = R_2 C_2/R_1 \equiv C_p$.

In the above analysis we have implicitly assumed a generator with zero source impedance (R_s). In practice, the infinite current cannot exist due to $R_s \neq 0$. Thus, the ideal step response can no longer be obtained from a practical compensated attenuator shown in Fig. 4.9-2. However, an improvement in rise time does result if the compensated attenuator is used. If $R_s \ll R_1 + R_2$, as is usually the case, the input to the attenuator will be an

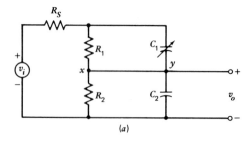

(a)

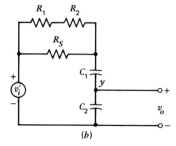

(b)

Figure 4.9-2 (a) Compensated attenuator including source impedance R_s. (b) Thevenin equivalent circuit: $V_i' = (R_1 + R_2)V/(R_s + R_1 + R_2)$. This equivalent circuit is obtained from the circuit in (a) when the branch xy is omitted.

exponential of time constant $\tau = R_s C_1 C_2/(C_1 + C_2)$. A cathode-ray-oscilloscope probe is an example of the bridge-type attenuator.

BIBLIOGRAPHY

Cobbold, R. S. C., *Transducers for Biomedical Measurements, Principle and Applications*. Wiley, New York, 1974, Chap. 6.

Cooper, W. D., *Electronic Instrumentation and Measurement Techniques*, 2nd ed. Prentice-Hall, Englewood Cliffs, New Jersey, 1978, Chap. 8.

Millman, J., *Microelectronics*. McGraw-Hill, New York, 1979, Chap. 17.

Operating and Service Manual of Hewlett-Packard Model 4265B Universal Bridge. Yokogawa-Hewlett-Packard, Ltd., Tokyo, Japan, 1975.

Webster, J. G., *Medical Instrumentation Application and Design*, Houghton Mifflin Company, Boston, 1978, Chap. 2.

Questions

4-1 Why is an electron-ray tube used in a capacitance bridge?

4-2 What type of bridge is best suited for measuring the capacitance of small-*D* capacitors?

4-3 What type of bridge is best suited for measuring the capacitance of large-*D* capacitors?

4-4 How is the power factor of a capacitor expressed?

4-5 What type of bridge is best suited for measuring the power factor of a low-loss capacitor?

4-6 What are the purposes of the source follower Q_1 and the direct-coupled amplifier $Q_2 - Q_4$ in Fig. 4.3-1?

4-7 Name several types of ac-bridge null detectors.

4-8 Describe the operation principle of phase detectors.

4-9 Explain the substitution method for measuring small values of capacitance with optimum accuracy.

4-10 Explain how the frequency is measured by the Wien bridge.

4-11 In Fig. 4.6-2*b*, we see an inverting op-amp circuit in which the non-inverting (+) input is grounded. The inverting (−) input is called the _____.

4-12 Describe the advantage of the transducer which is made of the parallel-plate differential capacitor.

4-13 Explain how the bridge-type attenuator is used as a cathode-ray-oscilloscope probe.

Problems

4.1-1 A balanced ac bridge has the following constants: $Z_{ab} = R_1 = 500$ Ω; $Z_{bc} = R_2 = 1K$; $Z_{cd} = Z_x$; $Y_{da} = Y_3 = j\omega$ (0.2 μF). A voltage of 10 V at 1 kHz is applied to the bridge at points a and c. (a) Calculate the unknown constants. (b) If the 1K resistor is changed to 1004 Ω, what is the voltage across the high-impedance detector? (c) Repeat (b), with the detector and the oscillator interchanged.

Answer.   (a) 0.1 μF; (b) 14 $\underline{/45°}$ mV; (c) 10 $\underline{/0°}$ mV.

4.1-2   In an ac bridge, $Z_{ab} = Z_1 = -j(10^6/0.01\omega)\Omega$, $Z_{bc} = R_2 = 2000\ \Omega$, $Z_{da}(\Omega) = Z_3 = 6000 - j(10^6/0.02\omega)$, and $Z_{cd} = Z_4 = R_x - j(1/\omega C_x)$ = unknown. The bridge is balanced at a frequency such that ω = 50,000 rad/s. Determine the unknown constants, R_x and C_x.

Answer. $R_4 = 1000\ \Omega$; $C_4 = 0.0033\ \mu$F.

4.1-3   Assume that the Q_2 stage in Fig. 4.1-1 has the following operating conditions: $h_{FE} = 40$, $V_{CEQ} = 12$ V, $I_c = 0.87$ mA, $V_{BE} = 0.7$ V, and $V_{cc} = 22$ V. Calculate R_{E2}, R_{C2}, R_1 and R_2.

Answer.   2.5K, 9K, 66K, 10K.

4.2-1   In Fig. 4.2-1, $R_1 = 1$K, $R_2 = 2$K, and $C_3 = 100$ pF. When the bridge is balanced, what is the value of the unknown capacitor?

Answer.   50 pF.

4.3-1   Calculate the voltage gain in the small-signal domain at the direct-coupled amplifier ($Q_2 - Q_4$) shown Fig. 4.3-1.

Answer.   197.

4.6-1   A bridge measurement at 1 kHz yields a series capacitance $C_s = 10$ μF with $D = 0.5$. Find the capacitor's (a) series resistance, (b) parallel resistance, (c) parallel capacitance.

Answer.   (a) $R_s = 8\ \Omega$; (b) $R_p = 40\ \Omega$; (c) $C_p = 8\ \mu$F.

4.6-2   An electrolytic capacitor has a capacitance of 25 μF and a power factor of 0.05 when measured on a 1-kHz bridge. The bridge uses the parallel approach to capacitor measurement. Find the capacitor's (a) parallel resistance, (b) series resistance, and (c) series capacitance.

Answer.   (a) $R_p = 127.3\ \Omega$; (b) $R_s = 0.3175\ \Omega$; (c) $C_s = 25.06\ \mu$F.

4.7-1   An ac bridge has the following constants: $Z_{ab} = Z_3(\Omega) = 500 - j\ 10^6/\omega$, $Z_{da} = Z_3(\Omega) = (-j\ 10^6/0.4\omega)(1200)/(1200 - j\ 10^6/0.4\omega)$, $Z_{bc} = R_2 = 1000\ \Omega$. $Z_{cd} = R_4 = $ unknown. (a) Calculate the frequency for which the bridge is at balance. (b) Calculate the value of R_4 required to produce balance.

Answer.   (a) 325 Hz, (b) 500 Ω.

4.7-2   In the ac bridge of Fig. 4.7-1, $R_3 = 1000\ \Omega$ in parallel with $C_3 = 0.16$ μF, $R_4 = 1000\ \Omega$, $R_2 = 500\ \Omega$, and $C_1 = 0.6\ \mu$F in series with R_1 = unknown. Find (a) the value of R_1 required to produce balance; (b) the frequency for which the bridge is in balance.

Answer.   (a) 233 Ω; (b) 1065 Hz.

4.8-1   Verify Eq. (4.8-4) for the differential three-terminal capacitor.

4.8-2   Show that the capacitance bridge in Fig. 4.8-1a provides an output voltage proportional to the fractional difference in capacitance given by Eq. (4.8-5). *Hint:* Use the voltage-divider relation.)

4.8-3   Verify Eq. (4.8-6) for the transformer ratio-arm bridge.

4.9-1   A 20-V step is applied to the bridge-type attenuator shown in Fig. P4.9-1. Compute and draw to scale the output waveform for $C_1 = 40$ pF, $C_1 = 60$ pF, and $C_1 = 20$ pF.

   Answer. $v_0(0+)$: 10 V for $C_1 = 40$ pF, 12 V for $C_1 = 60$ pF, 6.67 V for $C_1 = 20$ pF; time constant (for each case) $= ?$ v_0 waveform?

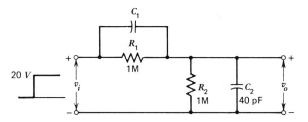

Figure P4.9-1

4.9-2   An oscilloscope test probe is shown in Fig. P4.9-2. The capacitance of the coaxial cable is 100 pF. The input impedance of the scope is 2 M in parallel with 10 pF. Calculate (a) the attenuation of the probe, (b) the capacitance C_1 for best response, and (c) the input impedance of the compensated probe.

   Answer.   (a) 1/20, (b) 5.75 pF, (c) 4.95 MΩ in parallel with 5.5 pF.

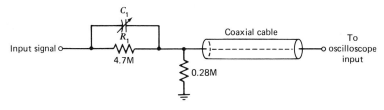

Figure P4.9-2

5
Inductance Bridges and Their Applications

5.1 SERIES-RESISTANCE INDUCTIVE BRIDGE

A series-resistance inductive bridge shown in Fig. 5.1-1 is referred to as an inductance-comparison bridge. In this bridge, the unknown inductance is determined by comparison with a known standard inductor. In addition to the four arms in Fig. 5.1-1b, there is one more resistor r, which gives the option of extending the adjustment range for the resistive balance equation.

The balance equations can be derived referring to Fig. 5.1-1a. At balance,

$$\frac{R_3 + j\omega L_3}{R_1} = \frac{R_x + j\omega L_x}{R_2} . \tag{5.1-1}$$

Separation of the real and j terms yields

$$R_x = \frac{R_2}{R_1} R_3, \tag{5.1-2}$$

and

$$L_x = \frac{R_2}{R_1} L_3. \tag{5.1-3}$$

In this bridge, R_2 is chosen as the inductive balance control and R_3 is the resistive balance control.

The variable resistor r shown in Fig. 5.1-1b is connected by means of switch S to either the standard arm (position 2) or the unknown arm (position

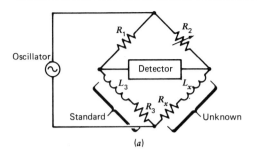

(a)

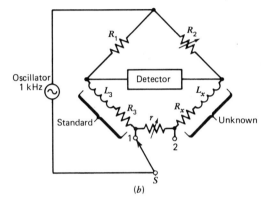

(b)

Figure 5.1-1 Series-resistance inductive bridge: (a) basic circuit; (b) resistor r added for extending measurement range.

1) so that Eq. (5.1-2) becomes either

$$R_x = \frac{R_2}{R_2} (R_3 + r) \tag{5.1-4}$$

or

$$R_x = \frac{R_2}{R_1} R_3 - r. \tag{5.1-5}$$

The resistive adjustment should be carried out first since it is more important than that of the inductive balance control.

Note that a practical inductor is essentially an *LCR* impedance, and its inductive component must be measured at a frequency well below its self-resonant frequency.

5.2 MAXWELL BRIDGE

The standard inductor of a series-resistance inductive bridge must be stable, accurate, and well shielded so that the circuit can provide very accurate measurements. However, such a standard inductor is comparatively expensive, and an iron-core standard inductor will not present its rated value of inductance unless the current flow is precisely adjusted. These disadvantages can be avoided by using a standard capacitor instead of a standard inductor, as explained next.

One of the inductance bridges with a standard capacitor is the Maxwell bridge, as shown in Fig. 5.2-1. This bridge measures an unknown inductance in terms of a known capacitance. Since the standard capacitor draws a leading current, while the unknown inductor draws a lagging current, the reactive arms are not adjacent, but are located opposite each other to obtain phase cancellation. The ratio arms are R_1 and R_2. The standard capacitance C_1 is connected in parallel with resistance R_1. At balance,

$$Z_1 Z_4 = Z_2 Z_3 \tag{5.2-1}$$

or

$$Z_x = Z_4 = Z_2 Z_3 Y_1, \tag{5.2-2}$$

where $Z_2 = R_2$; $Z_3 = R_3$; and $Y_1 = 1/R_1 + j\omega C_1$. Substitution of these values into Eq. (5.2-2) gives

$$Z_x = R_x + j\omega L_x = R_2 R_3 \left(\frac{1}{R_1} + j\omega C_1 \right). \tag{5.2-3}$$

Separation of the real and j terms yields

$$R_x = \frac{R_2}{R_1} R_3 \tag{5.2-4}$$

and

$$L_x = R_2 R_3 C_1. \tag{5.2-5}$$

The quality factor of the unknown inductor is

$$Q = \frac{\omega L_x}{R_x}. \tag{5.2-6}$$

Substitution of Eqs. (5.2-4) and (5.2-5) into Eq. (5.2-6) yields

$$Q = \omega R_1 C_1. \tag{5.2-7}$$

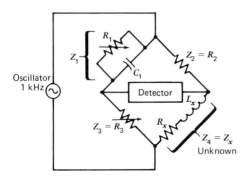

Figure 5.2-1 Maxwell bridge for the measurement of medium-Q coil $(1 < Q < 10)$.

The dissipation factor of the inductor is $D = 1/Q = 1/\omega R_1 C_1$. The Maxwell bridge is limited to the measurement of medium-Q coils ($1 < Q < 10$). This can be explained as follows.

Equation (5.2-1) is the general equation for balance of an ac bridge. If the impedance is written in the polar form $Z = |Z| \angle \theta$, where $|Z|$ represents the magnitude and θ the phase angle of the complex impedance, then Eq. (5.2-1) can be rewritten in the form

$$(|Z_1| \angle \theta_1)(|Z_4| \angle \theta_4) = (|Z_2| \angle \theta_2)(|Z_3| \angle \theta_3) \qquad (5.2\text{-}8)$$

or

$$|Z_1 Z_4| \angle \theta_1 + \theta_4) = |Z_2 Z_3| \angle \theta_2 + \theta_3). \qquad (5.2\text{-}9)$$

Equation (5.2-9) shows that the following two conditions must be met simultaneously when the bridge is balanced:

$$|Z_1 Z_4| = |Z_2 Z_3|, \qquad (5.2\text{-}10)$$

$$\angle \theta_1 + \angle \theta_4 = \angle \theta_2 + \angle \theta_3. \qquad (5.2\text{-}11)$$

Equation (5.2-11) indicates that the sum of the phase angles of one pair of opposite arms must be equal to the sum of the phase angles of the other pair. Since $Z_2 = R_2$ and $Z_4 = R_4$, Eq. (5.2-9) can be rewritten as

$$|Z_1 Z_4| \angle (\theta_1 + \theta_4) = R_2 R_3 \angle 0°. \qquad (5.2\text{-}12)$$

Hence, $\theta_1 + \theta_4 = 0°$. The phase angle (θ_4) of a high-Q coil will be very nearly $+90°$, which requires that the phase angle (θ_1) of the capacitive arm must be very nearly $-90°$. This means that the resistance R_1 must be very large. However, if the required value of R_1 becomes so large, residual leakages in the bridge circuit may impair the indication accuracy. Therefore a series RC standard arm is employed to measure high-Q coils, as described in Section 5.3.

5.3   HAY BRIDGE

When the parallel $R_1 C_1$ standard in the Maxwell bridge is replaced by a series $R_1 C_1$ standard, the configuration becomes a Hay bridge, as shown in Fig. 5.3-1. For large phase angles, the series resistance R_1 should have a very small value. The Hay bridge is therefore suitable for measuring high-Q coils ($Q > 10$).

The balance equation written in polar form is

$$|Z_1 Z_4| \angle (\theta_L - \theta_C) = R_2 R_3 \angle 0°, \qquad (5.3\text{-}1)$$

Hence, $\theta_L - \theta_C = 0°$, or $\theta_L = \theta_C$; then the quality factor of the unknown

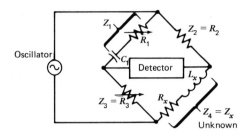

Figure 5.3-1 Hay bridge for measuring high-Q coils ($Q > 10$).

inductor is given by

$$Q = \tan \theta_L = \tan \theta_C = \frac{\omega L_x}{R_x} = \frac{1}{\omega C_1 R_1}, \qquad (5.3\text{-}2)$$

and the dissipation factor of the inductor is $D = 1/Q = \omega C_1 R_1$. The general balance equation is

$$Z_1 Z_4 = Z_2 Z_3, \qquad (5.3\text{-}3)$$

where $Z_1 = R_1 - j(1/\omega C_1)$; $Z_2 = R_2$; $Z_3 = R_3$; $Z_4 = R_x + j\omega L_x$. Substitution of these values into Eq. (5.3-3) gives

$$\left(R_1 - j\frac{1}{\omega C_1}\right)(R_x + j\omega L_x) = R_2 R_3.$$

Separation of the real and j terms yields

$$R_1 R_x + \frac{1}{C_1} L_x = R_2 R_3 \qquad (5.3\text{-}4)$$

and

$$\frac{1}{\omega C_1} R_x - \omega R_1 L_x = 0. \qquad (5.3\text{-}5)$$

Solving Eqs. (5.3-4) and (5.3-5) simultaneously, we obtain

$$R_x = \frac{\omega^2 C_1^2 R_1 R_2 R_3}{1 + \omega^2 C_1^2 R_1^2} = \frac{\omega^2 C_1^2 R_1 R_2 R_3}{1 + 1/Q^2} \qquad (5.3\text{-}6)$$

and

$$L_x = \frac{R_2 R_3 C_1}{1 + \omega^2 C_1^2 R_1^2} = \frac{R_2 R_3 C_1}{1 + 1/Q^2}. \qquad (5.3\text{-}7)$$

For $Q > 10$, $1/Q^2 \approx 0$, then

$$L_x \approx R_2 R_3 C_1. \qquad (5.3\text{-}8)$$

This is the same as the expression derived for the Maxwell bridge.

5.4   OWEN BRIDGE

As seen from the inductance bridges described above, the unknown inductor is represented by a series $R_x L_x$ circuit. Assume that the unknown is an iron-core inductor; the effective value of L_x shifts as the current flow changes, and the hysteresis loop for the iron-core is associated with a core loss that contributes the effective series resistance R_x. Thus, both the inductance and effective series resistance of the unknown shift as the null (balance) point is approached. Unless the Owen bridge of Fig. 5.4-1 is employed, inductance shift and resistance shift introduce a complication in bridge operation that is called chasing the null or sliding null. The Owen bridge has adjustable parameters R_3 and C_3 in the same arm, and thus eliminates the sliding null encountered in other inductance bridges. Hence, inductance measurements with the Owen bridge are speeded up considerably. The Owen bridge is balanced if $Z_1 Z_4 = Z_2 Z_3$ or

$$-j\,\frac{1}{\omega C_1}\,(R_x + j\omega L_x) = R_2 \left(R_3 - j\,\frac{1}{\omega C_3} \right) . \qquad (5.4\text{-}1)$$

Separation of the real and j terms yields

$$L_x = R_2 R_3 C_1 \qquad (5.4\text{-}2)$$

and

$$R_x = R_2 \frac{C_1}{C_3} . \qquad (5.4\text{-}3)$$

This bridge has a very wide range. It is typically rated for an accuracy of 0.1% with inductance values as low as 1 μH. Inductance values can be measured down to 0.001 μH and up to more than 1000 H. This bridge requires the use of a precision variable capacitor C_3 and a precision variable resistor R_3. The inductance measurement depends basically on the accuracy of R_3. Compared to Maxwell and Hay bridges, the Owen bridge is more expensive.

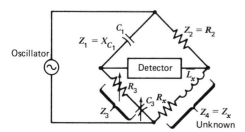

Figure 5.4-1 Owen bridge for wide-range inductance measurements.

5.5   INDUCTANCE-BRIDGE EQUIVALENT CIRCUITS

In order to change a bridge measurement to the most useful form for the problem at hand, the series $R_s L_s$ circuit may be converted into the equivalent parallel $R_p L_p$ circuit, or vice versa. In the series $R_s L_s$ circuit, its impedance and admittance are

$$Z_s = R_s + jX_{L_s}$$

and

$$Y_s = \frac{1}{Z_s} = \frac{1}{R_s + jX_{L_s}} = \frac{R_s - jX_{L_s}}{R_s^2 + X_{L_s}^2}.$$

Thus,

$$Y_s = \frac{R_s}{R_s^2 + X_{L_s}^2} - j\frac{X_{L_s}}{R_s^2 + X_{L_s}^2} = \frac{1}{R_p} - j\frac{1}{X_{L_p}} = \frac{1}{Z_p}. \qquad (5.5\text{-}1)$$

where Z_p is the impedance of the equivalent parallel circuit,

$$R_p = \frac{R_s^2 + X_{L_s}^2}{R_s}, \qquad (5.5\text{-}2)$$

$$X_{L_p} = \frac{R_s^2 + X_{L_s}^2}{X_{L_s}}. \qquad (5.5\text{-}3)$$

Equations (5.5-2) and (5.5-3) give the necessary conditions for the conversion of the series $R_s L_s$ circuit to its equivalent parallel $R_p L_p$ circuit.

The quality factor of a series $R_s L_s$ circuit is $Q = X_{L_s}/R_s$, thus

$$X_{L_s} = QR_s. \qquad (5.5\text{-}4)$$

Substitution of Eq. (5.5-4) into Eq. (5.5-2) yields

$$R_p = \frac{R_s^2 + Q^2 R_s^2}{R_s} = (1 + Q^2)R_s. \qquad (5.5\text{-}5)$$

Substituting $R_s = X_{L_s}/Q$ into Eq. (5.5-3) we find

$$X_{L_p} = \frac{(X_{L_s}/Q)^2 + X_{L_s}^2}{X_{L_s}} = \left(1 + \frac{1}{Q^2}\right) X_{L_s}. \qquad (5.5\text{-}6)$$

The parallel $R_p L_p$ circuit will be converted into the equivalent series $R_s L_s$ circuit if

$$R_s = \frac{R_p}{1 + Q^2} \qquad (5.5\text{-}7)$$

and

$$X_{L_s} = \frac{X_{L_p}}{1 + 1/Q^2}, \qquad (5.5\text{-}8)$$

where Q is the quality factor of the parallel $R_p L_p$ circuit and is given by

$$Q = \frac{R_p}{X_{L_p}}.$$ (5.5-9)

Substituting $X_{L_s} = \omega L_s$ and $X_{L_p} = \omega L_p$ into Eq. (5.5-8) we obtain

$$L_s = \frac{L_p}{1 + 1/Q^2},$$ (5.5-10)

where

$$Q = \frac{\omega L_s}{R_s} = \frac{R_p}{\omega L_p}.$$ (5.5-11)

For $Q > 10$, $L_s \simeq L_p$.

5.6   APPLICATIONS OF INDUCTANCE BRIDGES

Measurement of Incremental Inductance

An iron-core inductor is a nonlinear device because its inductance is greatest at low flux densities and is least at high flux densities. Thus, when an iron-core coil or audio-frequency choke is energized by a pulsating dc, the inductance decreases as the dc component is increased. The inductance that is offered to the superimposed ac on the dc component is known as the incremental inductance. When an incremental inductance is to be measured, the bridge we use must provide adjustable values of ac and dc. Since the quality factor Q_x is fairly large for most audio-frequency chokes, the modified Hay bridge shown in Fig. 5.6-1 is usually employed as an incremental inductance bridge.

Measurement of Mutual Inductance

The mutual inductance between two coils can be determined by taking two measurements on a conventional inductance bridge. Before these measure-

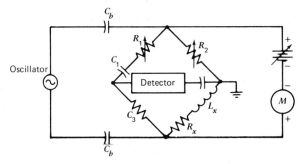

Figure 5.6-1 An incremental inductance bridge consisting of basic Hay configuration and adjustable dc and ac sources. $L_x \simeq R_2 R_3 C_1$ for $Q > 10$.

ments are taken, however, the location of the dots should be known. A dot is placed at the end of each coil showing induced voltages produced by the mutual flux, which are in phase. If the dots are not known, the circuit of Fig. 5.6-2 can be used to determine their location. A dc voltage applied to coil 1 induces a voltage in coil 2 only when the switch is closed or opened. When the switch is closed, the terminal a of coil 1 is driven positive. If at the instant of closure the voltmeter reads upscale, the induced voltage at the terminal a for coil 2 must also be positive. Therefore, dots would be placed at both terminals a. If, on the other hand, the terminal a of coil 1 is driven positive and the voltmeter reads downscale, the induced voltages at the terminal a of coil 1 and the terminal b of coil 2 are in phase and the dots would be placed at these terminals.

The dots are known at terminals a and c of Fig. 5.6-3a. Connect terminals b and d together as shown in Fig. 5.6-3b. Using an inductance bridge, measure the total inductance between terminals a and c. This net value is

$$L_A = L_1 + L_2 - 2M, \qquad (5.6\text{-}1)$$

where L_1, L_2, and M are the self-inductance of coil 1, the self-inductance of coil 2, and the mutual inductance between the coils, respectively. Equation (5.6-1) can be derived as follows. With terminals b and d connected, i flows into the undotted end of L_2, and the dotted end of L_1 is driven negative; also, i flows into the dotted end of L_1, and the dotted end of L_2 is driven positive—the induced voltages are represented by the dependent sources having polarities, as shown in Fig. 5.6-3c. The loop equation is given by

$$v_{ac} = L_1 \frac{di}{dt} - M \frac{di}{dt} - M \frac{di}{dt} + L_2 \frac{di}{dt}. \qquad (5.6\text{-}2)$$

Thus

$$\frac{v_{ac}}{di/dt} = L_A = L_1 + L_2 - 2M.$$

This equation indicates that the two coils are connected in a series-opposing configuration.

For the second measurement connect terminals b and c together as shown in Fig. 5.6-3d. An inductance bridge measurement between terminals a and d yields

$$L_B = L_1 + L_2 + 2M. \qquad (5.6\text{-}3)$$

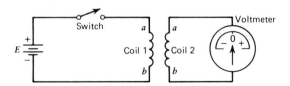

Figure 5.6-2   A test circuit to determine the ends of the coils that are in phase.

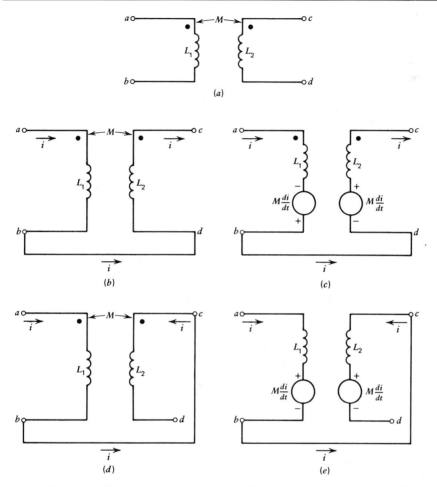

Figure 5.6-3 Circuit diagrams to measure mutual inductance. (a) An inductively coupled circuit: $M = K\sqrt{L_1 L_2}$, where K is the coefficient of coupling. (b) Undotted terminals connected together. (c) Circuit b drawn with dependent sources ($M\, di/dt$). (d) Terminals b and c connected together. (e) Circuit d drawn with dependent sources.

This equation indicates that the two coils are connected in a series-aiding configuration. Equation (5.6-3) can be derived in the same manner as for Eq. (5.6-1).

Subtracting Eq. (5.6-1) from Eq. (5.6-3) results in

$$L_B - L_A = 4M,$$

hence

$$M = \frac{L_B - L_A}{4}. \tag{5.6-4}$$

Measurement of Speaker Voice-Coil Impedance

An inductance bridge is useful for the measurement of speaker voice-coil impedance Z_x ($= R_x + j\omega L_x$). In a typical example, the measured inductance is $L_x = 100$ μH with $Q = 0.2$ at 1 kHz. These values correspond to $\omega L_x = 0.628$ Ω, $R_x = \omega L/Q = 3.14$ Ω, and $Z_x = 3.2$ Ω. Note that the voice-coil impedance is customarily rated at 1 kHz and has a much greater resistive component than an inductive-reactive component. The nominal impedance ratings for various speaker voice-coil are commonly 4, 8, and 16 Ω.

Measurement of CRT Deflection-Coil Inductance

Inductance and impedance of the deflection coil for a cathode ray tube (CRT) can be measured by using an inductance bridge. The horizontal-deflection coils for a small picture tube have a typical inductance of 8.2 mH and a winding resistance of 13 Ω, corresponding to a Q value of 4 at 1 kHz. The vertical deflection coils in the same yoke have an inductance of 4.8 mH and a winding resistance of 65 Ω, corresponding to a Q value of 0.46 at 1 kHz.

Operating Frequencies of ac (Impedance) Bridges

Most laboratory-type ac bridges have a basic accuracy rating (such as ±1% or ±0.05%) for 1-kHz operation. The same accuracy rating usually applies over the entire audio-frequency range. Another class of ac bridges is designed for operation at radio frequencies, for example, a typical rf bridge is rated for ±2% accuracy at frequencies from 400 kHz to 50 MHz.

5.7 BASIC Q METER

The Maxwell and Hay circuits contained in a conventional impedance bridge are restricted to operation in the audio-frequency range and thus are limited in their ability to measure Q and L. A typical high-quality Q meter is designed for operation at frequencies from 1 kHz to 300 MHz.

The operation of a Q meter shown in Fig. 5.7-1 is based on the charac-

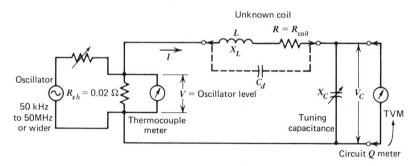

Figure 5.7-1   Basic Q-meter configuration.

teristics of a series-resonant circuit. At resonance,

$$X_C = X_L, \quad V_C = IX_C = IX_L, \quad V = IR,$$

$$Q = \frac{X_L}{R} = \frac{X_C}{R} = \frac{V_C}{V}, \tag{5.7-1}$$

where V = applied voltage,
X = circuit current,
V_C = voltage across the capacitor,
X_C = capacitive reactance,
X_L = inductive reactance,
R = ac resistance of the coil,
Q = magnification of the circuit.

Hence, if V is maintained at a constant and known level, an electronic voltmeter (such as TVM) connected across the capacitor can be calibrated directly in terms of the circuit Q.

In the Q-meter circuit of Fig. 5.7-1, the high-frequency oscillator with a practical range from 50 kHz to 50 MHz delivers current to a very small shunt resistance R_{sh} (on the order of 0.02 Ω). The voltage V across the shunt is measured with a thermocouple meter. The voltage V_c across the variable capacitor is measured with an electronic voltmeter whose scale is calibrated directly in Q values. The indicated or measured Q is called the circuit Q, which is commonly regarded as the Q of the coil. The effective or true Q is somewhat greater than the indicated Q. This difference is generally neglected, except in certain cases where $R_{coil} < R_{sh}$. When the error introduced by R_{sh} is not negligible, correction can be made by means of the formula:

$$Q_{true} = Q_{meas} \left(1 + \frac{R_{sh}}{R_{coil}}\right). \tag{5.7-2}$$

The measured Q of a coil with distributed capacitance is less than the true Q by a factor that depends on the value of the distributed capacitance and the total shunt capacitance. It can be shown that

$$Q_{true} = Q_{meas} \left(1 + \frac{C_d}{C_T}\right), \tag{5.7-3}$$

where C_d = distributed capacitance,
C_T = total shunt capacitance consisting of distributed capacitance, tuning capacitance, and input capacitance of TVM.

One method of finding the distributed capacitance C_d of a coil involves making two measurements at different frequencies f_1 and $f_2 = 2f_1$. The tuning capacitor is first fully meshed, and the oscillator tuned to resonance. We note the resonant frequency f_1 and the value C_1 of tuning capacitance. Then the oscillator is set to $f_2 = 2f_1$, and the tuning capacitor adjusted for resonance. We note the new value C_2 of tuning capacitance. The resonant frequencies $f_2 = 2f_1$ can be expressed as

$$f_2 = \frac{1}{2\pi\sqrt{L(C_2 + C_d)}} = \frac{2}{2\pi\sqrt{L(C_1 + C_d)}} = 2f_1.$$

This reduces to

$$\frac{1}{C_2 + C_d} = \frac{4}{C_1 + C_d}.$$

Solving for the distributed capacitance we obtain

$$C_d = \frac{C_1 - 4C_2}{3}. \tag{5.7-4}$$

The procedure for measuring C_d outlined above is known as the frequency-doubling method.

BIBLIOGRAPHY

Bell, D. A., *Electronic Instrumentation and Measurements*. Reston Publishing Company, Reston, Virginia, 1983, Chap. 7.

Carr, J. J., *Elements of Electronic Instrumentation and Measurement*. Reston Publishing Company, Reston, Virginia, 1979, Chap. 3.

Cooper, W. D., *Electronic Instrumentation and Measurement Techniques,* 2nd ed. Prentice-Hall, Englewood Cliffs, New Jersey, 1978, Chaps. 8 and 10.

Kantrowitz, P. G., Kousourou, and L. Zucker, *Electronic Measurements*. Prentice-Hall, Englewood Cliffs, New Jersey, 1979, Chapt. 6.

Questions

5-1   Why must the operating frequency of an inductance bridge be much less than the self-resonant frequency of the unknown inductor?

5-2   What type of bridge is best suited for measuring the inductance of low-Q coils?

5-3   What type of bridge is best suited for measuring the inductance of high-Q coils?

5-4   Explain the term sliding null (chasing the null) as it applies to an inductance bridge.

5-5   Describe the characteristics of an Owen bridge.

5-6   Describe the characteristics of an incremental inductance.

5-7   Describe the procedure for determining the mutual inductance of two inductors with an inductance bridge.

5-8   How do impedance bridges (capacitance and inductance bridges) usually indicate the resistive component of a capacitor or inductor?

5-9   Describe the operation principle of a Q meter.

5-10 Why is it necessary to measure the D or Q value at the frequency of
      interest and to specify the test frequency in the data report?

5-11 Describe the procedure for measuring the distributed capacitance of
      a coil.

Problems

5.1-1 Show that the resistive and inductive balance equations of the bridge
       in Fig. P5.1-1 is given by Eqs. (5.1-2) and (5.1-3).

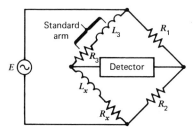

Standard arm L_3 R_1 R_3 Detector E L_x R_x R_2

Figure P5.1-1

5.1-2 In the bridge of Fig. P5.1-1, $R_2 = 3K$, $R_1 = 2K$, $L_3 = 20$ mH, and
       $R_3 = 40\ \Omega$. Calculate the values of the unknown inductance and
       resistance.
       Answer.   30 mH; 60 Ω.

5.2-1 In the balanced Maxwell bridge of Fig. 5.2-1, $R_1 = 300K$, $C_1 = 1000$
       pF, $R_2 = 1K$, and $R_3 = 5K$. The driving frequency is 1 kHz. Calculate
       R_x, L_x, and Q.
       Answer. $R_x = 16.7\ \Omega$; $L_x = 5$ mH; $Q = 1.88$.

5.3-1 The Hay bridge of Fig. 5.3-1 is balanced at 5 kHz and has the following
       constants: $R_1 = 10K$, $R_2 = 1K$, $R_3 = 2K$, and $C_1 = 500$ pF. Find:
       Q, $R_x (= R_s)$, $L_x (= L_s)$, and the equivalent R_p and L_p.
       Answer. $Q = 63.7$; $R_s = 4.9K$; $L_s = 1$ mH; $R_p = 19.888K$; $L_p =$
       1 mH.

5.3-2 The Hay bridge of Fig. 5.3-1 is balanced at 1 kHz and has the following
       parameters: $R_1 = R_3 =$ variable range from 100 Ω to 1000K; $C_1 =$
       500 pF; $R_2 = 1K$. Find the measurable maximum and minimum of
       L_p and D for the unknown inductor.
       Answer. $L_{p(max)} = 0.5$ H; $L_{p(min)} = 50\ \mu$H; $D_{max} = 3.14$; $D_{min} =$
       0.000314.

5.4-1 A modified Owen bridge is shown in Fig. P5.4-1. Suppose the values
       of the components in the bridge are $R_2 = 1K$, $R_3 = 2K$, $R_4 = 80\ \Omega$,

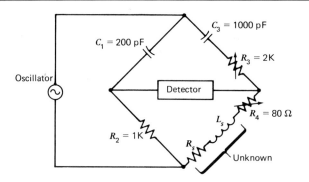

Figure P5.4-1

$C_1 = 200$ pF, and $C_3 = 1000$ pF. (a) Derive the resistive and inductive balance equations. (b) Find the values of R_s and L_s.

Answer. (a) $R_s = (C_1/C_3)R_2 - R_4$; $L_s = R_2R_3C_1$.
(b) $R_s = 120 \ \Omega$; $L_s = 400 \ \mu$H.

5.5-1 The series L_sR_s circuit has the values $L_s = 100 \ \mu$H and $R_s = 50 \ \Omega$. What are the element values in the equivalent parallel circuit at a frequency of 800 kHz?

Answer. $L_p = 101 \ \mu$H; $R_p = 5.102$K.

5.6-1 Suppose the bridge of Fig. 5.6-1 is balanced at 1 kHz when $R_1 = 100 \ \Omega$, $C_1 = 0.1 \ \mu$F, $R_2 = 1$K, and $R_3 = 10$K. What are the inductance and Q value of the unknown inductor?

Answer. $L_x = 1$ H; $Q = 15.9$.

5.6-2 In the inductively coupled circuit of Fig. 5.6-3a, $L_1 = 10$ mH, $L_2 = 2.5$ mH. L_A was measured as 2 mH and L_B as 20 mH. Calculate (a) the mutual inductance and (b) the coefficient of coupling.

Answer. (a) $M = 4.5$ mH; (b) $K = 0.9$.

5.7-1 Calculate the value C_d of the distributed capacitance of a coil when the following measurements are made: At frequency $f_1 = 2$ MHz, the tuning capacitor is set at $C_1 = 460$ pf. When the frequency is increased to $f_2 = 5$ MHz, the tuning capacitor is adjusted to resonance at $C_2 = 60$ pF. (*Hint:* Use the relation $f_2 = 2.5f_1$.)

Answer. $C_d = 16$ pF.

5.7-2 In checking the distributed capacitance C_d of a coil using the frequency-doubling method, the initial resonance is obtained with the tuning capacitor set to 460 pF. The resonance at twice the initial frequency is obtained with the tuning capacitor at 105 pF. Find C_d.

Answer. 13.3 pF.

6
Semiconductor Devices and Digital Systems

6.1 NETWORK THEOREMS APPLICABLE TO DIODE AND TRANSISTOR CIRCUITS

Kirchhoff's Laws

Kirchhoff's Current Law (KCL).

The algebraic sum of the currents into and out of each branch point in a given closed network is equal to zero. When nodal analysis is used, KCL is applied at each node of the network except the reference node (common point).

Kirchhoff's Voltage Law (KVL).

The voltages produced as a result of current(s) flowing in a closed path of a network are equal to the sum of all internal, applied, and induced voltages in the same path. KVL is applied to each closed path of the network when loop analysis is used.

Superposition Theorem

The response of a linear network containing several independent sources is found by considering each generator (source) separately and then adding the individual responses. When evaluating the response due to one source, one replaces each of the other independent generators by its internal impedance.

Thevenin's Theorem

Any linear network may, with respect to a pair of terminals, be replaced by a series combination of a voltage generator (source) and an impedance (see

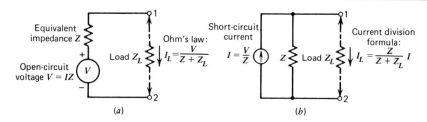

Figure 6.1-1 (a) Thevenin equivalent and (b) Norton equivalent for a linear two-terminal network.

Fig. 6.1-1a). The voltage generator equals the open-circuit voltage between the two terminals. The value of the equivalent impedance is equal to that impedance seen looking back into the network with all energy sources replaced by their respective internal impedance.

Norton's Theorem

Any linear network may, with respect to a pair of terminals, be replaced by a parallel combination of a current generator and an impedance (see Fig. 6.1-1b). The current generator equals the short-circuit current at the terminals. The value of the equivalent impedance is equal to that impedance seen looking back into the network with all energy sources replaced by their respective internal impedance.

6.2   COMPONENTS OF AN ELECTRONIC INSTRUMENT

An electronic instrument shown in Fig. 6.2-1 consists essentially of three chief sections: the input sensor, the signal processor, and the output device. The sensor is the instrument transducer. In general terms, a transducer is any device that converts energy in one form to energy in another. The function of an instrument transducer is to sense the presence, magnitude, change in, and/or frequency of some "measurand," and provide an electrical output that, when appropriately processed and applied to a readout device, will furnish accurate quantitative data about the measurand. The term "measurand" refers to the quantity, property, or condition that the transducer translates to an electrical signal. Different measurands require different input sensors. Thus we need microphones for acoustical, thermoelements for thermal, position switches for mechanical, or radiation detectors for nuclear applications. The results of the signal processing may be indicated optically or acoustically if the appropriate readout devices are used.

Since only measurable parameters can be dealt with, the processing of the electrical signals that represent these parameters is composed solely of arithmetical or logical operations. Signal processing can be elaborate, involving amplification, separation into frequency components, timing of var-

Figure 6.2-1   Block diagram of an electronic instrument.

ious segments, or conversion from ac to dc. Signal readout may consist of a panel meter or a digital display, or perhaps conversion into a form suitable for transfer to a digital computer. In an electronic instrument, its signal processor, readout (except the panel meter), and power supply usually contain semiconductor devices and integrated circuits.

6.3   SEMICONDUCTOR DIODES

p-n Junction Diode

In a *p-n* junction silicon diode, the *n*-type silicon serves as the cathode (K) and the *p*-type silicon serves as the anode (A). The arrow symbol used for the anode represents the direction of conventional current. When the positive terminal of an external battery is connected to A and the negative terminal to K, as shown in Fig. 6.3-1a, the space-charge region (depletion layer) at the junction becomes effectively narrower, and the energy barrier formed by the potential gradient (represented by the imaginary battery) decreases to an insignificant value. Excess (free) holes (majority carriers) from the *p*-type region then cross the junction into the *n*-type region, while excess (free) electrons (majority carriers) from the *n*-type region cross the junction into the *p*-type region. Both carrier flows continue as long as the external voltage is applied. Under these conditions the diode is forward biased. A small forward voltage is sufficient to yield a high current. The forward resistance is low. The holes (electrons) crossing the junction from the *p*-type (*n*-type) region become a minority current injected into the *n* side (*p* side).

The reverse-biased diode configuration is shown in Fig. 6.3-1b, where the positive terminal of the external battery is connected to K and the negative terminal to A. In this case, the space-charge region at the junction becomes effectively wider, and the potential gradient approaches the external voltage, and so a very small current results, which is the essentially constant minority carrier flow, often called the reverse saturation current (I_0 or I_s). Minority carriers such as holes in *n*-type material or free electrons in *p*-type material are due in part to those impurities that exist in every material not through design but because an absolutely pure material (Si or Ge) cannot be obtained. Other sources of minority carriers include thermally generated hole–electron pairs and carriers absorbing sufficient light energy in the form of photons to leave the parent atom.

For a *p-n* junction diode, the current I is related to the voltage V by the

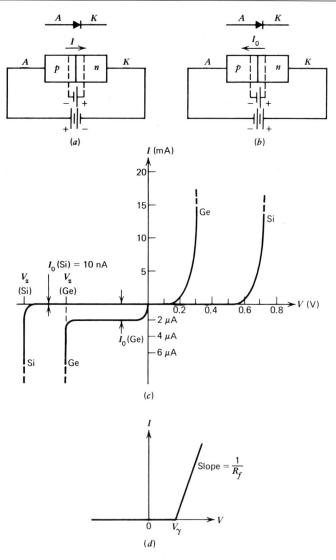

Figure 6.3-1 (a) Current flow in forward-biased p-n junction. (b) Current flow in reverse-biased diode. (c) Actual diode current–voltage curves. (d) Piecewise linear approximation for the diode characteristic.

equation

$$I = I_0(\epsilon^{V/\eta V_T} - 1), \qquad (6.3\text{-}1)$$

where I = current through diode, A;

V = voltage across diode, V;

I_0 = reverse saturation current, A;

V_T = volt equivalent of temperature = kT/q = $T/11{,}600$;

k = Boltzmann's constant = 1.38×10^{-23} J/K;
q = electron charge = 1.60×10^{-19} C;
T = absolute temperature, K;
η = empirical constant between 1 (for Ge) and 2 (for Si).

For most practical purposes, an appreciable current does not flow until 0.5–0.7 V forward bias (Si diode) is applied. In fact, there is a conduction threshold $V_\gamma \simeq 0.6$ V for silicon. Beyond V_γ, the current rises very rapidly. This fit to the actual exponential rise is referred to as the piecewise approximation, as shown in Fig. 6.3-1d. Germanium diodes have a less well-defined threshold $V_\gamma \simeq 0.2$ V, while GaAs diodes (light-emitting diodes) have a threshold $V_\gamma \simeq 1.6$ V.

Temperature can have a marked effect on the diode current. The forward current increases at fixed voltage when the temperature is increased. It is found that for either silicon or germanium (at room temperature),

$$\frac{dV}{dt} \simeq -2.5 \text{ mV/°C} \tag{6.3-2}$$

in order to maintain a constant value of forward current I. It is also found that the reverse saturation current I_0 approximately doubles for every 10°C rise in temperature for both silicon and germanium. Since $(1.07)^{10} \simeq 2$, we may also conclude that I_0 increases approximately 7%/°C. If $I_0 = I_{01}$ at $T = T_1$, then at a temperature T, I_0 is expressed by

$$I_0(T) = I_{01} \times 2^{(T - T_1)/10}. \tag{6.3-3}$$

The current in a reverse-biased diode is small (typically 10^{-8} A for silicon) and approximately independent of voltage until the breakdown region at high reverse voltages is reached (Fig. 6.3-1c). At this point, referred to as the avalanche or Zener breakdown region, the current rises rapidly with increasing (reverse) voltage.

For most applications, the only important parameters that distinguish various diodes are the maximum permissible operating voltage (peak reverse voltage) and current. For critical applications, the diode speed and reverse current leakage should be considered in addition to the maximum voltage and current, but otherwise the choice is arbitrary.

Power Diodes

The majority of rectifier or power diodes are constructed using silicon because of its higher current, temperature, and peak reverse voltage ratings. A typical small rectifier diode would be able to handle 1 A average (10 A surge) and have a peak reverse voltage of 100 V or more. Units rated well over 100 A and 1 kV are commercially available. Large units have a bolt (stud)-type case intended for mounting on a heat sink to dissipate the power produced by internal heating. When the power diodes are reverse biased, their current leakages can be rather high; their response times can be comparatively slow.

Zener Diodes

A Zener diode is also called an avalanche or voltage-regulator diode. It is designed with adequate power-dissipation capabilities to operate in the breakdown region, as shown in the I-V characteristic of Fig. 6.3-2*a*. A good regulator diode will exhibit a sharp knee and a high slope in the breakdown region. A voltage-regulator diode is operated through a series-limiting resistor R from an unregulated voltage source V, as shown in Fig. 6.3-2*b*. The source V ($>V_z$) and resistor R are chosen so that the diode is initially operating in the breakdown region. The diode will regulate the load voltage V_z against variations in load current and against variations in supply voltage V, since in the breakdown region large changes in diode current produce only small changes in diode voltage. For a regulator as shown in Fig. 6.3-2*b*, an empirical factor of 10% of the maximum load current is used as the minimum Zener-diode current.

At a specified operating point (V_z, I_z) a small-signal Zener resistance (= dynamic impedance) R_z can be defined as the inverse slope at that point. In a practical regulator circuit, R_s is much greater than R_z, and, therefore,

$$I_z \simeq \frac{V - V_z}{R_s}. \tag{6.3-4}$$

The output voltage approximates V_z.

Characteristics of the 18-V Zener diode 1N2816 are shown in Fig. 6.3-3

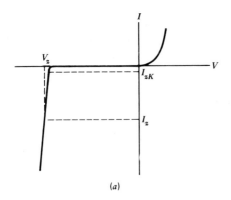

(a)

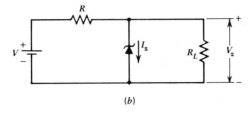

(b)

Figure 6.3-2 (a) Volt–ampere characteristic of a Zener diode. (b) A circuit in which a Zener diode is used as a source of reference potential V_z.

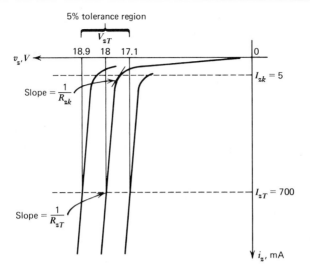

Figure 6.3-3 Reverse characteristic of 1N2816 Zener diode.

and specified as follows:

1. Nominal reference voltage $V_z\ (=V_{zT})\ =\ 18$ V.
2. Tolerance $=5\%$, so that V_z is between 17.1 and 18.9 V.
3. Maximum dissipation (at 25°C) $=50$ W.
4. Test current $I_z\ (=I_{zT})\ =\ 700$ mA.
5. Dynamic impedance ($=$ small-signal Zener resistance) at I_z,

$$R_z\ (=R_{zT})\ =\ 2\ \Omega.$$

6. Knee current $I_{zk}\ =\ 5$ mA.
7. Dynamic impedance at I_{zk}, $R_{zk}\ =\ 80\ \Omega$.
8. Maximum junction temperature $=150$°C.
9. Temperature coefficient

$$TC\ =\ \frac{\Delta V_z/V_z}{\Delta T}\ \times\ 100\%/°C\ =\ 0.075\%/°C.$$

High-Speed Diodes

A high-speed diode (known as computer diode) is a diode intended for high-speed switching applications. Diodes do not respond instantly since a finite time is required to input or remove the storage charge near the junction to produce the junction potential. The time to remove the charge is the storage time. It is referred to as the off time τ_{off}, because this process is responsible for the delay in the current or conductance drop when the potential across the diode is suddenly changed from forward to reverse bias. The on time τ_{on} when the bias is changed from reverse to forward is due to charge injection and is often shorter than τ_{off}. Switching times (both τ_{off} and τ_{on}) are typically 4 ns for a computer diode.

PIN Diodes

A PIN diode is constructed by adding a narrow insulating region between the *p*-type and *n*-type semiconductor region. Since the insulating region is very thin, the normal depletion region at the junction is extended only slightly and the dc diode characteristics are only slightly different from those of a standard diode; the high-frequency characteristics are better than those of standard high-speed diodes by one or two orders of magnitude, so that operation above 100 MHz or even 1 GHz is possible.

When the PIN diode is forward biased, it can be used as a current-controlled resistor for small signals. Its ac resistance can be varied from well above 10K to below 1 Ω. Attenuators for high-frequency operation often incorporate PIN diodes for this purpose.

Varactor Diodes

A varactor (voltage-variable capacitor) diode is operated in the reverse-bias region. A diode under reverse-biased conditions can be thought of as a capacitor with the depletion layer as the insulator. Increasing the bias voltage increases the width of the depletion layer and therefore decreases the capacitance. A varactor is only used at higher frequencies. The transition capacitance varies approximately as the inverse square root of voltage. A typical varactor diode (1N5452A or the equivalent) might have a capacitance of 50 pF at 4 V and vary $\pm 50\%$ in the range of 2–8 V. Varactor diodes are commonly used for frequency modulation and automatic tuning.

Tunnel Diodes

In a *p-n* junction diode, the depletion layer or space charge region constitutes a potential barrier at the junction. This potential barrier restrains the flow of majority carriers from one side of the junction to the other. The width of the depletion layer (junction barrier) varies inversely as the square root of the impurity concentration. A *p-n* junction diode of the conventional type has an impurity concentration of about 1 part in 10^8. With this amount of doping the width of the depletion layer is of the order of 5 μm (5×10^{-4} cm). If the concentration of impurity is greatly increased, say to 1 part in 10^3, then the width of the junction barrier is reduced from 5 μm to about 100 Å (10^{-6} cm). This thickness is only about one-fiftieth the wavelength of visible light. For a barrier as thin as this, quantum mechanics indicates that there is a great probability that an electron will penetrate through the barrier. The quantum-mechanical behavior is known as "tunneling," and therefore such a high-impurity-density p-n junction device is called the "tunnel diode."

With the tunnel-diode characteristic of Fig. 6.3-4 for small reverse voltages, the reverse current is high. Also, for small forward voltages (up to 50 mV for Ge) the resistance remains small (of the order of 5 Ω). At the peak current I_p corresponding to the voltage V_p, the slope dI/dV of the characteristic is zero. If V is increased beyond V_p, then the current decreases, and, therefore, the dynamic conductance $g = dI/dV$ is negative. Thus the tunnel diode exhibits a negative-resistance characteristic between the peak current I_p and the valley current I_v (corresponding to the valley voltage V_v). At V_v the conductance is again zero, and beyond this point the resistance becomes and remains positive. At the peak forward voltage V_f the current again reaches the value I_p. The explanation for the tunneling phenomenon is to be found in quantum mechanics. In order for an electron in the n region to cross the junction and enter a hole in the p region, it must have an energy that falls within a certain specified band of energies. Electrons with other energy levels are forbidden to enter these holes. Hence, as the forward voltage is increased, the energy of the electrons in the n region increases, first into a region that allows the electrons to cross the barrier and enter holes causing the current to increase, and then into a forbidden region causing the current to decrease; after this the current again begins to climb. Here the tunneling effect is no longer necessary, and current increases with increasing forward bias until the avalanche voltage is exceeded. Thus, the portion of the characteristic beyond V_v is caused by the injection current in

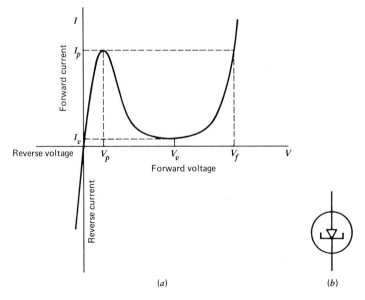

(a) (b)

Figure 6.3-4 (a) Tunnel diode characteristic and (b) symbol.

an ordinary *p-n* junction diode. The remainder of the characteristic is a result of the tunneling phenomenon in the highly doped diode. Since the current between the peak and the valley is decreasing with increasing voltage, the resistance of the tunnel diode is said to be negative. For currents whose values are between I_v and I_p the curve is triple valued because each current can be obtained at three different applied voltages. It is this multivalued feature that makes the tunnel diode useful in pulse and digital circuitry.

6.4 BIPOLAR JUNCTION TRANSISTORS

pnp and *npn* Transistors

A bipolar junction transistor (BJT) is a three-layer silicon (or germanium) device consisting of either two *p*- and one *n*-type layers of material (*pnp*) or two *n*- and one *p*-type layers of material (*npn*). The three portions of a transistor are called emitter (*E*), base (*B*), and collector (*C*).

Diodes, transistors, and other semiconductor devices are manufactured by using one or two of the basic techniques: grown, alloy, diffusion, or epitaxial. All microelectronic BJTs are fabricated by using the epitaxial technique, which consists of growing a very thin, high-purity, single-crystal layer of silicon on a heavily doped substrate of the same material. The term bipolar reflects the fact that holes and electrons participate in the injection process into the oppositely polarized material.

The basic amplifying action is produced by transferring a current from a low-resistance circuit (forward-biased *E-B* junction) to a high-resistance circuit (reverse-biased *C-B* junction). The combination of the two terms *transfer resistor* results in the name transistor. A transistor in each of the three basic configurations (common base, *CB*; common emitter, *CE*; common collector, *CC*) is normally operated in the active region for the amplification of signals with minimum distortion. In this region the *E-B* junction is forward biased, and the *C-B* junction is reverse biased.

Transistor Configurations

In a practical transistor, from 95 to 99.5% of the emitter current (I_E) reaches the collector (collector current $|I_C| < |I_E|$). The current gain (A_i) of the *CB* configuration shown in Fig. 6.4-1a is therefore always less than unity. The short-circuit amplification factor is $\alpha \simeq A_i = -I_C/I_E$. In the *CB* amplifier stage of Fig. 6.4-1b, the signal is applied to the emitter–base circuit and extracted from the collector–base circuit; the input or emitter–base circuit has a low impedance, and the output or collector–base circuit has a high impedance. There is no voltage phase reversal between the input and output signals.

The *CB* output characteristics are the collector static characteristics of collector current versus collector–base voltage for a BJT in the *CB* config-

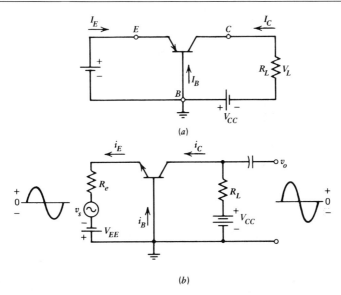

Figure 6.4-1 (a) A *pnp* transistor common-base (CB) (grounded base) configuration. Assume that I_E, I_C, and I_B all enter the transistor. $I_E + I_C + I_B = 0$. $A_i = -I_C/I_E \approx \alpha =$ current gain when R_L is 0 Ω. (b) Common-base amplifier stage. i_E, i_B, and i_C are all positive in the directions of the arrows.

uration. The collector characteristics have three basic regions: active, saturation, and cutoff regions. In the active region the input emitter-to-base diode is forward biased and the output collector-to-base diode is reverse biased. In the saturation region both emitter and collector junctions are forward biased. In the cutoff region the emitter and collector junctions are both reverse biased, hence the emitter current $I_E = 0$, and the collector current I_C equals the reverse saturation current I_{Co}.

In the common-emitter amplifier stage of Fig. 6.4-2a, the signal is applied to the *B-E* circuit and extracted from the *C-E* circuit. The input (base–emitter) impedance is low and the output (collector–emitter) impedance is high. The power gain is relatively high because the *CE* stage provides both current gain and voltage gain. The input signal voltage undergoes a phase reversal of 180°, as shown by the waveforms in Fig. 6.4-2a.

For the *CE* configuration (Fig. 6.4-2b) or amplifier stage operated in the active region, the dc and small-signal current gains are defined as

$$\beta_{dc} \equiv h_{FE} \equiv \frac{I_C}{I_B} \approx \beta \qquad (6.4\text{-}1)$$

and

$$\beta \equiv h_{fe} \equiv \frac{i_c}{i_b}, \qquad (6.4\text{-}2)$$

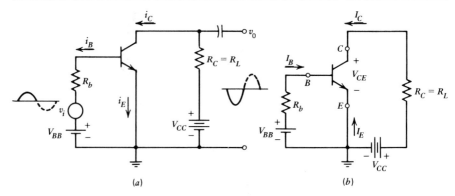

Figure 6.4-2 (a) Common-emitter amplifier stage. i_E, i_B, and i_C are all positive in the directions of the arrows. $i_E = I_E + i_e$; $i_B = I_B + i_b$; $i_C = I_C + i_c$; $v_B = V_B + v_b$; $v_C = V_C + v_c$. These expressions indicate that the instantaneous total value equals the sum of the dc and ac components. (b) CE configuration. I_E, I_B, I_C are all assumed to enter the transistor.

respectively. Strictly speaking β varies slightly with temperature, collector voltage, and current (Fig. 6.4-3), but for many purposes it can be considered constant for a given transistor. Typically, β is of the order of 100.

The *CE* output characteristics shown in Fig. 6.4-3 are the curves of collector current I_C versus collector-to-emitter voltage V_{CE}. The load line su-

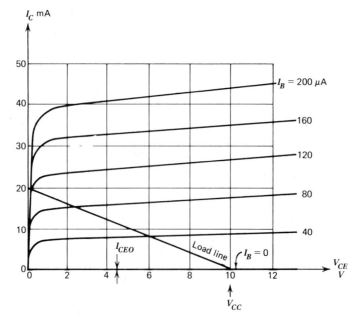

Figure 6.4-3 *CE* output characteristics of a typical *npn* Si transistor. A load line corresponding to $V_{CC} = 10$ V and $R_L = 500$ Ω is superimposed.

perimposed on the output characteristics is determined by the chosen values of a load $R_L = 500\ \Omega$ and a supply $V_{CC} = 10$ V. This load line passes through the point $I_C = 0$, $V_{CE}\ (= V_{CC}) = 10$ V, and $I_C\ (= V_{CC}/R_L) = 20$ mA, $V_{CE} = 0$. Its slope equals $-1/R_L$ independently of the output characteristics. The CE output characteristics also have three regions of interest: active, cutoff, and saturation regions. In the active region the E-B junction is forward biased and the C-B junction is reverse biased. When $I_E = 0$ and $I_C = I_{CO}$, the transistor is at cutoff. But when $I_B = 0$ or the base is open circuited, the transistor is not at cutoff. If $I_B = 0$, then $I_E = -I_C$ (the minus sign is due to the assumption that I_E, I_C, and I_B all enter the transistor), and

$$I_C = -I_E = I_{CEO} \simeq \beta I_{CO} \simeq \beta I_{CBO}, \qquad (6.4\text{-}3)$$

where I_{CEO} designates the actual collector current with collector junction reverse biased and base open circuited. I_{CBO} represents the current from collector to base with emitter open circuited. $|I_{CBO}|$ is slightly greater than $|I_{CO}|$ due to the surface leakage and some other factor.

In the saturation region, the collector and emitter junctions are forward biased by at least the cutin voltage (V_γ being a few tenths of a volt), resulting in the exponential change in collector current with a small change in voltage V_{CB}. Then $V_{CE} = V_{BE} - V_{BC}$ is only a few tenths of a volt at saturation.

A reasonable value for the temperature coefficient of $V_{BE(\text{active})}$, $V_{BE(\text{sat})}$, or $V_{BC(\text{sat})}$ is -2.5 mV/°C and that of $V_{CE(\text{sat})}$ is about -0.25 mV/°C.

In the CC configuration (emitter follower) of Fig. 6.4-4, the collector is at ac ground potential; the signal is applied to the base–collector circuit and extracted from the emitter–collector circuit. Since the emitter follower has a high input impedance and low output impedance, its voltage gain is less than unity and the power gain is normally lower than that obtained from

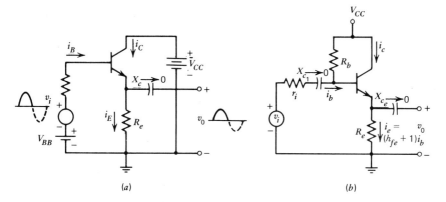

(a)   (b)

Figure 6.4-4 Common-collector (CC) configuration (EF). (a) EF with two bias sources; i_E, i_B, and i_C are all positive in the directions of the arrows. (b) EF with a single bias source. Since $i_e = (h_{fe} + 1)i_b$, the impedance is reflected from the emitter (or base) circuit to the base (or emitter) circuit by multiplying (or dividing) by $(h_{fe} + 1)$, so that the input (or output) impedance is high (or low).

Table 6-1   Typical *npn* Transistor-Junction Voltages (V) at 25°C

Material	$V_{CE \text{ (sat)}}$	$V_{BE \text{ (sat)}}$ $= V_\sigma$	$V_{BE \text{ (active)}}$	$V_{BE \text{ (cutin)}}$ $= V_\gamma$	$V_{BE \text{ (cutoff)}}$
Si	0.2	0.8	0.7	0.5	0.0
Ge	0.1	0.3	0.2	0.1	−0.1

either a *CB* or *CE* configuration. The *CC* configuration is used primarily for impedance-matching purposes. There is no phase reversal of the signal between the input and the output. The emitter follower can be designed using the *CE* characteristics.

A typical *npn* transistor-junction voltages at 25°C are listed in Table 6-1.

Transistor Switching Times

If a pulse waveform (v_i) is used instead of the input sinusoidal signal and the V_{BB} battery in Fig. 6.4-2*a*, then the *CE* stage becomes a transistor switch. The input waveform v_i operating between positive level and negative level drives the transistor from cutoff to saturation and back to cutoff. The collector current i_C does not immediately respond to input v_i. Instead, there is a turn-on time $t_{on} = t_d + t_r$, where t_d is the delay time required to bring i_C to $0.1I_{CS}$ ($\approx 0.1 V_{CC}/R_L$) and t_r is the rise time required to carry i_C from $0.1I_{CS}$ to $0.9I_{CS}$. When input v_i returns to its initial state at t = pulse duration T, current i_C again does not respond immediately. The time that elapses after the reversal of base current i_B before i_C falls again to $0.9I_{CS}$ is called the storage time t_s. The storage time is followed by the fall time t_f, which is the time required for i_C to fall from 90 to 10% of I_{CS}. The turn-off time $t_{off} = t_s + t_f$.

The delay time t_d results from the following three factors. First, when the input signal drives the base, a time interval is required to charge up the emitter-junction capacitance so that the transistor can be brought from cutoff to the active region. Second, a time is required before the minority carriers in the base can cross the base region to the collector junction; the minority carriers are those that have crossed the emitter junction into the base. Third, a time is required for current i_C to rise to $0.1I_{CS}$.

The rise time t_r and fall time t_f result from the following fact. When a base-current step is used to saturate the transistor or return it to cutoff from saturation, the collector current must traverse the active region. The collector current increases or decreases along an exponential curve whose time constant can be given by $\tau_r = h_{FE}(C_c R_c + 1/2\pi f_T)$, where C_c is the collector transition capacitance, and f_T is the frequency at which the *CE* short-circuit current gain drops to unity. The storage time results from the fact that a transistor in saturation has a saturation charge of excess minority carriers stored in the base; the transistor cannot respond until this saturation excess charge has been removed.

6.5   FIELD-EFFECT TRANSISTORS

Introduction

The field-effect transistor (FET) is available in two types: the junction field-effect transistor (abbreviated JFET, or simply FET) and the metal-oxide-semiconductor field-effect transistor (MOSFET, named after the process of forming the insulating layer). The FET is unipolar, operating wth either electron in an n-channel FET or hole current in a p-channel FET. The FET is essentially a semiconductor current path. The JFET conductance is controlled by an electric field that results from reverse biasing a p-n junction. Variation of a voltage between gate (G) and source (S) alters the resistance between drain (D) and source, resulting in the characteristics shown in Fig. 6.5-1. At lower drain voltages the drain current i_D (= source current i_S) increases linearly with the drain voltage v_{DS} for a fixed gate voltage V_{GS}. The slope, equal to the inverse of the FET drain-to-source resistance R_{DS}, decreases as the gate voltage is made more negative (for an n-channel JFET). When a sufficiently negative gate voltage is applied, no current flows and the FET is in the cutoff region. More positive gate voltages increase the drain current somewhat, but only to a certain limit, at which point the FET is turned fully on and R_{DS} has its minimum value, R_{on}. At high drain voltages the current levels off; that is, i_D becomes almost independent of v_{DS} (for small changes). The point above which i_D is almost constant is roughly equal to the pinchoff voltage. The depletion MOSFET has a behavior similar to the JFET characteristics.

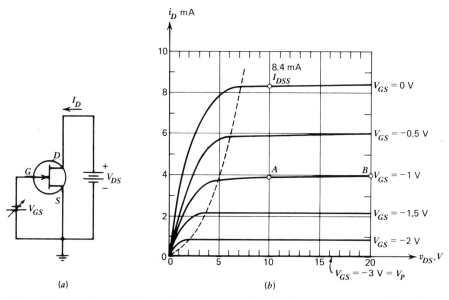

Figure 6.5-1 n-channel JFET: (a) circuit representation; (b) drain characteristics. When V_{GS} = 0 V, the drain current is I_{DSS}. At gate-source pinchoff voltage V_P, $i_D = 0$.

An n-channel JFET operated as a linear amplifier (common-source configuration) uses a negative voltage supply $(-V_{GG})$ on the gate side and a positive voltage supply $(+V_{DD})$ in series with a load resistor on the drain side. It is desirable for linear operation to have i_D independent of v_D, and thus normally the FET is operated beyond pinchoff (higher drain voltages). However, when the FET is used as a voltage-variable signal attenuator, linear operation in the low-voltage region well-below pinchoff is required.

An important parameter of the low-frequency small-signal model for a JFET or MOSFET is its gain or transconductance g_m, which is defined as

$$g_m \equiv \left. \frac{i_d}{v_{gs}} \right|_{v_{ds}=0} = \left. \frac{\partial i_D}{\partial v_{GS}} \right|_{V_{DS}} \simeq \left. \frac{\Delta i_D}{\Delta v_{GS}} \right|_{V_{DS}}. \qquad (6.5\text{-}1)$$

A typical value of g_m is 4 mA/V at $I_D = 4$ mA for the MOSFET with $K = 1$ mA/v^2. Since g_m varies considerably with i_D, the FET is a rather nonlinear device.

The experimental data on the current i_D as a function of gate voltage v_{GS} may be obtained from the output (drain) characteristics such as Fig. 6.5-1b or from the transfer characteristics indicated in Fig. 6.5-2. These data are required for proper biasing of the FET. The normal operating and cutoff regions of MOSFETs (Fig. 6.5-2) vary substantially with the FET type and may be classified into two general bias modes: depletion and enhancement. Normally the enhancement NMOSFET uses a positive gate voltage, while the enhancement PMOSFET uses a negative gate voltage.

The p-channel FET, like the pnp transistor, normally uses a negative voltage supply on the drain side. The JFET cannot be operated in the en-

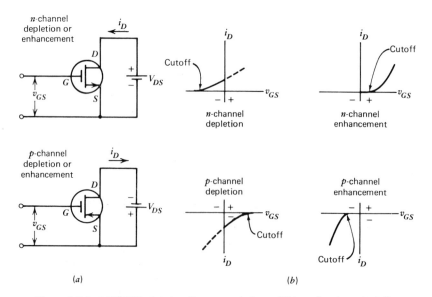

Figure 6.5-2 MOSFETs: (a) circuit representations; (b) transfer characteristics.

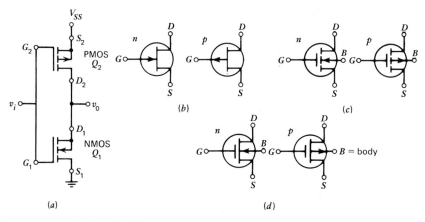

Figure 6.5-3   Circuit symbols for various FET types: (a) CMOS inverter; (b) JFETs; (c) enhancement MOSFETs; (d) depletion MOSFETs.

hancement mode since the gate would be forward biased and draw current. The FET has an extremely high input resistance (as high as 10^{14} Ω) (and a relatively high output resistance). It can dissipate high power and switch large currents in several nanoseconds. A complementary MOSFET (CMOS) connected as an inverter is fabricated as shown in Fig. 6.5-3a. When the CMOS inverter is used in digital circuits, the quiescent power dissipation is essentially zero at low frequencies. Several common FET symbols are shown in Fig. 6.5-3b–6.5-3d. For many FETs the source and drain are interchangeable.

Characteristics Analysis

It has been found that the relationship between i_D and v_{GS} can be approximated by the parabola

$$i_D = I_{DSS}\left(1 - \frac{v_{GS}}{V_P}\right)^2 \qquad (6.5\text{-}2)$$

for the drain current in the flat portion of the drain characteristic. The notation I_{DSS} is the drain current with the gate shorted to the source ($v_{GS} = 0$), and V_P is the gate–source pinchoff voltage at which the drain current is cut off.

If a slope is drawn to the drain characteristic at a point such as A on Fig. 6.5-1, the slope of the tangent defines the ac drain resistance r_d:

$$r_d \equiv \frac{\Delta v_{DS}}{\Delta i_{DS}}\bigg|_{v_{GS}} \qquad (6.5\text{-}3)$$

EXAMPLE 6.5-1   The value of I_{DSS} is 8.4 mA for the JFET used for Fig. 6.5-1b. The change in current from point A to point B is 100 μA. Determine i_D and r_d at point A.

Solution.   From Fig. 6.5-1 we see $V_P = -3$ V and $v_{GS} = -1$ V (at point A). Then i_D at point A is

$$i_D = I_{DSS}\left(1 - \frac{v_{GS}}{V_P}\right)^2 = 8.4 \text{ mA} \left(1 - \frac{-1 \text{ V}}{-3 \text{ V}}\right)^2 = 3.73 \text{ mA}.$$

Using Δv_{DS} and Δi_D between point B and point A we obtain the drain resistance at point A:

$$r_d = \frac{\Delta v_{DS}}{\Delta i_D} = \frac{20 - 10}{100 \times 10^{-6}} = 100,000 \ \Omega.$$

The slope of a tangent to the transfer characteristic at a point defines the transconductance g_m of the JFET at that point, as given by Eq. (6.5-1). If Eq. (6.5-2) is differentiated with respect to v_{GS}, a mathematical expression can be obtained for g_m:

$$i_D = I_{DSS}\left(1 - \frac{v_{GS}}{V_P}\right)^2,$$

$$\frac{di_D}{dv_{GS}} = 2I_{DSS}\left(1 - \frac{v_{GS}}{V_P}\right)\left(-\frac{1}{V_P}\right) = -\frac{2I_{DSS}}{V_P}\left(1 - \frac{v_{GS}}{V_P}\right),$$

$$g_m = -\frac{2I_{DSS}}{V_P}\left(1 - \frac{v_{GS}}{V_P}\right). \tag{6.5-4}$$

When v_{GS} is zero, g_m is g_{m0}:

$$g_{m0} = -\frac{2I_{DSS}}{V_P}. \tag{6.5-5}$$

And, substitution of Eq. (6.5-5) into Eq. (6.5-4) yields

$$g_m = g_{m0}\left(1 - \frac{v_{GS}}{V_P}\right). \tag{6.5-6}$$

Equations (6.5-1)–(6.5-6) are valid for both the JFET and the depletion MOSFET. The MOSFET itself can be used instead of a resistor in an IC. The gate of the MOSFET is placed either to the supply (switch position 1) or to the ground return (switch position 2) in Fig. 6.5-4.

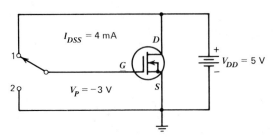

Figure 6.5-4   *n*-channel depletion-type MOSFET used as a resistor.

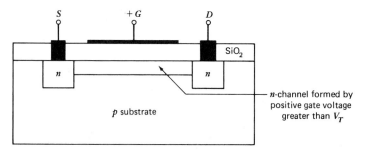

Figure 6.5-5   *n*-channel formed in enhancement MOSFET.

EXAMPLE 6.5-2   Determine the equivalent resistances of the FET circuit of Fig. 6.5-4 with the switch in position 1 and in position 2.

Solution.   When the switch is in position 1, $V_{GS} = +5$ V. Then

$$I_D = I_{DSS}\left(1 - \frac{V_{GS}}{V_P}\right)^2 = 4 \text{ mA}\left(1 - \frac{5 \text{ V}}{-3 \text{ V}}\right)^2 = 28.44 \text{ mA}$$

and

$$R_1 = \frac{V_{DD}}{I_D} = \frac{5 \text{ V}}{28.44 \text{ mA}} = 176 \text{ }\Omega.$$

When the switch is in position 2, $V_{GS} = 0$ and $I_D = I_{DSS} = 4$ mA. Then

$$R_2 = \frac{V_{DD}}{I_{DSS}} = \frac{5 \text{ V}}{4 \text{ mA}} = 1250 \text{ }\Omega.$$

I_{DSS} and V_P have no meaning for the enhancement MOSFET since it is constructed without a channel. If a positive voltage is applied to the gate of the *n*-channel unit (Fig. 6.5-5) at a sufficient positive voltage called the threshold voltage V_T, electrons are pulled into the substrate just below the layer of insulation beneath the gate. Now a virtual *n*-channel exists between the source and drain to permit a flow of drain current I_D. If the gate voltage is increased, the virtual channel deepens and I_D increases.

Figure 6.5-6 shows a set of typical drain characteristics and the corresponding transfer curve of a *n*-channel enhancement MOSFET. The transfer characteristic is described by

$$i_D = K(v_{GS} - V_T)^2, \tag{6.5-7}$$

where K, expressed in mA/V^2, is a property of the device construction. If Eq. (6.5-7) is differentiated with respect to v_{GS}, we obtain an equation for g_m:

$$i_D = K(v_{GS} - V_T)^2$$

$$\frac{di_D}{dV_{GS}} = 2K(v_{GS} - V_T)$$

$$g_m = 2K(v_{GS} - V_T). \tag{6.5-8}$$

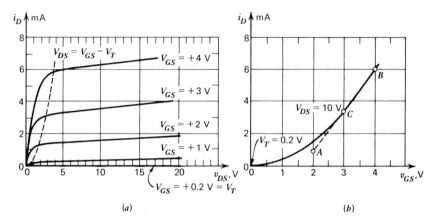

Figure 6.5-6 Characteristics for enhancement NMOSFET with $K = 0.445$ mA/V^2: (a) drain characteristic; (b) transfer curve.

EXAMPLE 6.5-3 The value of K for the enhancement NMOSFET shown in Fig. 6.5-6 is 0.445 mA/V^2. Determine the value of I_D and g_m at point C on the transfer curve.

Solution. $K = 0.445$ mA/V^2; $v_{GS} = +3$ V at point C; $V_T = +0.2$ V. Substituting these values into Eqs. (6.5-7) and (6.5-8) we obtain

$$I_D = K(v_{GS} - V_T)^2 = 0.455(3.0 - 0.2)^2 = 3.49 \text{ mA}$$

and

$$g_m = 2K(v_{GS} - V_T) = 2 \times 0.445(3.0 - 0.2) = 2.5 \text{ ms}$$
$$= 2500 \text{ μs}.$$

Another solution for g_m: at points B and A,

$$I_D = 6 \text{ mA}, v_{GS} = 4.0 \text{ V}$$

and

$$I_D = 1 \text{ mA}, v_{GS} = 2.0 \text{ V}.$$

Then by Eq. (6.5-1),

$$g_m = \frac{\Delta i_D}{\Delta v_{GS}} = \frac{6 - 1}{4 - 2} = 2.5 \text{ ms} = 2500 \text{ μs}.$$

6.6 SILICON-CONTROLLED RECTIFIERS AND UNIJUNCTION TRANSISTORS

Thyristors—SCR and TRIAC

A silicon-controlled rectifier (SCR) is a four-layer (*pnpn*) device but only three electrodes (cathode, anode, and gate) are exposed, as shown in Fig. 6.6-1. Basically, an SCR is a diode that will conduct in the forward direction only if turned on by a current pulse applied to the gate. Reverse current does not flow under any condition (except breakdown). Most SCRs are intended for power-switching applications, and so have a high current capacity, ranging from 1 to 100 A or more. They are mounted on cases that can dissipate the necessary power. The SCR is turned off by reversing the anode voltage or interrupting the anode current. Note that it cannot be turned off by simply removing the gate signal.

SCR gate-current requirements vary widely, ranging from about 0.1 mA to well over 100 mA. The voltage drop across the SCR in the conducting state is typically 1.5 V. Since the gate current need be applied only briefly (typically 2–20 μs), SCR gate drivers are frequently designed to deliver short current pulses, which have a high peak but a low average current so that power requirements are reduced. Note C_{AG}, the interelectrode capacitance between the anode and gate: If large inductive transients or radio-frequency (rf) interference is present on the anode, then a corresponding rf current will pass through C_{AG} from the anode to the gate, thus turning on the SCR. In this case, a filter may be used on the anode to avoid such gate triggering.

The DIAC is basically a two-terminal parallel–inverse combination of semiconductor layers that permits triggering in either direction. There is a breakover voltage in either direction, as shown in the characteristics of Fig. 6.6-2a. The TRIAC is fundamentally a DIAC with a gate terminal for controlling the turn-on conditions of the bilateral device in either direction. Characteristics of a TRIAC are shown in Fig. 6.6-3a. TRIACs are similar to SCRs except that they conduct current in either direction. They are intended for ac applications such as switching a motor or light connected to a 110-V (60-Hz) line. A TRIAC can be regarded as two SCRs back to back with the control voltage applied between the gate and only one anode, as shown in Fig. 6.6-4.

Both the SCR and TRIAC are high-current semiconductor switches often termed thyristors.

Figure 6.6-5a shows a light-activated SCR (LASCR), which is an SCR whose state is controlled by the light falling upon a silicon semiconductor

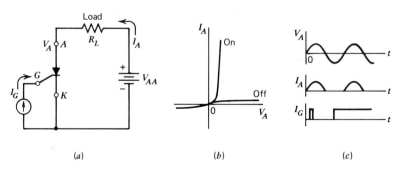

Figure 6.6-1 Silicon-controlled rectifier: (a) circuit representation; (b) output characteristics; (c) example of anode current controlled by gate current.

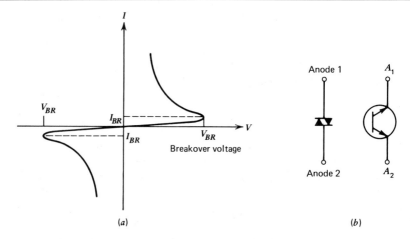

Figure 6.6-2   DIAC: (a) characteristics; (b) symbols.

layer of the device. Some of the main applications for the LASCR include optical light controls, relays, phase control, and motor control.

The Shockley diode shown in Fig. 6.6-5b is a four-layer *pnpn* diode with only two external terminals. The characteristics of the device are exactly the same as those encountered for the SCR with I_G = O. Shockley diodes are commonly used as a trigger switch for an SCR.

The silicon-controlled switch (SCS) is a commonly used low-power *pnpn* switch. It is essentially a miniature SCR with leads attached to all four semiconductor layers, as shown in Fig. 6.6-5c. The additional lead gives the designer access to the anode gate as well as the cathode gate. Less than 1 μA and 1 V are required for triggering at cathode gate. Less than 1 mA and 1 V are required to trigger at the anode gate.

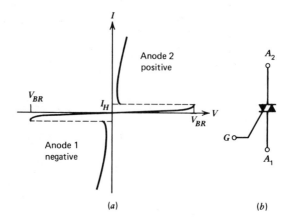

Figure 6.6-3   TRIAC: (a) characteristics; (b) symbol.

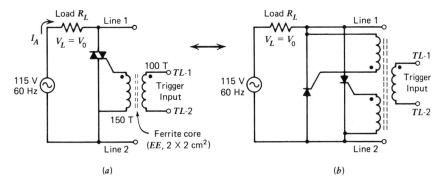

Figure 6.6-4 (a) TRIAC and (b) SCR circuits. Circuit (a) may be used instead of (b).

Unijunction Transistors

The unijunction transistor (UJT) is a three-terminal device having the basic construction of Fig. 6.6-6a. A slab of lightly doped n-type silicon material has two base contacts (B_1 and B_2) attached to both ends of one surface and an aluminum rod alloyed to the opposite surface. The emitter $p - n$ junction of the device is formed at the boundary of the aluminum rod and the n-type silicon slab, with the p-type region attached along (about halfway) the slab. The $p - n$ junction forms a reverse-biased emitter if the emitter voltage V_E is less than the voltage V_{RB1} at the attachment point A. Let R_{B_1} be the resistance from A to B_1 and R_{B_2} the resistance from A to B_2. When the $p - n$ junction is reverse biased, the emitter current I_E is zero, and the voltage V_{RB1} is given by

$$V_{RB1} = \frac{R_{B1} V_{BB}}{R_{B1} + R_{B2}} = \eta V_{BB} \big|_{I_E = 0}, \tag{6.6-1}$$

where

$$\eta = \frac{R_{B_1}}{R_{B_1} + R_{B_2}} \bigg|_{I_E = 0};$$

η is the intrinsic stand-off ratio (≈ 0.5–0.8). In fact, the slab acts as a resistive voltage divider. For applied emitter voltage V_E greater than ηV_{BB} by the forward-voltage drop of the diode, V_D ($= 0.35$–0.7 V), the junction diode will fire. The emitter firing potential (peak-point voltage) is given by

$$V_p = \eta V_{BB} + V_D. \tag{6.6-2}$$

If V_E is made sufficiently positive, the junction will be forward biased and current I_E will flow into the n region. As the charge carriers (holes) enter the n region, the conductivity will increase, predominantly at the negative B_2 end, where the positive charges are attracted. As the conduction increases, the region will become more forward biased in a regenerative man-

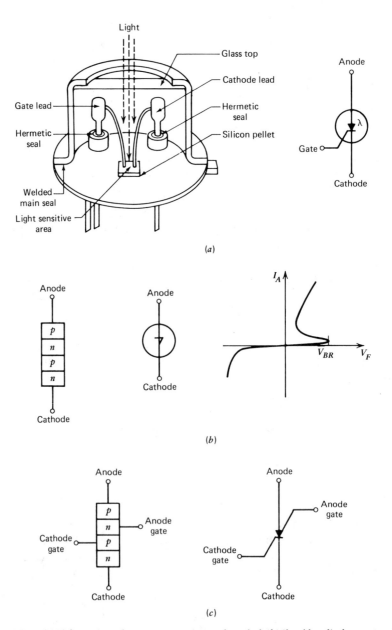

Figure 6.6-5 (a) Light-activated SCR construction and symbol; (b) Shockley diode construction, symbol, and characteristics; (c) Silicon-controlled switch construction and symbol.

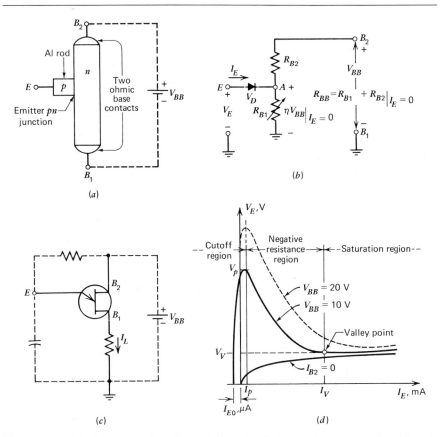

Figure 6.6-6   UJT: (a) construction; (b) equivalent circuit; (c) circuit representation; (d) static emitter characteristics.

ner. Hence a capacitor C connected to the emitter will rapidly discharge once V_E reaches $\eta V_{BB} + V_D$. If the emitter current falls sufficiently, the conduction will fall and the junction will return to the normal reverse-biased state. The UJT static emitter characteristics are shown in Fig. 6.6-6d.

The programmable unijunction transistor (PUT) is a *pnpn* silicon switch very similar to the SCS. The main differences are that the PUT anode gate operates at very low current levels, usually less than 1 μA, and only the anode gate is available for controlling the triggering of the PUT. The PUT is so named because it is functionally similar to the UJT. The major difference between the PUT and UJT is that the threshold voltage V_T is adjustable (programmable) by the external resistors (R_1 and R_2 in Fig. 6.6-7), whereas the emitter fire potential V_p is not adjustable. The PUT anode takes the place of the UJT emitter.

Both the UJT and the PUT are devices that switch to a high conductance, or on state, when voltage V_p or V_T is reached and switch back to a low conductance, or off state, when brought back to a lower current.

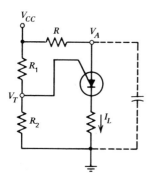

Figure 6.6-7 Programmable unijunction transistor (PUT).

UJT-SCRs Used in Timer and Regulated ac Power Supply

The UJT and SCR can be used in a timer circuit, as shown in Fig. 6.6-8a. When switch S_1 is turned on, SCR_2 conducts and SCR_1 is off at once. Then capacitor C_E starts charging. As soon as C_E is charged up to the emitter firing potential V_p, it starts discharging immediately. Under this condition, SCR_1 conducts and SCR_2 is off immediately. The time set is determined by

$$T \simeq R_E C_E \ln \frac{1}{1 - \eta}, \tag{6.6-3}$$

where η = intrinsic stand-off ratio $\simeq V_p/V$. Equation (6.6-3) is derived as follows.

The UJT stage in Fig. 6.6-8 is a relaxation oscillator. Its operation is explained referring to Fig. 6.6-9, where C_E is charged through resistor R_E toward supply voltage V. As long as the capacitor voltage V_E is below the emitter firing potential V_p, the UJT emitter lead appears as an open circuit.

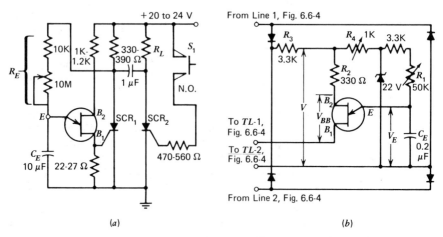

(a)   (b)

Figure 6.6-8 UJT-SCR circuits: (a) Timer; (b) UJT oscillator stage in the regulated ac power supply.

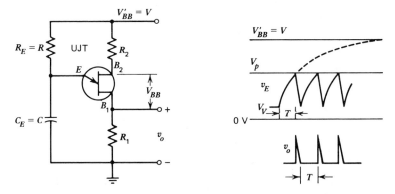

Figure 6.6-9 UJT oscillator and its waveforms.

From the waveforms shown in Fig. 6.6-9, we see that the peak-point voltage V_p is given by

$$V_p = (V - V_v)(1 - E^{-T/RC}).$$

Since $V \gg V_v$,

$$V_p \simeq V(1 - e^{-T/RC}).$$

Hence

$$T \simeq RC \ln \frac{V}{V - V_p} = RC \ln \frac{1}{1 - V_p/V},$$

or

$$\frac{1}{f} \simeq T \simeq R_E C_E \ln \frac{1}{1 - \eta},$$

where $R_E = R$, $C_E = C$, and $\eta \simeq V_p/V$. (Since $V_D \ll V_p$, $V_p = V_D + \eta V_{BB} \simeq \eta V_{BB} \simeq \eta V$.) The time set is approximately proportional to the time constant $R_E C_E$. The oscillator frequency $f \simeq 1/T$.

The voltage across the TRIAC or SCRs in Fig. 6.6-4 during blocking can be used for the interbase supply (V_{BB}) and synchronization for the UJT oscillator shown in Fig. 6.6-8b. The firing circuits is connected to the output of the bridge rectifier. When the emitter voltage equals the firing potential V_p, the UJT and SCR fire. As the line voltage increases, V_{BB} becomes higher, and so a higher V_p is needed (see Fig. 6.6-6d). Then the delay before the UJT and SCRs are triggered increases, causing the conduction angle of the SCRs to decrease. The decreased conduction angle reduces the power to the load, thus offsetting the increase of power to the load otherwise due to the increase in line voltage. By proper choice of the voltage divider ratio R_3 and R_4, it is possible to obtain perfect compensation of the circuit for small changes in line voltage. For a change in line voltage from 115 to 100

V, the circuit was adjusted so that the change in output voltage was less than 0.1 V with any output voltage setting from 10 to 30 V.

The resistor R_2 in base two (B_2) lead provides thermal stability for the oscillator of Fig. 6.6-9. An empirical formula for R_2 is given by

$$R_2 = 0.015VR_{BB}\eta, \tag{6.6-4}$$

where η = UJT standoff ratio,
$\quad\quad R_{BB}$ = internal B_1 to B_2 resistance.
$\quad\quad V$ = interbase supply = V'_{BB}.

EXAMPLE 6.6-1 A UJT relaxation oscillator shown in Fig. 6.6-9 has the following values: V'_{BB} = 12 V, R_2 = 5K, R_1 = 50 Ω, R_E = 50 K, η = 0.632. What value of C_E is required to obtain a frequency f = 400 Hz?

Solution. Substituting the values of f, R_E, and η into Eq. (6.6-3) gives

$$\frac{1}{400} = 50{,}000C_E \ln\frac{1}{1 - 0.632}$$

Hence

$$C_E = \frac{1}{(400)(50{,}000)(1.00)} = \frac{1}{20}\ \mu\text{F} = 0.05\ \mu\text{F}.$$

6.7   OPERATIONAL AMPLIFIERS (OP-AMPS)

Basic Characteristics of Op-amps

The basic block of an op-amp is symbolized in Fig. 6.7-1. The block usually consists of a differential amplifier input stage followed by one or more high-gain amplifier stages, which in turn drive some form of output stage. The direct-coupling technique used for the stages permits dc and ac amplification and ensures that the op-amp's quiescent point will not drift with temperature changes. The output voltage is proportional to the difference in voltage of two signal sources ($v+$ and $v-$), as indicated by

$$v_0 = A_v(v+ - v-) = A_v v_d, \tag{6.7-1}$$

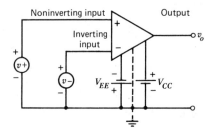

Figure 6.7-1   Basic block diagram of an op-amp.

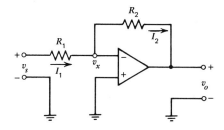

Figure 6.7-2 Inverting op-amp with voltage-shunt feedback. R_1 and R_2 are usually less than 500K, but greater than 1K.

where A_v is the open-loop gain. The op-amp can be used in a very special way as a result of several important characteristics, which are the following:

1. The gain, A_v, is extremely high; typically, $A_v = 100,000$.
2. The input impedance, R_i, is extremely high, meaning that very little current will flow into the input terminal; typically, $R_i = 5$ MΩ.
3. The output impedance, R_0, is extremely low, meaning that the output terminal can drive a fairly heavy load; typically, $R_0 = 150$ Ω.

Usually, the frequency response of an op-amp will range from dc to 50 kHz. In Fig. 6.7-1, the inputs labeled $(-)$ and $(+)$ are the inverting and noninverting inputs, respectively. Here the notation $(-)$ and $(+)$ refers to the sign of gain [Eq. (6.7-1)] and not to the polarity of the inputs ($v+$ or $v-$) with respect to ground. Either input may have a positive or negative applied voltage. It is the sign of the difference voltage that determines the sign of the output voltage. The op-amp is usually incorporated in the negative-feedback configuration to provide amplification with a lower but very stable gain. Such a circuit, called an inverting op-amp, is illustrated in Fig. 6.7-2.

Inverting Op-amp

In the inverting op-amp of Fig. 6.7-2, the output v_0 is connected to the inverting input $(-)$. As the input voltage v_s increases, v_x begins to increase; but, as v_x increases, output v_o decreases, owing to the fact that v_x is connected to the inverting input. Now, as v_o falls, it pulls the voltage at v_x back down via R_2. Whenever a change in output voltage is fed back to the inputs to oppose any further change, the circuit is considered to have negative feedback. Consider the currents in this inverting op-amp. The very high input impedance that is a characteristic of op-amps informs us that a negligible current flows into the $(-)$ or $(+)$ inputs. Since the $(+)$ input is grounded, we need to be concerned only with the condition of the $(-)$ input. Since practically no current flows into the $(-)$ input, we can consider I_1 to equal I_2. Then

$$I_1 = \frac{v_s - v_x}{R_1} = \frac{v_x - v_o}{R_2} = I_2. \tag{6.7-2}$$

From Eq. (6.7-1),

$$v_o = A_v(v+ \; - \; v-) = A_v(0 - v_x),$$

or

$$v_o = -Kv_x. \tag{6.7-3}$$

Hence $v_x = -(v_o/A_v)$. Replacing every occurrence of v_x with $-v_o/A_v$, we obtain

$$\frac{v_s + v_o/A_v}{R_1} = \frac{-v_o/A_v - v_o}{R_2} . \tag{6.7-4}$$

Since A_v is very large, $1/A_v$ is very small. Then Eq. (6.7-4) becomes

$$\frac{v_s}{R_1} = \frac{-v_o}{R_2} . \tag{6.7-5}$$

Hence the closed-loop gain is

$$A_{vf} = \frac{v_o}{v_s} = -\frac{R_2}{R_1} . \tag{6.7-6}$$

This equation indicates that the gain of the inverting op-amp is a function of only the two resistors R_2 and R_1, and is not related to the gain of the op-amp itself. Hence, we can obtain any gain that is necessary, within limits, by selecting the two resistors properly.

Noninverting Op-amp

Figure 6.7-3 shows that a circuit arrangement that has no phase reversal between output and input voltages. The resistors R_2 and R_1 are returned to the inverting input. The input voltage v_s is at the noninverting input. With this circuit arrangement, the feedback voltage v_f is in series and is subtracting from the input voltage v_s. Hence, this configuration is that of a voltage-series feedback amplifier. Since the input voltage is in series and is subtracting from the feedback voltage, the impedance seen by the source v_s is very high. The closed-loop gain of the noninverting op-amp is most easily determined if we assume an ideal op-amp. Then $R_o = 0$, R_i is infinite, and A_v is infinite, so that $v_d \simeq 0$. The resulting equivalent circuit is shown in Fig. 6.7-3b. Using this circuit, we have

$$v_s = v+ \; = v-$$

and

$$v_s = v- \; = \frac{R_1}{R_1 + R_2} v_o. \tag{6.7-7}$$

Hence,

$$A_{vf} = \frac{v_o}{v_s} = \frac{R_1 + R_2}{R_1} = 1 + \frac{R_2}{R_1} . \tag{6.7-8}$$

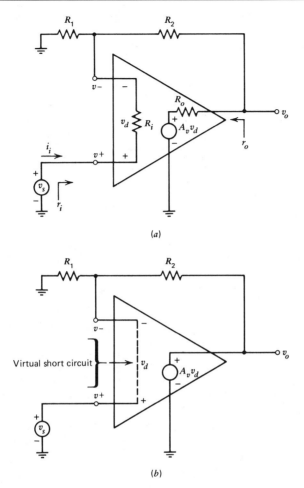

Figure 6.7-3 Noninverting op-amp: (a) circuit; (b) circuit with an ideal op-amp ($R_0 = 0$, $R_i = \infty$, $A_v = \infty$, and $v_d \simeq 0$).

If $R_2 \ll R_1$ (e.g., if R_1 is an open circuit and/or R_2 is a short circuit), then $A_{vf} = 1$ or $v_o = v_s$. This configuration, called the voltage follower, has a very high input impedance and a very low output impedance so that the source and load are in effect isolated. An emitter follower approximates the specifications of the noninverting op-amp.

Op-amps Used as the Main Parts of Butterworth Filters

Butterworth Filters.   Butterworth high- or low-pass filters exhibit a sharper cutoff than simple filters. A key design requirement for Butterworth filters is that the magnitude of the gain $|A|$ at the crossover (angular frequency ω_x) be $1/\sqrt{2}$ or -3 dB, just as it is for a simple single-section RC filter. Butterworth filters are composed of quadratic filters that have this property singly or in cascade. A quadratic low-pass filter has a transfer function, or gain, given by

$$A = \frac{1}{1 + jb\omega/\omega_x - (\omega/\omega_x)^2}. \tag{6.7-9}$$

Consider a single quadratic filter which is equivalent to a second-order Butterworth filter for which $b = \sqrt{2}$. $|A|^2 = AA^*$, where A^* is the complex conjugate. Using this relation and setting $b = \sqrt{2}$, we find

$$|A|^2 = AA^* = \frac{1}{1 - (\omega/\omega_x)^2 + jb(\omega/\omega_x)}$$

$$\times \frac{1}{1 - (\omega/\omega_x)^2 - jb(\omega/\omega_x)},$$

$$|A|^2 = \frac{1}{[1 - (\omega/\omega_x)^2]^2 + b^2(\omega/\omega_x)^2}, \tag{6.7-10}$$

$$|A|^2 = \frac{1}{1 + (\omega/\omega_x)^4}.$$

Hence,

$$|A| = \frac{1}{\sqrt{1 + (\omega/\omega_x)^4}}. \tag{6.7-11}$$

Higher-order low-pass Butterworth filters consist of two or more quadratic low-pass filters connected in cascade. While each filter is identical in form and ω_x is the same, the value of b is different from each filter. If the values of b are properly chosen, the magnitude of the total gain can be expressed as

$$|A| = \frac{1}{\sqrt{1 + (\omega/\omega_x)^{4n}}}, \tag{6.7-12}$$

where n is the number of quadratic filters and $2n$ is the order of the Butterworth filter. The frequency responses of Butterworth low-pass filters are plotted in Fig. 6.7-4.

Consider a fourth-order Butterworth low-pass filter consisting of two quadratic filters in series. The total gain is the product of the individual gains, A_1 and A_2, or

$$A = A_1 A_2 = \frac{1}{[1 - (\omega/\omega_x)^2 + jb_1\omega/\omega_x][1 - (\omega/\omega_x)^2 + jb_2\omega/\omega_x]}. \tag{6.7-13}$$

Let $\omega/\omega_x = \omega_n$, then

$$A = \frac{1}{(1 - \omega_n^2 + jb_1\omega_n)(1 - \omega_n^2 + jb_2\omega_n)}. \tag{6.7-14}$$

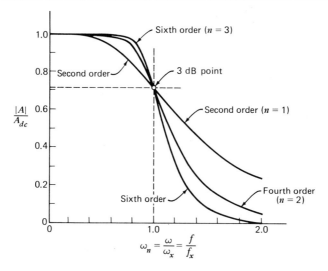

Figure 6.7-4 Frequency responses of Butterworth low-pass filters.

Multiplying by the complex conjugate to obtain $|A|^2$ yields

$$|A|^2 =$$

$$\frac{1}{[(1 - \omega_n^2 + jb_1\omega_n)(1 - \omega_n^2 - jb_1\omega_n)][(1 - \omega_n^2 + jb_2\omega_n)(1 - \omega_n^2 - jb_2\omega_n)]}$$

or

$$|A|^2 = \frac{1}{[(1 - \omega_n^2)^2 + b_1^2\omega_n^2][(1 - \omega_n^2)^2 + b_2^2\omega_n^2]} . \qquad (6.7\text{-}15)$$

Expanding, we obtain

$$|A|^2 = \frac{1}{(1 - \omega_n^2)^4 + (b_1^2 + b_2^2)\omega_n^2(1 - \omega_n^2)^2 + b_1^2 b_2^2 \omega_n^4}$$

or

$$|A|^2 = \frac{1}{\substack{1 + \omega_n^8 + \omega_n^2(b_1^2 + b_2^2 - 4) \\ \quad + \omega_n^4(6 - 2b_1^2 - 2b_2^2 + b_1^2 b_2^2) + \omega_n^6(b_1^2 + b_2^2 - 4)}} \qquad (6.7\text{-}16)$$

To fit the desired Butterworth form, all terms other than ω_n^8 and the constant ($=1$) must be set equal to zero. Thus

$$b_1^2 + b_2^2 - 4 = 0, \qquad (6.7\text{-}17)$$

$$6 - 2(b_1^2 + b_2^2) + b_1^2 b_2^2 = 0. \qquad (6.7\text{-}18)$$

Substituting Eq. (6.7-17) into Eq. (6.7-18) we get the relation in b_1 or b_2:

$$b^4 - 4b^2 + 2 = 0. \qquad (6.7\text{-}19)$$

Hence $b^2 = 2 \pm 1.414$, and the two positive roots of Eq. (6.7-19) are $b_1 =$ 1.8478 and $b_2 = 0.7654$, which are the fourth-order Butterworth filter constants. The procedure for finding the values of b for higher-order Butterworth filter is similar. The resulting values of b for even order up to the tenth are within the range from 0 to $+2$.

Butterworth high-pass filters can be designed with the same design criteria as used for the low-pass forms except that the variable ω/ω_x is inverted, that is, replaced by ω_x/ω in equations for A [Eq. (6.7-12)].

Quadratic Low-Pass Filter of Noninverting-Configuration Form.   A noninverting op-amp realization of the quadratic filter is shown in Fig. 6.7-5. It exhibits the response indicated by Eq. (6.7-11) except for a scaling factor or dc gain. Various combinations of resistors and capacitors will produce the same filter characteristics. While it is possible to set the dc gain A_{dc} to an even value (greater than unity), the specification of equal-valued resistors and capacitors was considered more desirable. The parameters ω_x and b are independently adjustable. Since capacitors are available in only a limited range of values, the capacitor value is usually chosen first and then R computed from $R = 1/\omega_x C$. Close-tolerance components are generally required for this filter.

It can be seen that part of the circuit in Fig. 6.7-5 is a standard noninverting op-amp of gain $K = 3 - b$, so that $V_o = KV_3$. Since b lies between 0 and 2, K is between 1 and 3. From Kirchhoff's current law (KCL) at junctions v_2 and v_3,

$$\frac{V_2 - V_i}{R} + \frac{V_2 - V_3}{R} + (V_2 - V_o)sC = 0$$

or

$$(2 + sCR)V_2 - V_1 - V_3 - sCRV_o = 0, \qquad (6.7\text{-}20)$$

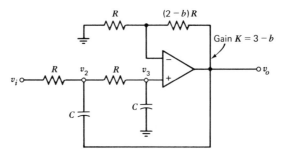

Figure 6.7-5   A quadratic low-pass filter of noninverting-configuration form. For Butterworth filters, $a = 1, b = 1.414, R = 5K$–$50K$ typically, and $\omega_x = 1/RC$.

and

$$\frac{V_3 - V_2}{R} + V_3 sC = 0$$

or

$$V_2 = V_3(1 + sCR). \tag{6.7-21}$$

$$V_o = KV_3 = (3 - b)V_3$$

or

$$V_3 = \frac{V_o}{3 - b}. \tag{6.7-22}$$

Variables V_2 and V_3 are eliminated from Eqs. (6.7-20), (6.7-21), and (6.7-22), and V_o is expressed as a function of V_i. The transfer function is

$$H(s) = \frac{V_o(s)}{V_i(s)} = \frac{3 - b}{1 + b(s/\omega_x) + (s/\omega_x)^2}, \tag{6.7-23}$$

where

$$\omega_x = 1/RC. \tag{6.7-24}$$

Expressed in terms of H_{dc} for $a = 1$, Eq. (6.7-23) becomes

$$H(s) = \frac{H_{dc}}{1 + bs/\omega_x + (s/\omega_x)^2}. \tag{6.7-25}$$

Substituting $s = j\omega$ yields $A(\omega)$. It is identical to Eq. (6.7-9) except for a multiplying constant of dc gain of H_{dc}. The more general form of the transfer function, where $a \neq 1$, is obtained by multiplying numerator and denominator by a, and redefining b and ω_x; hence

$$H(s) = \frac{H_o}{a + b's/\omega_x' + (s/\omega_x')^2} \tag{6.7-26}$$

where

$$\omega_x' = \frac{1}{\sqrt{a}\, RC}, \tag{6.7-27}$$

$$b' = b\sqrt{a}, \tag{6.7-28}$$

$$H_o = a\, H_{dc} = 3a - \frac{b'}{\sqrt{a}}. \tag{6.7-29}$$

Quadratic High-Pass Filter of Noninverting-Configuration Form. A noninverting op-amp realization of the quadratic high-pass filter is shown in

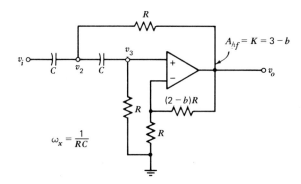

Figure 6.7-6 A quadratic high-pass filter of noninverting-configuration form. For Butterworth filters, $a = 1$ and $b = 1.414$.

Fig. 6.7-6. From *KCL* at junctions v_2 and v_3,

$$(V_2 - V_i)sC + (V_2 - V_3)sC + \frac{V_2 - V_o}{R} = 0, \qquad (6.7\text{-}30)$$

$$(V_3 - V_2)sC + \frac{V_3}{R} = 0 \qquad (6.7\text{-}31)$$

$$V_o = (3 - b)V_3 = KV_3. \qquad (6.7\text{-}32)$$

Eliminating variables V_2 and V_3 we obtain

$$H(s) = \frac{V_o(s)}{V_i(s)} = \frac{3 - b}{1 + b\omega_x/s + (\omega_x/s)^2} \qquad (6.7\text{-}33)$$

which is similar to Eq. (6.7-23) except that s/ω_x is replaced by ω_x/s.

6.8 BASIC DIGITAL INTEGRATED-CIRCUIT DEVICES

Introduction

The role of digital integrated circuits (ICs) in general instrument design is very important although their dominant application is in computers. The most extensive digital IC series are transistor–transistor logic (TTL) and CMOS. Digital IC devices have two stable states: a higher positive-voltage state termed high or logic 1 and a lower voltage state termed low or logic 0. This definition applies to conventional binary positive logic. Standard positive-logic notation refers to 0 V (approximately) as the low logic level or logic 0. Correspondingly the more positive voltage, for example, $+3$ to $+5$ V, is termed the high logic level or logic 1. From the mathematical logic point of view, the assignment of logic 0 or 1 to one of the two stable states

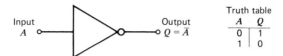

Figure 6.8-1   Inverter symbol and truth table.

of a particular logic device is, of course, arbitrary. Hence, for conventional negative logic, logic 0 is designated as the high state. However, nearly always, the more positive state is termed high. Positive logic is assumed for a signal, for example, A, and the bar added, for example, $\bar{A}$, whenever the signal is inverted. Usually a small circle is used as the inverting symbol.

Inverters and Drivers

The inverter is the digital equivalent of a unity-gain inverter. Its symbol and truth table are shown in Fig. 6.8-1.

A driver shown in Fig. 6.8-2 is the digital equivalent of a unity-gain amplifier with its output repeating the input logic state. The output presumably has a greater current (and/or voltage) capacity than the input source. Increased output-voltage capacity is required only for interfacing between different series of devices. Increased current capacity allows multiple loads to be connected. Two inverters in series are equivalent to a driver. Inverters are usually manufactured in IC form with six units per package, called the hex inverters.

OR, NOR, and Exclusive-OR (XOR) Gates

An OR gate has two or more inputs and one output. Its output will go to logic 1 if any input is at logic 1, as indicated by the truth table in Fig. 6.8-3.

A NOR gate is the equivalent of an OR gate followed by an inverter, as shown in Fig. 6.8-4. The small circle on the output side indicates inversion. The output is logic 1 only if all inputs are logic 0.

An exclusive-OR gate (XOR) has two or more inputs and one output. As indicated by the truth table in Fig. 6.8-5, the output will go to logic 1 if any input is at logic 1, except that the output will go to logic 0 if all inputs go to logic 1.

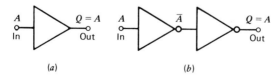

Figure 6.8-2   (a) Single driver unit; (b) driver formed from two inverters.

Inputs　　Output

A　　　$Q = A + B$

B

A	B	Q
0	0	0
1	0	1
0	1	1
1	1	1

Figure 6.8-3 Two-input OR gate: symbol and truth table.

AND and NAND Gates

An AND gate has two or more inputs and one output. Its output will go to logic 1 only if all inputs are at logic 1, as indicated by the truth table in Fig. 6.8-6.

A NAND gate shown in Fig. 6.8-7 is the equivalent of an AND gate followed by an inverter.

CMOS Transmission Device

A CMOS transmission device is shown in Fig. 6.8-8. It is equivalent to a voltage-controlled switch. It allows bidirectional flow of pulses when on and no flow when off. Typically, the resistance when on is $R_{on} = 400\ \Omega$, but devices with R_{on} below 25 Ω are available. These devices are symmetric, and either terminal can be connected to the input. For switching both analog and digital signals the input and output voltages are generally between the most negative terminal V_{SS} and most positive terminal V_{DD}. Voltages outside the range may cause the gate to lose control or damage the device. CMOS transmission devices are often manufactured with four units per package. When a transmission gate is connected in series with the output of a digital device, the three states are 0, 1, and off (or high impedance).

Set–Reset (SR) Flip-Flops (FFs)

The simplest type of FF is the S-R latch, which is a basic building block for other types of FFs. It has two inputs (S, R) and two outputs $(Q, \bar{Q})$. Application of a pulse to input S (i.e., momentarily bringing S to 1) will cause the output Q to go to 1 until reset by bringing input R to 1. The second output $\bar{Q}$ is the inverse of Q. The FF outputs remain in a state infinitely unless altered by an input pulse, however brief. SR flip-flops are usually composed of two NOR gates or two NAND gates, as shown in Figs. 6.8-9

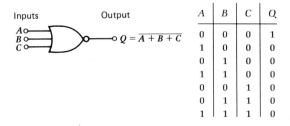

Inputs　　　　Output

A
B　　　$Q = \overline{A + B + C}$
C

A	B	C	Q
0	0	0	1
1	0	0	0
0	1	0	0
1	1	0	0
0	0	1	0
0	1	1	0
1	1	1	0

Figure 6.8-4 Three-input NOR gate: symbol and truth table.

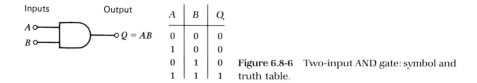

A	B	Q
0	0	0
1	0	1
0	1	1
1	1	0

Inputs   Output

$Q = A\bar{B} + \bar{A}B = A \oplus B$

Figure 6.8-5   Two-input XOR gate: symbol and truth table.

Inputs   Output

$Q = AB$

A	B	Q
0	0	0
1	0	0
0	1	0
1	1	1

Figure 6.8-6   Two-input AND gate: symbol and truth table.

Inputs   Output

$Q = \overline{AB}$

A	B	Q
0	0	1
1	0	1
0	1	1
1	1	0

Figure 6.8-7   Two-input NAND gate: symbol and truth table.

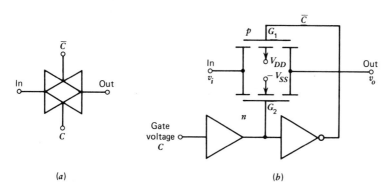

(a)   *(b)*

Figure 6.8-8   CMOS transmission device: (a) symbol; (b) CMOS analog or digital gate.

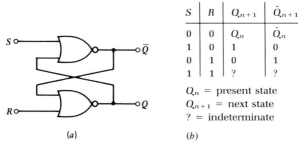

(a)   *(b)*

S	R	Q_{n+1}	$\bar{Q}_{n+1}$
0	0	Q_n	$\bar{Q}_n$
1	0	1	0
0	1	0	1
1	1	?	?

Q_n = present state
Q_{n+1} = next state
? = indeterminate

Figure 6.8-9   (a) SR latch (or FF) composed of two NOR gates; (b) truth table; logic 0 on both inputs produces no change.

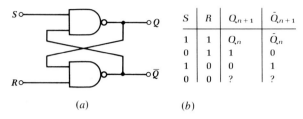

S	R	Q_{n+1}	$\bar{Q}_{n+1}$
1	1	Q_n	$\bar{Q}_n$
0	1	1	0
1	0	0	1
0	0	?	?

(a)   (b)

Figure 6.8-10 (a) SR flip-flop (or latch) composed of two NAND gates; (b) truth table, in which the input logic sense is the inversion of that for the NOR-type SR flip-flop (Fig. 6.8-9).

and 6.8-10. The NAND gate FF with active low inputs is sometimes called an $\bar{S}\bar{R}$ latch.

Most commonly used flip-flops are made in TTL. A TTL flip-flop has a single clock input. This single clock (CK) may be of the edge-triggered type or master–slave type. The SR edge-trigger flip-flop can be constructed from the NAND SR latch by introducing a clock input (CK), as shown in Fig. 6.8-11; note the function table and symbolic representation shown in Fig. 6.8-11b, where Q_n and Q_{n+1} are the states of the flip-flop before and after clocking, respectively.

There are three other types of commonly used flip-flops: JK FF, T-type FF, and D-type FF. These will be discussed in the next section.

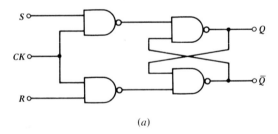

(a)

Input			Output	
CK	S	R	Q_{n+1}	$\bar{Q}_{n+1}$
1	0	0	Q_n	$\bar{Q}_n$
1	0	1	0	1
1	1	0	1	0
1	1	1	1*	1*

* Not allowed.

(b)

Figure 6.8-11 (a) Edge-triggered SR flip-flop. (b) Truth (function) table and symbolic representation.

6.9   *JK* FFS, *T*-TYPE FFS, *D*-TYPE FFS, AND COUNTERS

JK Flip-flop

The *JK* flip-flop shown in Fig. 6.9-1 has two data inputs, *J* and *K*, a single clock input (*CK*), and two outputs, *Q* and $\bar{Q}$. The operation of a *JK* flip-flop is identical to that of a *SR* flip-flop except that the former allows the input $J = K = 1$. When $J = K = 1$, the state *Q* will change state regardless of what the state *Q* was prior to clocking. Note that the operation of the *JK* flip-flop may be unstable: Because of the feedback connection *Q* ($\bar{Q}$) at the input to *K* (*J*), the input will change during the clock pulse (*CK* = 1) if the output changes state. Hence, for the pulse duration t_p (while *CK* = 1), the output will oscillate forth and back between 0 and 1. At the end of the pulse (*CK* = 0), the value of *Q* is ambiguous. This unstable situation is called a race-around condition.

T-Type Flip-Flop

When the two inputs, *J* and *K*, are connected together, a *T*-type flip-flop is formed, as shown in Fig. 6.9-2. This type of flip-flop has only one data input, *T*. When the input *T* is at a 0 level prior to a clock pulse, the *Q* output will not change with clocking. When the input *T* is at a 1 level, the *Q* output will be in the $\bar{Q}_n$ state after clocking (i.e., $Q_{n+1} = \bar{Q}_n$). This is called toggling,

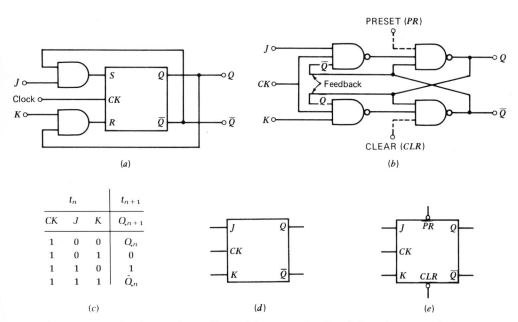

t_n			t_{n+1}
CK	J	K	Q_{n+1}
1	0	0	Q_n
1	0	1	0
1	1	0	1
1	1	1	$\bar{Q}_n$

(c)   (d)   (e)

Figure 6.9-1   *JK* flip-flop: (a) obtained by modifying an *SR* flip-flop; (b) logic diagram; (c) function table, where Q_{n+1} (Q_n) means the state of Q output after (before) clocking; (d) symbolic representation without CLEAR and PRESET; (e) symbolic representation with CLEAR and PRESET.

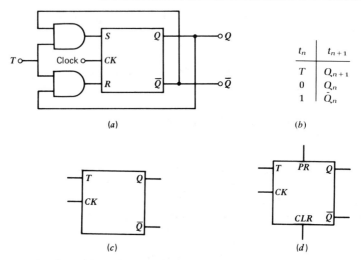

(a)

t_n	t_{n+1}
T	Q_{n+1}
0	Q_n
1	$\hat{Q}_n$

(b)

(c)

(d)

Figure 6.9-2 T flip-flop: (a) logic diagram; (b) function table; (c) symbol 1 (without CLEAR and PRESET); (d) symbol 2 (with CLEAR and PRESET).

thus the name T flip-flop. When the T input and the clock of a T flip-flop circuit are tied together, the flip-flop circuit generates two clock signals of the same frequency as the input clock signal. This circuit is useful in sequential circuit design and can be constructed from an SR flip-flop without using the two, AND gates, as shown in Fig. 6.9-3.

D-Type Flip-Flop

When the S (J) input of an SR flip-flop (JK flip-flop) is connected to the inverted R (K) input of the same flip-flop, a D-type flip-flop is formed, as shown in Fig. 6.9-4. This type of flip-flop has only one data input, D. The operation of this type of flip-flop may be described as follows. When the D input is at a 0 level prior to a clock pulse, the Q output will be 0 after clocking. When the D input is at a 1 level, the Q output will be 1 after clocking. In other words, $Q_{n+1} = D_n$, where Q_{n+1} and D_n denote the values of the output Q and the input D at t_{n+1} (after clocking) and t_n (before clocking), respec-

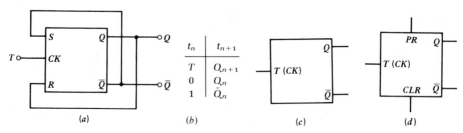

(a)

t_n	t_{n+1}
T	Q_{n+1}
0	Q_n
1	$\hat{Q}_n$

(b)

(c)

(d)

Figure 6.9-3 T flip-flop with its T input and clock tied together: (a) logic diagram; (b) function table; (c) symbol 1 (without CLEAR and PRESET); (d) symbol 2 (with CLEAR and PRESET).

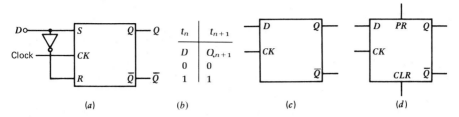

Figure 6.9-4 D flip-flop: (a) logic diagram; (b) function table; (c) and (d) symbols.

tively. For this type of flip-flop, the Q output is identical to the D input except with one pulse time delay; thus, the name D flip-flop.

Two Additional Inputs: CLEAR and PRESET

Most commercially available flip-flops have two additional inputs: CLEAR (CLR) and PRESET (PR). Their functions are as follows:

$$\text{PRESET} = \text{direct set} = 0, Q = 1.$$

$$\text{CLEAR} = \text{direct reset} = 0, Q = 0.$$

PRESET	CLEAR	Q	$\bar{Q}$	Remark
0	1	1	0	As PRESET = 0 (low), the flip-flop sets
1	0	0	1	regardless what the other inputs (e.g., S,
0	0	1*	1*	R, CK) are. Similarly, as CLEAR = 0, the
1	1	No change		flip-flop resets regardless what the other inputs are. Essentially, the PRESET or CLEAR input "overrides" the other inputs.

* Not allowed.

When both PRESET and CLEAR are active (i.e., they are at the low voltage level for a positive-logic flip-flop, or both are at logic 0), $Q = \bar{Q} = 1$, which is not allowed for any flip-flops. These inputs are asynchronous and independent of the clock. While they are present, all other operations are inhibited. An SR edge-triggered flip-flop with CLEAR and PRESET asynchronous inputs is shown in Fig. 6.9-5. Using this circuit as a building block, other types of flip-flops with CLEAR and PRESET can be constructed. The symbolic representations of the four types of flip-flops with CLEAR and PRESET are given in Figs. 6.9-1–6.9-4.

Master–Slave JK Flip-Flops

The circuit of a master–slave flip-flop is basically two latches connected serially. The first latch is called the master and the second is called the slave.

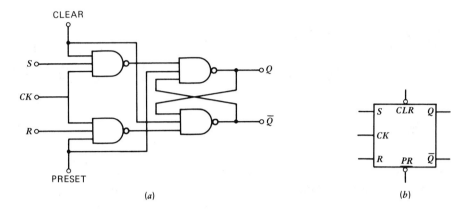

(a)   (b)

Inputs					Outputs	
PR	CLR	CK	S	R	Q_{n+1}	$\bar{Q}_{n+1}$
0	1	x^a	x	x	1	0
1	0	x	x	x	0	1
0	0	x	x	x	1^b	1^b
1	1	1	0	0	Q_n	$\bar{Q}_n$
1	1	1	0	1	0	1
1	1	1	1	0	1	0
1	1	1	1	1	1^b	1^b

[a] x = Regardless of state.
[b] Not allowed.

(c)

Figure 6.9-5 Edge-triggered SR flip-flop with CLEAR and PRESET asynchronous inputs: (a) logic diagram; (b) logic symbol; (c) function table.

A typical master–slave *JK* flip-flop and a master–slave *JK* flip-flop with CLEAR and PRESET asynchronous inputs are shown in Figs. 6.9-6*a* and 6.9-6*b*, respectively. Normal action in master–slave clocking consists of four steps, as shown in Fig. 6.9-7. The main feature of this type of clocking is that the data inputs are never directly connected to the outputs at any time during clocking; this provides total isolation of outputs from data inputs.

The function (or truth) table operation for both the master–slave and the edge-triggered *JK* flip-flops is identical; the primary difference is the clocking operation. For an edge-triggered type, the data on the inputs are entered into the flip-flop and appear on the outputs on the same edge of the clock pulse. For a master–slave type, the data on the *J* and *K* inputs are entered on the leading edge of the clock but do not appear on the outputs until the trailing edge of the clock.

Figure 6.9-6*a* exhibits the transitions for the flip-flop going from the RESET state to the SET state. Let us examine the operation of this circuit

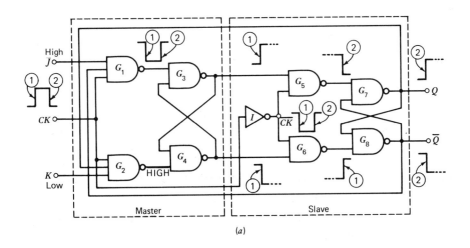

(a)

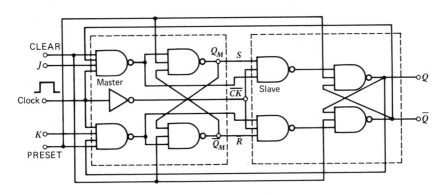

Function Table

Inputs					Outputs	
PR	CLR	CK	J	K	Q_{n+1}	$\bar{Q}_{n+1}$
0	1	x[a]	x	x	1	0
1	0	x	x	x	0	1
0	0	x	x	x	1[b]	1[b]
1	1	⊓	0	0	Q_n (No change)	$\bar{Q}_n$
1	1	⊓	0	1	0 (RESET)	1
1	1	⊓	1	0	1 (SET)	0
1	1	⊓	1	1	$\bar{Q}_n$ (TOGGLE)	Q_n

[a] x = Regardless of state.
[b] Not allowed.

(b)

Figure 6.9-6 (a) A typical master–slave JK flip-flop, making transition from the RESET state to the SET state. (b) Master–slave JK flip-flop with CLEAR and PRESET asynchronous inputs.

141

Figure 6.9-7 Master–slave clock. (1) Isolate slave from master; (2) enable data inputs to master; (3) disable data inputs; (4) transfer data from master to slave.

in detail. Assume that the flip-flop is RESET. Let J be HIGH and K be LOW, making a SET condition on the inputs. The circled number associated with each transition indicates when that transition occurs with respect to the clock pulse. A ① corresponds to the leading edge of the clock pulse and a ② corresponds to the trailing edge. On the leading edge of the clock pulse, the circuit operates as follows:

1. The output of gate G_1 goes from HIGH to LOW because all of its inputs are HIGH.
2. The output of gate G_3 goes from LOW to HIGH and the output of gate G_4 goes from HIGH to LOW because of the LOW on the output of gate G_1.
3. The inverted clock ($\overline{CK}$) goes LOW, disabling both the G_5 and G_6 gates. This ensures that their outputs remain HIGH.
4. The master section has been SET because the J input is HIGH and the K input is LOW. The slave section (SR) has not changed state, and so the Q and $\overline{Q}$ outputs remain in the RESET state.

On the trailing edge of the clock pulse, the circuit of Fig. 6.9-6a operates as follows:

1. The master section remains in the SET state.
2. The output of gate G_5 goes from HIGH to LOW because both of its inputs are now HIGH.
3. The output of gate G_7 goes from LOW to HIGH and the output of gate G_8 goes from HIGH to LOW because of the LOW on the output of gate G_5.
4. The flip-flop is now in the SET state because the Q output is HIGH. It did not become SET until the trailing edge of the clock pulse, although the master section was set on the leading edge.

In summary, during a clock pulse the output Q does not change but the output Q_M of the master section follows JK logic; at the end of the pulse, the value of Q_M is transferred to Q. Since Q_n (the output before clocking) is invariant for the duration t_p of the clock pulse, the unstable situation in the basic JK flip-flop of Fig. 6.9-1b will be eliminated.

Binary Counters

The JK flip-flop can be operated as a binary counter because it reverses states upon application of clock pulse when both JK inputs are at 1. For this application, as Fig. 6.9-8 indicates, the clock becomes the input and the *JK* terminals are permanently wired to a 1 voltage supply. Before counting, a pulse is applied to all the CLEAR inputs, so that all the outputs (Q_A, Q_B, etc.) are at 0. If a series of pulses is applied to the *A* input (clock), its output will reverse each time the clock pulse goes through one complete cycle, which for many types of flip-flops occurs on the falling edge of the clock pulse. Two complete input cycles are required to produce one complete output cycle. There is no requirement that the input pulse be periodic or symmetric. A package of four flip-flops with the *JK* inputs internally connected to 1 is referred to as a 4-bit binary or ripple counter (Fig. 6.9-8) with the clock input renamed the toggle or count input *T*. Decimal equivalents for several binary states are shown in Table 6-2. Note that a 4-bit binary counter ranges from 0 to 15 decimal and has an output or carry pulse that can be connected to the toggle input of additional stages.

Decimal Counters

A binary-coded decimal (BCD) counter is shown in Fig. 6.9-9. It is identical to the 4-bit binary counter except for an internal circuit that resets all the flip-flops to 0 on the tenth count (binary 1010) rather than the sixteenth. This may be done by an AND gate with its output connected to RESET (Clear) and its inputs connected to the Q_B and Q_D outputs. Each 4-bit counter is equivalent to one decimal digit and is separately decoded from the binary as a digit from 0 to 9. No 4-bit binary numbers higher than 1001 exist.

The decimal digit can be read by a seven-segment display after decoding. Each segment of the display can be independently illuminated and the decoder/driver selects the proper combinations of segments to form the desired digit, as determined by the BCD input. Illumination is provided by light-emitting diodes, neon bulbs, or incandescent lamps.

Counters with a modulus or base other than 10 can be constructed in a similar fashion. For example, if the AND-gate inputs are connected to the

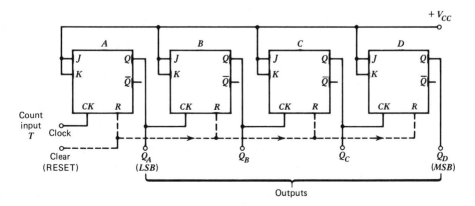

Figure 6.9-8 Basic 4-bit binary or ripple (asynchronous) counter.

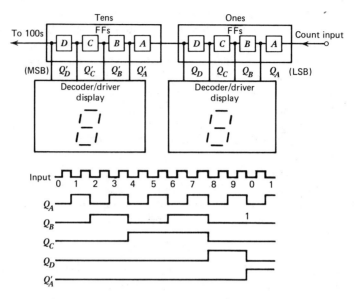

Figure 6.9-9 Two-digit BCD counter (asynchronous) and readout (BCD-to-seven-segment decoder/driver displays).

Table 6-2 States of the Binaries

Number of Input Pulses	State of Binary			
	B_3	B_2	B_1	B_0
0	0	0	0	0
1	0	0	0	1
2	0	0	1	0
3	0	0	1	1
4	0	1	0	0
5	0	1	0	1
6	0	1	1	0
7	0	1	1	1
8	1	0	0	0
9	1	0	0	1
10	1	0	1	0
11	1	0	1	1
12	1	1	0	0
13	1	1	0	1
14	1	1	1	0
15	1	1	1	1

Q_C and Q_B counter outputs and the gate output to RESET, the counter acts as a base 6 or divide-by-6 counter (the Q_D output is not used unless a divide-by-12 counter is desired).

Two of many available decade counters are indicated in Fig. 6.9-10. The LSTTL counter (74LS90) is pin compatible and functionally identical to the popular TTL counter SN7490. It features a flip-flop *A* with a separate clock input. In normal BCD operation, input *A* is the pulse-count input, and the clock inputs to the *B* section must be externally connected to the *A* output (the remaining clocks are internally connected). Following the decade counting stage inputs are tied to the *D* outputs, which function as carry pulses. Note that the following stage count is incremented on the falling edge of the *D* output, that is, on the tenth count of the driving stage. When functioned as a frequency divider (divide-by-10), the separate stage *A* is sometimes placed last (input frequency to *B* clock) so that the output is a symmetric square wave. In this case the unit acts as a decade counter, but it is not BCD encoded.

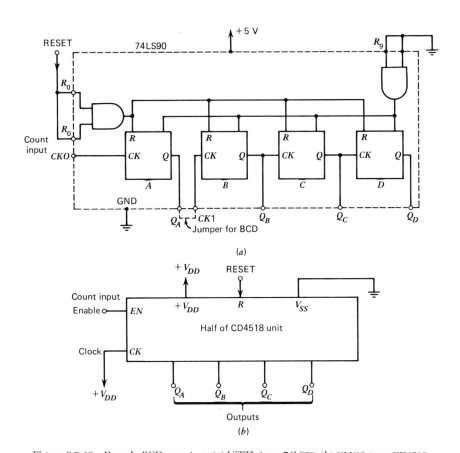

Figure 6.9-10   Decade BCD counters: (*a*) LSTTL-type 74LS90; (*b*) CMOS-type CD4518.

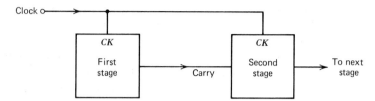

Figure 6.9-11   Synchronous counter.

Two BCD counters per package are available in CMOS-type CD4518. Half of this unit is shown in Fig. 6.9-10*b*. Normally pulses to be counted are connected to the enable input because the flip-flops change on the enable falling edge. Counting is inhibited (turned off) by bringing the enable (strobe) input to 1. Clocking on the rising edge is possible by interchanging the clock and enable inputs (inhibit on 0).

Synchronous Counters

Asynchronous, or ripple-type, counters discussed above, will exhibit a delay. The carry propagation delay is the time required for a counter to complete its response to an input pulse. The carry time of a ripple counter is longest when each stage is in the 1 state, because, in this situation, the next pulse must cause all previous flip-flops to change state. Any particular binary will not respond until the preceding stage has nominally completed its transition. The clock pulse effectively "ripples" through the chain. Thus, the carry time will be of the order of magnitude of the sum of the propagation delay times of all the binaries. If the chain is long, the carry time may well be longer than the interval between input pulses. In such a case, it will not be possible to read the counter between pulses.

If the asynchronous operation of a counter is changed so that all flip-flops are clocked simultaneously (synchronously) by the input pulses, the propagation delay time may be reduced considerably. Another advantage of the synchronous counter is that no decoding spikes appear at the output since all flip-flops change state at the same time. Thus no strobe pulse is required when decoding a synchronous counter. In order to obtain these advantages, the clock must be connected to all stages, as shown in Fig. 6.9-11, but each stage is inhibited unless the carry output of the previous stage occurs. Carry occurs just after the ninth pulse of a BCD counter, thus enabling the subsequent stage on the tenth count of the driving stage.

6.10   TTL AND CMOS INTEGRATED CIRCUITS

Totem-Pole TTL Circuits

Almost all TTL circuits used in industry have a two-state totem-pole output circuit. A typical TTL circuit shown in Fig. 6.10-1 consists of a NAND-gate input section (Q_1-Q_2) and a two-state totem-pole output (R-Q_3-D-Q_4). As-

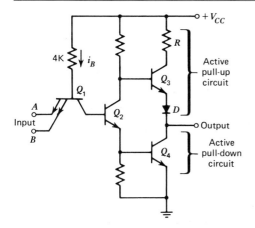

Figure 6.10-1 A typical TTL circuit—two-input NAND gate (e.g., $\frac{1}{4}$ 7400, Texas Instruments).

sume that either input (A or B) is made low by grounding it through a sufficiently low resistance R_1. Then the base current i_B flows through the 4K resistor, the emitter diode, and R_i to ground. The base voltage V_B is $0.6 + i_B R_i \simeq 0.6$ V. Current i_B does not flow through the collector diode and into the base of Q_2 unless $V_B \geq 1.2$ V (equal to two diode drops). If both inputs are made sufficiently positive, V_B will rise to the point where current flows into the base of Q_2 and the emitter diodes are reverse biased. Then Q_2 and, so, Q_4 goes into saturation. Hence, the output from the Q_4 collector is low (≈ 0.2 V) when both inputs are high. Under these conditions, the base of Q_3 is slightly reverse biased. If no current flows into the base of Q_2 (with input A or B low), Q_3 becomes forward biased and pulls the output up to high state (about $+3.8$ V). In fact, the totem-pole driver shorts the output and acts as a low-impedance emitter follower on high output, thus allowing rapid charging and discharging of any output-line capacitance. One important rule must be remembered when using two-state totem-pole outputs: Never connect their outputs together.

Open-Collector TTL Circuits

Open-collector circuits are used primarily as interfacing devices; but when open-collector outputs are connected together, they also produce wired-logic DOT-AND/OR equivalent gate forms. The major difference between an open-collector output and a totem-pole output is the absence of the active pull-up circuit. When the active pull-down transistor (or open-collector output transistor) in Fig. 6.10-2 turns off, external resistor $R_{\text{pull-up}}$ pulls the output logic line to a high level ($H = V_{\text{pull-up}}$). The open-collector output transistor can be obtained with a maximum breakdown voltage of either 15 or 30 V, depending on the design. This allows $V_{\text{pull-up}}$ to be selected for interfacing with LEDs, lamps, or other logic families. When open-collector output circuits are used in interfacing applications, they are generally referred to as buffer or driver circuits.

To explain how wired logic works let us take a closer look at the circuit

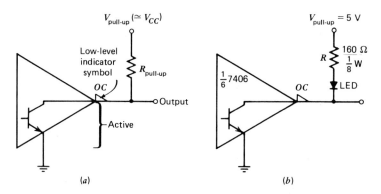

Figure 6.10-2 (a) Inverter with an open-collector output; (b) an example of driver circuit.

illustrated in Fig. 6.10-3 with only a single transistor inverter occupying each inverter logic symbol. Note that $R_{pull-up}$ is an external pull-up resistor in an active circuit. With either logic line A or B or both connected to a high level (V_{CC}), a low-level output results on logic line F (≈ 0.2 V). As both input logic lines A and B are connected to a low level (GND), output logic F is pulled to a high level (V_{CC}). These two situations correspond to the DOT-OR form and the DOT-AND form, as illustrated in Fig. 6.10-4. The DOT-

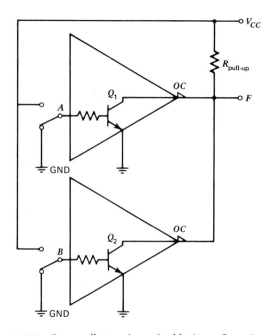

Figure 6.10-3 Open collectors in a wired-logic configuration.

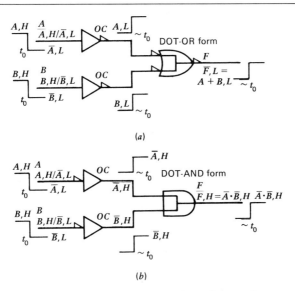

Figure 6.10-4 Wired logic: (a) open-collector DOT-OR form; (b) equivalent open-collector DOT-AND form for the same wired-logic circuit.

OR form for TTL open-collector outputs is always a low-level-input–low-level-output OR form. The equivalent DOT-AND form for TTL open-collector outputs is always a high-level-input–high-level-output AND form. The notation $A,H/\overline{A},L$ ($B,H/\overline{B},L$) may be read, "if $A(B)$ is high, then NOT $A(B)$ is low," or "$A(B)$ is high, but $A(B)$ bar is low." In the level-indicator system, small right triangles are used rather than small circles. A small right triangle at the input to any logic symbol indicates that the relatively low-voltage level (L) must be applied at that input to enable the function. The absence of a small right triangle at the input indicates that the relatively high voltage level (H) must be applied to enable the function. A right triangle appearing at the output means that the output is relatively low (L) when the function is enabled.

EXAMPLE 6.10-1 Show a wired-logic implementation for the following equation using only inverters with open-collector outputs (SN7405); assume a positive-logic assignment for inputs:

$$\overline{F},L = A + B + \overline{C}, L$$

Solution. A two-step procedure is used. The first step involves drawing the required open-collector DOT-OR form, as shown in Fig. 6.10-5a. The second step only requires us to add inverters where necessary to achieve matching Boolean level functions and level indicators, as shown in Fig. 6.10-5b. As a reminder, the dotted lines show where the external pull-up resistors should be included in the actual circuit.

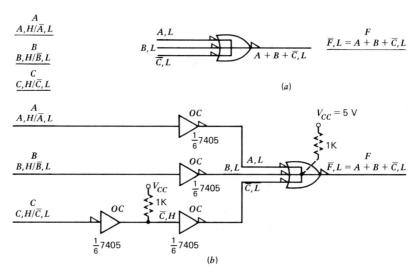

Figure 6.10-5 Example 6.10-1: Wired logic.

CMOS Integrated Circuits

CMOS IC devices contain *p*-channel and *n*-channel FETs on the same chip. CMOS logic devices are compatible with TTL devices; that is, they have approximately the same 1 (high) and 0 (low) voltage levels when used with a 5-V power supply, which TTL requires. CMOS is used more often in SSI (small-scale integration) and MSI (medium-scale integration) devices. CMOS74COO has several major advantages over low-power-series

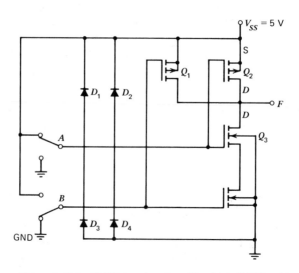

Figure 6.10-6   CMOS two-input positive-logic NAND gate.

TTL74LOO. With V_{SS} ($=V_{CC}$) = 5 V, CMOS (TTL) has P_{diss}/package = 40 nW (4 mW), V_{supply} = 3–15 V (4.75–5.25 V), dc noise immunity = 1 V (0.4 V), and t_{pD} (= signal propagation-delay time) = 50 ns (33 ns).

The CMOS circuit of Fig. 6.10-6 is a two-input positive-logic NAND gate. It consists of two *p*-channel transistors Q_1 and Q_2 connected in parallel and two *n*-channel transistors Q_3 and Q_4 connected in series. By using only one upper *p* channel and one lower *n* channel enhancement-type device, we obtain an inverter including Q_1 and Q_4 or Q_2 and Q_3. To turn on either PMOS connected in parallel (Q_1 or Q_2) to obtain a high-level output on logic line *F* requires that either input logic line *A* or *B* or both be connected to ground. To turn on both NMOS FETs connected in series (Q_3 and Q_4) to obtain a low-level output on logic line *F* requires that both input logic lines *A* and *B* be simultaneously connected to V_{SS}. Diodes D_1–D_4 are used only to provide protection to the extremely high input impedance (typically 10^{12} Ω shunted by 5 pF) inherent in MOS devices. These diodes are biased OFF for applied input voltages between V_{SS} and ground.

6.11 INTRODUCTION TO DIGITAL COMPUTER SYSTEMS

Comparison of Large Computers, Minicomputers, and Microcomputers

A digital computer is a programmable device capable of operating on numerical data with basic arithmetic and logic processes. It deals with three types of information, namely, data, instructions, and control signals. A microprocessor is the central processing unit (CPU) of a microcomputer and is a single large-scale integrated circuit (LSIC). A microprocessor in combination with memory and input/output (I/O) circuits constitutes a microprocessor-based system called a microcomputer. Compared to the large mainframe computer systems, the microprocessor-based systems have the following features: (1) slower clock speed, (2) smaller memory size, (3) smaller word size (data paths), (4) more-limited instruction set, (5) more-basic input/output handling, (6) low cost, (7) small size (often portable), (8) little power consumption (sometimes battery operated). Minicomputers fall somewhere between the two extremes of microcomputers and large mainframe computers. Both the minicomputer and the microcomputer are small computers (Fig. 6.11-1) with word lengths usually ranging from 12 to 32 bits and 4 to 16 bits, respectively. The typical minicomputer has a memory of the size of 4K to 64K words (1K word = one kiloword).

Programmable calculators constitute another part of the fast-growing computer-systems field. They are intended to solve complex equations with direct contact between the user and the calculator. By comparison with the minicomputers and microcomputers, these calculators are quite slow. They have limited main memory. They often include microprocessors. They are

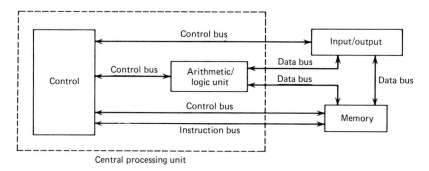

Figure 6.11-1   Organization of a typical small computer.

usually designed with a built-in display or printer and a keyboard so that peripheral equipment is not needed. Their languages are implemented as hardware rather than software.

Software

Software consists of (1) the programs that a computer user writes a process information and (2) the routines that manage the computer's operation. Most computers represent numbers in binary, or base 2. The chief problem of representation occurs in the handling of negative numbers. Nearly all mini-computers and microcomputers represent negative numbers by complementing the positive number (i.e., changing 0's to 1's and 1's to 0's) and adding 1. This representation is called 2's complement. Because octal (base 8) or hexadecimal (base 16) numbers are easier for us to read and write, many computer manufacturers show instructions in octal or hexadecimal rather than binary. We need instructions to tell the computer what to do. We must also be able to represent floating-point numbers, that is, numbers expressed in scientific notation, such as 3.2×10^7, and nonnumeric characters such as letters and punctuation symbols. Nonnumeric characters are represented by any of several codes that assign a binary number to each lower-case and upper-case letter and to each special symbol. Some microprocessors represent numbers in binary-code decimal (BCD) form rather than as binary numbers. In BCD, each decimal digit is represented by four binary bits.

Computer languages allow us to direct the computer's operations. High-level languages, such as FORTRAN or PL/1, use instructions that look something like the English words and arithmetic expressions we use to describe the operations of solving a problem. A program called a translator is needed to convert instructions given in a high-level language to the patterns of 0's and 1's that direct the computer's operations. There is a strong trend toward increased use of high-level languages on both minicomputers and microcomputers. Programs for minicomputers and microcomputers are often written in assembly language. Assembly language depends on the specific hardware configuration and thus is different for each computer. Instructions in assembly language correspond directly to the machine language instructions that direct the basic operations of the computer, but are represented as abbreviations instead of as binary numbers. The abbreviations are usually three or four letters, such as STO for store or SUB for subtract. The advantage of user programs written in assembly language is that they can be converted to machine language by a special translator called an assembler, which requires less space in memory than the translators needed for high-level languages. In addition to machine and assembly language, some minicomputers and microcomputers operate at an even more basic language level known as microprogramming. In microprogramming, a smaller computer inside the main computer executes the control functions for the instruction set of the main computer. Microprogramming uses microinstructions that require only a small part of a computer word.

Software contains not only user programs designed to achieve specific tasks, but also various routines that simplify use of the computer. Loaders load programs into the computer. Editors simplify the task of preparing and changing a program. Debug routines help to detect errors. Simulators allow the testing of a model of a complete system. Operating systems control the use of other software. The major cost for most computer systems, especially those used primarily for general-purpose computing, is software, not hardware.

Hardware

As shown in Fig. 6.11-1, the main parts of the computer are the central processing unit (CPU) (including the central unit and the arithmetic/logic unit), the memory, and the input/out unit.

The memory stores instructions and data. It can be classified into two general categories, namely, random access memory (RAM) and serial access memory (SAM). Most modern RAM devices consist of transistor (bipolar or MOS) arrays, and are characterized by the fact that time needed to retrieve data from any location within the memory is approximately equal. Hence, we would say that the access time for RAM devices is independent of the data's location within the device. On the other hand, SAM devices frequently consist of a movable physical medium, such as magnetic tape or punched paper tape, and require that data be read from the medium only in the sequence in which they were recorded. Consequently, the time needed to gain access to a particular memory location in a SAM device depends on how much intervening data must be read before the desired memory location is reached. For example, in a spool of tape, the access time to data that are wound deeply in the spool will be much longer than that for data that are located closer to the outer rim. RAM can be both read, to retrieve information, and written into, to store information. The name random access means that any word in RAM may be reached or accessed in the same amount of time. RAM is used to store data and results and can be reused repeatedly for new data. It is also called read/write memory to emphasize that it has both functions. By contrast, read-only memory (ROM) provides permanent or semipermanent storage only; its contents can be read but cannot be written during normal computer operation. ROM is usually used for permanent storage of instructions. Most semiconductor read/write RAMs consist of an array of flip-flop memory cells that require a constant power source to maintain data. Should power be removed, voluntarily or not, any stored data will be lost. Such memory circuits are referred to as being volatile. If nonvolatile read/write RAM is desired, battery backup must be provided for the semiconductor circuit, or older magnetic-core RAMs must be used (magnetic cores do not require power to maintain data). Another alternative is to use electrically programmable read-only memories (PROMs), in which the recorded data are nonvolatile.

Storage of information in memory requires two auxiliary devices, usually called the memory address register (MAR) and the memory data register (MDR). A register is a device that can hold several bits; it can be read or written into relatively quickly. Figure 6.11-2 shows the organization of memory and the two registers. The array of memory shown in Figure 6.11-2*a* includes addresses and contents. To write a word in memory, the control unit sends the address of the desired location in memory to the MAR, such as address 2 sent to the MAR in Fig. 6.11-2*b*; at the same time, the control unit sends the word to be written into memory to the MAR; then the contents of the MDR are stored in the location whose address is given by the contents of the MAR; in this case the contents of the MDR becomes the contents of location 2, and the previous contents of that location are lost. Reading a word is done similarly; the address of the desired location is placed in the

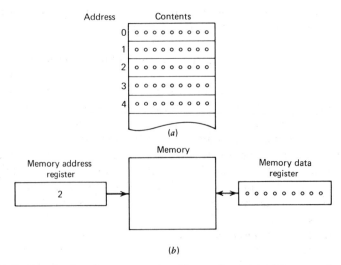

Figure 6.11-2 Organization of memory and auxiliary registers. (*a*) Addresses and contents. (*b*) Register contents while location 2 is being addressed.

MAR; then the contents of that memory location are sent to the memory data register, from which they can be sent to any other register.

The operation of the arithmetic/logic unit (ALU) is related to 2's complement addition. The rules for binary addition are $0 + 0 = 0$; $0 + 1 = 1 + 0 = 1$; $1 + 1 = 10$ (sum of 0 with carry of 1). To see whether the sum is correct, we examine carries into and out of the sign bit (the leftmost bit of a computer word: 0 represents $+$; 1 represents $-$). If there are two carries or no carries, the sum is correct. If there is just one carry (either into or out of the sign bit), we know that the sum is incorrect, that is, that its magnitude is too large to be represented correctly. This condition is called overflow. The ALU consists of one or more registers and circuits for arithmetic and logic operations. At least one register of the ALU is called an accumulator, since it accumulates or holds the results of arithmetic and logical operations. Most small computers have a special one-bit register, usually called a carry register or flag, that holds carries out of the sign bit, so that they can be inspected for overflow; the carry register acts as a high-order extension to an accumulator, but does not hold part of the sum. Logic circuits in the ALU usually permit shifting the contents of a register one or more cells to the right or left, clearing the register (setting its contents to 0), and complementing each bit or all bits of the register, that is, interchanging 0's and 1's. The arithmetic/logic unit nearly always includes an adder, a device that can add the contents of some other register or memory location to the contents of the accumulator and store the results in the accumulator. The ALU may perform other arithmetic operations as well; with an adder and a shifter all arithmetic operations can be performed.

The control unit contains logic and timing circuits that control the operation of the computer. A basic unit has the following devices: (1) one or more clocks, which provide the basic timing of operation; (2) decoders, which decode the instruction to be executed and activate the circuits to execute the instruction; (3) an instruction register (IR), which holds the current instruction; (4) a program counter (PC), which holds the address of the next instruction to be executed. A program is stored in consecutive locations in memory. Instructions are usually executed in sequence. Hence, most of the time the PC can be advanced to the address of the next instruction by increasing it (adding 1). The execution of instructions requires two phases—fetch and execute. During fetch phase, the address of the current instruction is sent from the PC to the MAR, the current instruction is read into the MDR and then sent to the IR of the CPU, and the PC is incremented to hold the address of the next instruction. During execute phase, the instruction is decoded, and control signals to perform the operations of the instruction are sent to the appropriate registers, memory, or input/output devices.

The input/output unit controls the transfers of information between a computer and its peripheral devices. There is more variety in input/output systems than any other part of the computer.

A major consideration in systems is the interface between the computer and the input/output devices connected to the computer. The interface must handle any differences in voltage levels between the computer and the peripheral device. It must ensure that the signals transferred are coded in a way that both the computer and the peripheral device can use.

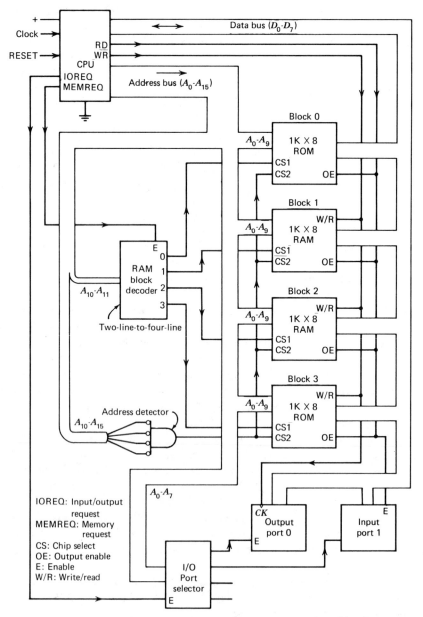

Figure 6.11-3 Simplified block diagram of a typical microprocessor-based hardware system.

The block diagram of a typical microprocessor-based system (also called a microprocessor system) is shown in Fig. 6.11-3, where eight-bit-wide data and sixteen-bit-wide address paths are assumed. Only the most important connections are shown.

6.12   MC6800 MICROCOMPUTER SYSTEM

Motorola M6800 Chip Set

The computing power contained in a US$1 million computer in the early 1950s is now available in a single LSI chip called a microprocessor unit (MPU) and typically at a cost of less than US$20. There are many types of MPU, such as Intel 8085, Motorola MC6800, Zilog Z80, etc. The MC6800 is one of the leading MPUs in use today. It provides an excellent example of microprocessors in general. The motorola M6800 chip set is a family of LSI devices designed to work together as the small MC6800 microcomputer system shown in Fig. 6.12-1.

MC6800 MPU

The package and pin assignment of MC6800 MPU are shown in Fig. 6.12-2. The symbolic diagram of MC6800 is shown in Fig. 6.12-3, where arrows pointing into the diagram represent externally generated signals, and arrows pointing out represent signals originating inside the MPU. There are several registers inside the MPU chip. The MC6800 actually has two accumulators. The index register is a 16-bit register that is used as a pointer to point to a memory location. The stack pointer (SP) is also a 16-bit register, similar to the index register. (The stack is a section of RAM set aside by the programmer as a temporary storage area.) The other register shown as separable

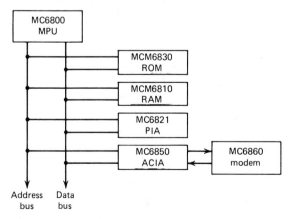

Figure 6.12-1   A small Motorola MC6800 microcomputer system. MPU = microprocessor unit; ROM = read only memory; RAM = random access memory; PIA = peripheral interface adaptor; ACIA = asynchronous communications interface adaptor; modem = modulator–demodulator.

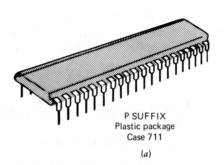

P SUFFIX
Plastic package
Case 711

(a)

PIN ASSIGNMENT

1	V_{SS}	$\overline{\text{Reset}}$	40
2	$\overline{\text{Halt}}$	TSC	39
3	ϕ_1	N.C.	38
4	$\overline{\text{IRQ}}$	ϕ_2	37
5	VMA	DBE	36
6	$\overline{\text{NMI}}$	N.C.	35
7	BA	R/W	34
8	V_{CC}	D_0	33
9	A_0	D_1	32
10	A_1	D_2	31
11	A_2	D_3	30
12	A_3	D_4	29
13	A_4	D_5	28
14	A_5	D_6	27
15	A_6	D_7	26
16	A_7	A_{15}	25
17	A_8	A_{14}	24
18	A_9	A_{13}	23
19	A_{10}	A_{12}	22
20	A_{11}	V_{SS}	21

(b)

Pin 1, 21. V_{ss}: to +5V power supply.
    2. $\overline{\text{Halt}}$: signal, active low.
    3, 37. ϕ_1, ϕ_2: signals, two nonoverlapping phases of a two-phase clock.
    4. $\overline{\text{IRQ}}$: INTERRUPT $\overline{\text{REQUEST}}$ input.
    5. VMA: a control line called VALID MEMORY ADDRESS, active HIGH.
    6. NMI: $\overline{\text{NONMASKABLE INTERRUPT}}$ input.
    7. BA: BUS AVAILABLE signal, active HIGH.
    8. V_{CC}: to +5 V power supply.
    9–20 & 22–25. A_0–A_{15}: 16-Bit address bus.
    26–33. D_7–D_0: 8-Bit Data Bus.
    34. R/W::READ/$\overline{\text{WRITE}}$ control line.
    35, 38. N. C.: no connection.
    36. DBE: DATA BUS ENABLE.
    39. TSC: THREE-STATE CONTROL.
    40. $\overline{\text{RESET}}$: active low.

Figure 6.12-2 MC6800 MPU: (a) package; (b) pin assignment.

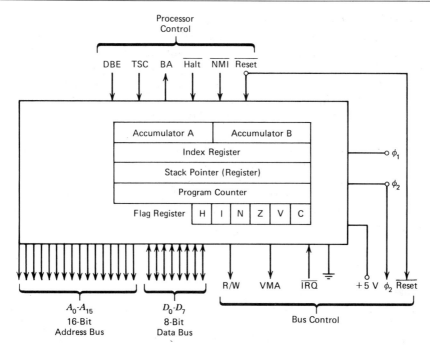

Figure 6.12-3 Symbolic diagram of MC6800 MPU. Within the MPU, but not shown, are the memory address register connected to the address bus, the memory buffer register connected to the data bus, and all op code decoding and control circuitries. Part of the timing distributor circuitry is included within the chip.

blocks H, I, N, Z, V, C is called the condition or flag register. The little squares represent individual flip-flops that act as flag or status bits, which are examined by the MPU during program execution. Also within the MPU, but not shown in Fig. 6.11-3, are the memory address register connected to the address bus, the memory buffer register connected to the data bus, and all op code decoding and control circuitry. Part of the timing distributor circuitry is included within the chip, but the clock signals fed to the MPU should be generated externally.

The MPU can address up to 2^{16} or 65,536 memory locations. Both address bus and data bus have three-state output drivers capable of driving one standard TTL load each. Note that a capacitance on any line must not exceed 130 pF. The R/$\overline{\text{W}}$ (READ/$\overline{\text{WRITE}}$) control line (pin 34) is fed to memory and to other chips to signal whether the MPU is doing a memory read or write operation.

Whenever the TSC (three-state control) pin 39 is driven high, the address bus and the R/$\overline{\text{W}}$ control lines are placed in their high-impedance states. Similarly, driving DBE (data bus enable) pin 36 high places the data bus in its high-impedance state. When a valid address appears on the address bus, the VMA (valid memory address) control line (pin 5) goes active high. When the $\overline{\text{HALT}}$ line (pin 2) is driven low, the MPU halts operation after completing its present instruction. When the MPU is halted, all three-state (tristate) outputs go to their high-impedance state (off). Whenever the MPU is halted, the BA (bus available) signal (pin 7) goes active high, indicating that the address bus is in its high-impedance state and available for use by an external device. Whenever the external devices want to interrupt the current program, they signal the MPU by the $\overline{\text{IRQ}}$ (interrupt request, pin 4) and $\overline{\text{NMI}}$ (nonmaskable interrupt, pin 6) inputs.

As the RESET input (pin 40) is driven low, the MPU commences its restart sequence. Proper registers are initiated internally, and the program counter (PC) is loaded with the address of the first instruction to be decoded. As the MC6800 is reset, the contents of memory locations FFFE and FFFF are loaded into the PC. Then the MPU effectively does a jump to the location now loaded in the PC and starts execution of the program.

ROM and RAM and Two-Phase Clock Signals

Normally, a ROM is connected in the highest memory locations and a RAM in the lowest for the MC6800 microcomputer system. The programmer should place the address of the starting location of his or her program in locations FFFE and FFFF, with the most significant byte in FFFE.

Signals ϕ_1 and ϕ_2 to pins 3 and 37 are two nonoverlapping phases of a two-phase clock timing waveform (shown in Fig. 6.12-4) generated externally by a clock circuit, which can be built with SSI components. In order to obtain higher speeds, the leading and trailing edges are used to generate the desired control signals. The total time required to complete one MPU cycle depends on the clock frequency and is typically 1 μs.

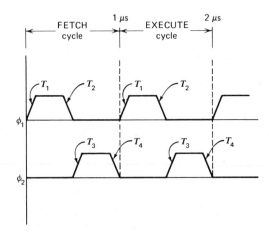

Figure 6.12-4   Two-phase clock timing waveforms.

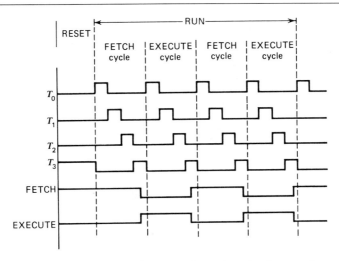

Figure 6.12-5 Timing distributor waveforms in a simple computer.

The MPU Cycle Compared to that of a Simple Computer

The two-phase clock signals shown in Fig. 6.12-4 are needed to perform a relatively simple instruction, such as load accumulator. For that instruction, the timing of events is similar to that of a simple computer.

Since instructions and data are both stored in the same memory, a timing distributor (TD) must be used to keep track of whether an instruction or data byte is being read from memory. The TD generates two timing cycles called the FETCH cycle and the EXECUTE cycle. For a simple computer, each of these cycles is further subdivided into four time periods, or T states— T_0, T_1, T_2, and T_3—as indicated in Fig. 6.12-5. During the FETCH cycle, the control circuitry fetches, or reads, an instruction from memory. Then during the EXECUTE cycle, the instruction is carried out. The sequence of events shown in Fig. 6.12-5 is governed by the TD. After a system RESET, the computer is placed in the RUN mode. Thereafter, the TD generates a FETCH signal and T_0, T_1, T_2, T_3 in that order followed by EXECUTE signal and T_0, T_1, T_2, T_3. The TD continues this operation over and over, until it is told to HALT. In Fig. 6.12-4, the rising edge of ϕ_1 corresponds to T_1 in Fig. 6.12-5, and the falling edge of ϕ_1 corresponds to T_2, etc.

MC6800 MPU Instruction Set

The list of instructions that a computer can interpret is called the instruction set. Instructions generally have two parts: an operation code (op code) and operand address. The op code tells the computer what to do. The operand address identifies the memory location holding the data that are to be operated on.

Table 6-3 Motorola MC6800 Accumulator and Memory Instructions (Other Addressing Modes

Accumulator and Memory Operations	Mnemonic	IMMED			DIRECT			INDEX			EXTND			INHER		
		OP	~	#	OP	~	#	OP	~	#	OP	~	#	OP	~	#
Add	ADDA	8B	2	2	SB	3	2	AB	5	2	BB	4	3			
	ADDB	CB	2	2	OB	3	2	EB	5	2	FB	4	3			
Add Acmltrs	ABA													1B	2	1
Add with Carry	ADCA	89	2	2	99	3	2	A9	5	2	B9	4	3			
	ADCB	C9	2	2	09	3	2	E9	5	2	F9	4	3			
And	ANDA	84	2	2	94	3	2	A4	5	2	B4	4	3			
	ANDB	C4	2	2	D4	3	2	E4	5	2	F4	4	3			
Bit Test	BITA	85	2	2	95	3	2	A5	5	2	B5	4	3			
	BITB	C5	2	2	D5	3	2	E5	5	2	F5	4	3			
Clear	CLR							6F	7	2	7F	6	3			
	CLRA													4F	2	1
	CLRB													5F	2	1
Compare	CMPA	81	2	2	91	3	2	A1	5	2	B1	4	3			
	CMPB	C1	2	2	D1	3	2	E1	5	2	F1	4	3			
Compare Acmltrs	CBA													11	2	1
Complement, 1's	COM							63	7	2	73	6	3			
	COMA													43	2	1
	COMB													53	2	1
Complement. 2's (Negate)	NEG							60	7	2	70	6	3			
	NEGA													40	2	1
	NEGB													50	2	1
Decimal Adjust. A	DAA													18	2	1
Decrement	DEC							6A	7	2	7A	6	3			
	DECA													4A	2	1
	DECB													5A	2	1
Exclusive OR	EORA	88	2	2	98	3	2	A8	5	2	B8	4	3			
	EORB	C8	2	2	D8	3	2	E8	5	2	F8	4	3			
Increment	INC							6C	7	2	7C	6	3			
	INCA													4C	2	1
	INCB													5C	2	1
Load Acmltr	LOAA	86	2	2	96	3	2	A6	5	2	B6	4	3			
	LOAB	C6	2	2	D6	3	2	E6	5	2	F6	4	3			
Or. Inclusive	ORAA	BA	2	2	9A	3	2	AA	5	2	BA	4	3			
	ORAB	CA	2	2	DA	3	2	EA	5	2	FA	4	3			
Push Data	PSHA													36	4	1
	PSHB													37	4	1
Pull Data	PULA													32	4	1
	PULB													33	4	1

Boolean/Arithmetic Operation	Cond. Code Reg.					
(All register labels	5	4	3	2	1	0
refer to contents)	H	I	N	Z	V	C
$A + M \rightarrow A$	↕	•	↕	↕	↕	↕
$B + M \rightarrow B$	↕	•	↕	↕	↕	↕
$A + B \rightarrow A$	↕	•	↕	↕	↕	↕
$A + M + C \rightarrow A$	↕	•	↕	↕	↕	↕
$B + M + C \rightarrow B$	↕	•	↕	↕	↕	↕
$A \cdot M \rightarrow A$	•	•	↕	↕	R	•
$B \cdot M \rightarrow B$	•	•	↕	↕	R	•
$A \cdot M$	•	•	↕	↕	R	•
$B \cdot M$	•	•	↕	↕	R	•
$00 \rightarrow M$	•	•	R	S	R	R
$00 \rightarrow A$	•	•	R	S	R	R
$00 \rightarrow B$	•	•	R	S	R	R
$A - M$	•	•	↕	↕	↕	↕
$B - M$	•	•	↕	↕	↕	↕
$A - B$	•	•	↕	↕	↕	↕
$\overline{M} \rightarrow M$	•	•	↕	↕	R	S
$\overline{A} \rightarrow A$	•	•	↕	↕	R	S
$\overline{B} \rightarrow B$	•	•	↕	↕	R	S
$00 - M \rightarrow M$	•	•	↕	↕	①	②
$00 - A \rightarrow A$	•	•	↕	↕	①	②
$00 - B \rightarrow B$	•	•	↕	↕	①	②
Converts Binary Add. of BCO Characters into BCD format	•	•	↕	↕	↕	③
$M - 1 \rightarrow M$	•	•	↕	↕	④	•
$A - 1 \rightarrow A$	•	•	↕	↕	④	•
$B - 1 \rightarrow B$	•	•	↕	↕	④	•
$A \bullet M \rightarrow A$	•	•	↕	↕	R	•
$B \bullet M \rightarrow B$	•	•	↕	↕	R	•
$M + 1 \rightarrow M$	•	•	↕	↕	⑤	•
$A + 1 \rightarrow A$	•	•	↕	↕	⑤	•
$B + 1 \rightarrow B$	•	•	↕	↕	⑤	•
$M \rightarrow A$	•	•	↕	↕	R	•
$M \rightarrow B$	•	•	↕	↕	R	•
$A + M \rightarrow A$	•	•	↕	↕	R	•
$B + M \rightarrow B$	•	•	↕	↕	R	•
$A \rightarrow M_{SP}, SP - 1 \rightarrow SP$	•	•	•	•	•	•
$B \rightarrow M_{SP}, SP - 1 \rightarrow SP$	•	•	•	•	•	•
$SP + 1 \rightarrow SP, M_{SP} \rightarrow A$	•	•	•	•	•	•
$SP + 1 \rightarrow SP, M_{SP} \rightarrow B$	•	•	•	•	•	•

Table 6-3   (*Continued*)

Addressing Modes

Accumulator and Memory Operations	Mnemonic	IMMED			DIRECT			INDEX			EXTND			INHER		
		OP	~	#	OP	~	#	OP	~	#	OP	~	#	OP	~	#
Rotate Left	ROL							69	7	2	79	6	3			
	ROLA													49	2	1
	ROLB													59	2	1
Rotate right	ROR							66	7	2	76	6	3			
	RORA													46	2	1
	RORB													56	2	1
Shift left, Arithmetic	ASL							68	7	2	78	6	3			
	ASLA													48	2	1
	ASLB													58	2	1
Shift right, Arithmetic	ASR							67	7	2	77	6	3			
	ASRA													47	2	1
	ASRB													57	2	1
Shift Right, Logic	LSR							64	7	2	74	6	3			
	LSRA													44	2	1
	LSRB													54	2	1
Store Acmltr	STAA				97	4	2	A7	6	2	B7	5	3			
	STAB				D7	4	2	E7	6	2	F7	5	3			
Subtract	SUBA	80	2	2	90	3	2	A0	5	2	80	4	3			
	SUBB	C0	2	2	D0	3	2	E0	5	2	F0	4	3			
Subtract Acmltrs	SBA													10	2	1
Subtr. with Carry	SBCA	82	2	2	92	3	2	A2	5	2	B2	4	3			
	SBCB	C2	2	2	D2	3	2	E2	5	2	F2	4	3			
Transfer Acmltrs	TAB													16	2	1
	TBA													17	2	1
Test, Zero or Minus	TST							6D	7	2	7D	6	3			
	TSTA													40	2	1
	TSTB													50	2	1

Legend:

OP	Operation code (hexadecimal);	M_{SP}	Contents of memory location pointed to be Stack Pointer;
~	Number of MPU cycles;	+	Boolean Inclusive OR;
#	Number of program bytes;	$\oplus$	Boolean Exclusive OR;
+	Arithmetic plus;	$\overline{M}$	Complement of M;
−	Arithmetic minus;	→	Transfer Into;
•	Boolean AND;	0	Bit = Zero

The MC6800 has an instruction set of 72 instructions, which may be expanded to nearly 200 different possible operations by using various addressing modes. The addressing mode refers to how the PMU obtains the operand. To decode almost 200 instructions, the MC6800 uses an 8-bit op code in hexadecimal form.

Table 6-3 lists the MC6800 instruction set. The leftmost column lists the

Boolean/Arithmetic Operation (All register labels refer to contents)		Cond. Code Reg.					
		5	4	3	2	1	0
		H	I	N	Z	V	C
M A B	$C \leftarrow [b_7 \leftarrow b_0]$	• • •	• • •	↕ ↕ ↕	↕ ↕ ↕	⑥ ⑥ ⑥	↕ ↕ ↕
M A B	$C \rightarrow [b_7 \rightarrow b_0]$	• • •	• • •	↕ ↕ ↕	↕ ↕ ↕	⑥ ⑥ ⑥	↕ ↕ ↕
M A B	$C \leftarrow [b_7 \cdots b_0] \leftarrow 0$	• • •	• • •	↕ ↕ ↕	↕ ↕ ↕	⑥ ⑥ ⑥	↕ ↕ ↕
M A B	$[b_7 \cdots b_0] \rightarrow C$	• • •	• • •	↕ ↕ ↕	↕ ↕ ↕	⑥ ⑥ ⑥	↕ ↕ ↕
M A B	$0 \rightarrow [b_7 \cdots b_0] \rightarrow C$	• • •	• • •	R R R	↕ ↕ ↕	⑥ ⑥ ⑥	↕ ↕ ↕
A → M		•	•	↕	↕	R	•
B → M		•	•	↕	↕	R	•
A − M → A		•	•	↕	↕	↕	↕
B − M → B		•	•	↕	↕	↕	↕
A − B → A		•	•	↕	↕	↕	↕
A − M − C → A		•	•	↕	↕	↕	↕
B − M − C → B		•	•	↕	↕	↕	↕
A → B		•	•	↕	↕	R	•
B → A		•	•	↕	↕	R	•
M − 00		•	•	↕	↕	R	R
A − 00		•	•	↕	↕	R	R
B − 00		•	•	↕	↕	R	R

00	Byte = Zero;	R	Reset always
H	Half-carry from bit 3;	S	Set always
I	Interrupt mask	↕	Test and set if true, cleared otherwise
N	Negative (sign bit)	•	Not affected
Z	Zero (byte)	CCR	Condition code register
V	Overflow, 2's complement	LS	Least significant
C	Carry from bit 7	MS	Most significant

instructions, such as add, in alphabetical order. Next to the instructions is listed the assembly language mnemonic for each. ADDA means add the operand to accumulator A, etc. There are five columns under the heading of addressing modes. These are IMMED (immediate), DIRECT, INDEX, EXTEND (extended), and IMPLIED.

Now, look at the column labeled DIRECT. DIRECT addressing means that the address of the operand is given directly following the op code. For example, to instruct the MC6800 to add the contents of memory location 14 to the accumulator, we would list the instruction in two bytes, the first byte being 9B, the op code in hexadecimal, and the second byte being 14, which is the operand address. The symbol ~ represents the number of MPU cycles (clock cycles) required to perform the operation. For the add instruction, three MPU cycles are needed. Hence, it would take 3 μs to complete if a 1-MHz clock were being used. The symbol # indicates the number of program bytes necessary for that mode, which we know to be 2.

Finally, let us look at the last two columns of Table 6-3. In the column labeled Boolean/arithmetic operation, the notation $A + M \rightarrow A$ signifies that when the operation is carried out, the present contents of the accumulator plus the contents of the indicated memory location will be placed in the accumulator. The last column shows how the various flag bits are affected by a particular instruction. For the ADDA instruction, we see that there is a dot (.) listed for the I flag. This means that the I flag is not affected by that instruction. All other flags showing double-headed arrows mean that each flag may change according to the result of the operation. For example, if the addition causes a carry to occur, the C flag will set (↑). If no carry occurs, the C flag will reset (↓). Likewise, if the result of the addition causes the number in the accumulator to be negative, the N flag will set (↑); otherwise it will reset (↓).

6.13   DIGITAL-TO-ANALOG CONVERTERS (DAC) AND ANALOG-TO-DIGITAL CONVERTERS (ADC)

D/A Converters

A digital-to-analog converter (DAC) is designed to convert binary digital information to analog form; for example, it can convert the binary digital word 1010001_2 to the decimal format 81, since

$$1 \times 64 + 0 \times 32 + 1 \times 16 + 0 \times 8 + 0 \times 4 + 0 \times 2 + 1 \times 1 = 81_{10},$$

and the DAC contains the weighting factors (1, 2, 4, 8, 16, . . .), which can be summed or not depending on the state of the input digit.

A basic DAC shown in Fig. 6.13-1 is constructed in a resistive $(R\text{-}2R)$ ladder arrangement, with a noninverting op-amp as the analog buffer amplifier. In practice, the digital input signals would come from a logic circuit (such as latch), but not through the mechanical switches as shown. In this basic DAC, each signal $(V_0 - V_{n-1})$ is low (ground potential) or high ($= V_{\text{ref}}$) depending on the position of its respective switch. The analog output (V_a) of an $R\text{-}2R$ ladder is generally expressed by

$$V_a = \frac{V_0 + 2V_1 + 4V_2 + 8V_3 + 16V_4 + \cdots + 2^{n-1}V_{n-1}}{2^n}, \quad (6.13\text{-}1)$$

where n is the number of digital inputs and V_0 represents the LSB.

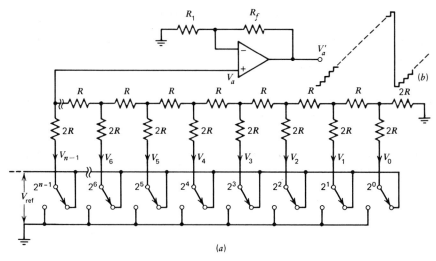

Figure 6.13-1 (a) Basic DAC using a resistive (R-2R) ladder and an analog buffer amplifier. (b) Analog output voltage of a staircase waveform if the switches in (a) replaced by logic signal outputs of an n-bit (e.g., 8-bit) binary counter.

Table 6-4 lists some typical single-chip DACs and ADCs. The faster the speed, the higher the cost.

When a microprocessor is used for digital-to-analog conversion, the data output (data lines) drives the data input port of a digital-to-analog converter. Such an application is shown in Fig. 6.13-2, where two DACs generate sig-

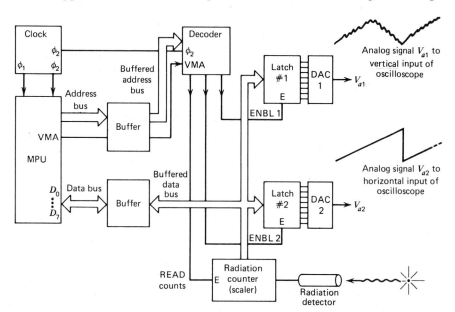

Figure 6.13-2 Application of DAC in microprocessor-based system. ROM and RAM are not shown.

Table 6-4   Typical Single-Chip DACs and ADCs

Manufacturer	Type	Resolution	Speed
Motorola	DAC, MC1408	8	300 ns
PMI	DAC-08	8	100 ns
Analog Devices	DAC, AD7520	10	500 ns
Datel	DAC-4Z12D	12	1 μs
Burr-Brown	DAC70/CSB	16	75 μs
PMI	SA, ADC, AD-02	8	8 μs
Analog Devices	SA, ADC, AD7570	10	18 μs
Datel	Integrat., ADC-EK12B	12	24 ms
Analog Devices	Integrat., ADC, AD7550	13	40 ms
National	SA, ADCO816	8	114 ms

nals for analog devices (such as a cathode-ray oscilloscope). The digital output of the radiation counter is applied to the latches.

Analog-to-Digital Converters (ADCs)

An analog-to-digital converter (ADC) is able to receive a smoothly changing voltage, called an analog signal voltage, and convert it to a series of digital equivalent signals. Analog signals can be amplified and applied to the input of an ADC. The ADCs output can be read by an MPU, as directed to do so with appropriate software. There are three major methods for analog-to-digital conversion, namely, successive-approximation (SA), integration (dual-slope), and direct comparison (for very high speed). Some commercially available single-ship ADCs are listed in Table 6-4. Perhaps the SA-type ADC is the most popular one. Let us discuss the SA type starting from a counter-type ADC.

A counter-type ADC is shown in Fig. 6.13-3. It consists of a comparator, a clock, a NAND gate, a group of BCD counters, and a digital-to-analog converter (DAC). The DAC allows comparison of the converter analog output v_A with the analog input v_i, assumed positive. Initially the counter is cleared by the convert pulse, so that $v_A = 0$ and the NAND gate allows clock pulses to pass. As time progresses, v_A increases in steps. At the point where v_A exceeds v_i the comparator output switches negative and the pulses to the counter cease. At this point the digital output is essentially equal to the equivalent analog input.

Clock frequency determines the rate of conversion but has no effect on accuracy. While the ADC illustrated in Fig. 6.13-3 shows only four bits, this can be increased to any number although the time required grows rapidly with the number of bits, an important drawback for many applications.

A successive-approximation analog-to-digital converter is shown in Fig. 6.13-4. It is similar to the counter type except that the bits are tested in

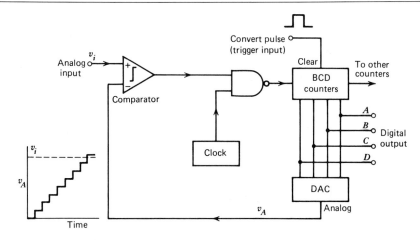

Figure 6.13-3 Block diagram of a counter-type ADC, showing only four bits.

succession, that is, incremented in large steps starting with the most significant bit (MSB), rather than incrementing by the smallest step, as done when counting pulses. To do this, the bit is switched on during a test phase. If the bit causes v_A to exceed v_i as detected by the comparator, it is switched off and kept off; otherwise it is left on. Sequentially the control logic tests the next most significant bit until the conversion is completed. Obviously the process is much faster than counting pulses, especially if the number of bits is large. The chief disadvantage is the rather involved control logic required.

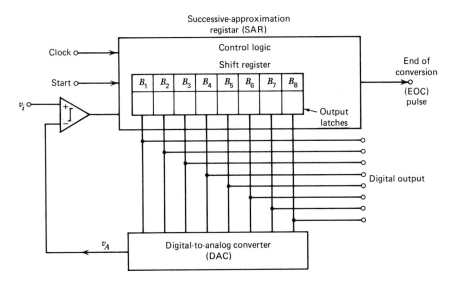

Figure 6.13-4 Block diagram of a successive-approximation ADC.

A successive-approximation register (SAR) in Fig. 6.13-4 contains the control logic, a shift register, and a set of output latches, one for each register section. The outputs of the latches drive a DAC. A start pulse sets the first bit of the shift register high, so the DAC will see the word 10000000_2, and there produces an output voltage equal to one-half of the full-scale output voltage ($V_{fs}/2$). If the input voltage v_i is greater than $V_{fs}/2$, then the $B1$ latch is set high. On the next clock pulse, register $B2$ is set high for trial 2. The output of the DAC is now three-quarter scale. If, on any trial, it is found that $v_i < v_A$, then that bit is reset low.

Here is an example, illustrating a three-bit SAR through a sample conversion. Assume that the full-scale potential is 1 V, and v_i is 0.625 V. Consider the timing shown in Fig. 6.13-5.

Time t_1: The start pulse is received, so register $B1$ goes high. The output word is now 100_2, so $v_A = 0.5$ V. Since v_A is less than v_i, latch $B1$ is set to "1," so that at the end of the trial the output word remains 100_2.

Time t_2: On this trial (which starts upon receiving the next clock pulse), register $B2$ is set high, so the output word is 110_2. Voltage v_A is now 0.75 V. Since v_i is less than v_A, the B_2 latch is set to "0," and the output word reverts to 100_2.

Time t_3: Register $B3$ is set high, making the output word 101_2. The value of v_A is now 0.625 V, so $v_i = v_A$. The $B3$ register is latched to "1," and the output word remains 101_2.

Time t_4: Overflow occurs, telling the control logic to issue an end of conversion (EOC) pulse. In some cases the overflow pulse is the EOC pulse.

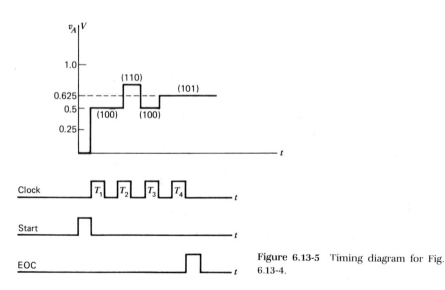

Figure 6.13-5 Timing diagram for Fig. 6.13-4.

Note that in the example, we had a three-bit SAR, and required four $(=n + 1)$ clock pulses to complete the conversion. Hence, an SA-type ADC takes $(n + 1)$ clock pulses for a full-scale conversion. The SA-type ADCs are normally purchased as complete units or implemented by microprocessor software. The National Semiconductor ADC 0816/ADC 0817 (a CMOS device) is a SA-type eight-bit ADC designed for use with a microprocessor.

BIBLIOGRAPHY

Boylestad, R. and L. Nashelsky, *Electronic Devices and Circuit Theory,* 3rd ed. Prentice-Hill, Englewood Cliffs, New Jersey, 1982.

Floyd, T. L., *Digital Logic Fundamentals,* 2nd ed. Charles E. Merrill Publishing Company, Columbus, Ohio, 1982.

Lurch, E. N., *Fundamentals of Electronics,* 3rd ed. Wiley, New York, 1981.

McKay, C. W., *Digital Circuits.* Prentice-Hall, Englewood Cliffs, New Jersey, 1978.

McWane, J., *Introduction to Electronics and Instrumentation.* Breton Publishers, North Scituate, Massachusetts, 1981.

Millman, J., *Microelectronics.* McGraw-Hill, New York, 1979.

Oleksy, J. E. and G. B. Rutkowski, *Microprocessor and Digital Computer Technology.* Prentice-Hall, Inc., Englewood Cliffs, New Jersey 07632, 1981.

Sandige, R. S., *Digital Concepts Using Standard Integrated Circuits.* Prentice-Hall, New Jersey, 1978.

Schilling, D. L. and C. Belove, *Electronic Circuits,* 2nd ed. McGraw-Hill, New York, 1979.

Strangio, C. E., *Digital Electronics.* Prentice-Hall, Englewood Cliffs, New Jersey, 1980.

Stout, D. F. and M. Kaufman, *Handbook of Operational Amplifier Circuit Design.* McGraw-Hill, New York, 1976, pp. 10-1–10-12 and 11-1–11-11.

Taub, H., *Digital Circuits and Microprocessors.* McGraw-Hill, 1982.

Taub, H. and D. Schilling, *Digital Integrated Electronics.* McGraw-Hill, New York, 1977.

Questions

6-1 Define "transducer" and "sensor" in your own words.

6-2 Describe the difference between majority and minority carriers.

6-3 How does the reverse saturation current of a $p - n$ diode vary with temperature?

6-4 How does the diode voltage (at constant current) vary with temperature?

6-5 Explain physically why a $p - n$ diode acts as a rectifier.

6-6 What are the only important parameters which distinguish various diodes for most applications?

6-7 Describe the operation of a Zener diode when it is used as a voltage regulator.

6-8   What are the off time and on time of a computer diode?

6-9   Describe the characteristics of a PIN diode.

6-10   What are the main application and characteristic of a voltage-variable capacitor or varactor diode?

6-11   What is the major difference between a bipolar and a unipolar device?

6-12   How must the transistor junctions be biased for BJT amplifier operation?

6-13   Are I_{CO} and I_{CEO} related in a BJT? If they are, how and why?

6-14   What is the fact from which the transistor storage time results.

6-15   Describe the major characteristics of the FETs.

6-16   Describe the major characteristics of a Triac.

6-17   What are the differences between a SCS and a SCR?

6-18   Compare the characteristics of a UJT and a PUT.

6-19   Describe the major characteristics of a basic op-amp.

6-20   Draw the schematic diagram of an ideal inverting op-amp with voltage-shunt feedback.

6-21   Draw the schematic diagram of an ideal noninverting op-amp with voltage-series feedback.

6-22   Define an EXCLUSIVE OR gate and give its truth table.

6-23   Define a NAND gate and give its truth table.

6-24   Draw the circuit of a TTL gate and explain its operation.

6-25   Draw the circuit of a CMOS transmission device for switching both analog and digital signals; explain its operation.

6-26   (a) Sketch the logic system for a clocked *SR* flip-flop. (b) Give the truth table.

6-27   (a) Draw the logic diagrams of *JK*, *T*, and *D* flip-flops. (b) Give their truth tables.

6-28   (a) Draw a system of *JK* master–slave flip-flop with CLEAR and PRESET asynchronous inputs. (b) Give its function table.

6-29   (a) Draw a basic four-bit binary or ripple (asynchronous counter). (b) Explain its operation.

6-30   (a) Draw a decimal (BCD) counter and (b) explain its operation.

6-31   What are the advantages of the synchronous counter over the asynchronous counter?

6-32   What are the four main units of a computer? What purposes do they have?

6-33   Compare minicomputers and microcomputers.

Problems

6.1-1   A digital-to-analog converter (DAC) accepts a number in binary form and gives an output voltage proportional to the number. Figure P6.1-

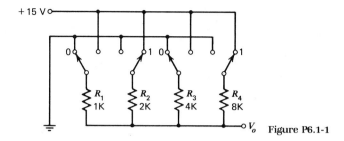

Figure P6.1-1

1 shows a four-bit DAC. The four switches, which in practice would be transistors, store the binary number; $+15$ V represents 1 and ground is 0. The switches are shown in the position corresponding to the number 5 ($=0101$ binary). Calculate the output voltage V_o.

Answer.   5 V.

6.1-2   Figure P6.1-2 shows the basic elements of an analog adding circuit. Show that the output voltage V_o is approximately proportional to the sum of the input voltages V_1-V_3.

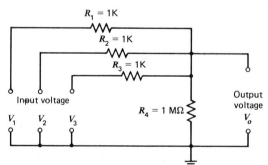

Figure P6.1-2

6.1-3   A current pulse of amplitude I is applied to a parallel RC circuit, as shown in Fig. P6.1-3. (a) Convert the circuit to Thevenin form and then show that the capacitor current is given by $i_C = Ie^{-t/RC}$. (b) Draw and scale the waveforms of the current i_C for the cases (1) $t_p < RC$, (2) $t_p = RC$, (3) $t_p > RC$. (*Hint:* When $RC > t_p$, C charges slowly.)

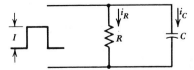

Figure P6.1-3

6.3-1 (a) Assume a reverse saturation current $I_o = 1$ nA in a $p-n$ junction silicon diode. Find the applied voltage for a forward current of 0.5 μA. (b) Assume $I_o = 20$ μA in a germanium diode. What current would result if the voltage found in part (a) were applied across this device in the forward direction?

Answer. (a) 0.323 V; (b) 4.97 A.

6.3-2 (a) For what voltage will the reverse current in a $p-n$ junction silicon diode reach 100% of its saturation value at room temperature? (b) What is the ratio of the current for a forward bias of 0.2 V to the current for the same magnitude of reverse bias? (c) If the reverse saturation current is 10 nA, find the forward currents for voltages of 0.5, 0.6, and 0.7 V, respectively?

Answer. (a) -0.036 V; (b) -46.89; (c) 0.150, 1.026, 7.020 nA.

6.3-3 (a) Find the factor by which the reverse saturation current of a silicon diode is multiplied when the temperature is increased from 25 to 175°C. (b) Repeat part (a) for a germanium diode over the range 25–100°C.

Answer. (a) 3.273×10^4; (b) 181.

6.3-4 In the Zener voltage regulator of Fig. 6.3-2 the Zener voltage $V_Z = 20$ V. For the Zener diode, $I_{ZK} = 5$ mA and $I_Z = 200$ mA. The unregulated dc voltage is $V = 30$ V. Determine (a) the value of R; (b) the permissible variation in R_L over which the load voltage is still regulated at the Zener voltage.

Answer. (a) 50 Ω; (b) R_L may vary from infinity to 102.6 Ω.

6.3-5 The circuit shown in Fig. P6.3-5 represents a dc voltmeter which reads 25 V full scale. The meter resistance is 560 Ω, and the full scale is 0.2 mA. The Zener diode is used to prevent overloading of the sensitive meter movement without affecting meter linearity. If the diode is a 20-V Zener, find R_1 and R_2 so that, when $V_i > 25$ V, the Zener diode conducts and the overload current is shunted away from the meter.

Answer. $R_1 = 25$ Ω; $R_2 = 99.44$K.

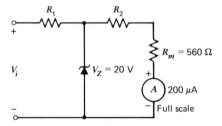

Figure P6.3-5

6.3-6 A 10-mA dc meter whose resistance is 20 Ω is calibrated to read rms volts when used in the diode bridge as shown in Figure P6.3-6. The effective resistance of each diode may be considered to be zero in the forward direction and infinite in the inverse direction. The sinusoidal input voltage is applied in series with a 5K resistance. What is the full-scale reading of this meter?

Answer. 55.74 V.

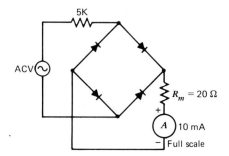

Figure P6.3-6

6.4-1 A common-base equivalent circuit with large-signal behavior is shown in Fig. P6.4-1, where the input impedance $r_d \approx 0$, $I_{CBO} \approx 0$, $V_{EE} = 2$ V, $R_e = 1K$, $V_{CC} = 48$ V, $R_c = 20K$, and a 1-V peak sinusoidal source is connected in series with V_{EE}. Calculate (a) the current i_E, (b) the voltage v_{CB}, and (c) the voltage gain A_v.

Answer. (a) $i_E = 1.3 + 1.0 \cos \omega t$ mA; (b) $v_{CB} = -22 + 20 \cos \omega t$ V; (c) $A_v = V_{cbm}/V_{im} = 20/1 = 20$.

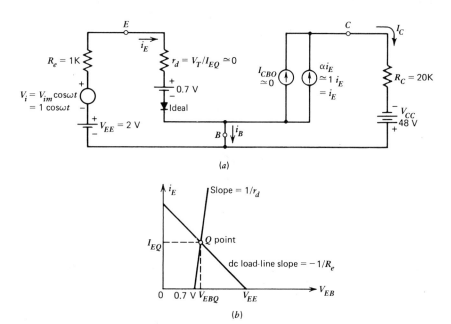

Figure P6.4-1 Large-signal common-base *pnp* equivalent circuit: (a) basic model; (b) emitter–base volt–ampere characteristic.

6.4-2   Two transistors are direct coupled with the first emitter connected
to the second base and the two collectors connected together, as
indicated in Fig. P6.4-2. For this circuit shown, $\alpha_1 = 0.99$, $\alpha_2 = 0.98$,
$V_{CC} = 22$ V, $R_C = 120$ Ω, and $I_E = -120$ mA. Neglect the reverse
saturation currents and assume both transistors are in the active re-
gion. Find: (a) the currents I_{C_1}, I_{B_1}, I_{E_1}, I_{C_2}, I_{B_2}, and I_C; (b) the current
ratios I_C/I_B and I_C/I_E; (c) the voltage V_{CE}.

Answer.   (a) $I_{C_1} = 2.376$ mA; $I_{B_1} = 0.024$ mA; $I_{E_1} = -2.4$ mA;
$I_{C_2} = 117.6$ mA; $I_{B_2} = 2.4$ mA; $I_C = 119.98$ mA; (c) $V_{CE} = 7.6$ V.

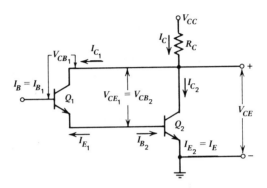

Figure P6.4-2

6.4-3   Calculate the value of the feedback resistor R_f in the circuit of Fig.
P6.4-3 for $V_{CC} = 20$ V, $R_C = 5$K, $R_e = 100$ Ω, $V_{CE} = 5$ V, $V_{EE} =$
0.7 V, and $\beta = 100$. Neglect the reverse saturation current.

Answer.   148 K.

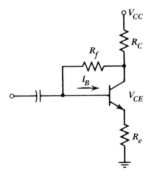

Figure P6.4-3

6.4-4   Calculate the base and collector currents in the transistor circuit of
Fig. P6.4-4 for $R_1 = 90$K, $R_2 = 10$K, $R_C = 5$K, $R_e = 1$K, and $\beta =$
50. (*Hint:* Apply Thevenin's Theorem looking to the left of the base
terminal.)

Answer. $I_B = 0.13$ mA; $I_C = 6.5$ mA.

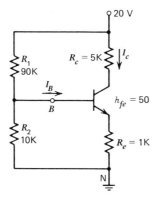

Figure P6.4-4

6.5-1   (a) Obtain the operating point for the fixed-bias circuit of Fig. P6.5-1a and n-channel JFET with drain characteristic shown in Fig. P6.5-1b. (b) Calculate the operating drain current using Eq. (6.5-2); does the result correspond to the operating point?

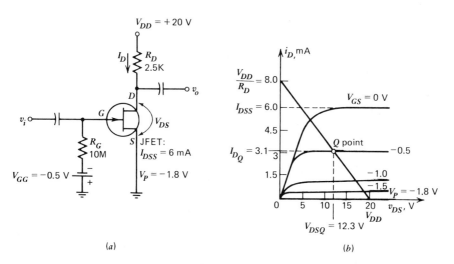

Figure P6.5-1

6.5-2   (a) The equation for drain current in an n-channel JFET is

$$I_D = 8.4 \left(1 - \frac{V_{GS}}{-3.0}\right)^2 \text{ mA} \qquad (V_{DS} = 10 \text{ V}).$$

Find I_D at each of the following values of V_{GS}: 0, -0.5, -1.0, -1.5, -2.0, and -3.0 V. (b) By using the results of part a, sketch the drain characteristic.

6.5-3   In the circuit of Fig. P6.5-3a the p-channel enhancement MOSFET

has the characteristics as shown in Fig. P6.5-3b. (a) Find i_{SD} and v_{SD}.
(b) If it is desired that $v_{SD} = 25$ V, what should be the value of V_{GG}?
Answer. (a) $i_{SD} = 20$ mA, $v_{SD} = 16.7$ V; (b) $V_{GG} = -10.8$ V.

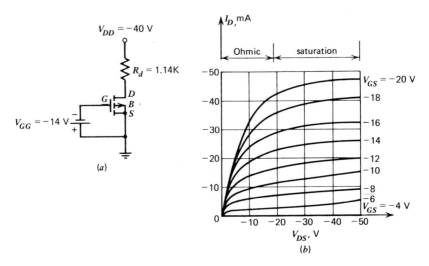

Figure P6.5-3

6.5-4 Determine the equivalent resistances of the enhancement NMOSFET
circuit of Fig. P6.5-4 with the switch in position 1 and in position 2.
Answer. $R_1 = 860 \ \Omega$; $R_2 = \infty$.

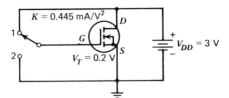

Figure P6.5-4

6.5-5 The equation for an enhancement NMOSFET is

$$I_D = 0.6(V_{GS} - 0.6)^2 \text{ mA} \qquad (V_{DS} = 10 \text{ V}).$$

(a) Find I_D at each of the following values of V_{GS}: 1.0, 2.0, 3.0, and
4.0 V. (b) Draw the drain characteristic and the transfer characteristic. (c) Find g_m for each value of V_{GS} in part (a). Sketch a curve of
g_m against V_{GS}.

6.5-6 A p-channel depletion-type MOSFET has $I_{DSS} = 6$ mA and $V_P = 3$
V. Find V_{GS} and g_m when I_D is 4 mA.
Answer. 0.55 V; 3267 μs.

6.6-1 In the circuit of Fig. P6.6-1 an external inductance is used in the anode circuit to limit di/dt. If $V_{ac} = 120$ V rms, $R_L = 50\ \Omega$, and Q_1 is a G.E. C106B1 SCR where di/dt maximum is 50 A/μs, determine the value of L to protect the SCR. [*Hint:* Since some allowance must be made for variations in line voltage and tolerance on value of L, $(di/dt)_{max} = 80\%$ of the specified value may be used.]

Answer. 4.25 μH.

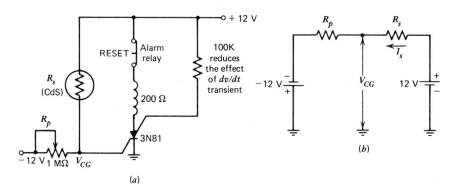

Figure P6.6-1

6.6-2 The SCS circuit of Fig. P6.6-2a is a resistance-sensitive alarm actuator. The resistor R_s (e.g., CdS) is sensitive to light. The voltage at the cathode gate (V_{CG}) will respond on the voltage divider circuit of R_s, R_p, and the two dc supplies (see Fig. P6.6-2b). (a) Show that the cathode-gate voltage is given by

$$V_{CG} = 12\,\frac{R_p - R_s}{R_p + R_s}.$$

(b) Discuss the situations when $R_s = R_p$, $R_s < R_p$, and $R_s > R_p$. In which case will the alarm be actuated?

Figure P6.6-2

6.6-3 It is necessary to vary the frequency of the circuit of Example 6.6-1 between 60 Hz and 12 kHz using a fixed value of 0.01 μF for C_E. What is the range of R_E that is required?

Answer. From 8.33K to 1.667 MΩ.

6.7-1   Find the value of output v_0 obtained from the circuit of Fig. P6.7-1.
Use the superposition theorem.

Answer. $+2$ V.

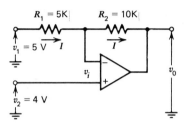

Figure P6.7-1

6.7-2   A μA741C op amp is used in the circuit of Fig. P6.7-2. Calculate the
output voltage for the ideal op amp.

Answer. -2 V.

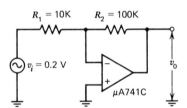

Figure P6.7-2

6.7-3   Derive the transfer function of the quadratic low-pass Butterworth
filter of inverting-configuration form shown in Fig. P6.7-3.

Answer. $H(s) = \dfrac{-1}{1 + bs/\omega_x + (s/\omega_x)^2}$,   where $\omega_x = \dfrac{1}{RC}$.

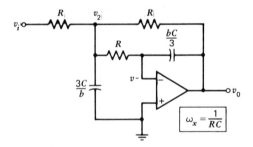

Figure P6.7-3

6.7-4   Derive the transfer function of the quadratic high-pass Butterworth
filter of inverting-configuration form shown in Fig. P6.7-4.

Answer. $H(s) = \dfrac{-1}{1 + b\omega_x/s + (\omega_x/s)^2}$,   where $\omega_x = \dfrac{1}{RC}$.

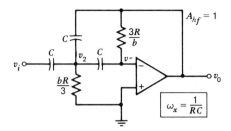

Figure P6.7-4

6.8-1   For the two-input OR gate in Fig. P6.8-1, determine the output waveform in proper relation to the inputs.

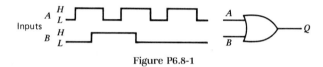

Figure P6.8-1

6.8-2   If the three waveforms shown in Fig. P6.8-2 are applied to the NOR
gate, what is the resulting output waveform?

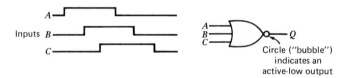

Figure P6.8-2

6.8-3   Sketch the output waveform for the three-input NAND gate in Fig.
P6.8-3, showing its proper relationship to the inputs.

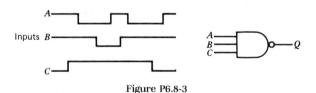

Figure P6.8-3

6.8-4   For the four-input NOR gate operating in the negative-AND mode
in Fig. P6.8-4, determine the output relative to the inputs.

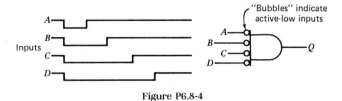

Figure P6.8-4

6.8-5   For the four-input NAND gate in Fig. P6.8-5 operating in the negative-OR mode, determine the output with respect to the inputs.

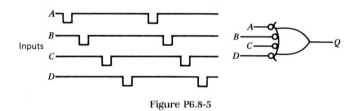

Figure P6.8-5

6.8-6   Determine the output waveform in Fig. P6.8-6.

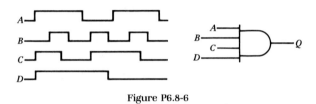

Figure P6.8-6

6.8-7   Determine the output waveform in Fig. P6.8-7.

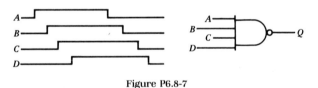

Figure P6.8-7

6.8-8   Implement the expression $AB + BCD + EFGH$ with logic gates.

6.8-9   Construct the following function with logic gates: $(A + B)(C + D + E)(F + G + H + I)$.

6.8-10   Sketch the logic symbol for a negative edge-triggered SR FF. If the waveforms in Fig. P6.8-10 are applied to the inputs of this flip-flop, determine the Q output. Assume the flip-flop is initially RESET.

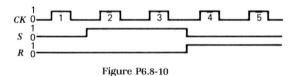

Figure P6.8-10

6.9-1   The waveforms in Fig. P6.9-1 are applied to the J, K, and clock inputs
        as indicated in Fig. 6.9-1. Determine the Q output, assuming the J
        − K flip-flop is initially RESET.

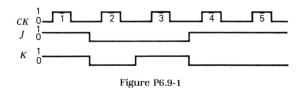

Figure P6.9-1

6.13-1 If in the circuit of Fig. 6.13-1, $n = 8$ and $V_{ref} = 5$ V, find the buffered
        analog output V_a' with each of the following hexadecimal equivalent
        inputs: (a) 02_H; (b) 10_H; (c) 80_H; (d) $4E_H$; (e) FF_H. Assume that the
        buffer amplifier has a voltage gain of 2. (*Hint:* $02_H = 00000010_2$;
        $10_H = 00010000_2$; $80_H = 10000000_2$; $4E_H = 01001110_2$; $FF_H =
        11111111_2$.)

        Answer.   (a) 78 mV; (b) 625 mV; (c) 5 V; (d) 3.047 V; (e) 9.96 V.

6.13-2 If in the circuit of Fig. 6.13-1, $n = 8$, $V_{ref} = 10$ V, and the amplifier
        gain is 1, find the analog output V_a' with each of the following inputs:
        (a) 169_{10}; (b) 251_{10}; (c) 33_H; (d) CC_H ($= 11001100_2$).

        Answer.   (a) 6.6 V; (b) 9.805 V; (c) 1.99 V; (d) 7.969 V.

7
Transducers

A transducer is a device that converts nonelectrical physical parameters into electrical signals which are proportional to the value of the physical parameter being measured. Transducers take many forms. Some of commonly used transducer types will be discussed.

7.1 THERMISTORS

A thermistor is a temperature-sensitive resistor. It is a semiconductor in various geometrical configurations to which two leads are attached. It is not a junction device but a mixture of oxides of cobalt, nickel, strontium, or manganese. Its terminal resistance is somehow related to its body temperature. It has a negative temperature coefficient, indicating that its resistance will decrease with increasing temperature. Its negative temperature coefficient α is defined as

$$\alpha \equiv \frac{\Delta R/R_T}{\Delta T} . \tag{7.1-1}$$

α is quite high but nonlinear in the resistance–temperature variation. Over most of its range the resistance decreases exponentially with temperature. Thus the resistance of a thermistor can be expressed by

$$R_t = R_0 \, e^{\beta[1/T - 1/T_0]} \tag{7.1-2}$$

where R_t = the resistance of the thermistor at temperature T,
  R_0 = the resistance of the thermistor at a reference temperature (usually 25°C or 0°C),
  e = 2.71828,
  T = thermistor temperature (K),
  T_0 = reference temperature (K),

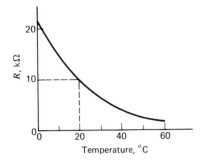

Figure 7.1-1 Typical thermistor resistance versus temperature.

β = a property of the material used to make the thermistor (K).

Usually, β will have a value between 1500 and 7000 K,

For most commercial thermistors the temperature coefficients are in the range of 3–5%°C. The typical set of a thermistor characteristics is shown in Fig. 7.1-1. For precision units the specified resistance R_0 at a reference temperature T_0 is closely controlled (e.g., ±1%), but for standard units only α is controlled while R_0 may alter by 10% or more from unit to unit.

There are two additional parameters, namely, the time constant τ_t and the internal temperature rise due to power dissipation η. They may be 3–10 times higher for a thermistor in still air compared with a stirred liquid. Typical values in a liquid are τ_t = 0.5 s and η = 0.1°C/mW. The thermistor resistance does not jump immediately to the new value when the temperature changes, but requires a small amount of time to stabilize at the new resistance value; this is expressed in terms of the time constant of the thermistor in a manner that is reminiscent of capacitors in *RC* circuits. To minimize η or internal (self-) heating it is necessary to control the power dissipation of the thermistor. A thermistor transducer may be used as one part of a voltage divider or an arm of a resistance bridge.

Instead of a thermistor, it is possible to use a temperature-sensitive resistor with a positive temperature coefficient, such as a metal, or the sensistor manufactured by Texas Instruments. The temperature coefficient of the sensistor is +0.7%/°C over the range from −60 to +150°C. A heavily doped semiconductor can exhibit a positive temperature coefficient of resistance.

7.2 THERMOCOUPLES

A thermocouple is a junction made from two dissimilar metals joined together. When the junction is heated, a small thermoelectric potential is generated. This phenomenon is due to different work functions for the two metals. The potential generated by the junction is proportional to the junction temperature and is approximately linear over wide temperature ranges. Actually two junctions are employed for a practical thermocouple, as shown

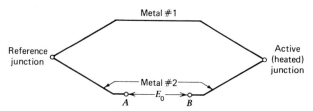

Figure 7.2-1 Thermocouple measurement junction and reference junction employed for a practical thermocouple.

in Fig. 7.2-1. They are the measurement junction used as a thermometry probe and the reference junction kept at a reference temperature (e.g., 0°C).

There is an inverse thermocouple phenomenon, in which an electrical potential applied across the two terminals A and B in Fig. 7.2-1 will cause one junction to absorb heat (get hot) and the other to lose heat (get cold). Semiconductor thermocouples have been used in small-scale environmental temperature chambers.

There are many kinds of thermocouple wires. They generate different thermoelectric potentials in millivolts with reference at 0°C. For example, at 20°C, the copper–constantan, platinum–platinum plus 10% rhodium, and Chromel–Alumel thermocouples generate 0.78, 0.113, and 0.80, respectively. At 60°C, they produce 2.47, 0.364, and 2.43, respectively. Their temperature–voltage relation is more linear than thermistors.

7.3   SEMICONDUCTOR TEMPERATURE TRANSDUCERS

Ordinary p-n diodes can be used as temperature transducers in less exacting applications. When they are operated at a constant current, the forward-biased diode voltage changes with temperature. Figure 7.3-1 shows a semiconductor temperature transducer, which is a pair of diode-connected transistors. In each of the transistors, the base–emitter voltage is given by

$$V_{BE} = \frac{kT}{q} \ln \frac{I_c}{I_0},\qquad(7.3\text{-}1)$$

where V_{BE} = base–emitter voltage, V,
k = Boltzmann's constant = 1.38×10^{-23} J/K,
T = absolute temperature, K,
q = electronic charge = 1.6×10^{-19}/C,
I_c = collector current, A,
I_0 = reverse saturation current, A.

Since both currents can be made to be constant, the only variable is the temperature T. The differential output voltage ΔV_{BE} is

$$\Delta V_{BE} = V_{BE_2} - V_{BE_1} = \frac{kT}{q}\left(\ln \frac{I_{c2}}{I_0} - \ln \frac{I_{c1}}{I_0}\right)$$

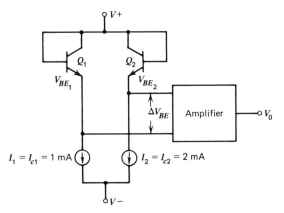

Figure 7.3-1 A pair of diode-connected transistors used as a temperature sensor operating at constant currents.

or

$$\Delta V_{BE} = \frac{kT}{q} \ln \frac{I_{c_2}}{I_{c_1}}. \tag{7.3-2}$$

Designers usually set $I_{c_2} = 2$ mA and $I_{c_1} = 1$ mA, that is, the current ratio is $2:1$. Since I_{c_1} and I_{c_2} are supplied from constant current sources, the ratio I_2/I_1 is a constant. Hence Eq. (7.3-2) can be expressed as

$$\Delta V_{BE} = kT = \frac{59.8 \ \mu V}{K} T, \tag{7.3-3}$$

where $K = (k/q) \ln (I_{c_2}/I_{c_1})$
$\qquad = (1.38 \times 10^{-23}/1.6 \times 10^{-19}) \ln (2/1) = 59.8 \ \mu V/K.$

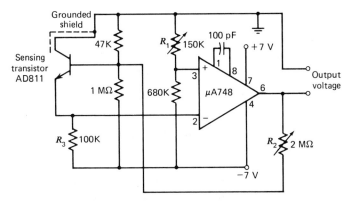

Figure 7.3-2 A temperature thermometer consisting of a transistor sensor and an op-amp connected in a feedback circuit. [FROM Koch, C. J., "Diode or Transistor Makes Fully Linear Thermometer," *Electronics*, 110–112 (May 13, 1976). [Copyright © McGraw-Hill, Inc., 1976. All rights reserved.]

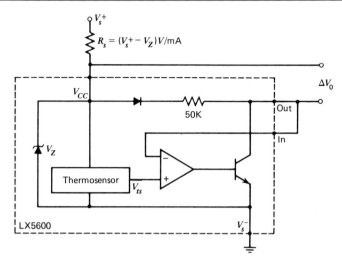

Figure 7.3-3 An IC temperature transducer with its op-amp connected as a unity-gain amplifier. $\Delta V_0 = 10$ mV/°C (2.98 V at 25°C).

The most common scale factor of 10 mV/K must be employed so that the temperature can be easily read from a digital voltmeter. Hence, the amplifier in Fig. 7.3-1 requires a gain of

$$A_V = \frac{10 \text{ mV/K}}{59.8 \text{ } \mu\text{V (1 mV/10}^3 \text{ } \mu\text{V)}} = 167.$$

Figure 7.3-2 shows a temperature transducer consisting of a sensing transistor and an op-amp connected in a feedback circuit. Use of bipolar supply for the op-amp makes the transducer circuit fully linear even at low temperatures. Accuracy is within 0.05°C. The zero point is set by R_1 and the gain is set by R_2.

An improved version of the semiconductor thermosensor is the IC temperature transducer (e.g., LX5600), as shown in Fig. 7.3-3. The temperature response is linear within the range of -55 to $+125$°C. The sensor output V_{ts} has a sensitivity of 10 mV/°C and a voltage range of 2.98 V (at 25°C) to 3.98 V.

7.4 SEMICONDUCTOR PHOTOCONDUCTIVE CELLS

A semiconductor photoconductive cell is shown in Fig. 7.4-1; it is also known as a photoconductor or photoresistor. Its conductance increases (linearly) with the intensity of the incident light. The photoconductive materials most frequently used are cadmium sulfide (CdS) and cadmium selenide (CdSe). The peak spectral response of CdS occurs at nearly 5100 Å and for CdSe at nearly 6150 Å. (1 angstrom $= 1$ Å $= 10^{-4}$ μm $= 10^{-8}$ cm.) The response time of CdS cells is about 100 ms and it is 10 ms for CdSe units. These cells

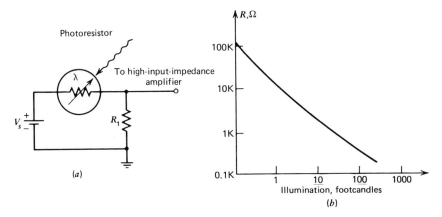

Figure 7.4-1 Semiconductor photoconductive cell (photoresistor): (a) circuit; (b) characteristics.

can be made to operate at light levels of about 10^{-3}–10^3 foodcandles. The resistance of a photoresistor may change from 10 MΩ in the dark to under 10 Ω in a bright light.

7.5 PHOTODIODES AND PHOTOTRANSISTORS

The photodiode is a photojunction device. It is formed from a *p-n* junction and is usually made of silicon. If a photon is incident in the junction with sufficient energy to jump the energy-band gap, hole–electron pairs are produced that modify the junction characteristics, as shown in Fig. 7.5-1. From

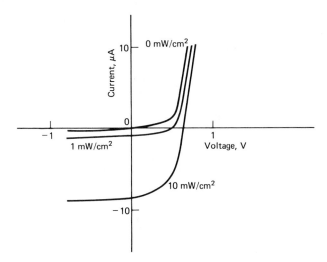

Figure 7.5-1 Voltage–current characteristics of an irradiated silicon *p-n* junction.

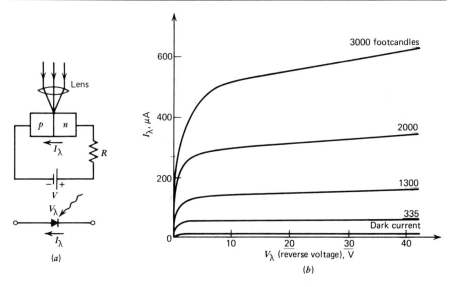

Figure 7.5-2   (a) Photodiode and (b) its characteristics.

this figure, we see the three different curves. For zero irradiance, both forward and reverse characteristics are normal. For 1 mW/cm^2, the open-circuit
voltage is 500 mV and the short-circuit current is 0.8 μA. For 10 mW/cm^2,
the open-circuit voltage is 600 mV and the short-circuit current is 8 μA. If
the junction is reverse biased, the reverse photocurrent (μA) is proportional
to the light intensity (mW/cm^2). Typical photodiode characteristics are
shown in Fig. 7.5-2.

Sometimes the photodiode junction is coincident with the base–collector
junction of a transistor, which then becomes a phototransistor. It is equivalent to a photodiode connected to the base of a transistor which provides
a current gain. The phototransistor symbol and equivalent are shown in Fig.
7.5-3. The radiation–current characteristics have a monlinearity of about 2%
since β varies with collector current. The response time is about 10 μs.

Light intensity can be defined as radiant flux, expressed in watts per
square meter. The intensity can also be expressed in photometric units, such

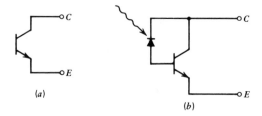

Figure 7.5-3   (a) Phototransistor symbol; (b) phototransistor equivalent.

as candelas (cd) or lumens (lm), which are based on effective brightness as perceived by the eye. The sensitivity of the eye depends on the wavelength. At 550 nm (green, where the eye is most sensitive), the luminous efficiency or the ratio between the photometric unit and fundamental unit is about 680 lm/W. A one-candela point source results in a flux of one lumen through a one-square-foot area on a sphere with a radius of one foot. An illumination of 1 footcandle equals 1 lm/ft^2.

7.6 LIGHT-EMITTING DIODES

The light-emitting diode (LED) will give off visible light when it is energized. Since it is forward biased, electrons from the n region and holes from the p region cross the p-n junction. Some of the electrons in the p region recombine with holes, and similarly, some holes in the n region recombine with electrons. Because of recombination, radiant energy is released. This is the basis of operation for the LED. The released energy is in the form of photons with a photon energy equal to that of the band-gap energy. Semiconductors with higher gap energies result in higher photon energies, that is, light of shorter wavelengths. For silicon diodes, the maximum wavelength is about 900 mm, which lies in the near infrared. A popular gallium arsenide phosphide (GaAsP) LED emits red light. Its emission spectrum [relative intensity (%) versus wavelenths (μm)] is within the region of 0.62–0.76 μm with a peak (100%) at 0.66 μm. LEDs are also available in orange, yellow, and green; for these three colors gallium phosphide is frequently used. Current–voltage characteristics shown in Fig. 7.6-1 are similar to those of any forward-biased diode except that appreciable current does not flow until the threshold voltage of 1.4–1.8 V is applied. Both the conduction threshold voltage and photon energy increase with band-gap energy. When an LED is used, as in the current of Fig. 7.6-2, a series resistor is ordinarily required to limit the current flow to a safe value. The reverse voltage is typically 3–10 V.

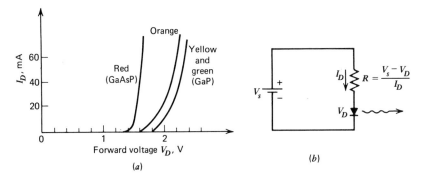

Figure 7.6-1   LED current–voltage characteristics.

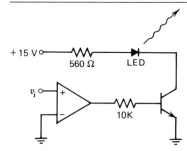

Figure 7.6-2   LED driven by op-amp and transistor.

The seven-segment LED display shown in Fig. 7.6-3 is the commonly used display. Each segment can be individually illuminated. Any digit from 0 to 9 can be displayed by turning on the proper combination of segments. One lead is brought out for each segment in a standard display. Most units also include a decimal point.

7.7   PHOTOMULTIPLIER

A photomultiplier schematic is shown in Fig. 7.7-1. It is a phototube combined with an electron multiplier. When an incoming photon strikes the photocathode, a photoelectron is released if its energy exceeds that of the work function of the cathode surface. The energy of the photon E_p given by

$$E_p = h\nu = \frac{hc}{\lambda},\qquad(7.7\text{-}1)$$

where h = planck's constant = 6.62×10^{-34} J·S,
   λ = wavelength of light,
   c = velocity of light = 3×10^8 m/s.
If the photon energy is expressed in electron volts (1 eV = 1.6×10^{-19} J) and the wavelength in nanometers (nm), then the constant hc equals 1240 eV/nm. The electron emitted from the cathode is accelerated toward the first dynode (i.e., secondary-emission electrode), which is 100 V more positive than the cathode. The impact liberates several electrons by secondary emission. They are accelerated toward the second dynode, which is 100 V

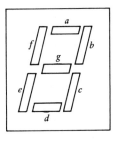

Figure 7.6-3   LED seven-segment array for numerals (0–9), used as a display.

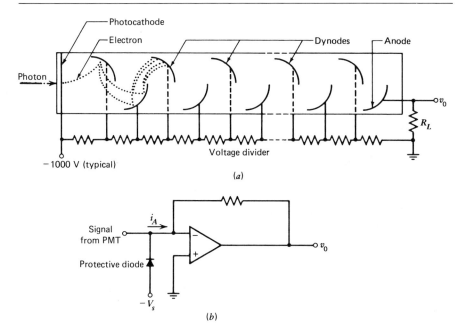

Figure 7.7-1   (a) Photomultiplier tube (PMT) dynode voltage divider; (B(PMT amplifier.

more positive than the first dynode. This electron multiplication continues until the anode, where currents of about 1 μA flow through the load R_L. Current gains of 10^5 or 10^6 are common and account for the high sensitivity of the photomultipliers.

Photomultipliers are the most-sensitive photodetectors. They will respond to single photons with high efficiency. At low light levels, where the single-photon pulses do not overlap, pulse-counting techniques can be used to measure light intensity. If R_L is from 50 to 1000 Ω, the pulse width is typically 5–50 ns. In this case, $R_L C_A$ is small, where C_A is the anode capacitance to ground.

7.8   VACUUM GAGES

Pressures are usually expressed in torrs; 1 torr is the pressure required to raise a column of mercury 1 mm. Hence 760 torrs = 760 mm Hg = 1 atm. For the 10^{-3}–1 torr range, a popular sensor is the thermocouple gage. It conveniently covers the lower operating range of mechanical vacuum pumps. Usually it is a monitor of proper pump operation where only a rough idea of the pressure is needed. The gage consists of a thermocouple attached to a heater, as illustrated in Fig. 7.8-1. Although the amount of power dissipation in the heater is a constant, the temperature depends on the degree

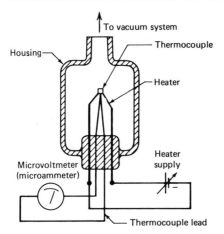

Figure 7.8-1 Thermocouple vacuum gage.

of cooling, which in turn depends on the gas pressure. Since high vacuum is an insulator, the thermocouple heats up. The voltage produced by the thermocouple is read by a microammeter. When the meter scale is calibrated, a good vacuum ($\ll 10^{-3}$ torr) is desirable, at which point the heater power is adjusted so that the meter reads full scale electrically. A more-accurate thermoconductive gage uses a resistance thermometer as the temperature sensor.

Perhaps the most-accurate high vacuum gage is the hot-cathode-ionization gage, as shown in Fig. 7.8-2. Usually it is intended for use below 10^{-3} torr, where the mean free path of electrons is fairly long. The filament (cathode) is heated to produce electrons, as in a vacuum tube. Screen and plate voltage

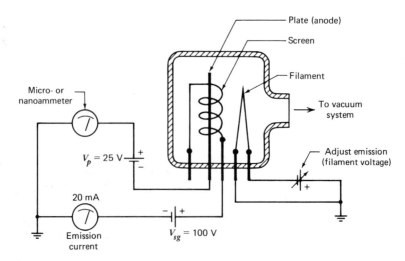

Figure 7.8-2 Hot-cathode-ionization-type vacuum gage.

supplies are typically $+100$ V and -25 V, respectively. The emission current is typically 20 mA. Emitted electrons are accelerated by the positive screen. The energetic electrons collide with neutral gas molecules, causing ionization. The negative plate collects the positive ions. Because the electron current and the path length are fixed, the number of ions produced, and thus ion current hitting the plate, will be proportional to the pressure. Ion currents are below 10^{-6} A at lower pressures. The lower pressure limit is below 10^{-10} torr. Note that the filament may burn out if it is accidentally exposed to air at higher pressures.

7.9   pH METERS

Meaning of pH

A solution is acid when the hydronium-ion (H^+) activity is greater than the hydroxyl-ion (OH^-) activity and is alkaline when the reverse is true. The ion activity a is related to the concentration of ions, and can be expressed as

$$a = \gamma C,$$

where C is the concentration and γ is the activity coefficient. When the solution is infinitely dilute, $\gamma = 1$, and the activity is equal to the concentration. Since the activities may vary by many orders of magnitude, it is convenient to compress the scale by using as a measure of the acidity the base-10 logarithm of the hydrogen-ion activity. The pH is a measure of the hydrogen-ion concentration defined as

$$pH = -\log [H^+], \tag{7.9-1}$$

where $[H^+]$ is the hydrogen-ion concentration. For pure water at 25°C, $[H^+] \simeq 1 \times 10^{-7}$, and so

$$pH = \log \frac{1}{[H^+]} = \log 10^7 = 7.$$

Thus, for an acid, $pH < 7.0$, and for an alkaline solution, $pH > 7.0$. The values of pH in milk and egg white are 6 and 8, respectively. The values of pH can vary from 0 (purely acidic) to 14 (purely alkaline). When the pH is 7, the solution is neutral. The normal range of pH in arterial blood is 7.38 to 7.44.

Measurement of pH

A pH meter shown in Fig. 7.9-1 consists of a glass electrode, a reference electrode (usually calomel), and an electrometer (high-input-impedance voltmeter) dc amplifier. The glass electrode is the primary detecting element in pH measurements. It is made from a special glass that is permeable to the

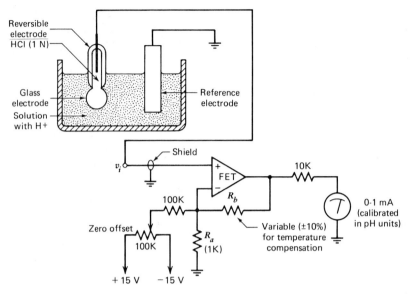

Figure 7.9-1 The pH meter containing a glass electrode, a reference electrode, and an electrometer amplifier (noninverting FET op-amp.

hydrogen ion (only), and the thin bottom acts as a membrane. Hydrochloric acid ($1N$) inside the sealed electrode is in contact with the glass membrane as well as with a reversible electrode at the top. Both the reversible electrode inside the glass electrode and the reference electrode are identical, and thus the potential differences due to the reference electrodes are zero. The net potential across the leads connected to the amplifier is just the membrane voltage.

The potential develops across the membrane when the concentrations of a permeable ion are unequal on either side of a membrane. A positive ion (cation) diffusion from the higher- to the lower-concentration side leaves the anion (assumed incapable of penetrating the membrane) behind, and the resulting charge separation produces the membrane potential, which can be measured by a pair of reference electrodes. The voltage difference continues to build up until at equilibrium the electric field within the membrane is strong enough to inhibit further diffusion.

Most chemical transducers are based on membrane voltage. Practically all biological potentials are associated with membranes, especially the membranes surrounding cells.

BIBLIOGRAPHY

Carr, J. J., *Elements of Electronic Instrumentation and Measurement*. Reston Publishing Co. Inc., Reston, Virginia, 1979, Chapter 13.

Cobbold, R. S. C., *Transducers for Biomedical Measurements, Principles and Applications*. Wiley, New York, 1974.

Koch, C. J., "Diode or Transistor Makes Fully Linear Thermometer." *Electronics*, 110–112 (May 13, 1976).

Thomas, H. E., *Handbook of Biomedical Instrumentation and Measurement*. Reston Publishing Co. Inc., Reston, Virginia, 1974.

Webster, J. G., *Medical Instrumentation Application and Design*. Houghton Mifflin, Boston, 1978, Chapter 2.

Questions

7-1   Define, in your own words, "transducer".

7-2   List three types of temperature transducers.

7-3   The temperature coefficients for most elemental metals are _____.

7-4   What is the main difference between the thermistor and the sensistor?

7-5   A thermocouple is formed of two _____ metals, and works because the metals have different _____.

7-6   Explain the operation of the temperature thermometer shown in Fig. 7.3-3.

7-7   What is the main use of a CdS photoresistor?

7-8   How do the photodiode and phototransistor work?

7-9   Describe the operating principle of a light-emitting diode.

7-10 Describe the operating principle of a photomultiplier tube.

7-11 What must be observed for the safe operation of the hot-cathode-ionization-type vacuum gage?

7-12 How does the membrane potential develop?

Problems

7.1-1 Calculate the resistance of a thermistor at 100°C if the resistance at 0°C was 20K. The material of the thermistor has a value of 2200 K. *Answer.*   2330 Ω.

7.2-1 A thermocouple thermometer is shown in Fig. P7.2-1. The voltage applied to the millivoltmeter is given by

$$V_i = \frac{R_m + R_c}{R_m + R_c + R_t + R_w} V_t.$$

where V_i = the voltage applied to the millivoltmeter,
V_t = the potential generated by the thermocouple,
R_m = the series resistor of the millivoltmeter,
R_c = the resistance of the moving coil,

R_t = the thermocouple resistance,
R_w = the resistance of the wire connected to the thermocouple.

Assume R_m = 550 Ω, R_c = 50 Ω, and R_t = 2.5 Ω. Let the value of R_w increase from 0.15 to 5 Ω due to the temperature increasing. Calculate the measurement error.

Answer. 0.41%

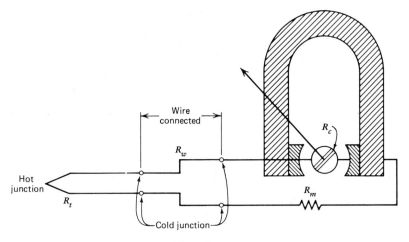

Figure P7.2-1

7.3-1 Calculate the output voltage ΔV_{BE} from the circuit of two transistors connected as diodes shown in Fig. 7.3-1. Assume the temperature equal to 35°C.

Answer. 0.0184 V.

7.4-1 Design a light-controlled switching circuit consisting of two CdS photoresistors in series and a common-emitter transistor.

7.5-1 What is the "dark current" of a photodiode?

7.5-2 Assume that a phototransistor can deliver a current of 50 μA when a light beam is interrupted. It is used as a sensor in an amplifier which will turn on a relay (12 V, 50 mA) with the light on. Design this amplifier circuit. (*Hint:* The main amplifier may be a noninverting op-amp with a phototransistor as the input section. The output section may be a transistor stage with the relay as a load.)

7.6-1 Suppose you wanted to light a common-anode seven-segment LED display (Fig. P7.6-1a) and form your own special symbols for digits 1 and 9 as shown in Fig. P7.6-1b. Show the output level required for each segment of your special BCD-to-seven-segment decoder/driver.

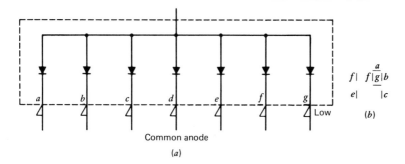

Common anode

(a)

(b)

Figure P7.6-1

7.7-1 Draw a dynode voltage divider and an amplifier circuit for the photomultiplier tube.

7.8-1 A resistance-thermometer-type thermoconductive vacuum gage is
shown in Fig. P7.8-1. It is one arm of a bridge whose output is a
readout op-amp. Draw the basic bridge readout circuit.

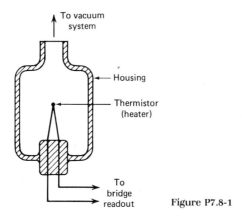

Figure P7.8-1

7.9-1 Design a MOSFET electrometer amplifier for a pH meter.

8

General Description of Oscilloscopes

8.1 INTRODUCTION TO THE CATHODE-RAY OSCILLOSCOPE

The cathode-ray oscilloscope is also called the oscilloscope, or simply scope. It can present visual representations of many dynamic phenomena by means of transducers. Although many of other instruments are very useful and can be more accurate than the oscilloscope, their application is mainly limited to the measurement of one parameter of the signal. With an oscilloscope one can visualize the signal of interest and also observe whether the signal contains properties that would not be made apparent by most other instruments (e.g., whether the signal is superimposed on a dc level). In the oscilloscope there is one screen on which a part of the signal is written during a certain time span. In order to obtain a stable picture, the scope will draw a great many of those parts of the signal, one part after the other and each one covering the previous one exactly. Therefore, the eye observes one steady picture of the waveform. The speed of a chart recorder is limited to a few transients per second, while the electron beam in the scope can visualize transients in the nanosecond (10^{-9}s) range. Hence, the scope is able to visualize much faster phenomena than the recorder. Since the oscilloscope has the above advantages, it is a more-valuable instrument and widely used today.

The oscilloscope is much like a television set in which it produces a picture of the shape of an input signal on a light-emitting phosphor screen. The heart of the scope is the cathode-ray tube (CRT), since it performs the basic functions to convert a signal into an image; it is the output device, or display portion, of the instrument. The CRT is a vacuum tube similar in shape to a TV picture tube.

In an $X-Y$ oscilloscope, the beam can be positioned anywhere on the CRT screen by controlling the potentials applied to the vertical and horizontal deflection plates (see Fig. 8.2-1). In most commercial oscilloscopes vertical and horizontal amplifiers are used to drive the deflection plates. Most service and laboratory oscilloscopes sweep the horizontal plates with a time base signal, and so are ''Y-time'' designs. Many of these models can be used in the $X-Y$ mode because a selector switch allows the user to apply either the internal time base signal or an external signal to the horizontal deflection plates.

8.2   THE CATHODE-RAY TUBE

Operation of the CRT

The symbol of a basic CRT is shown in Fig. 8.2-1. The main components of the CRT are the electron gun, the vertical and horizontal deflection plates, the fluorescent phosphor screen, the glass envelope, and the base of the tube. The electron gun in the narrow part of the tube includes a heated cathode, a control electrode (grid), a preaccelerating anode (a_1), a focusing anode (a_2), and an accelerating anode (a_3). Anodes a_1 and a_3 are connected together. Some CRT models also have a postdeflection accelerating electrode (sometimes called a second anode) in the next section toward the screen, called the postacceleration area. The heated cathode gives off electrons by thermionic emission. The cathode is surrounded by a negatively charged cylindrical control electrode which has a hole in one end. Electrons also carry a negative charge, so are repelled from the walls of the cylinder, and will stream out through the hole toward a less negatively charged accelerating electrode. In such a case the electron gun furnishes a controllable source of electrons and focuses these electrons into a beam with the focus point (spot) on the screen. The beam is deflected vertically and horizontally in the deflection section before striking the layer of phosphor at the screen

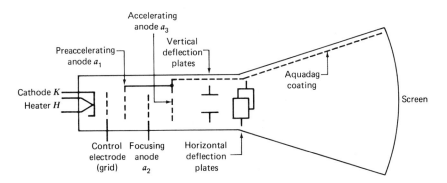

Figure 8.2-1   CRT symbol.

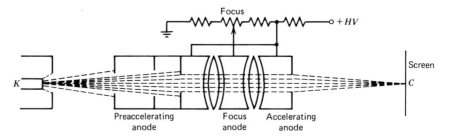

Figure 8.2-2   Focusing arrangement of a CRT. The curved equipotential surfaces form a double convex lens system.

to produce light. The inside of the tube from the deflection plates to almost as far as the screen is covered with a conductive coating of Aquadag (i.e., a graphite coating). The Aquadag coating is connected to anodes a_1 and a_3. Hence, secondary electrons emitted by the screen when it is struck by the electron beam are attracted to the comparatively positive Aquadag coating, and the return circuit to the cathode is thereby completed. This knid of tube is called the monoaccelerator tube, which has time coefficients up to 0.1 μs/div; thus 10-MHz sine waves can be displayed with 1 period per division.

Electrostatic Focusing

The electrostatic focusing system of a CRT is shown in Figure 8.2-2. The preaccelerating and accelerating anodes are connected together to a high positive potential supply. The focusing anode, located between the two accelerating anodes, is connected to a lower positive potential. The potential difference between the focusing anode and the accelerating anodes creates electric fields between the cylindrical elements. Since the field lines between the cylinder ends are nonuniformly spaced, as in Fig. 8.2-3, the equipotential surfaces are curved to form a double concave lens system, as indicated in Figure 8.2-3.

Note that electrons are emitted by the cathode as a slightly divergent beam. Those electrons that enter the electric field between the preaccelerating anode and the focusing anode at an angle other than normal to the equipotential surface will be refracted toward the normal. The electron beam

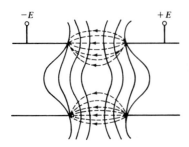

Figure 8.2-3   Equipotential surface for two cylinders placed end to end. The lateral repulsion of the electric field lines causes the spreading of the space between lines. The force on an electron in the field is in the direction normal to the equipotential surfaces, or in the direction opposite to the direction of the field: $f_e = -1.6 \times 10^{-19} \epsilon$ N, where ϵ = electric field intensity (V/m).

thus tends to become parallel to the CRT axis, as shown. This approximately parallel beam then enters the second convex lens and is refracted once again to become slightly convergent and focused on the screen at the center of the CRT axis. The focal point of the beam can be moved along the CRT axis by varying the voltage on the focusing anode.

Refraction of an electron ray at an equipotential surface is explained as follows. Consider the regions on both sides of an equipotential surface, S, as shown in Fig. 8.2-4. The potential to the left of the surface S is V^- and to the right it is V^+. An electron, which is moving in the direction AB and entering the area to the left of surface S with a velocity u_1, experiences a force at surface S. Because of this force, the velocity of the electron increases to a new value, u_2, after it has passed surface S. Thus, the normal component of the velocity, u_n, is increased by the force at surface S to a new value, u_n'. But the tangential component, u_t, of the velocity on both sides of S remains the same. Then

$$u_t = u_1 \sin \theta_i = u_2 \sin \theta_r, \qquad (8.2\text{-}1)$$

or

$$\frac{\sin \theta_i}{\sin \theta_r} = \frac{u_2}{u_1}, \qquad (8.2\text{-}2)$$

where θ_i is the angle of incidence and θ_r is the angle of refraction of the electron ray. Equation (8.2-2) is identical to the expression relating the refraction of a light beam in geometrical optics.

Electrostatic Deflection

The path of an electron, traveling through an electric field of constant intensity and entering the field at right angles to the lines of flux, is parabolic in the $X - Y$ plane, because the electron moves with a constant acceleration in the Y direction of the uniform electric field.

The two parallel deflection plates shown in Fig. 8.2-5 are placed a distance d apart and are connected to a source of potential difference E_d so that an

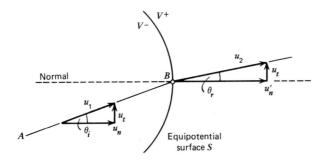

Figure 8.2-4 Refraction of an electron ray at an equipotential surface in the focusing system.

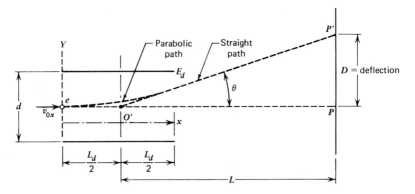

Figure 8.2-5　Deflection of the CRT.

electric field (E_d/d) exists between the plates. An electron entering the field with an initial velocity v_{0x} is deflected toward the positive plate following the parabolic path as shown. When the electron leaves the region of the deflection plates, the deflecting force no longer exists, and the electron travels in a straight line toward point P', a point on the screen. The straight line of travel of the electron is tangent to the parabola at $x = L_d$, and this tangent intersects the X axis at point O', the apparent origin. From Fig. 8.2-5 we see that the deflection D is proportional to L, $L_d/2$, and E_d; but D is inversely proportional to d and E_a. Hence,

$$D = \frac{LL_dE_d}{2\,dE_a}\quad(\text{m}),\tag{8.2-3}$$

where　D = deflection on the screen, m;
　　　L = distance from center of deflection plates to screen, m;
　　L_d = effective length of the deflection plates, m;
　　　d = distance between the deflection plates, m;
　　E_d = deflection voltage, V;
　　E_a = accelerating voltage, V.

In high-frequency oscilloscopes, the vertical deflection plates may be segmented to increase the bandwidth of the CRTs. By segmenting these plates, a specific drive per deflection segment may be obtained, resulting in a proper deflection of high-frequency signals. In this case, the stray capacitance between the plates are used to form a delay line together with the externally connected coils. The deflection sensitivity of a CRT is defined by

$$S \equiv \frac{D}{E_d} = \frac{LL_d}{2\,dE_a}\quad\left(\frac{\text{m}}{\text{V}}\right).\tag{8.2-4}$$

The deflection factor of a CRT is defined by

$$G \equiv \frac{1}{S} = \frac{2\,dE_a}{LL_d}\quad\left(\frac{\text{V}}{\text{m}}\right).\tag{8.2-5}$$

Typical values of deflection sensitivities are from 0.1 to 1 mm/V, corresponding to deflection factors of 100–10 V/cm. The high sweep frequencies to be supplied to the deflection plates means the high writing speed to be displayed.

Postdeflection Acceleration System

The next section toward the screen is the postacceleration area (see Fig. 8.2-2). For time coefficients up to 0.1 μs/div no postacceleration is needed. In order to raise the writing speed, the preaccelerating and accelerating anodes in the monoaccelerator can be brought to a potential of approximately 4 kV. But increasing E_a from 2 to 4 kV causes the deflection sensitivity to decrease proportionally. In order to overcome the problem of the writing speed, postdeflection acceleration (PDA) is applied. This method allows the focusing and deflection sections to be operated at even lower voltages than the monoaccelerator tube. These lower operating voltages reduce the velocity of the beam in the deflection system and are an aid to better deflection sensitivity. After the beam has been deflected, it is then accelerated in the postacceleration area to give a high light output.

The recently developed CRT with the PDA system is shown in Fig. 8.2-6. Instead of the Aquadag coating used in the monoaccelerator, a conductive coating covers the inside surface of the envelope within the postacceleration area. A domed mesh is located in the CRT just beyond the deflection plates. The mesh is a metal electrode at a positive potential, such as 2 kV. The gain in the deflection sensitivity (DS) is even such that the CRT can be shortened and still retain an acceptable DS. The conductive coating for the PDA system is called the postaccelerator electrode, which is at a potential up to about 20 kV.

Magnetic Deflection System

In the magnetic deflection system there are no internal deflection plates, but an external electromagnet surrounds the glass neck of the CRT. There are two separate electromagnet coils, one each for vertical and horizontal de-

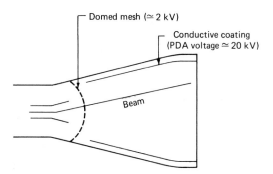

Figure 8.2-6 The domed mesh in the CRT provides shorter tube while maintaining the increase in deflection sensitivity.

flection. These two coils are constructed inside of a single assembly called a deflection yoke. The yoke can create the deflection since magnetic fields will bend the path of the electron beam emitted by the gun assembly.

The inductance of the yoke coils limits magnetic deflection to low-frequency applications. For example, a TV receiver using magnetic deflection creates a lighted area on the screen called a raster by sweeping the vertical at approximately 60 Hz and the horizontal at 15,734 Hz (Note: 15,750 Hz was used in the days before color TV.) Medical oscilloscopes deal with very slow signals (up to either 100 or 1000 Hz), and thus often use magnetic deflection in order to take advantage of the possibly shorter CRT length. But all service, engineering laboratory, and most scientific oscilloscopes employ electrostatic deflection. For high values of anode voltage, a greater deflection can more readily be obtained with magnetic deflection than with electrostatic deflection.

8.3   FREE-RUNNING OSCILLOSCOPE

The block diagram of a basic free-running oscilloscope is shown in Fig. 8.3-1. In the vertical channel, the push–pull output vertical signal amplifier drives the vertical deflection plates. The vertical amplifier has a high gain, so large signals must be passed through an attenuator. The horizontal amplifier connects to an internal time base signal, and is controlled by a horizontal gain control and two sweep frequency controls: sweep selector and sweep vernier.

The time base generator produces a sawtooth waveform that deflects the beam horizontally. The voltages between the CRT horizontal deflection plates are arranged so that the electron beam (or spot) is positioned on the left side of the screen when the sawtooth ramp voltage is zero. The beam is pulled to the right as the ramp voltage rises. If adjustments are proper, the ramp will reach maximum just as the beam disappears off the right edge of the screen. Once the ramp portion is completed, the sawtooth waveform drops rapidly back to zero, snapping the beam back to the left edge of the screen; in this case, the spot on the screen reaches the end position and is quickly brought back to the start position. This action results in a retrace (flyback) line being printed on the screen. To overcome this problem, a retrace blanking pulse is produced, which extinguishes the beam during the flyback time. This eliminates the retrace line on the screen.

The free-running oscilloscope is a low-cost instrument. Its time base generator must be synchronized to the vertical amplifier signal for a stable CRT display. Otherwise, the waveform will march across the screen and remain unstable. Synchronization means that the time base signal sweeps across the screen in a time that is equal to an integer number of vertical waveform periods. Thus, the vertical waveform can appear locked on the CRT screen provided that the frequency of the vertical input signal is a whole multiple of the sweep frequency ($f_v = nf_s$).

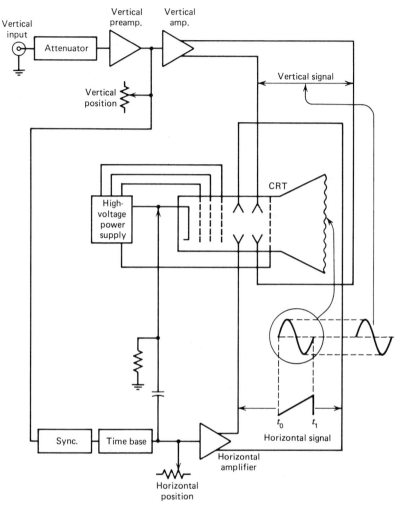

Figure 8.3-1 Block diagram for a free-running oscilloscope.

8.4   TRIGGERED-SWEEP OSCILLOSCOPES

The low-cost free-running oscilloscope provides only limited usefulness. For example, the rise time of a pulse cannot be measured with the free-running scope, but can be measured with the triggered-sweep scope. The triggered-sweep scope is considered far more versatile, and is the standard of the industry. In triggered-sweep modes, the sawtooth generator does not develop a ramp voltage unless told to do so by a trigger pulse. The triggered sweep allows CRT displays of vertical input signals of very short duration, stretched out over an appreciable area of the screen, simply because the sweep is initiated by a triggered pulse derived from the waveform under observation.

Figure 8.4-1 shows a block diagram of a basic triggered-sweep oscillo-scope, including the power supplies, CRT, delay line, vertical amplifier system, trigger pick-off amplifier, trigger circuit, sweep generator, horizontal amplifier, and the Z-axis circuit. When a signal is applied to the vertical input, it is immediately passed to the preamplifier (A), where it is converted to a push–pull signal. The signal then goes on to the vertical output amplifier (C) via the delay line (B). The signal from the output amplifier is used to drive the CRT electron beam vertically, causing the spot on the screen to move vertically. A sample of the vertical signal is taken from the vertical preamplifier just prior to the delay line by the trigger pickoff amplifier (D), which serves to isolate the trigger circuits (E). This signal will be used by the time base system (E, F, G). The trigger signal is used to force a time relation between the vertical signal and the time base. The trigger pick-off signal is shaped into a trigger signal by the trigger circuits (E). This triggers the ramp in the sweep generator (F), causing it to run up at a predetermined rate. The ramp is amplified and changed to a push–pull signal by the horizontal amplifier (G), which is connected to the horizontal plates of the CRT and causes the trace to sweep across the screen horizontally as the ramp voltage increases. A gate from the sweep generator (F) turns the beam on during the ramp's rise time and off during retrace. This is done by the Z-axis portion of the CRT and high-voltage circuit (H). Meanwhile, the am-plified input signal is further amplified by the vertical output amplifier (C)

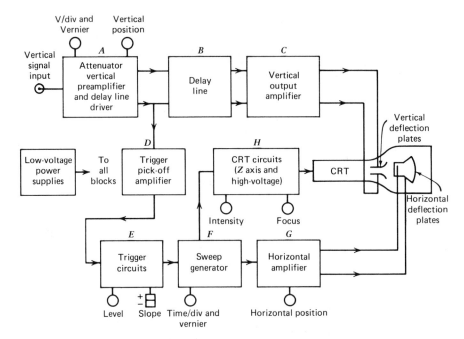

Figure 8.4-1 Block diagram of a basic triggered-sweep oscilloscope.

after being delayed by the delay line. The trigger point of this waveform arrives at the vertical deflection plates at, or slightly after, the start of the sweep since the delay line delays the signal for a time equal to or greater than the trigger delay. Hence, on the screen there appears a trace which represents the input-signal-voltage change on the vertical axis, and the time relationships of the various events of the signal shown as a horizontal distance on the CRT display.

Many oscilloscopes contain circuits to vary the intensity through an external jack that is usually labeled Z axis, Z input, or intensity modulation. There are two methods for accomplishing Z-axis modulation: (1) The CRT cathode is held constant by a bias supply, and the Z-axis signal is applied to the control grid. (2) The negative grid bias is contant, and the Z-axis signal is applied to the cathode. In most oscilloscopes using a Z-axis input, it is merely necessary to reduce the intensity so that the beam is just barely extinguished. If this is done when the Z-axis signal is zero, then applying a signal will brighten the CRT trace.

Figure 8.4-2 shows the approximate time relationship of the signals at the outputs of each of the blocks in Fig. 8.4-1 to a vertical input signal. This time relationship involves quite a few time delays since all electronic circuits have a certain propagation time. The delays through each block, other than the delay line, are assumed to be the same, although in actuality, the delay may vary considerably from one block to the next. Actual delays in a typical oscilloscope may range from a few nanoseconds per block in faster oscilloscopes to a few hundred nanoseconds per block in slower oscilloscopes. Delay line delay times are usually longer than delays for other blocks and might range from about 100 ns or less to nearly 1 μs. Faster oscilloscopes usually have the shortest delay line times. The effects of the delay line are usually visible only on the fastest sweep speeds. Waveforms 1 through 4 show the successive delays that occur between the vertical input and the CRT. Since the trigger signal is passed through some amplification on its way to the trigger circuit, some delay also occurs between the input and the trigger circuit. This delay can be seen by comparing waveforms 1 and 5. Additional delays occur between the triggering signal, waveform 5, and the start of the sweep, waveform 7. Further delays of the unblanking signal and horizontal signal occur in the Z-axis circuit, waveform 9, and horizontal amplifier, waveform 10, respectively. The delay established by the delay line, waveform 3, must be just slightly greater than the total of the delays shown in waveforms 1, 2, 5, 6, 7, and 10. This causes the sweep to be unblanked and moving across the screen by the time the part of the vertical signal that triggered the sweep deflects the beam vertically. In other words, the triggering event can be viewed on screen. The most important time relations to remember are waveforms 1, 4, and 10.

The delay lines in scopes should not be confused with delayed triggering, or delayed sweep, found in some models. The delayed trigger feature creates a small time delay between the beginning of a waveform and the initiation of a sweep cycle. The delayed sweep feature allows viewing of a small feature, or section, of a larger waveform. For example, we may see several cycles of a 1 MHz ($= 1/1$ µs) oscillation riding on the peak of a 10-kHz ($= 1/100$ µs) sine wave, as shown in Fig. 8.4-3. It can be observed that the 1-MHz oscillation begins at approximately 20 µs past the trigger point. If the trigger delay control is adjusted for 20 µs, and the sweep set for 1 µs/div, then the screen will display only the 1-MHz signal rather than the 10-kHz wave.

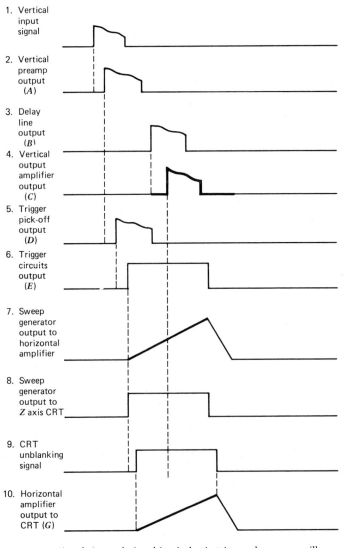

1. Vertical input signal
2. Vertical preamp output (*A*)
3. Delay line output (*B*)
4. Vertical output amplifier output (*C*)
5. Trigger pick-off output (*D*)
6. Trigger circuits output (*E*)
7. Sweep generator output to horizontal amplifier
8. Sweep generator output to *Z* axis CRT
9. CRT unblanking signal
10. Horizontal amplifier output to CRT (*G*)

Figure 8.4-2 Signal time relationships in basic triggered-sweep oscilloscopes.

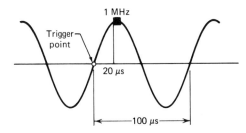

Figure 8.4-3 Using delayed sweep allows one to see a small oscillation riding on a larger waveform. For example, one may see several cycles of a 1-MHz oscillation riding on the peak of a 10-kHz sine wave. In order to view the 1-MHz signal, set the horizontal sweep to show several cycles of the 1-MHz waveform, and then trigger the sweep on the 10-kHz signal.

The main controls are shown in Fig. 8.4-1. The attenuator and the vertical amplifier system enables signals, ranging from a few millivolts to several hundred volts, to be displayed on the screen. The V/div (= volts/division) and vernier controls select vertical deflection factor and calibrator signal. The time/div and vernier controls select sweep rate and external horizontal input. The slope control determines whether sweep is triggered on + or − slope of trigger signal. The level control selects a point on trigger signal at which sweep is triggered. The focus and intensity controls permit a well-focused display of convenient brightness to be set on the screen.

8.5   BASIC OSCILLOSCOPE SPECIFICATIONS

Vertical Sensitivity

The vertical sensitivity of an oscilloscope is the smallest deflection factor on the attenuator knob, indicating a measure of the deflection for a specified input signal. If, for example, the most-sensitive position of the vertical attenuator is 2 mV/div, then the vertical sensitivity of that oscilloscope is 2 mV/div. The CRT screen has a grid pattern called a graticule. Each division of the graticule is between 0.75 and 1.3 cm apart.

Bandwidth

An oscilloscope with a dc-coupled amplifier in the vertical channel may have a bandwidth or frequency response from dc to some specified upper frequency limit, such as 10 MHz, indicated in Fig. 8.5-1. The bandwidth is defined as the frequency where the amplitude of the output voltage is −3 dB with respect to the reference (0 dB). In order to improve the same amplifier's bandwidth, the amplification at 10 MHz may be somewhat boosted (e.g., by RC networks in feedback loops). This results in an extra gain of +3 dB at 10 MHz, while at the same time the −3 dB level is shifted to 50 MHz, as indicated by the line of dashes in Fig. 8.5-1. But the extraamplification at the higher frequencies results because the display of a pulse shows an overshoot due to high-frequency peaking.

A low-frequency scope usually has a vertical amplifier with a bandwidth of dc to 10 MHz, while a high-frequency scope may go as high as 350 MHz.

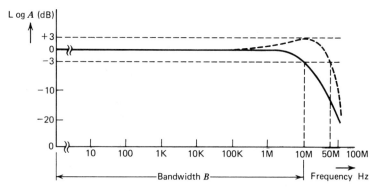

Figure 8.5-1 Frequency characteristic of a dc-coupled low-pass, broad-band amplifier, showing the frequency response from dc to some specified upper frequency limit.

Rise Time

With high-frequency oscilloscopes often intended for pulse measurements, the rise time (t_r) capability is often specified. The rise time of a pulse is defined as the time required for its leading edge to rise from 10% of its final amplitude to 90%. For a dc-coupled or low-pass RC circuit, the rise time equals $2.2RC$, and the upper 3-dB frequency equals the bandwidth (BW), which is given by

$$BW = \frac{1}{2\pi RC} = \frac{2.2}{2\pi t_r} = \frac{0.35}{t_r},$$

or

$$BW = 0.35/t_r, \qquad\qquad (8.5\text{-}1)$$

where BW = bandwidth in megahertz,
t_r = rise time in microseconds.

A high-frequency oscilloscope with a rise time of 10 ns has a bandwidth of 35 MHz.

Horizontal Sweep Rate

Most oscilloscopes calibrate the horizontal section in units of time per division. Suppose a scope with 10 horizontal graticule divisions is set to sweep at 0.1 μs/div. Then the sweep time across the CRT screen is

$$\frac{0.1\ \mu s}{\text{div}} \times 10\ \text{div} = 1\ \mu s.$$

Using this information, we can determine the approximate period or frequency of displayed waveforms.

8.6 DUAL-TRACE OSCILLOSCOPES

Today most oscilloscopes have the feature of dual-trace displays, which enables the operator to perform time and amplitude comparisons between two waveforms. To obtain the two traces on the screen, we may use one of the two techniques: (1) a single beam is subjected to two channel signals by means of electronic switching (dual trace); (2) two beams are provided to display one channel signal each (dual beam). Since the construction of dual-beam and split-beam CRTs is more expensive, the technique of dual trace is usually employed.

With the dual-trace oscilloscope shown in Fig. 8.6-1, a single-gun CRT with one pair of vertical plates is used. The final vertical amplifier, providing the deflection voltage to the plates, is alternately connected to the two input channels by an electronic switch. The electronic switch is operated either by a free-running multivibrator or by a pulse coming from the time base circuits, respectively, in the chopped mode or in the alternate mode.

The alternate mode displays one vertical channel for a full sweep and the other channel on the next sweep. The result is that the output of each vertical channel is alternately displayed. The sweep speed above which the mode is switched from chopped to alternate is about 0.1 ms/div. The alternate mode is commonly used for viewing high-frequency signals, where sweep speeds (faster than 0.1 ms/div) are much faster than CRT phosphor decay time, but the chopped mode of operation at high repetition rates is objectionable since the chopper frequency might interfere with the signal frequency. When low-frequency signals are to be viewed, requiring a slow sweep (from 10 to 0.1 ms/div), the alternate mode of operation is objectionable due to the extreme flicker. This is overcome by utilizing the chopped mode. In this mode, a small time segment of the sweep will be allotted to one vertical channel, with

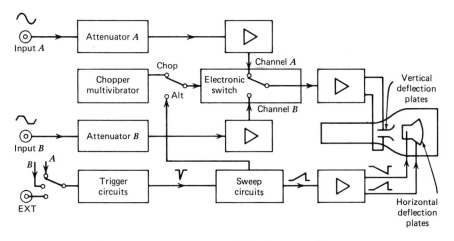

Figure 8.6-1 Block diagram of a dual-trace oscilloscope.

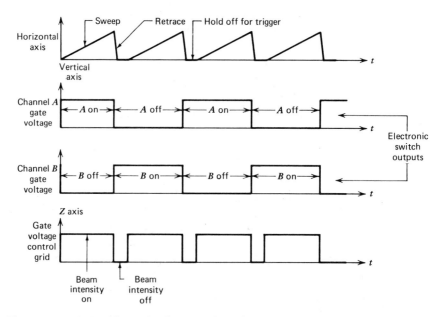

Figure 8.6-2 Timing relationship for a two-channel vertical amplifier system in alternate mode operation.

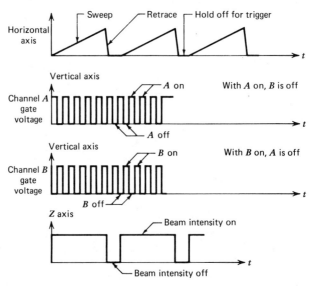

Figure 8.6-3 Timing relationship for a two-channel vertical amplifier system in chopped mode operation.

the next time segment allotted to the second channel. The result is that each vertical channel is composed of small chopped segments, which merge to appear continuous to the eye.

In the alternate mode, the electronic switch alternates between channel amplifiers every sweep period, while in the chopped mode, the electronic switch oscillates between channels at a fixed rate of approximately 100 kHz. Figures 8.6-2 and 8.6-3 show timing relationships for a dual channel vertical amplifier system in alternate and chopped modes of operation.

If a dual-trace oscilloscope is used to display the two sinusoidal signals simultaneously so that one signal is used for the EXT sync input, the two waveforms will appear in proper time perspective and the oscilloscope can be used to measure the amount of time between the start of one cycle of each of the waveforms. This amount of time can be used to calculate the phase angle between the two signals.

8.7   STORAGE CRT AND DIGITAL STORAGE OSCILLOSCOPES

A storage oscilloscope is a scope equipped with a storage feature. This scope is employed in those applications where the display time at the screen is too short to examine the signals to be measured. There are two basic types of storage oscilloscope, one type using a special storage CRT and the other using digital storage techniques.

Storage Cathode Ray Tubes

If a single-shot signal is applied to the vertical input, only one sweep is generated by the sweep circuit. During this sweep the CRT screen is excited by a high-energy electron beam. At the end of the single sweep the beam is suppressed and a phosphorescence remains for some time. The time that the phosphorescence remains visible is called the persistence of the tube and depends upon the type of phosphor used. The persistence time of a storage CRT can be extended in excess of minutes. The storage principle is based on the phenomenon of secondary electron emission. The composition used for the storage layer possesses not only good secondary emission but also good insulation against cross migration of electric charge of the particles loaded by secondary emission. When the high-energy electron beam hits the layer of particles of the composition, more secondary electrons are emitted from the particles than are captured from the primary electrons that have hit them. Hence the bombarded particles become positively charged, and so the pattern written by the high-energy beam remains as a charge pattern at the layer. This charge pattern is later employed to visualize the pattern again. This application is the purpose of storage CRTs, most of which are bistable tubes and mesh tubes.

In the bistable storage CRT of Fig. 8.7-1, the phosphor used for the screen is applied in scattered particles, so that cross migration of the stored charge

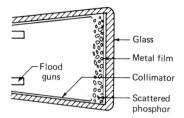

Glass

Metal film

Flood
guns

Collimator

Scattered
phosphor

Figure 8.7-1 Construction of the bistable storage
CRT.

is avoided. The layer of scattered phosphor may be more than one particle
thick to provide a homogeneous display. Between the phosphor layer and
glass surface, there is a transparent metal layer, which acts as a controlling
electrode. The writing beam hits the particles with such an energy that a
charge pattern is caused. In order to visualize this pattern again after writing,
a cloud of low-energy electrons is sent toward the screen from a pair of flood
guns in the tube. The energy of the flood-gun electrons is sufficient to illuminate the phosphor on those parts of the screen where the particles have
a positive charge. But the low-energy electrons do not have enough energy
to illuminate the phosphor on all the other parts of the screen. These electrons are repelled and should be collected by the metal film in the cone of
the collimator, which is also used to shape flood beam.

The metal film backplate of the bistable tube can be split into upper and
lower halves. For comparison purposes a waveform can be stored in the
upper half of the screen, while the lower half can be used in the normal
mode for the operator's measurements. With the bistable storage CRT, no
halftones are possible.

The mesh storage CRT is widely used. The mesh tube makes use of a
dielectric material deposited on a storage mesh as a target. The mesh is
located between the deflection plates and the normal standard phosphor
screen of the CRT, about 6 mm in front of the screen when viewed from
the deflection plates. Positively charged areas of the storage target allow
electrons from the flood guns to pass through the mesh toward the standard
phosphor, thereby reproducing the stored replica. As shown in Fig. 8.7-2,
the mesh tube includes the writing system and the flood system. The writing
system is the same as that of a conventional CRT. The flood system contains
a pair of flood guns operated in parallel, both having a cathode K, a control
grid G_1, and an accelerator grid G_2. Common to both flood guns are the
flood-beam collimator G_7, the collector mesh G_8, the storage mesh G_9 carrying the storage layer, and the phosphor viewing screen G_{10} with its alu-
minized layer. A cloud of electrons is emitted by each flood-gun cathode.
These clouds are combined, shaped, and accelerated by the two control
grids, G_1 and G_2, and by the collimator G_7, which consists of a coating on
the inside of the tube. The positive collimator voltage is adjusted such that
the flood-gun electron cloud just fills the CRT viewing screen. The cloud is
further accelerated toward the storage mesh and viewing screen by the voltage of collector mesh G_8. After passing through the collecter mesh, the flood

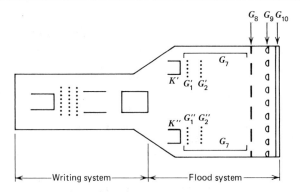

Figure 8.7-2 The construction of a mesh storage CRT.

electrons are further controlled by the voltages of the storage mesh and storage layer. Since the cathode side of the storage mesh is coated with the nonconductive storage material, only a capacitive coupling exists between the storage layer and the storage mesh. This capacitive coupling is essential for the storage and erase functions. The rest potential of the storage mesh is about $+1$ V with respect to the flood-gun cathodes. The potential of the storage layer varies from 0 V to negative in the write and erase routines. The postaccelerator voltage (about 6 kV with respect to the storage mesh) is connected to the transparent aluminizing layer of the phosphor viewing screen. The flood electrons passing the storage mesh are accelerated and strike the phosphor, causing it to emit light. When the storage layer is at negative potential, the number of electrons passing the storage mesh is reduced considerably.

Digital Storage Oscilloscope

A digital storage oscilloscope is shown in Figure 8.7-3, where an analog-to-digital (A/D) converter is used to digitize the input waveform. It will sample the waveform at many points, and then convert the instantaneous amplitude at each point to a binary number value proportional to the amplitude. The

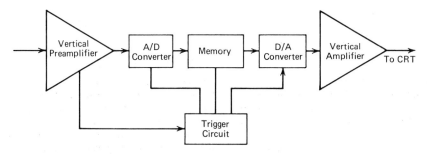

Figure 8.7-3 Block diagram of digital storage oscilloscope.

binary numbers are then stored in memory. A digital-to-analog converter at the memory output converts the binary words back to analog voltages, which are capable of driving the vertical output amplifier. Because the memory is scanned many times per second, the CRT screen is constantly being refreshed by the data stored in memory before it can fade out. The digital storage technique is very expensive for some high-frequency oscilloscopes designed as "transient catchers." However, the digital storage technique does find extensive use in medical and physiological oscilloscopes, where the sweep time is typically 25 mm/s.

8.8   SAMPLING OSCILLOSCOPES

The vertical amplifier bandwidths of the costliest conventional oscilloscopes extend to only 250 MHz or so. Similarly, sweep speeds only go up to possibly 0.1 μs/div. Hence, it is hardly possible to view pulse waveforms in the over-500-MHz region except using a sampling oscilloscope.

The sampling scope permits measuring of signals in gigahertz (10^9 Hz) range. The input waveform is not continuously measured, but discrete samples of signal amplitude at related intervals are measured and synthetically reproduced into a complete signal. The signal must be repetitive (not necessarily periodic) with thousands of samples required in order for the signal to appear continuous on the CRT trace.

Figure 8.8-1 shows a simple view of sampling and the waveform recon-

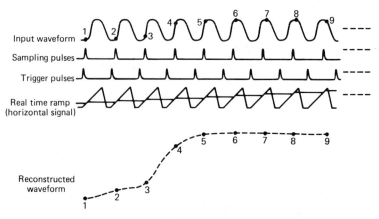

Figure 8.8-1 Waveforms illustrating the operation of a sampling oscilloscope. The horizontal displacement of the beam is synchronized with the trigger pulses, which also determine the moment of sampling. A staircase generator produces the horizontal staircase signal (V_{hs}). It is triggered by a voltage comparator, which compares the ramp voltage to a fraction of V_{hs}. When both the voltage values are equal, the generator is permitted to advance one step, and simultaneously a sampling pulse is applied to the vertical input sampling gate; at this moment, a sample of the vertical input voltage is taken.

structed. The input signal is too fast (or its frequency is too high) to be displayed directly on the CRT screen. But the input repetitive waveform can be reconstructed from many amplitude samples that are taken from the different pulses in the wave train. In reconstructing the waveform, the sampling pulse turns the sampling circuit on for an extremely short time interval. The waveform voltage at that instant is measured. The CRT spot is then positioned vertically to the corresponding voltage input. The next sample is taken from the successive pulse in the wave train at a slightly later position. The CRT spot is moved horizontally over a very short distance and is re-positioned vertically to the new value of the input voltage. In this way the scope plots the waveform point by point, using 1000 samples or more to reconstruct the original waveform. The sample frequency may be as low as one-hundredth of the input periodic signal frequency. If the input signal has a frequency of 1000 MHz, the required bandwidth of the amplifier would be only 10 MHz, a very reasonable figure.

8.9   THE PRACTICAL CIRCUIT OF A LOW-COST FREE-RUNNING OSCILLOSCOPE

This section discusses the whole circuit for the vertical deflection system, horizontal channel, and the cathode-ray tube circuit in a low-cost oscilloscope with free running sweep.

Vertical Deflection System

The vertical amplifier is shown in Fig. 8.9-1. It consists of a source follower (SF, Q_1), a first paraphase amplifier ($Q_2 - Q_3$), an output paraphase amplifier ($Q_4 - Q_5$), and two diodes (D_1, D_2) used to protect the input FET. The vertical input circuit contains an ac/dc switch, an attenuator, and a T-type resistor network. Across the T-network output resistor is a capacitor (C_1) connected for high-frequency compensation. If a bipolar transistor (BJT) is operated in the active region for amplification, the emitter-to-base junction should be forward biased, and the collector-to-base junction reverse biased. These requirements must be satisfied for the two stages of the dc-coupled paraphase amplifier ($Q_2 - Q_5$). The n-channel FET source follower is used for impedance matching. Its small negative gate bias is supplied through the gate-to-source network.

The paraphase amplifier produces a push–pull output signal. A single signal from the source follower (Q_1) must be phase inverted through the grounded-base transistor Q_3 (note the base capacitor C_2 connected to ground), so that the push–pull signal output can be obtained. The bias voltages applied to the paraphase circuit should be balanced. Hence, a dc balance control is used at the base of Q_3. At each paraphase amplifier stage, there are a pair of the same emitter-biasing resistors and a pair of the same collector resistors. At the first paraphase stage, a vertical gain control is in the emitter circuit, and a vertical position control is between the pair of collector resistors. When an ac signal path is traced, all the bias-voltage source lines must be regarded as at ac ground potential. Therefore, a push–pull signal from a paraphase amplifier output is clearly a double of the single collector output amplitude. (The two collector output signals are connected in series aiding.)

The equivalent circuit of a dc-coupled amplifier can be simplified as a low-pass RC circuit, where C is the effective base-shunt capacitance. In order to obtain high-frequency compensation, the emitter resistor (1K) at the output stage is shunted by a 270-pF capacitor (C_3), so that the 1K resistor is by-passed at high-frequencies. But at low frequencies, the by-pass effect is negligible. This causes the emitter resistor to produce negative feedback, resulting in a broader bandwidth.

The vertical deflection system has a bandwidth of dc to 1 MHz, a vertical sensitivity of 20 mV/div (1 div = 6 mm), and an input impedance of 1 MΩ shunted by 40 pF. The maximum vertical input can be 600 V through a proper attenuator. The vertical amplifier output signal is not only applied between the two vertical deflection plates of the CRT, but also partially used as a synchronous signal applied to the sweep generator input.

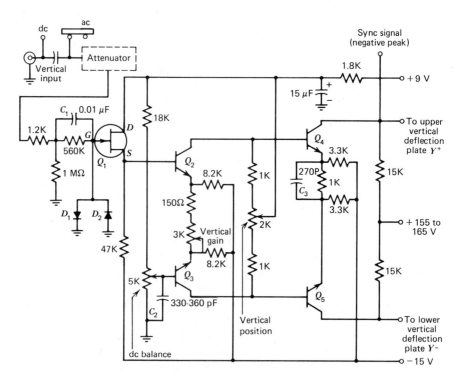

Figure 8.9-1   Vertical amplifier circuit for a free-running scope.

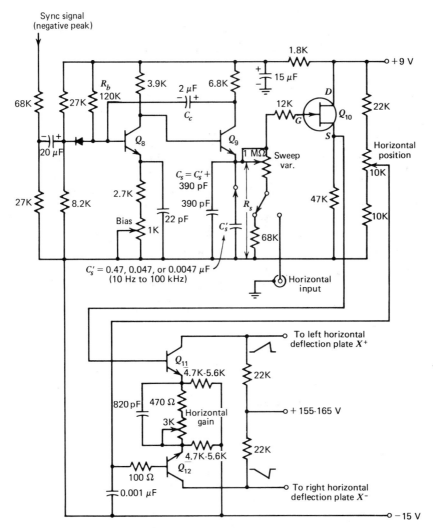

Figure 8.9-2   Free-running mode horizontal circuit.

Horizontal Channel

The horizontal channel shown in Fig. 8.9-2 consists of a sweep generator ($Q_8 - Q_9$), a source follower (Q_{10}), and a paraphase amplifier ($Q_{11} - Q_{12}$). The negative peak of the sync signal taken from the vertical amplifier output is applied to the sweep generator input.

The sweep generator is a modified run-down-type multivibrator. It produces a run-down waveform, that is, a negative going sweep. When no sync signal is applied, the generator operates to cause the free-running sweep. This initially unsynchronized generator falls into synchronization shortly after the application of a synchronizing signal. When the bias and sweep controls are properly adjusted, the sweep capacitor (C_s) is charged until Q_9 is saturated while Q_8 at cutoff. Q_8 is maintained at cutoff by the negative voltage of capacitor C_c (which was previously charged due to the regenerative process) until C_c discharges through R_b toward the collector-supply voltage. When the junction of C_c and R_b reaches a slight positive voltage, however, Q_8 begins to conduct while the current through Q_9 begins to decrease from saturation and the sweep capacitor C_s begins to discharge through the sweep resistor R_s. Since the regenerative process reverses, Q_8 then becomes saturated while Q_9 is off and C_s discharges continuously until the base of Q_9 is slightly positive. C_s charges fast and discharges slowly, resulting in a repetitively negative going sweep. The discharging period is the sweep or trace time, and the duration from starting charge to starting discharge is the flyback or retrace time.

The run-down waveform from the sweep generator goes to source follower Q_{10} which operates for impedance match. To obtain a push–pull signal from the output paraphase amplifier, a single signal from the source follower is phase inverted by means of the effective base-grounded transistor Q_{12}. Across the emitter circuit of the output stage is a 820-pF capacitor connected

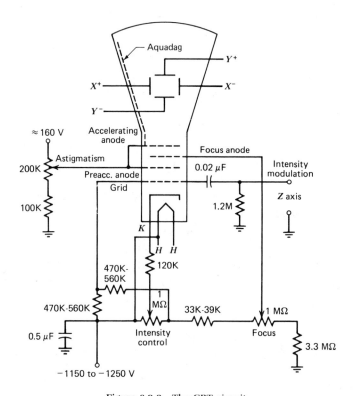

Figure 8.9-3 The CRT circuit.

for high-frequency compensation. The horizontal position control is at the Q_{12} base circuit.

The horizontal channel has a bandwidth of dc to 250 kHz, a horizontal sensitivity of 300 mV/div, and an input impedance of 1 MΩ shunted by 40 pF. The sweep control dial is calibrated from 10 Hz to 100 kHz in several ranges.

CRT Circuit

The cathode-ray-tube circuit is shown in Fig. 8.9-3. Accelerating voltage is about 1200 V. Z-axis modulation is over 20 V peak-to-peak.

The focus control adjusts the size and "roundness" of the CRT beam. The astigmatism control varies the voltage to focus the spot in both dimension simultaneously. It is used to ensure the "roundness" of the beam across the entire width of the CRT. For the Z-axis modulation, a third signal is applied to the control grid (used as Z axis) so that the beam is modulated. By means of Z modulation, a time interval can be indicated.

BIBLIOGRAPHY

Cameron, D., *Advanced Oscilloscope Handbook*. Reston Publishing, Reston, Virginia, 1977.
Erk, R. E., *Oscilloscopes*. McGraw-Hill, New York, 1978.
Herrick, C. N., *Oscilloscope Handbook*. Reston Publishing, Reston, Virginia, 1974.

Questions

8-1 What are the advantages of the oscilloscope over the most other instruments?

8-2 Name two different types of CRT, classified by the deflection method used.

8-3 List the main components of an electrostatic CRT.

8-4 Describe the construction of an electron gun.

8-5 The mesh in a CRT provides _____ .

8-6 What is mean by "synchronization" in a free-running oscilloscope?

8-7 A _____ sweep oscilloscope holds off the beginning of a sweep until the input signal begins.

8-8 What is the reason for using a delay line in the vertical deflection system of a triggered sweep oscilloscope?

8-9 What advantages are given by delay sweep?

8-10 Define vertical sensitivity in your own words.

8-11 Define vertical bandwidth.

8-12 Define the rise time of a pulse.

8-13   Electronic _____ is employed to create a dual-trace oscilloscope, even though the CRT has but one electron gun.

8-14   What are the differences between the chop and alternate modes of a dual-trace oscilloscope?

8-15   A flood gun is used in _____ CRTs.

8-16   Describe how a digital storage oscilloscope works.

8-17   Describe the paraphase amplifier.

8-18   In a free-running oscilloscope the horizontal deflection system is driven by a _____ waveform.

8-19   Explain the high-frequency compensation for the output horizontal stage shown in Fig. 8.9-2.

8-20   Describe the Z-axis modulation.

8-21   What is the difference between focus and astigmatism?

Problems

8.2-1   A CRT has a deflection factor of 16 V/cm. Calculate the amount of deflection seen on the screen for deflection voltages of 32 and 96 V.
Answer. 2 cm; 6 cm.

8.2-2   The deflection sensitivity of a CRT is given as 0.2 mm/V. What voltage should be applied to the deflection plates to obtain a deflection of 2 in.?
Answer. 254 V.

8.2-3   If a deflection of 5 cm is obtained from a deflection voltage of 50 V between the plates of the CRT, (a) calculate the deflection factor of the tube. (b) How much will the beam be deflected for a voltage of 80 V for the same tube?
Answer. (a) 10 V/cm; (b) 8 cm.

8.3-1   A sinusoidal waveform at 5 kHz is applied to the vertical input of a free-running oscilloscope. (a) The horizontal sweep speed is set so that a full cycle takes 0.4 msec. Show the resulting display for one sweep of beam. (b) Repeat part (a) for a sweep frequency of 10 kHz.
Answer. (a) Two cycles of the input signal; (b) one-half cycle.

8.5-1   A sinusoidal waveform is observed on an oscilloscope as having a peak-to-peak amplitude of 6 cm. If the vertical sensitivity setting is 5 V/cm, calculate the peak-to-peak and the rms values of the voltage.
Answer. $V_{p-p} = 15$ V; $V_{rms} = 10.6$ V.

8.5-2   A pulse is measured as having a peak amplitude of 18 V. If the vertical dial setting was 10 V/cm, how many centimeters of signal amplitude were observed (peak-to-peak)?
Answer. 3.6 cm.

8.5-3   Two pulse signals are observed on a dual-trace oscilloscope using a sweep scale setting of 5 ms/cm. If the pulse widths are measured as 3.6 and 4.4 cm, respectively, and the distance between the start of each pulse is 2.5 cm, calculate the time measurements for the pulse widths and delay between pulses.

Answer.   18 ms, 22 ms, 12.5 ms.

8.5-4   A square-wave signal is observed on the oscilloscope to have one cycle measured as 6 cm at a scale setting of 30 s/cm. Find the signal frequency.

Answer.   5.55 kHz.

8.5-5   An amplifier bandwidth is improved by an *RC* network in the feedback loop. Draw the amplifier circuit.

8.5-6   Calculate the bandwidth in hertz required of an oscilloscope vertical amplifier if the rise time of the input pulse is 20 ns.

8.6-1   In measuring phase shift between two sinusoidal signals on a dual-trace oscilloscope the scale setting is adjusted so that one full cycle is 10 boxes. Find the scale factor (degrees/box) for this adjustment and the amount of phase shift for a reading of 0.8 boxes.

Answer.   360°/box; 28.8°.

8.6-2   One full cycle is set to 8 boxes. The phase displacement is measured as 0.5 boxes. Find (a) the scale factor and phase shift in degrees and (b) the number of boxes to observe for a phase shift of 45°.

Answer.   (a) 22.5°; (b) 1 box.

9
Solid-State Electronic Voltmeters and Multimeters

9.1 BJT VOLTMETER BRIDGE CIRCUITS

Basic TVM Bridge Circuit

The basic transistor voltmeter (TVM) arrangement is analogous to the basic vacuum-tube voltmeter (VTVM) circuit. A basic BJT voltmeter bridge circuit is shown in Fig. 9.1-1. This bridge with substantial negative feedback minimizes drift due to variations in β, I_{CO}, V_{CC}, and V_{EE}. Variations of β and I_{CO} usually result from ambient temperature change, and variations of V_{CC} and V_{EE} result from power-supply fluctuation. Since the variations of the four parameters are the same on both transistors, the bridge remains balanced. When a positive voltage under test is applied to the base of Q_1, the emitter current I_{E1} is increased and I_{E2} is decreased due to the increased voltage drop across R_E. Thus, the bridge is imbalanced, and the current flows through meter movement. Maintenance of calibration accuracy requires a substantial amount of negative feedback provided by R_E and R_N in addition to a stable balance condition. This can be explained as follows.

Suppose R_E and R_N are replaced by short circuits and the TVM remains operative. If the ambient temperature increases, the β of Q_1 and Q_2 increase. When a positive voltage is applied to the base of Q_1, its collector current is greater than at the original temperature and the meter reads higher. Thus, R_E and R_N are needed in the circuit to provide the negative feedback for Q_1 and Q_2. Since the amount of negative feedback is mostly given by R_E, the value of R_E should be very high. Then the TVM operates effectively from a constant-current source provided that R_E is sufficiently high in value.

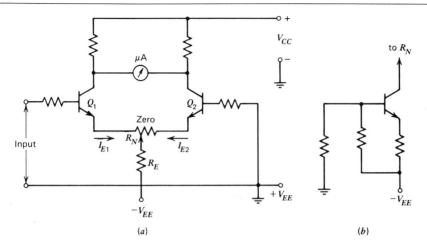

Figure 9.1-1 (a) A basic TVM bridge circuit. (b) Constant current source instead of R_E in (a).

In a more-elaborate arrangement, R_E may be replaced by a transistor used as a constant current device (Fig. 9.1-1b). The ac resistance of a constant-current device (source) is ideally infinite and practically from 100K to about 1 MΩ.

Typical BJT Voltmeter Bridge Circuit

The range of input voltages for a TVM can easily be extended by an input attenuator or range switch, as shown in Fig. 9.1-2a. The unknown dc input voltage is applied through a large resistor in the probe body to a resistive voltage divider. The main disadvantage of the TVM bridge shown in Fig. 9.1-1a is its comparatively low input resistance. If the high-resistance input voltage divider is directly connected to the low-input-resistance terminal of the basic TVM bridge, then the base of Q_1 will present a large load to the high-resistance voltage attenuator. In order to solve this problem, two em-itter followers are employed as the bridge preamplifier transistor, as shown in Fig. 9.1-2b. The emitter followers Q_3 and Q_4 provide very high input resistances and thus present a minimum load to the high-resistance input voltage divider. In addition, Q_3 and Q_4 provide very low output resistances and thus suitably drive the bases of the bridge transistors Q_1 and Q_2, re-spectively. Because of the unbypassed emitter resistors of Q_1 and Q_2, their input resistances are very high, thus preventing loading of the Q_3 and Q_4 emitters. The output of the TVM bridge is indicated on the 200-μA meter movement. Note the zero control for operation, the internal adjustments for calibration, and the bypass capacitors C_1 and C_2 for preventing ac signals from reaching the amplifier and affecting the meter reading.

Note also that the bridge preamplifier stages Q_3 and Q_4 are used merely for the impedance match, which involves characteristics of the approximate transistor equivalent circuit. Any resistor in the emitter leg will appear as a much larger resistance in the base circuit owing to the factor $1 + h_{fe}$ ($\approx h_{fe}$

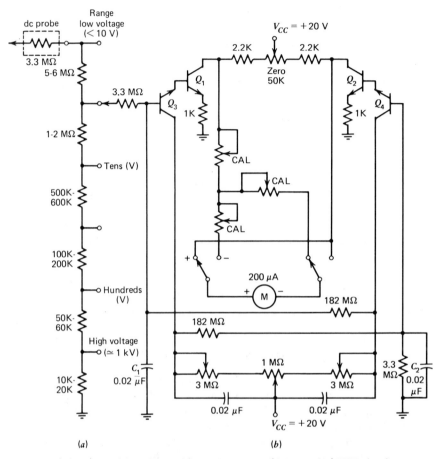

Figure 9.1-2 (a) An input voltage attenuator. (b) A practical TVM circuit.

$= \beta$). The resistance reflected from the base circuit to the emitter is normally small in magnitude owing to division by the factor $1 + h_{fe}$ ($\approx h_{fe}$). The emitter current equals the base current multiplied by the factor $1 + h_{fe}$ ($\approx h_{fe}$). An emitter follower has an input resistance of hundreds of kilohms (very high) and an output resistance of tens of ohms (very low).

A typical TVM usually has an ohmmeter function, which is accomplished when a battery voltage is applied across a voltage divider composed of the unknown resistance and a known resistance.

9.2 FET VOLTMETER BRIDGES

The FET Voltmeter (VM) on DCV Function

A high input resistance of a TVM comparable to that of a VTVM can be obtained by employing two field-effect transistor (FETs) as bridge-circuit components. Figure 9.2-1 shows a typical FET VM for dcV and acV measurements. On the dcV function, an input resistance of 15 MΩ is provided on all ranges. The basic bridge circuit is made up of Q_1, Q_2, R_3, and R_4. Since this FET bridge is also a difference amplifier, control voltages are applied to the gates of both Q_1 and Q_2, and the meter reading is proportional to differences between these gate voltages. The zero adjust is an operating control and set to make zero meter reading when no voltage is applied to the input of Q_1. The dc balance control is a maintenance control and adjusted to compensate for production tolerances and component aging. When a voltage under test is applied to the input of Q_1, the unbalanced current flows through the meter movement and the pointer deflection are proportional to the value of the voltage drop. Diode D_7 operates in the Zener mode to protect Q_1 against overload damage. Diode D_5 provides temperature compensation for maintenance of calibration accuracy over a wide temperature range. D_3 and D_4 provide protection against meter overload. A dc calibration control

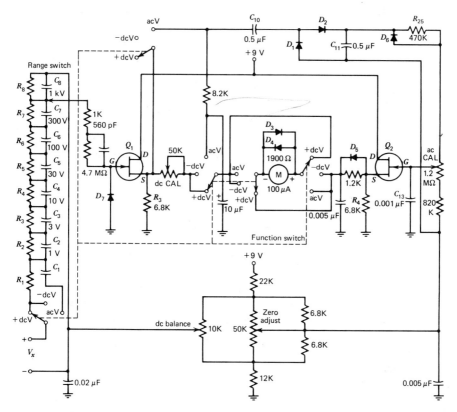

Note: $R_1 = 5$ MΩ, $R_2 = 6.67$ MΩ, $R_3 = 2.33$ MΩ, $R_4 = 667$K, $R_5 = 233$K, $R_6 = 66.7$K, $R_7 = 23.3$K, $R_8 = 10$K. $C_1 = 0.05$ μF, $C_2 = 20$ pF, $C_3 = 47$ pF, $C_4 = 120$ pF, $C_5 = 560$ pF, $C_6 = 2100$ pF, $C_7 = 5600$ pF, $C_8 = 0.015$ μF.

Figure 9.2-1 Simplified bridge circuits of FET VM on dcv and acv functions (adapted from Sencore Model FE14 FET meter). (Courtesy of Sencore, Inc.)

is provided in series with the lead between the source electrodes. In the FET difference amplifier, Q_2 conducts less if Q_1 conducts more owing to a positive voltage impressed on its gate. Since the gate of Q_2 draws negligible current, the ac calibration control has no effect on the dcV function.

The FET VM on acV Function

The attenuator (multiplier) in Fig. 9.2-1 is a resistance voltage divider. Precision resistors are used for R_1 through R_8, and most of these are of the wire-wound variety. Such resistors have significant inductances, so if the meter is to be used on ac, some type of compensation is required. This compensation takes the form of capacitors C_2 through C_8. The proper capacitor values may be determined by adjusting them with a square-wave input, until best "squareness" is viewed on an oscilloscope.

The unknown ac voltage V_{xac} is measured when the TVM in Fig. 9.2-1 is on acV function. V_{xac} is applied to Q_1, which operates as an electronic impedance transformer to drive the peak-to-peak–type detector that is composed of D_1, D_2, C_{10}, and C_{11}. The positive dc output voltage from the detector is applied to the gate of Q_2 through a voltage divider consisting of R_{25}, R_{26}, and R_{27}. Thus Q_2 conducts more, whereas Q_1 conducts less. The unbalanced current then flows through the meter movement. On the acV function of the TVM, diode D_6 provides temperature compensation, and Q_1 maintains its high input impedance. The practical input impedance is 10 MΩ shunted by 29 pF. The frequency response is flat from about 25 Hz to 1 MHz. The rated accuracy on the acV function is $\pm 5\%$ of full scale, compared with $\pm 3\%$ of full scale on the dcV function. Peak-to-peak scales are provided in addition to the rms scale.

9.3   MOSFET AC VOLTMETER

The ac TVM circuit shown in Fig. 9.3-1 is made up of four MOSFET stages. The circuit is designed as usual: that is, the first and final stages Q_1 and Q_4 are employed for matching impedance, and thus operated as source followers; the middle stages Q_2 and Q_3 are operated as common-source amplifiers, which provide the required voltage gain. The source follower Q_1 presents a very low input capacitance to the conventional 1-MΩ input-signal voltage divider. With this stage operating at a drain current of only 230 μA and a drain-to-source voltage of 0.5 V, the effective input capacitance is only 0.5 pF. The source of Q_1 is coupled to the gate of Q_2 by a 0.33-μF ceramic capacitor. Q_2 provides a voltage gain from 16 to 20, and also has a quiescent drain current of 230 μA. Stage Q_3 is similar to Q_2 except that an unbypassed 1000-Ω potentiometer is added in series with the bypassed 10K source resistance. This potentiometer, providing negative feedback, can be adjusted to vary the voltage gain of Q_3 from 10 to 20. With a 10-mV signal at the input of Q_1, the maximum output-signal voltage at the drain of Q_3 is about 2.8 V rms. The source follower Q_4 provides the necessary impedance trans-

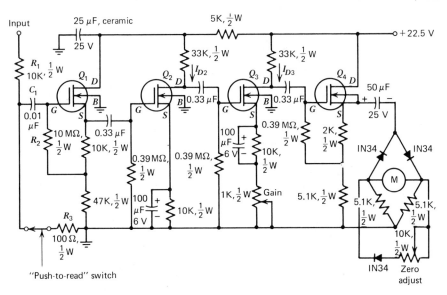

Figure 9.3-1   MOSFET ac voltmeter circuit (Courtesy RCA Solid State.)

formation between the high output impedance (about 30,000 Ω) of Q_3 and low impedance of the meter–rectifier bridge circuit. In the meter–rectifier circuit, two diodes are operated as a full-wave rectifier, and a third diode is used in conjunction with a potentiometer to compensate for the nonlinear rectification characteristic of the rectifier diodes at the low end of the meter scale. A "push-to-read" switch controls the 100-to-1 voltage divider ($R_1 : R_3$) placed ahead of the input-coupling capacitor of Q_1. When the range switch is in the 10-mV position (on the lowest range), this 100-to-1 attenuation network is necessary to protect the gate of Q_1 from overload in the event that an excessively large signal is accidentally applied to the input terminals.

This MOSFET ac voltmeter has an input impedance of 1 MΩ, a full-scale sensitivity of 10 mV on the lowest range, a flat frequency response over the audio range of 20–20,000 Hz, and a total current drain of only 2.5 mA, which permits fully portable operation.

9.4   MOSFET ELECTROMETERS

An electrometer is a very high-input-impedance voltmeter. The ultimate in voltmeter design requires that the input impedance be infinite for each scale so that the response of a circuit is not changed when the meter is introduced. The very high input impedance of the MOSFET amplifier is employed toward this end in the MOSFET electrometer. An electrometer has its ability to measure potentials that are present across extremely high impedances such

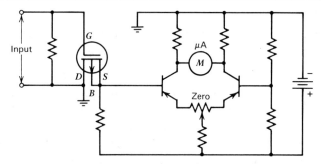

Figure 9.4-1 An electrometer consisting of a MOSFET source follower and a BJT differential stage.

as crystals, photocells, multiplier tubes, and counters. Thus voltages generated in solutions, ion chambers, and other places with very high impedances and extremely low current capabilities can be measured by employing the electrometers. Sensitivites of MOSFET electrometers can be made better than 0.25 μA/mV, and input-impedance figures of over 10^{15} Ω are obtainable. An electrometer shown in Fig. 9.4-1 comprises a MOSFET source follower and a BJT differential stage. This circuit can be designed into a highly stable and sensitive electrometer.

A more stable and sensitive electrometer circuit is shown in Fig. 9.4-2. The circuit is composed of two emitter followers Q_2 and Q_5 in addition to

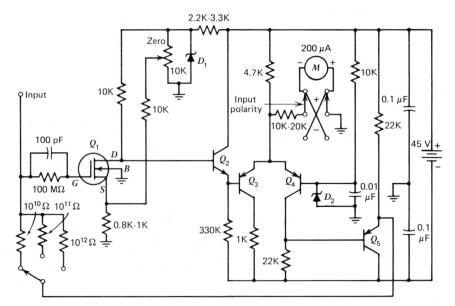

Figure 9.4-2 An electrometer composed of two emitter followers as the second and final stages in addition to MOSFET and BJT differential amplifier stages.

MOSFET and BJT differential amplifier stages (Q_1 and Q_3–Q_4). The second stage Q_2 is operated for matching impedance between Q_1 and Q_3. The negative feedback voltage is delivered from the collector of Q_4 and returned to the input terminal through the high-ohm resistor (10^{10}–10^{12} Ω) and the final stage Q_5, which provides a high input resistance and thus prevents loading of the Q_4 collector. The differential amplifier stage Q_3–Q_4 is used as a means of stabilizing the output and input impedances of the dc amplifier, as well as for reducing the equivalent input drift. Diodes D_1 and D_2 operate in the Zener mode to regulate the V_{GS} bias for Q_1 and base voltage of Q_4, respectively.

Note that the high-ohm resistors at the input terminal should be supported by Teflon or another excellent insulator so that the input impedance of the electrometer can be maintained at a very high value. This supporting insulator should be kept clean and dry.

9.5   OP-AMP MILLIVOLTMETERS

dc Millivoltmeter

Since an inverting op-amp has a very low output resistance and its voltage gain with feedback depends only on the resistor values, it is suitable for use as the basic amplifier in a dc millivoltmeter, as shown in Fig. 9.5-1. Scale factors of the millivoltmeter depend only on the resistor values. The meter reads millivolts of signal V_i at the circuit input. Since $R_s \ll R_f$,

$$I_0 = I_s - I = \frac{V_s}{10} - \frac{V_s}{100K} \simeq \frac{V_s}{10},$$

or

$$I_0 = \frac{V_s}{R_s} = \frac{IR_f}{R_s} = \frac{V_i}{R_i}\left(\frac{R_f}{R_s}\right).$$

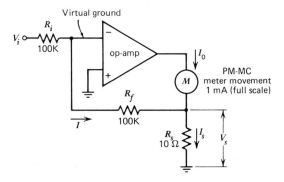

Figure 9.5-1   Dc millivoltmeter using inverting op-amp.

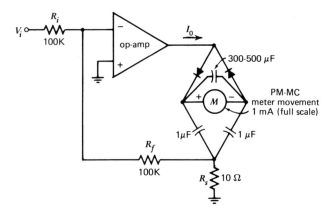

Figure 9.5-2   Ac millivoltmeter using inverting op-amp.

Thus, the circuit transfer function is

$$\frac{I_0}{V_i} = \frac{R_f}{R_i}\left(\frac{1}{R_s}\right) = \frac{100\text{K}}{100\text{K}} \times \frac{1}{10} = \frac{1\text{ mA}}{10\text{ mV}}.$$

Hence, an input of 10 mV will result in a current through the meter of 1 mA. An input of 5 mV will result in a current of 0.5 mA, which is half-scale deflection of the pointer.

ac Millivoltmeter

An op-amp ac millivoltmeter circuit is shown in Fig. 9.5-2. The circuit transfer function for an ac signal is

$$\frac{I_o}{V_i} = \frac{R_f}{R_i}\left(\frac{1}{R_s}\right) = \frac{100\text{K}}{100\text{K}} \times \frac{1}{10} = \frac{1\text{ mA}}{10\text{ mV}},$$

which shows that the meter indication provides a full-scale deflection for an ac input voltage of 10 mV.

9.6   CHOPPER-TYPE ELECTRONIC DC VOLTMETER

Direct-coupled amplifiers have a certain inherent drift and tend to be noisy. For example, a 50-μV/°C drift figure in an $\times 100$ amplifier produces an output of 5 mV/°C, which is not too important. But in an $\times 100{,}000$ amplifier, the output is 5 V/°C; this amount of drift will probably obscure any real signals in a very short time.

High-sensitivity dc voltmeters often employ the chopper-type dc amplifier to avoid the drift problems usually associated with direct-coupled dc amplifiers. A chopper is an electronic switch used to turn a dc signal on and off at an ac rate. In the chopper amplifier the dc input voltage is converted into an ac voltage, amplified by an ac amplifier, and then converted back into a dc voltage, proportional to the original input signal. The ac chopping signal comes from a relaxation oscillator.

Figure 9.6-1 shows the simplified circuit of a dc voltmeter with the chopper amplifier. Photoconductors (such as *CdS*) are used as choppers for modulation and demodulation. Modulation means conversion from dc to ac. Demodulation means conversion from ac back into dc. A photoconductor has a low resistance, from a few hundred to a few thousand ohms, when it is illuminated by a neon lamp. The photoconductor resistance increases sharply, usually to several megohms, when it is darkened. The first two

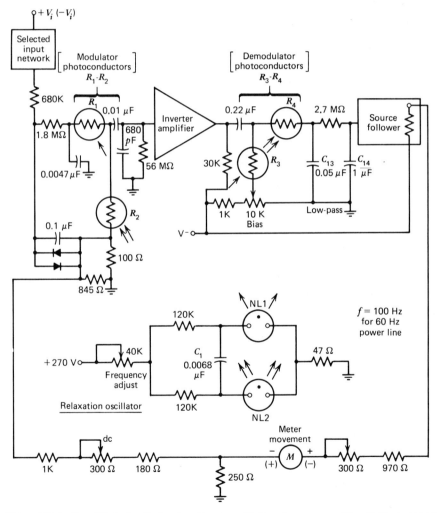

Figure 9.6-1 Simplified circuit of a dc voltmeter with the chopper amplifier. (Referring to HP 410C electronic voltmeter.)

photoconductors in the amplifier input circuit form a series-shunt half-wave modulator–chopper. The other two photoconductors in the amplifier output circuit form a shunt-series demodulator–chopper. A relaxation oscillator drives two neon lamps into illumination on alternate half-cyles of oscillation. Each neon lamp illuminates one photoconductor in the modulator and one in the demodulator. The two photoconductors of the modulator act like a switch across the input of the amplifier, alternately opening and closing at a rate determined by the frequency of the relaxation oscillator.

The dc signal through the input network is applied to the modulator, which converts the dc to ac for amplification. The input to the amplifier is a non-sinusoidal ac voltage with an amplitude proportional to the level of the input dc voltage and a frequency equal to the oscillator frequency. This frequency is limited to a few hundred Hertz since the transition time between the high- and low-resistance states of the photoconductors limits the chopping rate. As the photoconductors are alternately illuminated by the neons, the input voltage and feedback voltage are alternately applied to the ac amplifier. The amplitude of the resultant signal to the amplifier is the voltage difference between the input and feedback voltages. The amplified signal to the input of the demodulator is 180° out of phase with the output of the modulator. An approximately 150-mV square wave is usually applied to the demodulator from the amplifier. Since the same neon lamps illuminate both the modulator and demodulator photoconductors (R_1–R_4), operation of the two choppers is synchronous. Therefore, when R_1 is sampling the input voltage, R_3 is clamping the amplified and inverted difference voltage to ground. Alternately, when R_2 is sampling the feedback voltage, R_4 is charging capacitors C_{13} and C_{14} to the peak value of the square wave. These capacitors maintain this charge as long as the input voltage remains constant since they have no discharge path and they are being repetitively recharged by the demodulator. Therefore, a dc potential, proportional to the difference between the input and feedback voltages, is applied to the gate of the source follower and subsequently to the meter circuit. A portion of the meter circuit voltage is feed back to the modulator. The meter stabilizes when the feedback voltage and input voltage are nearly equal. In a practical case, the input network applies approximately 15 mV dc for full-scale meter deflection to the modulator–chopper; meanwhile, applied to the opposite side of the chopper is the amplifier feedback voltage, which is of the same polarity and approximately 5 μV lower in amplitude than the input voltage. The input impedance of the chopper–amplifier dc voltmeter is usually in the order of 10 MΩ or more.

The operation of the relaxation oscillator shown in Fig. 9.6-1 can be explained as follows. The two neon lamps cannot start to fire at the same time. Suppose lamp NL_1 fires first. Then C_1 charges until NL_2 fires. As soon as NL_2 fires, NL_1 extinguishes. The reason is that the negative charge is delivered from C_1 toward the +270 V source so that the voltage across NL_1 is decreased to the extinguishing level. Now C_1 is charged in the other direction until NL_1 fires again. The circuit is thus oscillating. The oscillator frequency is normally set to 100 Hz for operation from the 60-Hz power line. This frequency is selected so that it is not harmonically related to the power-line frequency, precluding possible beat indications on the meter.

A pair of CMOS or JFET electronic switches driven out of phase with each other will serve as a chopper. Some monolithic or hybrid function module chopper amplifiers use a varactor switching bridge for the chopper. The chopper amplifier is used not only to avoid drift problems, but also to limit noise.

9.7 SOLID-STATE ELECTRONIC MULTIMETER

Comparision between EMM and VOM

The chief advantages of the electronic multimeter (EMM) over the volt–ohm–milliammeter (VOM) are the following. The input resistance of the EMM is usually very high and constant over all voltage ranges, thus lessening the problem of loading for most voltage-measuring situations. On current-measuring functions, the resistance of the EMMs is usually quite low, thus loading effects are again avoided for most current-measuring conditions. Resistance scales on the EMM deflect in the same direction as the rest of the scales, thereby avoiding confusion. In addition, a lower voltage is used in an actual resistance-measuring process and thereby allows one to measure the junction resistance of a BJT without damaging the transistor.

The electronic voltmeter (EVM) can have an input resistance as low as 10 MΩ or as high as 100 MΩ, and the input resistance is the same for all voltage ranges. On the other hand, the input resistance of the VOM is different for all voltage ranges. On low-voltage ranges, the resistance of the VOM tends to be low. In the case of the 20,000 Ω/V meter with a 0–1-V range, the input resistance is only (20,000 Ω/V) (1 V) = 20K.

The solid-state EVMs, however, cannot be used in the presence of strong electric or electromagnetic fields such as those produced by TV flyback transformers, radio transmitters, etc. The field will tend to bias the transistors or integrated circuits used in the EVM to a point where they will not operate properly. The VOM on dc function, on the other hand, is practically immune to such effects.

Analog Electronic Multimeter

The basic analog (continuous) electronic multimeter can be subdivided into three main blocks: the measuring networks, the amplifier circuits, and the analog meter movement (such as a PM–MC type). In the case of voltage and current measurements the measuring network is a voltage divider that limits the voltage to the amplifier and essentially sets the instrument range.

The Philips PM2505 electronic (VAΩ) multimeter is shown in Fig. 9.7-1. It measures full-scale dc and ac voltages down to 100 mV. The dc and ac

Figure 9.7-1 Philips PM2505 electronic multimeter. (Courtesy of Phillips Test and Measuring Instruments.)

current ranges are from 1 μA full scale to 10 A full scale. The ohmmeter ranges are from 100 Ω to 30 MΩ (FSD). A function switch provides selection for volts, ampere, and ohm ranges. This multimeter is designed around a monolithic IC amplifier with FET input, giving a high input impedance (10–20 MΩ), which eliminates possible errors caused by loading the test circuit.

EMM on dcV Function

EVMs use an analog meter movement that is commonly driven by an electronic balance circuit, such as those shown in Figs. 9.1-2, 9.2-1, or 9.7-2.

The circuit in Fig. 9.7-2 employs a differential amplifier composed of transistors Q_2 and Q_3 to form a balanced bridge circuit. JFET Q_1 operates as a source follower and is used to provide impedance transformation between the input and the base of Q_2. The constant current source $I_s = I_2 + I_3 = K$. When the unknown voltage V_x is zero, $I_2 = I_3$, $V_{E2} = V_{E3}$ and so the current through the meter movement (M) is zero: $I_m = 0$. This condition is maintained by the fixed bias on Q_3 and the bias on Q_2, which is a function of the voltage drop across R_s. When an unknown positive voltage V_x is applied, the bias on Q_2 base increases, so that V_{E2} increases to a value greater than V_{E3}, and the current I_m rises; hence the deflection of the movement is proportional to V_x. A compensated attenuator is used on acV function since precision resistors are used for the attenuator, and most of these are of the wire-wound variety. Such resistors have significant inductances; the effect of these inductances can be balanced out by the shunting capacitors.

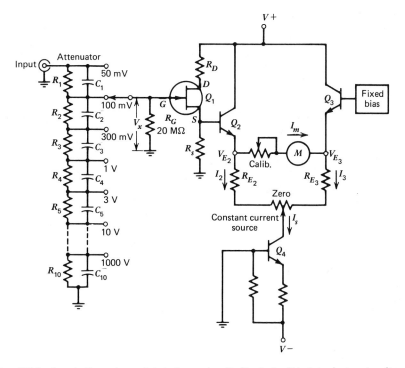

Figure 9.7-2 Input attenuator and dc balance circuit of typical solid-state electronic voltmeter.

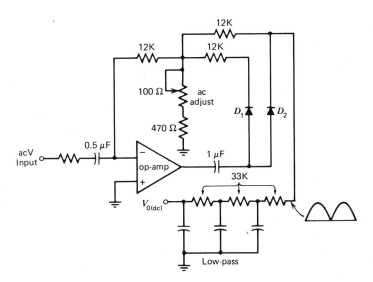

Figure 9.7-3 Precision-rectifier and averager circuit.

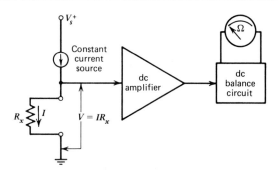

Figure 9.7-4   Electronic ohmmeter circuit in a solid-state VOM.

EMM on ACV Function

The basic EVM circuit such as that in Fig. 9.7-2 is for use with dc voltages only. To accomodate ac voltage some sort of ac-to-dc converter must be added.

A rectifier-averager circuit is shown in Fig. 9.7-3. The op-amp circuit serves as a precision rectifier. The nonlinear characteristics of the p-n junction diodes D_1 and D_2 in the forward direction are "servoed out" by the op-amp and its negative-feedback loop. The pulsating dc is smoothed out (averaged) by the low-pass filter. The smoothed dc output from the filter $[V_{0(dc)} \propto V_{i(rms)}]$ is applied to an analog balance circuit such as Fig. 9.7-2, or a dc digital voltmeter. Most ac voltmeters are calibrated in terms of volts rms, yet they will not actually read the rms values unless the input signal has a pure sine-wave shape.

EMM on Ohm Function

If the current through an unknown resistor is kept constant, then the value of the voltage drop across the resistor will give the data needed to calculate the resistance value; $R_x = V/I = V/k \ \Omega$. Accordingly, an electronic ohmmeter circuit can be formed as shown in Fig. 9.7-4. The output current of the constant-current source and the voltage gain of the dc amplifier are set by the range selector switch (not shown) so that full-scale resistances from milliohms to megohms can be accommodated. Ordinary ohmmeters using a battery of 1.5 V or more will forward bias p-n junctions when these instruments are used in solid-state circuits, whereas the circuit of Fig. 9.7-4 using low-voltage levels will not usually forward bias p-n junctions when such electronic ohmmeters are used in solid-state circuits. Some EVM manufacturers include a high-power ohmmeter scale so that a quick test for diodes and transistors can be made.

9.8 DIGITAL MULTIMETERS

Introduction

A typical digital multimeter (DMM) is shown in Fig. 9.8-1. It displays a measured value in terms of discrete numerals, rather than as a pointer deflection on a continuous scale as used in analog instruments. The DMM is becoming increasingly more popular as the prices of these instruments become more competitive. It has some important advantages over analog multimeters. The direct numerical readout reduces the human error and the tedium involved in long periods of measurement. The increase of speed in the measurement and the elimination of the parallax error are additional advantages. Features such as autoranging and automatic polarity further reduce measurement error and any possible instrument damage through overload or reversed polarity. In some cases even hard copy in the form of printed cards or punched tape is also available. However, the digital multimeters are as of now still limited in that nonlinear parameters cannot be measured. Furthermore, for comparable accuracy they are at this time not competitively priced.

A digital voltmeter (DVM) employs an analog-to-digital converter (ADC) to convert the dc input to a binary-coded-decimal (BCD) digital word that is used to drive a digital display device. Most digital voltmeters or digital multimeters employ a dual-slope integrator as their ADC circuit, since the dual-slope DVMs or DMMs offer relative immunity to noise riding on the input voltage, and relative immunity to error due to inaccuracy or long-term drift in the clock frequency, etc. The dual-slope waveform is a dual-ramp voltage.

Dual-Ramp Digital Voltmeter

The dual-ramp (or dual-slope) DVM system is shown in Fig. 9.8-2a, while associated waveforms are shown in Fig. 9.8-2b. The op-amp A_1 and R_1–C_1

Figure 9.8-1 Simpson 467 digital multimeter. (Courtesy of Simpson Electric Company.)

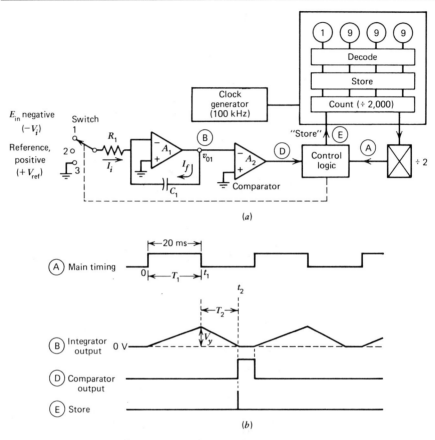

Figure 9.8-2 (a) System and (b) waveforms of the dual-ramp DVM. Note: the ÷ 2000 counter (for 1999 display) is further divided by two to drive the voltmeter and capacitance meter (which is not included in the system). (From Waddington, D. E. O'N., Wireless World, P. 157, April, 1973.)

combination make up an integrator, where

$$I_f = -I_i, \quad I_i = \frac{E_{in}}{R_1}, \quad \text{and} \quad I_f = C_1 \frac{dV_{01}}{dt}.$$

Thus,

$$\frac{C_1 \, dV_{01}}{dt} = \frac{-E_{in}}{R_1}. \tag{9.8-1}$$

Integrating both sides,

$$\int \frac{C_1 \, dV_{01}}{dt} \, dt = -\int \frac{E_{in}}{R_1} \, dt,$$

$$C_1 V_{01} = -\int \frac{E_{in}}{R_1} \, dt,$$

hence

$$V_{01} = \frac{-1}{R_1 C_1} \int_0^t E_{in}\, dt. \qquad (9.8\text{-}2)$$

Equation (9.8-2) is the transfer equation for the op-amp integrator circuit. The dual-ramp DVM system operates as follows.

At the beginning of the measurement cycle the capacitor C_1 in Fig. 9.8-2a is fully discharged. The input to the integrator is connected to the negative input voltage $(-V_i)$ so that the capacitor C_1 begins to charge due to current $-V_i/R_1$. While the integrator output V_{01} begins rising from zero, the counter starts to count the clock pulses from the 100-kHz clock generator. The charging is continued until the counter has counted 2000 (i.e., for 2 K/100 K or 20 ms). At the end of this period, the voltage V_y across this capacitor will be

$$V_y = \frac{-1}{R_1 C_1} \int_0^{t_1} (-V_i)\, dt = \frac{V_i T_1}{R_1 C_1}. \qquad (9.8\text{-}3)$$

Then the control logic section switches the input of the integrator to the positive reference voltage V_{ref} so that the capacitor discharges under the influence of current V_{ref}/R_1. Since the reference voltage magnitude is larger than the voltage to be measured, the charge on the capacitor decreases more rapidly than it is built up, and at a time t_2 it will be zero; that is,

$$0 = V_y - \frac{1}{R_1 C_1} \int_{t_1}^{t_2} V_{ref}\, dt = V_y - \frac{V_{ref} T_2}{R_1 C_1}$$

$$= \frac{V_i T_1}{R_1 C_1} - \frac{V_{ref} T_2}{R_1 C_1},$$

thus

$$V_i T_i = V_{ref} T_2,$$

or

$$V_i = \frac{T_2}{T_1} V_{ref}. \qquad (9.8\text{-}4)$$

The zero-voltage condition is sensed by the comparator, which causes the control logic to switch the input of the integrator to zero volts (ground

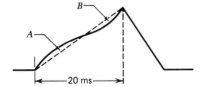

Figure 9.8-3 Integrator output when measuring a dc voltage with a superimposed 50- or 60-Hz signal. The area A cancels the area B.

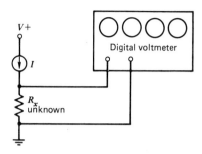

Figure 9.8-4   Resistance measurement by means of digital voltmeter.

potential), thus preventing any further change in the charge on the capacitor. At the same time, the control logic commands the counter to store the count. As indicated by Eq. (9.8-4), the time displayed gives a direct measure of the input voltage in terms of the reference voltage. Hence, the reference voltage can be chosen to give a suitable basic range for the digital voltmeter. For example, with a reference voltage of 2 V, the basic range will be 2 V, although it will only be possible to display 1.999 V. The counter continues counting until it reaches the all-zero state, when the measurement cycle is repeated. The choice of a measuring period of 20 ms gives good rejection of 50 or 60-Hz interference. Figure 9.8-3 shows the integrator output when measuring a dc voltage with a superimposed 50- or 60-Hz signal. Since the area A cancels the area B, the interfering signal is effectively rejected.

We may see terms such as "$3\frac{1}{2}$-digit" and "$4\frac{1}{2}$-digit" in the advertisements for DVM/DMM products. This term refers to the fact that the most significant digit can only be a "0" or a "1," while all other digits can be anything between "0" and "9." Such terminology indicates that the meter can read 100% overrange from its basic range. For example, a $3\frac{1}{2}$-digit voltmeter will read 0–1999 mV, while its basic range is only 0–999 mV. If this range is exceeded, then the "1" lights up; otherwise, it remains dark.

A DVM usually has an input resistance larger than 10 MΩ and an accuracy better than $\pm 0.2\%$ of reading.

Resistance Measurement

The resistance-measuring system is shown in Fig. 9.8-4. The method is to pass a known current through the unknown resistor and to measure the voltage drop across it. However, this system can only be used to measure the range from 100 Ω to 100K with adequate accuracies.

BIBLIOGRAPHY

Bell, D. A., Electronic Instrumentation and Measurements, Reston Publishing, Reston, Virginia, 1983, Chapters 8 and 10.

Carr, J. J., *Electronic Instrumentation and Measurement*. Reston Publishing, Reston, Virginia, 1979, Chapter 7.

Millman, J., *Microelectronics*. McGraw-Hill, New York, 1979, Chaps. 11 and 16.

Waddington, D. E. O'N., "Digital Multimeter, Part 1," *Wireless World*, 155–158 (April, 1973).

Questions

9-1   Why is the basic TVM bridge circuit such as Fig. 9.1-1*a* impractical?

9-2   How is a high-input resistance maintained in the circuit of Fig. 9.1-2?

9-3   Draw a simplified FET dc voltmeter circuit, referring to the FET bridge circuit shown in Fig. 9.2-1. Explain the controls, temperature compensation, and protective devices.

9-4   Draw a simplified FET ac voltmeter circuit, referring to the FET bridge circuit shown in Fig. 9.2-1. Explain the compensation attenuator and the peak-to-peak detector.

9-5   Explain design consideration to a MOSFET ac voltmeter such as Fig. 9.3-1. How is the impedance matched in its input and output sections?

9-6   What are the requirements of the electrometer circuit? How is this instrument maintained in good condition when it is not used?

9-7   Write the transfer function for an op-amp dc millivoltmeter.

9-8   How does an analog EVM differ from a VOM?

9-9   What technique is used to measure resistance in many EVMs?

9-10   What is the relationship between the input voltage (V_i) and the reference voltage (V_{ref}) in the dual-ramp digital voltmeter?

9-11   What is a "$3\frac{1}{2}$-digit" voltmeter? What is meant by a "$\frac{1}{2}$-digit"?

Problems

9.1-1   The transistor voltmeter bridge circuit shown in Fig. 9.1-1*a* is basically a differential amplifier. If R_E is replaced by a transistor circuit, then the bridge configuration becomes a differential amplifier with a constant-current source; one example of this arrangement is shown as in Fig. P9.1-1. calculate: (a) constant current I_E, (b) dc collector current I_C of Q_1 or Q_2; (c) dc collector voltage V_C of Q_1 or Q_2; (d) ac voltage gain A_v (neglecting $1/h_{oe}$); (e) input impedance R_i; and (f) derive an expression for the Q_3 circuit ac output impedance R_0 and calculate R_0.

Answer.   (a) $I_E = 4$ mA; (b) $I_C = 2$ mA; (c) $V_C = 7.2$ V; (d) $A_v =$

$$\dfrac{h_{fe}\,R_C}{2\,(h_{ie} + h_{fe}r_E)} = 30;\quad \text{(e)}\ R_i = 10.4\text{K};\quad \text{(f)}\ R_0 \simeq \dfrac{1}{h_{oe}} \times$$

$$\left(1 + \dfrac{h_{fe}\,R_E}{R_E + h_{ie} + R_1\|R_2}\right) = 3.84\ \text{M}\Omega.$$

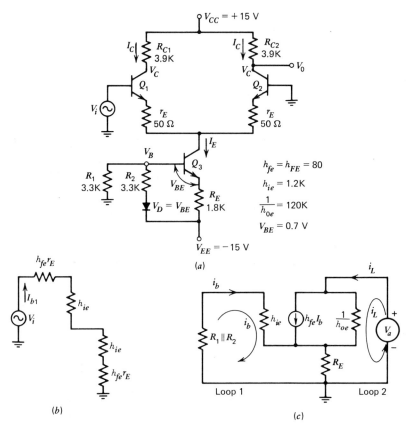

Figure P9.1-1   Difference amplifier with constant current source: (a) practical circuit; (b) input ac equivalent circuit; (c) Q_3 equivalent circuit.

9.1-2   In order to match the input impedance of an ac BJT TVM circuit, an emitter follower shown in Fig. P9.1-2 may be used as the preamplifier stage. Calculate: (a) input impedance Z_i; (b) output impedance

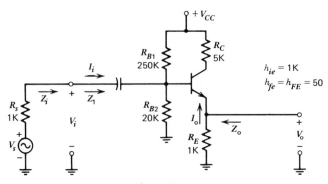

Figure P9.1-2

Z_0; (c) voltage gain $A_{v_1} = V_0/V_s$, and $A_{v_2} = V_0/V_i$; (d) current gain $A_I = I_0/I_i$.

Answer. (a) $Z_i = 14.5K$; (b) $Z_0 \simeq 40 \ \Omega$; (c) $A_{v_1} \simeq 0.98$, $A_{v_2} \simeq 1$; (d) $A_i = -14.9$.

9.2-1 The Q_1 stage of the FET differential amplifier shown in Fig. 9.2-1 is virtually operated as a source follower. Calculate A_v, V_0, R_i, and R_0 for the basic source follower shown in Fig. P9.2-1. *Hint:* Use

$$g_m = -\frac{2I_{DSS}}{V_P}\left(1 - \frac{V_{GS}}{V_P}\right) = g_{m0}\left(1 - \frac{V_{GS}}{V_P}\right),$$

$$A_V = \frac{V_0}{V_i} = \frac{g_m R_s}{1 + g_m R_s}, \quad \text{and} \quad R_0 \simeq \frac{1}{g_m}.$$

Answer. $A_v = 0.835$; $V_0 = 0.334$ V; $R_i = 1$ MΩ; $R_0 \simeq 435 \ \Omega$.

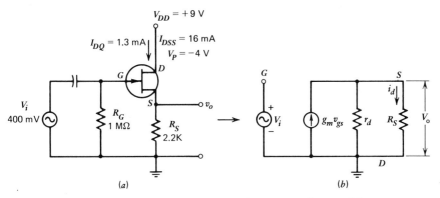

Figure P9.2-1 (a) Practical and (b) equivalent circuit of source follower.

9.2-2 Assume that the input voltage attenuator for the TVM of Fig. 9.2-1 is not connected to the dc balance control and the input gate circuit of Q_1. Calculate the voltage that can be applied to the gate circuit of Q_1 when a voltage of (a) 1000 V, (b) 300 V, or (c) 3 V is to be measured. *Answer.* (a), (b) 0.667 V; (c) 0.664 V.

9.3-1 In the MOSFET ac voltmeter circuit of Fig. 9.3-1, the gain control potentiometer is set to 0 Ω so that Q_2 is similar to Q_3, and $I_{D_2} = I_{D_3} \simeq 230$ μA. When a signal of 10 mV is applied to the input terminals of the voltmeter, the maximum output-signal voltage at the drain of Q_3 is 2.8 V rms. Assume that the voltage gain of Q_1 is about 1. Calculate (a) the gate-to-source bias voltage of Q_2 or Q_3; (b) the voltage gain of Q_2 or Q_3; (c) the input impedance of Q_1. *Answer.* (a) $V_{GS_2} = V_{GS_3} = -2.3$ V; (b) $A_{v_2} = A_{v_3} = -16.7$; (c) $R_i = 10.047$ MΩ.

9.3-2   Figure P9.3-2 shows a low-frequency small-signal *CS* amplifier with
a depletion MOSFET. Derive the expression for (a) transconductance
(g_m) and (b) voltage gain (A_v). (c) Find the values of A_v and V_0 if
$R_{S1} = 0$.

Answer.   (b) $A_v = \dfrac{-g_m R_D}{1 + g_m R_{S1}}$; (c) $A_v = -3.38$; $V_0 = -50.7$ mV.

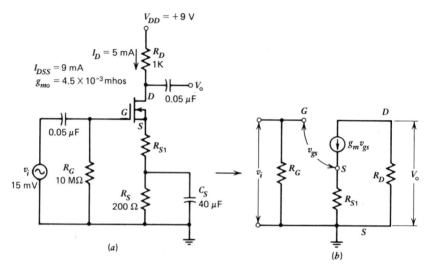

Figure P9.3-2

9.3-3   (a) For the circuit of Fig. P9.3-3 calculate the output voltage (V_0)
developed across a load resistance $R_L = 10K$. (b) Repeat if an output
resistance, $r_0 = 50K$, of the FET is considered.

Answer.   (a) $V_0 = -233.3$ mV; (b) $V_0 = -225.7$ mV.

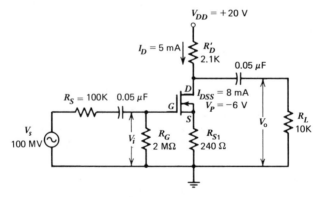

Figure P9.3-3

9.4-1   Assume in Fig. 9.4-2, the emitter of Q_2 disconnected from the Q_3 base. If each of Q_2 and Q_5 has $h_{FE} = 80$, what are the approximate input-resistance values of Q_2 and Q_5?

   Answer.   26.73 MΩ; 1.78 MΩ.

9.5-1   Suppose R_f in Fig. 9.5-1 is changed to (a) 200K, (b) 250K, and (c) 400K. Calculate and explain the resulting circuit scale factor in each case.

   Answer.   (a) $I_0/V_i = 1$ mA/5 mV; (b) $I_0/V_i = 1$ mA/4 mV.

9.6-1   Suppose each of the two neon lamps in Fig. 9.6-1 has a firing voltage of 50 V, an operating voltage of 45 V, and an extinguishing voltage of 40 V, and the potentiometer in the oscillator circuit is set to maximum value. (a) Find the approximate value of operating current of each neon lamp. (b) Explain the oscillator operation.

   Answer.   (a) 1.41 mA.

9.7-1   A 100-μA constant current is passed through an unknown resistance R_x. An EVM with its input resistance R_m is connected across R_x and reads 800 mV. (a) Find R_x if $R_m = 10$ mΩ. (b) Find R_x if R_m is infinite.

   Answer.   (a) 8.008K; (b) 8K.

9.7-2   A 50-μA constant current is passed through a 1-MΩ resistor. An EVM with its input resistance R_m is connected across the resistor. (a) Find the voltage reading V_x if $R_m = 15$ MΩ. (b) Find V_x if R_m is infinite.

   Answer.   (a) 46.88 V; (b) 50 V.

9.8-1   An operational amplifier integrator uses a 0.69 μF capacitor and an 820K. Assume a 100-mV dc level is applied to the input for 4 s. Calculate the output voltage.

   Answer. -0.707 V.

10

Oscillators
and Signal Generators

10.1 AUDIO GENERATORS

Introduction

The oscillator is the heart of most signal generators. Audio generators usually cover the range from 20 Hz to 20 kHz or more. They always produce sine waves, and most also produce square waves from the sine waves. These instruments typically use either output metering, or calibrated output attenuators, and produce controllable signal levels down to approximately -40 or -50 VU (volume unit) (1 VU $=$ 10 log P). The VU reference (zero VU) is a 1-mW power dissipated in a 600-Ω resistive load. Audio generators are typically rated to have a 600-Ω output impedance. Figure 10.1-1 shows the block diagram of a typical audio-signal generator.

Sinusoidal Generators

If the input to a linear network is sinusoidal, the output is also sinusoidal of the same frequency, although the magnitude and phase may be varied. Consequently, if a linear feedback amplifier (without input-signal excitation) oscillates, the output waveform is sinusoidal.

In the block diagram of Fig. 10.1-2a the amplifier has transfer gain $A = x_0/x_i$ and the feedback network ($\beta = x_f/x_0$) is not yet connected to form a closed loop. The output of the feedback network is $x_f = \beta x_0 = A\beta x_i$, and the output of the mixing circuit is $x_f' = -x_f = -A\beta x_i$. The loop gain is

$$-\beta A = \frac{x_f'}{x_i} = \frac{-x_f}{x_i}. \tag{10.1-1}$$

250

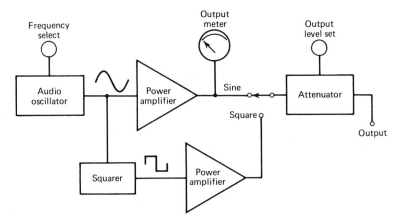

Figure 10.1-1   Block diagram of an audio-signal generator.

Assume that the signal x'_f is adjusted to be identically equal to the externally applied input signal x_i. Then, if the external source were removed and if terminal 2 were connected to terminal 1, the amplifier would continue to provide the same output signal x_0 as before. Note that the statement $x'_f = x_i$ means that the instantaneous values of x'_f and x_i are exactly equal at all times. The condition $x'_f = x_i$ is equivalent to $-A\beta = 1$.

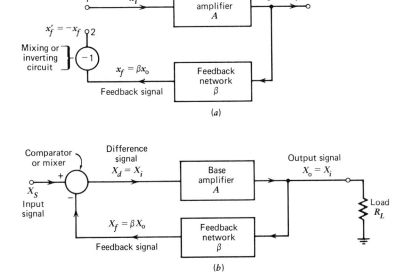

Figure 10.1-2   (a) Block diagram of an amplifier with its feedback network not yet connected to form a closed loop. (b) Block diagram of a single-loop feedback amplifier; the gain with feedback is $A_f = A/(1 + \beta A)$.

For a sinusoidal waveform the equation $x'_f = x_i$ is equivalent to the condition that the amplitude, phase, and frequency of x_i and x'_f must be identical. Consequently, the frequency of a sinusoidal oscillator is determined by the condition that the loop-gain phase shift is zero. Oscillations will not be sustained if, at the oscillator frequency, the magnitude of the product of the transfer gain of the amplifier and the magnitude of the feedback factor of the feedback network (the magnitude of the loop gain) are less than unity. The condition of unity loop gain $-A\beta = 1$ is called the Barkhausen criterion. This condition implies both that $|A\beta| = 1$ and that the phase of $-A\beta$ is zero. If $-\beta A = 1$, then $A_f = A/(1 + \beta A) = \infty$, which means that there exists an output voltage even in the absence of an externally applied signal voltage.

Referring to the feedback amplifier of Fig. 10.1-2b, it appears that if $|\beta A|$ at the oscillator frequency is precisely unity, then, with the feedback signal connected to the input terminals, the removal of the external source will make no difference. If $|\beta A|$ is less than unity, the removal of the external source will result in a cessation of oscillations. If $|\beta A|$ is larger than unity, the oscillation amplitude will continue to increase until it is limited by the onset of nonlinearity of operation in the active devices associated with the amplifier.

In reality, no input signal is needed to start the oscillator going. Only the loop gain $|\beta A| = 1$ must be satisfied for self-sustained oscillations to result. In practice, $|\beta A|$ is made larger than 1, and the system is started oscillating by amplifying noise voltage which is always present. Nonlinearity factors in the practical circuit provide an average value of $|\beta A|$ of 1. The resulting waveforms are never exactly sinusoidal. However, the closer the value of $|\beta A|$ is to exactly 1, the more nearly sinusoidal is the waveform.

Low Frequency Sine- and Cosine-Wave Generators

A quadrature oscillator is used to generate low-frequency sine and cosine waves with less distortion compared to the Wien-bridge oscillator. This sine- and cosine-wave generator operates on the basic principle of op-amp integrator.

The sine wave is expressed as

$$y = A \sin \omega_0 t. \tag{10.1-2}$$

Differentiating Eq. (10.1-2) with respect to time t, we obtain

$$y' = \omega_0 A \cos \omega_0 t \quad \text{or} \quad y'/\omega_0 = A \cos \omega_0 t. \tag{10.1-3}$$

Differentiating Eq. (10.1-3) gives

$$y'' = -\omega_0^2 y = -\omega_0^2 A \sin \omega_0 t. \tag{10.1-4}$$

Solving Eqs. (10.1-4) and (10.1-3) for y' and y we get Eqs. (10.1-3) and (10.1-2), respectively. To simulate these equations we may design a circuit as shown in Fig. 10.1-3, where there are two integrators and one inverter.

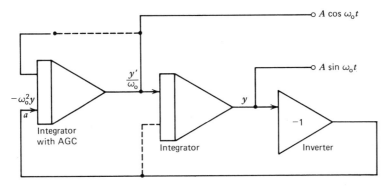

Figure 10.1-3 Sine- and cosine-wave oscillator.

The integrator in Fig. 10.1-3 is similar to that of Fig. 9.8-2a is redrawn in Fig. 10.1-4 as a basic integrator, where

$$i = -\frac{v_s}{R_1} = -C_1 \frac{dv_0}{dt} ,$$

and the output voltage is

$$v_0 = -\frac{1}{R_1 C_1} \int v_s \, dt. \tag{10.1-5}$$

Equation (10.1-5) shows that the output voltage of the integrator is the integral of the input, with an inversion and scale multiplier of $1/R_1 C_1$.

In order to sustain oscillations, the signal $-\omega_0^2 y$ must be applied to input a of the first integrator in Fig. 10.1-3, where the dotted lines indicate the return paths of the voltage stabilization network. The voltage amplitude can be automatically stabilized by an error signal resulting from a comparison between the output signal and a reference voltage, as shown in Fig. 10.1-5. The amplitude can also be stabilized by means of over excitation and nonlinearity. Although the comparison network is more complicated and expensive, it is usually preferred since it causes less distortion. In the circuit of Fig. 10.1-5, the multiplier has an output voltage equal to the product of the two input voltages. (See Section 10.7 and Fig. 10.7-1.) The comparison

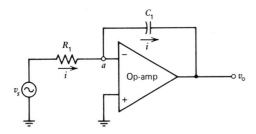

Figure 10.1-4 Op-amp integrator.

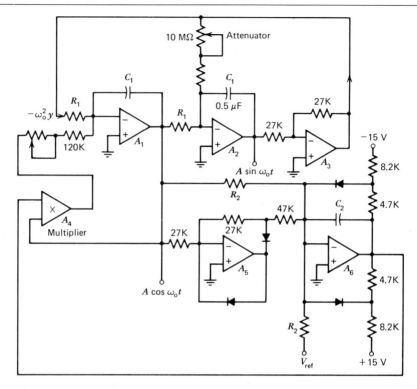

Figure 10.1-5 Sine- and cosine-wave oscillator with automatic gain control (AGC).

network is composed of A_5 and A_6. When the input signal of A_5 is positive, its output is negative, which is suitable to compare with the reference voltage.

From Eq. (10.1-5) we see that the oscillation frequency depends on the time constant $R_1 C_1$ and the voltage gain of the integrator.

10.2 WIEN-BRIDGE OSCILLATORS

Introduction

The Wien-bridge oscillator is one of the standard circuits used to generate sine-wave signals at frequencies up to about 1 MHz. This oscillator has a relatively pure waveform, and enjoys excellent frequency stability. It is essentially a feedback amplifier with a Wien bridge as the feedback network between the output and input terminals of the amplifier, as shown in Fig. 4.7-2. The amplifier will oscillate when the Barkhausen criterion ($-A\beta = 1$) has been satisfied, as described in Section 4.7.

A basic signal generator consists of an oscillator connected with an output meter and a calibrated attenuator so that a known output level is provided.

Four-Transistor Wien-Bridge Oscillator

A practical circuit of Wien bridge oscillator is shown in Fig. 10.2-1. The bridge consists of R_1 ($=R$) and C_1 ($=C$) in series, R_3 ($=R$) and C_3 ($=C$) in parallel, R_2 and R_4. Feedback is applied from the output of the circuit through the coupling capacitor C_4 and variable resistor R_5 to the top of the bridge. C_4 is large enough to introduce no phase shift at the lowest frequency of oscillation. R_5 is an adjustment for a suitable amount of feedback. Resistor R_4 serves the dual purpose of emitter resistor of Q_1 and element of the Wien bridge.

In a practical oscillator, the amplitude of oscillation is determined by the extent to which loop gain $\mid \beta A \mid$ is greater unity. If feedback factor is fixed, the amplitude is determined by A, increasing as A increases until further increase is limited by the nonlinear behavior of the transistor. Regulation of the amplitude is provided by resistor R_4, which provides a variable $\beta [= V_{OB}/V_{iB} = (V_{R_3} - V_{R_4})/V_{iB}]$. Resistor R_4 can be a tungsten-filament light bulb, acting as a variable resistance element. If the output of the amplifier tends to increase (decrease), the increased (decreased) current through R_4 raises (reduces) its temperature and increases (decreases) its resistance. Thus β would decrease (increase) and would tend to keep the product $A\beta$ constant, thereby regulating the amplifier output to a constant level. The thermal lag of the filament of the tungsten lamp causes its resistance to remain almost constant during the course of the alternating cycle of output voltage or current except at very low frequencies. Since the resistance of the lamp may be changed during the cycle at very low frequencies,

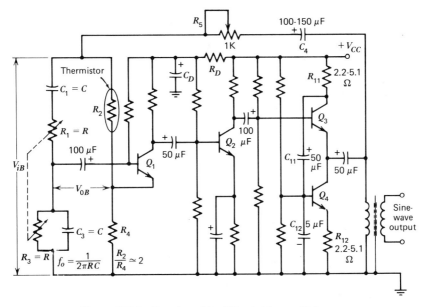

Figure 10.2-1 Four-transistor Wien-bridge oscillator.

the lamp may be substituted by a thermistor which has sufficient bulk to provide adequate thermal lag. However, since the thermistor has a negative temperature coefficient, it should be used as the bridge element R_2 rather than R_4. Although the tungsten lamp can be replaced by a sensistor as the element R_4 for amplitude stabilization over a range extended to very low frequencies, its positive temperature coefficient (40.7%/°C) in magnitude is small compared to that of the thermistor (-1 to -5%/°C). Note that the negative feedback path includes R_2 and R_4 in series with $R_2 = 2R_4$. The positive feedback path includes R_1 and C_1 in series, and R_3 and C_3 in parallel, with the oscilation frequency $f_0 = 1/2\pi RC$.

In the circuit of Fig. 10.2-1, the output stage is composed of two *npn* transistors Q_3 and Q_4 in cascode. Basically, Q_3 is an emitter follower with Q_4 in the emitter circuit as a control resistor. The two low-ohm resistors R_{11} and R_{12} (2.2–5.1 Ω, each) are operated for temperature compensation. When the currents through Q_3 and Q_4 increase with temperature, the voltage drops across R_{11} and R_{12} are increased. These will decrease the forward bias between the base and emitter of Q_4, thus decreasing the transistor currents to balance out the temperature effect.

Op-amp Wien-Bridge Oscillator

Figure 10.2-2 shows a circuit of an op-amp Wien-bridge oscillator with FET amplitude stabilization. A chief advantage of this circuit is that the traditional

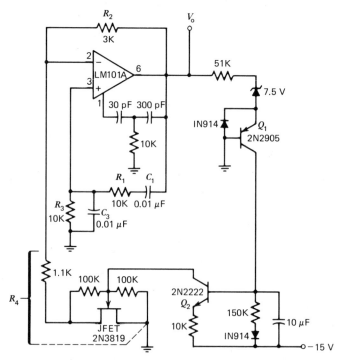

Figure 10.2-2 Op-amp Wien-bridge oscillator with FET amplitude stabilization. (From Hnatek, E. R., *Applications of Linear Integrated Circuits*, Wiley, New York, 1975.)

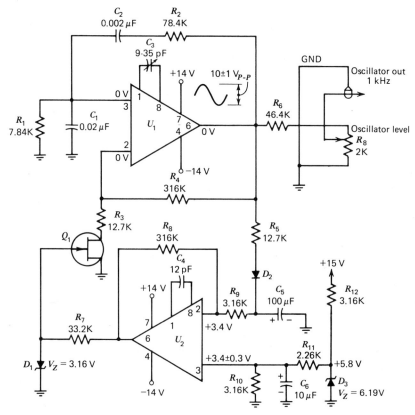

Figure 10.2-3 Typical circuit of the op-amp Wien-bridge oscillator contained in the Hewlett-Packard Model 4265B Universal Bridge. (Courtesy of Yokogawa-Hewlett-Packard, Ltd.)

tungsten-filament-lamp amplitude regulator is eliminated along with its time constant and linearity problems. In addition, the reliability problems associated with a lamp are eliminated. When the Wien-bridge network is used as a positive-feedback element around the operational amplifier, oscillation occurs at the frequency at which the phase shift is zero. Additional negative feedback is provided to set loop gain to unity at the oscillation frequency, to stabilize the oscillation frequency, and to reduce harmonic distortion.

The main problem in producing a low-distortion constant-amplitude sine-wave is obtaining the correct loop gain ($|A\beta| = 1$). This can be achieved by using the 2N3819 FET as a voltage-variable resistor in the negative-feedback loop. A Zener diode at the output terminal provides the voltage reference for the peak sine-wave amplitude. This is rectified and fed to the gate of the FET through Q_1 and Q_2 stages, thus varying the FET channel resistance and, hence, the loop gain. Note that the gain $A = 1 + R_2/R_4$ (see Section 4.7). When the output V_0 is positively increased, the FET channel resistance will be increased, resulting in a larger value of R_4 and, thus, a smaller value of A. The bridge elements are $R_1 = R_3 = R$, and $C_1 = C_3 = C$. The oscillation frequency is $f_0 = \frac{1}{2}\pi RC$.

Typical Wein-Bridge Oscillator Contained in HP Universal Bridge 4265B

Figure 10.2-3 shows the typical circuit of Op-amp Wien-bridge oscillator contained in Hewlett-Packard Universal Bridge 4265B. The oscillator generates a 1 kHz $\pm$ 1.5% sinusoidal waveform applied to bridge network and detector. The 1 kHz oscillator consists of four networks, namely, a frequency selective network, a negative feedback network, an amplifier, and a peak comparator. Positive feedback is accomplished through the frequency-determining network consisting of C_1, C_2, R_1, and R_2. To provide low impedance for the input of the amplifier U_1, R_1 is taken at one-tenth of R_2; C_1 is also 10 times C_2. Therefore, the negative feedback network of R_4, R_3, and Q_1 is designed with a divider ratio of 21 to 1. The oscillating frequency, which is determined by the square root of the product of all four elements, is the same as the product of each pair, R_1 and C_1 or R_2 and C_2. The amplifier has a low output impedance and R_6 isolates and protects the oscillator from output circuit. The input of the peak comparator U_2 is the difference between the voltage detected by positive peak detector D_2 and the reference voltage produced by R_{10}, R_{11}, and the Zemer diode D_3. The output (6) of U_2 controls the dynamic resistance of Q_1 to maintain a constant output of U_1 at about 10 V peak-to-peak.

10.3   A TYPICAL WIDE-BAND SIGNAL GENERATOR

Figures 10.3-1a and 10.3-1b show the block diagram and the simplified circuit of a typical wide-band signal generator. It provides a wide frequency range from 5 Hz to 1.2 MHz. The output sine wave has an open-circuit amplitude range from 150 μV_{rms} to 5 V_{rms}. With a 600-Ω load, the maximum available output voltage is reduced to 2.5 V_{rms}. Distortion is less than 1%. The oscillator can be synchronized by an external 5-V source for optimum locking action.

The circuit is composed of nine transistors. An oscillator is formed by the Wien bridge connected to an amplifier containing transistors Q_1, Q_3–Q_8 in addition to a peak comparator Q_9 and an automatic-gain-control (agc) stage Q_2. The oscillator output from the final stage Q_7–Q_8 is attenuated by a 600-Ω variable attenuator. The required loop gain ($|A\beta| = 1$) is provided by positive feedback through the RC network combined with negative feedback through the R_2–R_4 network. The output voltage from the Wien bridge to the + input of the amplifier at f_0 is one-third the amplitude of the positive-feedback voltage. In order to maintain $-A\beta = 1$ and oscillation, the negative feedback network including agc is designed with the R_2/R_4 ratio of 2 so that $A = 3$. The agc stage Q_2 is driven by comparator Q_9. The comparator com-

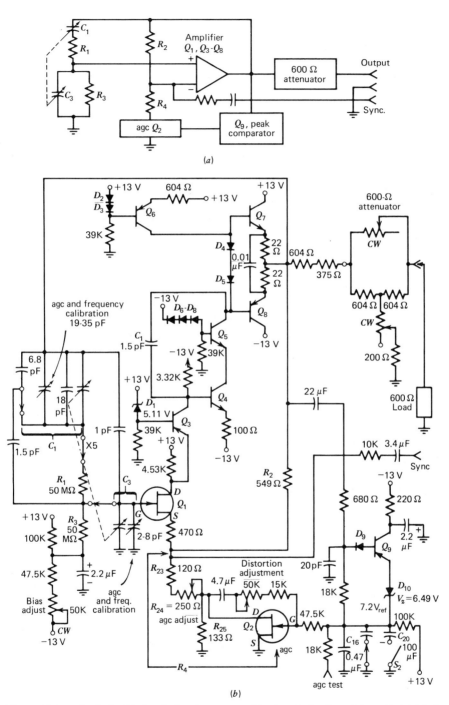

Figure 10.3-1 Typical wide-band signal generator (Hewlett-Packard Model 204C): (a) block diagram; (b) simplified circuit. $R_1 = R_3 = R$; $C_1 = C_3 = C$; $R_2/R_4 = 2$; $f_0 = 1/2\pi RC$. (Courtesy of Hewlett-Packard.)

pares the negative peak of the oscillator output to a 7.2 V reference. The difference signal from the comparator controls the gate of the agc (FET) stage and, thus, the dynamic resistance of the agc circuit. Hence, the oscillator output is maintained at 7.2-V peak.

When the generator is first turned on, the agc stage provides an amplifier gain $A > 3$, so that noise is greatly amplified. The frequency selective network in the bridge provides positive feedback at the selected noise frequency, and the system starts to oscillate at the frequency to which it is tuned. As the oscillator–amplifier output approaches the 7.2 V peak, the agc action decreases the amplifier gain to $A = 3$, thus maintaining stable oscillation. The agc stage maintains optimum negative feedback for the amplifier by means of impedance variation. The generator has a constant output impedance of 600 Ω on any setting of the 600-Ω attenuator and the following step attenuator (not shown).

The gate of FET Q_1 is connected to the positive-feedback network, and its source circuit combined with the agc system. Zener diode D_1 establishes proper bias for Q_3. Diodes D_6, D_7, and D_8 establish proper bias for Q_5. Capacitor C_9 is used to provide a stable roll-off at high frequencies. Q_6 is a current source for Q_4 and Q_5. D_4 and D_5 provide proper bias for the complementary output transistors Q_7 and Q_8. The negative-feedback action of the Wien bridge depends on the R_2/R_4 ratio. R_4 is the total impedance of R_{23}, R_{24}, R_{25}, and Q_9. R_{25} reduces the effect of the FET Q_2, and thereby increases operating stability. The conduction of the FET Q_2 is controlled by the peak comparator Q_9. Note that Q_9 conducts during the most negative portion of each half cycle, thus developing a negative charge in C_{16} and its shunt capacitor(s). When the amplifier output increases, Q_9 conducts more current and C_{16} becomes more negatively charged. Hence, the input voltage to the FET Q_2 becomes more negative, and so its impedance is increased. This results in an increase in the amount of negative feedback, thus decreasing the amplifier output voltage. C_{20} with switch S_2 provides a longer agc time constant for minimum distortion at low operating frequencies.

10.4 *LC* OSCILLATOR USED IN RF SIGNAL GENERATORS

Introduction

An *LC* tank is an *LC* tuned (resonant) circuit. In an *LC* oscillator, the *LC* resonant circuit is connected as a whole or a section of the feedback network between the input and output of an op-amp, FET, or BJT amplifier. Part of the amplifier output voltage is fed back to the input by inductive or capacitive coupling to compensate for the power losses in the tank circuit. This regenerative or positive feedback causes an output voltage of constant amplitude at the resonant frequency

$$f_0 = \frac{1}{2\pi\sqrt{LC}}. \tag{10.4-1}$$

LC oscillators can operate at high frequencies, up to several hundred MHz.

Most rf (radio-frequency) signal generators use an *LC* oscillator to produce sinusoidal signals. This type of instrument produces output frequencies in the "over 30 kHz" region. Most high-quality rf signal generators have a precision output attenuator that allows setting of output levels from under 1 μV to 100,000 μV. Rf signal generators typically have a 50-Ω output impedance and a 1-mW power level in a load of 50 Ω is 0 dB$_m$.

A General Form of *LC* Oscillators

Many *LC* oscillators are of the general form shown in Fig. 10.4-1a. In the following analysis we assume an active device with very high input resistance such as a FET or an op-amp. Figure 10.4-1b shows the linear equivalent circuit of Fig. 10.4-1a, using an amplifier with an open-circuit negative gain $-A_v$ (e.g., $-\mu = -g_m r_d$ for FET) and output resistance R_0. The load impedance Z_L consists of Z_2 in parallel with the series combination of Z_1 and Z_3, or $Z_L = (Z_1 + Z_3)Z_2/(Z_1 + Z_2 + Z_3)$. The gain without feedback is

$$A = \frac{V_0}{V_{13}} = \frac{-A_v R_L}{R_0 + R_L} = \frac{-A_v Z_2(Z_1 + Z_3)}{R_0(Z_1 + Z_2 + Z_3) + Z_2(Z_1 + Z_3)}.$$

The feedback factor is $\beta = -V_f'/V_0 = -Z_1/(Z_1 + Z_3)$. The loop gain is

$$-A\beta = \frac{-A_v Z_1 Z_2}{R_0(Z_1 + Z_2 + Z_3) + Z_2(Z_1 + Z_3)}. \tag{10.4-2}$$

If the impedances are pure reactances (either inductive or capacitive), then $Z_1 = jX_1$, $Z_2 = jX_2$, and $Z_3 = jX_3$. Thus,

$$-A\beta = \frac{+A_v X_1 X_2}{jR_0(X_1 + X_2 + X_3) - X_2(X_1 + X_3)}. \tag{10.4-3}$$

For the loop gain to be real or zero phase shift, the *j* term must be zero, that is,

$$X_1 + X_2 + X_3 = 0; \tag{10.4-4}$$

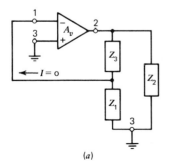
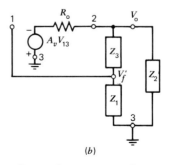

(a)   (b)

Figure 10.4-1 (a) Basic configuration for many *LC* oscillators. (b) Linear equivalent circuit of Fig. 10.4-1a.

thus,

$$-A\beta = \frac{-A_vX_1}{X_1 + X_3}.\qquad(10.4\text{-}5)$$

Equation (10.4-4) indicates that the circuit will oscillate at the resonant frequency of the series combination of X_1, X_2, and X_3. Using Eq. (10.4-4) in Eq. (10.4-5) we find

$$-A\beta = \frac{+A_vX_1}{X_2}.\qquad(10.4\text{-}6)$$

For operation as an oscillator, $-A\beta$ must be positive and at least unity in magnitude. Therefore, X_1 and X_2 must have the same sign. In other words, they must be either both inductive or both capacitive. Then from Eq. (10.4-4), $X_3 = -(X_1 + X_2)$, and X_3 must be capacitive if X_1 and X_2 are inductive, or vice versa.

When X_1 and X_2 are capacitors and X_3 is an inductor, the circuit is called a Colpitts oscillator. When X_1 and X_2 are inductors and X_3 is a capacitor, the circuit is called a Hartley oscillator.

Practical Colpitts and Hartley Oscillators

A practical version of a FET Colpitts oscillator is shown in Fig. 10.4-2. The circuit is basically the same form as shown in Fig. 10.4-1a, with the addition of the components needed for dc bias of the FET amplifier. The oscillation frequency derived from Eq. (10.4-4) is given by

$$f_0 = \frac{1}{2\pi\sqrt{LC_{eq}}}.\qquad(10.4\text{-}7)$$

where

$$C_{eq} = c_1c_2/(c_1 + c_2).$$

A BJT Colpitts oscillator and a BJT Hartley oscillator are shown in Figs.

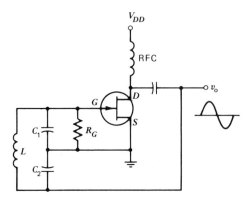

Figure 10.4-2 FET Colpitts oscillator.

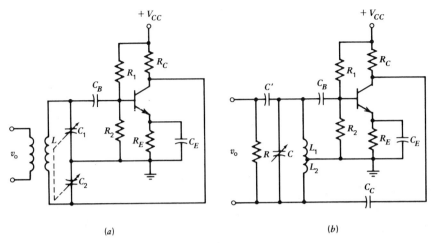

Figure 10.4-3 (a) A BJT Colpitts oscillator. (b) A BJT Hartley oscillator.

10.4-3a and 10.4-3b. Qualitatively, these two circuits operate in the manner described above. However, the detailed analysis of a BJT oscillator circuit is more difficult because the low input impedance of the BJT shunts Z_1 in Fig. 10.4-1a and the low-frequency h-parameter model is no longer valid at high frequencies.

The operating frequency of a Hartley oscillator is approximated by

$$f_0 = \frac{1}{2\pi\sqrt{L_{eq}C}}, \tag{10.4-8}$$

where $L_{eq} = L_1 + L_2 + 2M$.

The Colpitts and Hartley oscillators are usually employed in signal generators that cover the frequency ranges up to 250 MHz.

10.5 AM SIGNAL GENERATORS

General Description

A standard AM signal generator is a complete and highly accurate sine-wave signal source. It is often used to measure circuit properties, such as gain, bandwidth, signal to noise ratio, etc. It is extensively used to test radio or television receivers and transmitters. The plan of a basic standard AM signal generator is shown in Fig. 10.5-1. The carrier frequency is generated by a very stable LC oscillator, which delivers a good sinusoidal waveform and has no appreciable hum or noise modulation. Amplitude modulation (AM) is provided from an internal, fixed-frequency sine-wave oscillator, or from an external source. Usually the modulating frequency is 400 or 1000 Hz. Modulation takes place in the output amplifier circuit, which delivers the

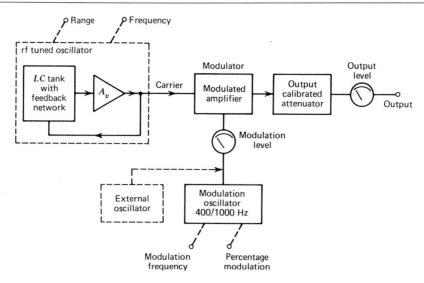

Figure 10.5-1   Plan of a basic standard AM signal generator.

modulated carrier to the output attenuator. Shield compartments are provided to prevent radiation of rf energy into surrounding space.

Accuracy Rating and Frequency Stabilization

A service-type signal generator may be rated for $\pm 2\%$ accuracy, whereas a
laboratory-type or standard generator may be rated for $\pm 0.25\%$ or better
accuracy. The accuracy rating of a signal generator depends directly on the
frequency stability of its oscillator. In order to attain a given accuracy assignment the designer must observe the mechanical and electrical stability
requirements. Thus rigid construction is employed, the effects of expansion
and contraction resulting from temperature changes are taken into consideration, fixed capacitors with suitable temperature coefficients are used for
optimizing frequency stabilization, a temperature-compensating capacitor is
mounted at a point where it provides optimum compensation during both
warm-up and sustained operation, and the load condition must have practically no effect on the oscillator frequency.

Amplitude Modulation (AM)

The variation of a high-frequency carrier characteristic proportional to a
lower-frequency signal is known as modulation. In radio-voice transmission
by an AM transmitter, the rf wave is varied in accordance with the audio
signal being transmitted. Similarly, the rf signal from an AM signal generator
is modulated by the audio signal (400 or 1000 Hz). Thus, the operating principle of an AM signal generator is basically the same as that of a radio AM
transmitter. Amplitude modulation for a sinusoidal carrier can be generated
by a nonlinear amplifier or by an analog multiplier, as will be described.

10.6   AM WAVEFORM AND NONLINEAR MODULATOR

The energy medium by which the lower-frequency signal is to be transferred is referred to as the carrier. The lower-frequency signal is often termed the modulating signal v_m. A carrier signal may be represented by

$$v_0 = V_{0p} \cos \omega_0 t = V_{0p} \cos 2\pi f_0 t, \qquad (10.6\text{-}1)$$

where f_0 is the frequency of the carrier wave. In amplitude modulation, the magnitude of the carrier V_{0p} is varied in accordance with the modulating signal v_m. Amplitude modulation produces a modulation envelope on the carrier waveform. This modulation envelope follows the modulating signal waveform. The amplitude of the modulation envelope is a fraction m of the amplitude of the unmodulated wave. This fraction m given in percent is termed percent modulation. Based on this definition for modulation, the equation for the modulating signal v_m can be written as

$$v_m = mV_{0p} \cos \omega_m t = mV_{0p} \cos 2\pi f_m t. \qquad (10.6\text{-}2)$$

When a carrier is amplitude modulated by a sine-wave signal, the amplitude of the carrier contains the sinusoidal variations as expressed by $(1 + m \cos \omega_m t)V_{0p}$. The instantaneous voltage of the resultant wave is

$$v = (1 + m \cos \omega_m t)V_{0p} \cos \omega_0 t$$

$$= V_{0p} \cos \omega_0 t + mV_{0p} \cos \omega_0 t \cos \omega_m t. \qquad (10.6\text{-}3)$$

Using the trigonometric formula

$$\cos x \cos y = \tfrac{1}{2} \cos(x + y) + \tfrac{1}{2} \cos(x - y),$$

we have

$$v_0 = V_{0p} \cos \omega_0 t + \frac{mV_{0p}}{2} \cos(\omega_0 + \omega_m)t + \frac{mV_{0p}}{2} \cos(\omega_0 - \omega_m)t. \qquad (10.6\text{-}4)$$

The above is the equation of an amplitude-modulated wave containing three terms. The first term is identical with Eq. (10.6-1), which is the unmodulated wave. The frequency of the second term is $f_0 + f_m$, and the frequency of the third term is $f_0 - f_m$. As an example, when the carrier is 1000 kHz and the modulating signal is 1 kHz, the frequencies of the three terms are 1000 kHz, 1001 kHz, and 999 kHz. The term $f_0 + f_m$ (at 1001 kHz) is called the upper side-band, the term $f_0 - f_m$ (at 999 kHz) is called the lower side-band. In this example, the modulating signal is 1 kHz, but the total bandwidth required is from 999 kHz to 1001 kHz, or 2 kHz. Thus the bandwidth required in amplitude modulation is twice the frequency of the modulating signal. When the modulation is 100% ($m = 1$), the maximum instantaneous voltage is $2V_{0p}$ and the minimum instantaneous voltage of the envelope is zero, as shown in Fig. 10.6-1c.

The coefficients of the terms of Eq. (10.6-4) are

$$V_{0p} \qquad mV_{0p}/2 \qquad mV_{0p}/2$$

and are in the ratio

$$1 \qquad m/2 \qquad m/2.$$

Since power may be expressed as V^2/R, these terms can be converted to a power ratio:

$$1 \qquad m^2/4 \qquad m^2/4.$$

The total power is in the ratio

$$1 + m^2/4 + m^2/4 = 1 + m^2/2.$$

This can be stated in the form of an equation

$$P_T/P_0 = 1 + m^2/2, \tag{10.6-5}$$

where P_0 is the carrier power, and P_T is the total power for a modulation m. If R is the resistance,

$$\frac{P_T}{P_0} = \frac{I_T^2 R}{I_0^2 R} = 1 + \frac{m^2}{2}$$

$$I_T = I_0\sqrt{1 + m^2/2}. \tag{10.6-6}$$

When the modulation is 100%, m is unity and the total power ratio becomes 1.5. The two halves of the additional 50% represent the energy content rates of the upper and lower side-bands, respectively.

The amplitude-modulated waveforms of Figs. 10.6-1b and 10.6-1c can be generated by using a nonlinear device, such as the balanced modulator shown in Fig. 10.6-2. The output of a nonlinear device is represented by the usual

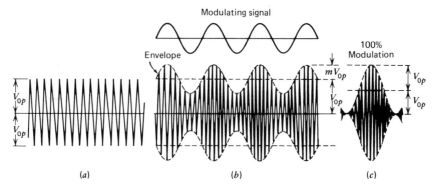

Figure 10.6-1 Amplitude modulation (AM). (a) Unmodulated wave (carrier). (b) Modulated waveform. (c) 100% modulation waveform.

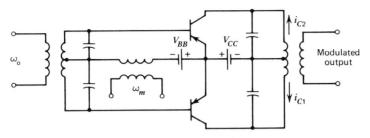

Figure 10.6-2 A balanced modulator circuit. Its transfer characteristic is nonlinear.

power series,

$$i = a_0 + a_1 v + a_2 v^2 + a_3 v^3 + \cdots.$$

With the balanced modulator circuit, either the carrier or the modulating frequency, as well as many of the intermodulation terms, are automatically eliminated. The operation of the circuit depends on the transfer characteristic of the circuit being nonlinear. Specifically, suppose that the collector currents i_{c1} and i_{c2} for the identical driving elements are related to the base voltages v_{b1} and v_{b2} by the first two terms of a power series relation, respectively; then

$$i_{c1} = a_1 v_{b1} + a_2 v_{b1}^2 \quad \text{and} \quad i_{c2} = a_1 v_{b2} + a_2 v_{b2}^2, \qquad (10.6\text{-}7)$$

and the input voltages have the form

$$v_{b1} = V_{0p} \cos \omega_0 t + V_{mp} \cos \omega_m t$$

and

$$v_{b2} = V_{0p} \cos \omega_0 t + V_{mp} \cos \omega_m t. \qquad (10.6\text{-}8)$$

The currents in the collector circuits oppose each other, and the effective current is

$$i_{c1} - i_{c2} = a_1 2 V_{0p} \cos \omega_0 t + a_2 4 (V_{0p} \cos \omega_0 t)(V_{mp} \cos \omega_m t),$$

which is

$$i_{c1} - i_{c2} = 2a_1 V_{0p} \cos \omega_0 t \left(1 + \frac{2a_2 V_{mp}}{a_1} \cos \omega_m t \right), \qquad (10.6\text{-}9)$$

which is precisely in the form of Eq. (10.6-3) for the amplitude-modulated wave without approximation.

10.7   ANALOG MULTIPLIER FOR AMPLITUDE MODULATION

By multiplying any carrier waveform by a modulating signal v_m an amplitude-modulated signal is obtained because the instantaneous value of the carrier is proportional to v_m. An analog multiplier, such as the transconductance multiplier of Fig. 10.7-1, is used in this application for a sinusoidal carrier, and thus can be used as the modulated amplifier in Fig. 10.5-1. In the circuit of Fig. 10.7-1, transistors $Q_1–Q_3$ comprise a differential amplifier with a constant current stage. Operation of the transductance multiplier can be understood by examining the transfer characteristic (I_c versus $V_{B1} - V_{B2}$) of the differential amplifier. Suppose V_{B2} is constant. When V_{B1} is below cutoff, all the current I_0 flows through Q_2. As V_{B1} carries Q_1 above cutoff, I_{E1} increases while I_{E2} decreases, and

$$I_{E1} + I_{E2} = -I_0 = \text{constant.} \tag{10.7-1}$$

The total range ΔV_0 over which the output can follow the input is $R_c I_0$ and thus ΔV_0 is adjustable through an adjustment of I_0. In the base circuits of Q_1 and Q_2,

$$V_{B1} - V_{B2} = V_{BE1} - V_{BE2}. \tag{10.7-2}$$

Each transistor has its emitter current I_E related to the voltage V_{BE} by a diode volt–ampere characteristic of the form

$$I_E = I_S e^{V_{BE}/V_T}, \tag{10.7-3}$$

where I_S = Ebers–Moll parameter, and

$$V_T = \bar{k}T/q = T/11,600. \tag{10.7-4}$$

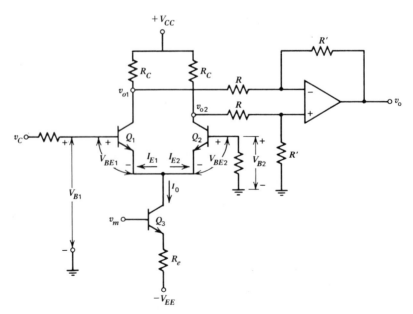

Figure 10.7-1 Variable transconductance multiplier. Signal v_m controls I_0. $v_0 = kv_c v_m$, where k is constant.

In Eq. (10.7-4), $\bar{k}$ is the Boltzmann constant in joules per degree Kelvin. At room temperature (300 K), the volt equivalent of temperature is $V_T = 0.0259$ V. Assume that Q_1 and Q_2 are matched, then

$$-I_0 = I_{E1} + I_{E2} = I_{E1} + I_{E1}e^{(V_{B2} - V_{B1})/V_T} = I_{E1}(1 + e^{-(V_{B1} - V_{B2})/V_T}).$$

Hence

$$I_{C1} \simeq -I_{E1} = \frac{I_0}{1 + e^{-(V_{B1} - V_{B2})/V_T}}, \tag{10.7-5}$$

and I_{C2} is given by the same expression with V_{B1} and V_{B2} interchanged. Differentiating Eq. (10.7-5) with respect to $V_{B1} - V_{B2}$, we have the transconductance g_{md} of the differential amplifier with respect to the differential input voltage, or

$$\frac{dI_{C1}}{d(V_{B1} - V_{B2})} = -\frac{I_0 e^{-(V_{B1} - V_{B2})/V_T}}{(1 + e^{-(V_{B1} - V_{B2})/V_T})^2} \cdot \frac{-d(V_{B1} - V_{B2})/V_T}{d(V_{B1} - V_{B2})}$$

$$= \frac{I_0}{V_T} \cdot \frac{1}{(1 + 2e^{-(V_{B1} - V_{B2})/V_T} + e^{-2(V_{B1} - V_{B2})/V_T})e^{(V_{B1} - V_{B2})/V_T}}$$

$$= \frac{I_0}{V_T} \cdot \frac{1}{e^{(V_{B1} - V_{B2})/V_T} + 2 + e^{-(V_{B1} - V_{B2})/V_T}}.$$

g_{md} is evaluated at $V_{B1} = V_{B2}$; therefore,

$$g_{md} = \frac{dI_{C1}}{d(V_{B1} - V_{B2})} = \frac{I_0}{4V_T}. \tag{10.7-6}$$

Note that, for the same value of I_0, the effective transconductance of the differential amplifier is one-fourth that of a single transiscor $(g_m = |I_C|/V_T)$. Substituting $V_T \simeq 25$ mV into Eq. (10.7-6), we get

$$g_{md} = 10I_0. \tag{10.7-7}$$

The variation of g_{md} starts from zero, reaches a maximum of $I_0/4V_T$ when $I_{C1} = I_{C2} = \frac{1}{2}I_0$, and again approaches zero. Thus the value of g_{md} is proportional to I_0. Since the output voltage change V_{02} is given by

$$V_{02} = g_{md}R_c\Delta(V_{B1} - V_{B2}) \simeq 10I_0R_c(V_{b1} - V_{b2}), \tag{10.7-8}$$

it is possible to change the differential gain by varying the value of the current I_0. This means that automatic gain control is possible with the differential amplifier.

Equation (10.7-8) indicates that the output voltage of a differential amplifier depends on the current source I_0. If the high-frequency carrier V_c is applied to one input and the low-frequency modulating signal v_m is used to vary I_0, as in Fig. 10.7-1, the output will be proportional to the product of the two signals, or $v_0 = kv_cv_m$. Figure 10.7-2 shows the equivalent circuit of the Fairchild monolithic double-balanced modulator/demodulator μA796

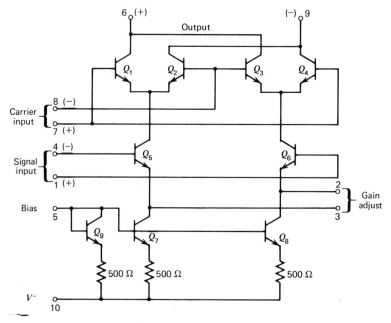

Figure 10.7-2   Equivalent circuit of μA796 double-balanced modulator/demodulator. (Courtesy Fairchild Camera and Instrument, Corp.)

(equivalent to MC1496); this circuit produces an output voltage which is the product of an input voltage (signal) and a switching function (carrier). Communications applications include modulation and demodulation of AM, single side-band (SSB), double side-bands (DSB), FM, etc.; the maximum transadmittance bandwidth is 300 MHz when the load R_L is 50 Ω, carrier input port has a sine wave of 60 mV$_{rms}$, and modulating sine wave is of 300 mV$_{rms}$ at 1 kHz; or, the transadmittance bandwidth is 80 MHz when the signal input port has a sine wave of 300 mV$_{rms}$ and $V_7-V_8 = 0.5$ V$_{dc}$.

10.8   QUARTZ CRYSTALS FOR RF STANDARDS

Quartz is a crystalline combination of silicon and oxygen atoms (SiO_2). The reason for the concerned with quartz is twofold: (1) It possesses a piezoelectric effect, which means pressure electricity; when a mechanical force is applied to a quartz crystal, an electrical charge will appear on its surface; the charge is proportional to the force. (2) Quartz possesses very stable physical properties that do not vary radically with temperature or time.

Usually the crystal is cut in a plane perpendicular to the Z axis so that an X-Y cut (Fig. 10.8-1) is formed. Note that there is threefold symmetry in the characteristics of the cross section in different directions in the X-Y plane. Rotate the crystal 120° about the Z axis and observe the characteristic

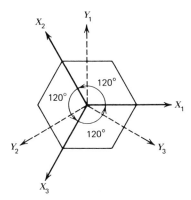

Figure 10.8-1 $X - Y$ cut.

as initially. There are three X axes—X_1, X_2, and X_3. The corresponding Y axes—Y_1, Y_2, and Y_3—can be found by rotating the positive $X_{1,2,3}$ counterclockwise by 90°. There are three directions in the X-Y plane—X_1, X_2, and X_3—with similar characteristics, as is true with Y axes—Y_1, Y_2, and Y_3.

If we cut a wafer of quartz, oriented as shown in Fig. 10.8-2, it will deform with external fields. If we apply a static field ($\mathscr{E}_x$) and release it, the crystal will oscillate between the position shown in Figs. 10.8-2b–10.8-2d. Since the wafer is flexing in its X-Y plane, it is called an X-Y flexure mode. Placing a

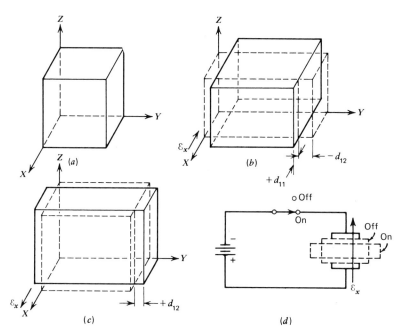

Figure 10.8-2 (a) A wafer of quartz (b)–(d) deformed with external fields applied.

cover over the wafer and sealing it to the base completes our X-Y flexure reasonator. Figures 10.8-3a and 10.8-3b show the resonator equivalent circuit and reactance versus frequency curve, and reactance versus resistance curve, respectively. In the equivalent circuit, C_0 is the capacitance between the electrodes with quartz as the dielectric; C_m, L_m, and R_m are the motional capacitance, inductance, and resistance of the flexure mode, respectively. The series resonant frequency $\omega_s^2 = 1/L_m C_m$, anti- or parallel resonant frequency $\omega_a^2 = 1/[L_m C_m C_0/(C_m + C_0)] = (1/L_m C_m)[(C_m + C_0)/C_0]$, and the natural bandwidth $BW = \omega_a - \omega_s$. The impedance of the crystal is highest at ω_a and lowest at ω_s. Typically, values for a 32.768-kHz crystal are $C_0 \simeq$ 2.2 pF, $C_m \simeq 6.5 \times 10^{-15}$ F, $L_m \simeq 4000$ H, and $R_m \simeq 25$K, corresponding to the quality factor $Q = 2\pi f_0 L_m/R_m \simeq 33{,}000$. The dimensions of such a crystal bar are approximately 8.9 mm by 10.16 mm by 1.5 mm.

If we neglect the resistance R_m, the impedance of the crystal is a reactance jX whose dependence on frequency is given by

$$jX = -\frac{j}{\omega C_0}\frac{\omega^2 - \omega_s^2}{\omega^2 - \omega_a^2}. \tag{10.8-1}$$

Since $C_0 \gg C_m$, $\omega_a \simeq \omega_s$. For the crystal whose parameters are specified above, the parallel frequency is only 0.3% higher than the series frequency. For $\omega_s < \omega < \omega_a$, the reactance is inductive, and outside this range it is capacitive, as indicated in Fig. 10.8-3a.

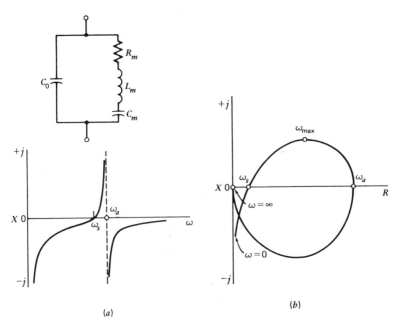

(a)

(b)

Figure 10.8-3 (a) Equivalent circuit of quartz crystal with its reactance versus frequency curve; (b) resistance versus frequency curve.

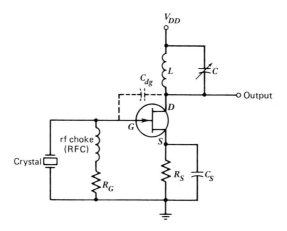

Figure 10.8-4   Miller crystal-controlled FET oscillator.

If in the basic configuration of Fig. 10.4-1a a crystal is used for Z_1, a tuned LC combination for Z_2, and the capacitance C_{dg} between drain and gate for Z_3, the resulting circuit is a Miller crystal-controlled FET oscillator, as shown in Fig. 10.8-4. From Eqs. (10.4-4) and (10.4-6), we see that both the reactances of the crystal and the LC network must be inductive. For the loop gain to be greater than unity, we also see from Eq. (10.4-6) that X_1 cannot be too small. Therefore, the circuit will oscillate at a frequency that lies between ω_s and ω_a but close to the ω_a value. Because $\omega_a \simeq \omega_s$, the oscillator frequency is essentially determined by the crystal.

Quartz-crystal-controlled oscillators can be formed into other circuits, as discussed next.

10.9   QUARTZ-CRYSTAL-CONTROLLED OSCILLATOR CIRCUITS

If in the basic configuration of Fig. 10.4-1a a crystal is used for Z_3, a capacitor C_1 for Z_1, and another capacitor C_2 for Z_2, then the resulting circuit is a modified Pierce crystal-controlled oscillator, as shown in Figs. 10.9-1a and 10.9-1b. From Eq. (10.4-4) we see that the crystal must operate in its inductive region, that is, between ω_s and ω_a. Since $X_3 = -(X_1 + X_2)$, then

$$\omega_0 L_3 = \frac{C_1 + C_2}{\omega_0 C_1 C_2} \quad \text{or} \quad \omega_0^2 = \frac{C_1 + C_2}{L_3 C_1 C_2}.$$

If $C_1 = 150$ pF and $C_2 = 41$ pF, then the effective capacitance, called the crystal load capacity, is $C_L = C_1 C_2/(C_1 + C_2) = (150)(41)/(150 + 41) \simeq 32$ pF. The feedback element $2\pi f_0 L_3 = 2\pi f_0 L_{x,\text{eff}} = \frac{1}{2}\pi f_0 C_L$. Thus $f_0 = \frac{1}{2}\pi\sqrt{C_L L_{x,\text{eff}}}$. The reactance versus frequency plot is shown in Fig. 10.9-2a. What makes the oscillator operate at f_0 since the reactive restrictions

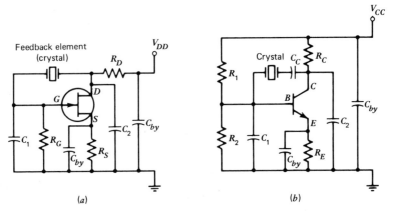

Figure 10.9-1   Modified Pierce crystal-controlled oscillators: (*a*) FET circuit. (*b*) BJT circuit.

are also satisfied at f'_0? The answer to the question is implicit in the R vs X plot shown in Fig. 10.9-2*b*. Note that R'_3 at f'_0 is much larger than R_3 at f_0. The f_0 oscillation will capture the circuit to the exclusion of f'_0.

If we desire to change frequency slightly, we can place a trimmer (C_t) in parallel with C_1 or C_2 (Fig. 10.9-3*a*), or in series with the crystal (Fig. 10.9-

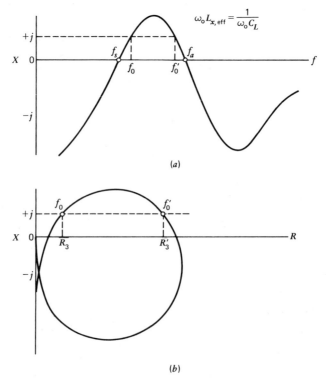

Figure 10.9-2   (*a*) X vs f plot; (*b*) X vs R plot.

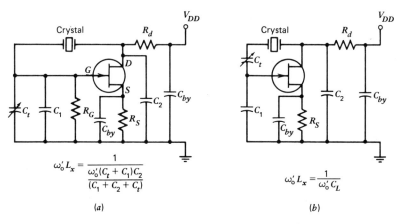

$$\omega_o' L_x = \frac{1}{\frac{\omega_o'(C_t + C_1)C_2}{(C_1 + C_2 + C_t)}}$$

(a)

$$\omega_o' L_x = \frac{1}{\omega_o' C_L}$$

(b)

Figure 10.9-3 (a) C_t in parallel with C_1. (b) C_t in series with the crystal.

3b). As C_t is varied, C_L varies. For the parallel case, $C_L = (C_t + C_1)C_2/(C_1 + C_2 + C_t)$:

$$\frac{dC_L}{dC_t} = \frac{(C_1 + C_2 + C_t)C_2 - (C_t + C_1)C_2}{(C_1 + C_2 + C_t)^2} = \frac{C_2^2}{(C_1 + C_2 + C_t)^2}$$

For the series case, $C_L = C_1 C_2 C_t/(C_1 C_2 + C_2 C_t + C_1 C_t)$:

$$\frac{dC_L}{dC_t} = \frac{(C_1 C_2 + C_2 C_t + C_1 C_t)C_1 C_2 - C_1 C_2 C_t(C_1 + C_2)}{(C_1 C_2 + C_2 C_t + C_1 C_t)^2}$$

$$= \frac{(C_1 C_2)^2}{(C_1 C_2 + C_2 C_t + C_1 C_t)^2} = \frac{(C_1 C_2)^2}{[C_1 C_2 + C_t(C_1 + C_2)]^2} .$$

Changing C_L will affect the frequency, as shown in Fig. 10.9-4.

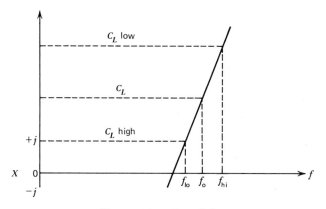

Figure 10.9-4 C_L vs f plot.

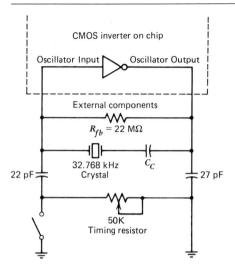

Figure 10.9-5   CMOS-type crystal oscillator.

CMOS-Type Crystal Oscillator

A CMOS-type crystal oscillator is shown in Fig. 10.9-5. The CMOS inverter on a chip provides the necessary gain. The external components provide startup, bias shift, and attenuation. This circuit, which is currently being used in electronic wrist watches, can be employed as a reference to design a timer for instrumentation.

In the CMOS inverter of Fig. 10.9-6, Q_1 (n channel) and Q_2 (p channel) are the enchancement MOSFETs constructed on the same chip. Q_1 is the driver and Q_2 acts as the load. The input V_i varies from $V(0) = 0\ V$ to $V(1) = V_{DD}$. When $V_i = 0$, then $V_{GS1} = 0$ and Q_1 is off, whereas $V_{GS2} = -V_{DD}$ and Q_2 is on. In this case, $V_{DS2} = 0$, and so $V_0 = +V_{DD}$ when $V_i = 0$. When $V_i = +V_{DD}$, then Q_1 is on but Q_2 is off, hence $V_0 = 0$.

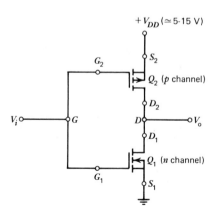

Figure 10.9-6   CMOS inverter.

10.10 FREQUENCY MODULATION AND VARACTOR-DIODE MODULATOR

Frequency-Modulation Signal

Frequency modulation (FM) is a method of keeping the amplitude V_p constant while incorporating the modulating signal (usually, audio) into variations of the carrier frequency f_c. The FM reduces most natural and man-made electrical noise in the form of amplitude-modulated signals.

A numerical example to illustrate frequency modulation is presented next. Let the carrier frequency f_c be 1000 kHz, the modulating frequency f_m be 1 kHz, and the amplitude of the modulating signal V_{mp} be 1 V. At the instant the modulating signal is zero, the FM wave is 1000 kHz. When the modulating signal increases in a positive direction, assume that the output wave increases its frequency, and, when the modulating signal is negative, the output wave decreases its frequency. Assume that at the instant the signal V_{mp} is $+1$ V, the instantaneous frequency of the output is 1010 kHz, and that, at the instant the signal V_{mp} is -1 V, the output frequency is 990 kHz. This concept is shown in Fig. 10.10-1. For each complete modulating cycle,

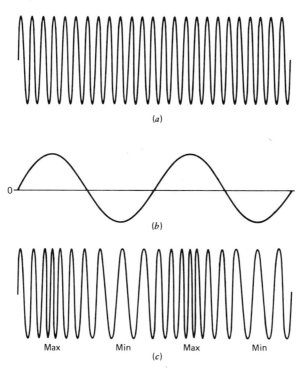

Figure 10.10-1 Frequency modulation. (a) Unmodulated wave (carrier). (b) Modulating signal. (c) Frequency-modulated wave.

the instantaneous frequency of the output follows:

V_{mp} (V)	0	+1	0	-1	0
f_c (kHz)	1000	1010	1000	990	1000

If this relation is linear, 2-V and 0.5-V signals change these figures to

V_{mp} (V)	0	+2	0	-2	0
f_c (kHz)	1000	1020	1000	980	1000
V_{mp} (V)	0	+0.5	0	-0.5	0
f_c (kHz)	1000	1005	1000	995	1000

The amplitude of the modulating signal determines the frequency deviation f_d from the carrier. For 0.5-V, 1-V, and 2-V signals, the deviations are 5 kHz, 10 kHz, and 20 kHz, respectively. Note that f_d is measured one way from the carrier. For FM broadcasting, the Federal Communications Commission (FCC) limits f_d to a maximum of 75 kHz and, in television broadcasting, it limits f_d to 25 kHz for the sound portion of the program. Note also that f_d contains the information on the amplitude or volume of the signal f_m. The deviation f_d is proportional to amplitude V_{mp} of the modulating signal.

If f_m is 1 kHz and V_{mp} is 1 V, as in the original case, the frequency deviation is $f_d = 10$ kHz. This indicates that the output frequency is changing between 1010 kHz and 990 kHz at the rate of 1000 times a second. If V_{mp} is kept at 1 V, and f_m is changed from 1 kHz to 2 kHz, the deviation stays the same at 10 kHz, but the output frequency changes between 1010 kHz and 990 kHz 2000 times a second instead of 1000 times a second. Thus, the modulating frequency is the rate of change of the output frequency. The ratio of frequency deviation to modulating frequency is termed the index of modulation, m_f, or

$$m_f \equiv \frac{f_d}{f_m}. \tag{10.10-1}$$

In the original example, m_f is 10 kHz/1 kHz or 10. When V_{mp} is increased from 1 V to 2 V, m_f changes from 10 to 20. When V_{mp} is 0.5 V, m_f is 5. If a 1-V signal at 1 kHz produces a deviation of 10 kHz, the index of modulation is 10. A 2-V signal at 2 kHz produces a deviation of 20 kHz. The index of modulation is still 10.

In FM broadcasting, the maximum allowed audio frequency is 15 kHz. The maximum allowed deviation is a swing of 75 kHz above and below the carrier frequency. These two limiting figures give a particular index of modulation that is 75 kHz/15 kHz or 5. This index of modulation is known as the deviation ratio. When the deviation ratio is larger than unity, wide-band frequency modulation is employed; when the deviation ratio is less than unity, narrow-band frequency modulation is used.

The general equation of a sine wave, neglecting the phase angle, is

$$v = V_p \cos \omega t.$$

In frequency modulation, as shown above, the instantaneous frequency is a function of f_c, f_d, f_m, and V_{mp}. Since the index of modulation $m_f\ (= f_d/f_m)$ unites f_d, f_m, and V_{mp}, we can reduce the variables to f_c, f_m, and m_f. The instantaneous frequency f_i of the FM wave may be given by

$$f_i = f_c + f_d \cos 2\pi f_m t,$$

$$2\pi f_i = 2\pi f_c + 2\pi f_d \cos 2\pi f_m t,$$

$$\omega_i = \omega_c + \omega_d \cos \omega_m t. \qquad (10.10\text{-}2)$$

The equation for ω_i may be converted to an expression for instantaneous voltage:

$$v = V_p \cos \left(\int \omega_i(t)dt \right) = V_p \cos \left(\int (\omega_c + \omega_d \cos \omega_m t)dt \right)$$

$$= V_p \cos \left(\omega_c t + \frac{\omega_d}{\omega_m} \sin \omega_m t \right) = V_p \cos(\omega_c t + m_f \sin \omega_m t), \quad (10.10\text{-}3\text{a})$$

or

$$\frac{v}{V_p} = \cos(\omega_c t + m_f \sin \omega_m t). \qquad (10.10\text{-}3\text{b})$$

Referring to the formula

$$\cos(x + y) = \cos x \cos y - \sin x \sin y,$$

Eq. (10.10-3b) can be rewritten as

$$\frac{v}{V_p} = \cos \omega_c t \cos(m_f \sin \omega_m t) - \sin \omega_c t \sin(m_f \sin \omega_m t). \quad (10.10\text{-}4)$$

When Eq. (10.10-4) is expanded,* we find

$$\frac{v}{V_p} = J_0(m_f) \cos \omega_c t$$

$$+ J_1(m_f) \cos(\omega_c + \omega_m)t - J_1(m_f) \cos(\omega_c - \omega_m)t$$

$$+ J_2(m_f) \cos(\omega_c + 2\omega_m)t + J_2(m_f) \cos(\omega_c - 2\omega_m)t$$

$$+ J_3(m_f) \cos(\omega_c + 3\omega_m)t - J_3(m_f) \cos(\omega_c - 3\omega_m)t$$

$$+ J_4(m_f) \cos(\omega_c + 4\omega_m)t + J_4(m_f) \cos(\omega_c - 4\omega_m)t$$

$$+ \cdots.$$

$$(10.10\text{-}5)$$

* Jahnke, E. and F. Emde, *Tables of Functions*. Tuebner, Leipzig, 1938 (Dover Publications, New York, 1943).

From Eq. (10.10-5), it is evident that in frequency modulation there are many side-bands, whereas in amplitude modulation there were only two side-bands. The side-bands in frequency modulation occur in pairs. There are an upper side-band and a lower side-band for the modulating frequency, for the second harmonic of the modulating frequency, for the third harmonic of the modulating frequency, for the fourth harmonic of the modulating frequency, and so on. As in amplitude modulation there is a term that represents energy at the carrier frequency—the J_0 term. Since the equation is for v/V_p, the vector sum of the coefficients of the carrier and side-band terms must add to unity. The coefficients $[J_0(m_f), J_1(m_f), J_2(m_f), \cdots]$ of the terms are called Bessel functions of the first kind. The subscript of the J is called the order. Thus the third-order Bessel function is the coefficient of the third side-bands, which are located at $(f_c + 3f_m)$ and at $(f_c - 3f_m)$. The bandwidth required in FM is $2(f_d + f_m)$.

One of the chief advantages of the FM radio transmitter is that, unlike the AM transmitter, the output power is always constant. The process of frequency modulation reduces the carrier power and puts this decreased energy into useful signal-carrying side-band energy. The FM generator is basically operated on the same principle as the FM transmitter.

Varactor-Diode Modulator

Figure 10.10-2 shows a varactor-diode modulator used to generate an FM signal. The capacitance across a varactor diode varies inversely with the reverse voltage across the diode. A dc reverse voltage is applied to the diode through the R_1–R_2 divider network and is bypassed by C_1. R_3 isolates the rf and can be large since the leakage current of the diode is very small. The instantaneous voltage across the diode is the sum of the dc bias plus the audio input. Consequently, a variable capacitance ΔC of the diode is placed in parallel with the L-C tank of the oscillator and, since ΔC varies with amplitude v_{mp} of the modulating signal, deviation f_d produced in the oscillator is proportional to amplitude v_{mp}. Therefore, the frequency-modulated oscillator is a voltage-to-frequency converter.

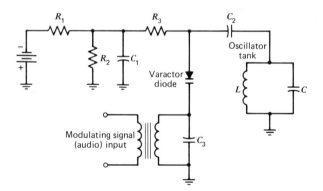

Figure 10.10-2   Varactor-diode (variable-voltage capacitor diode) modulator.

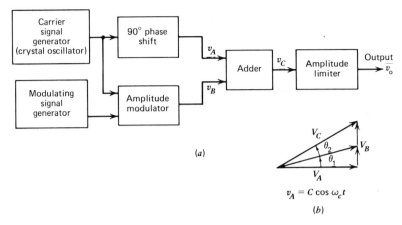

$$v_A = C \cos \omega_c t$$

(b)

Figure 10.10-3   A small-deviation phase modulator; (a) block diagram; (b) vector diagram.

The Frequency-Modulation System Based on a Phase Modulator

The phase modulator is used to form the basis for some frequency-modulation systems. A small-deviation phase modulator is shown in Fig. 10.10-3. The carrier with a 90° phase shift is added to a much smaller AM waveform that has the same carrier frequency. The resultant waveform is the vector sum of the two components, as shown in Fig. 10.10-3b. As the AM waveform varies in amplitude, the output voltage v_C varies in phase and slightly in amplitude. If the phase deviation is limited to approximately 30° ($\pm 15°$), a reasonably linear variation of phase with modulating voltage results. The limiter is used to keep the amplitude variations of the output waveform at a minimum level. The expression for the output voltage is then

$$v_0 = C \cos(\omega_c t + \phi_0 + m_p \sin \omega_m t), \qquad (10.10\text{-}6)$$

which corresponds to a phase-modulated wave. The quantity m_p is called the phase-modulation index, and the constant ϕ_0 is the phase in the absence of modulation. Neglecting ϕ_0, the form of Eq. (10.10-6) is the same as that of Eq. (10.10-3a).

The relationship between phase and frequency modulation can be derived mathematically starting with the expression for a phase-modulated waveform given by Eq. (10.10-6). The frequency of this signal is not constant since the total instantaneous phase or angle

$$\theta = \omega_c t + \phi_0 + m_p \sin \omega_m t$$

does not vary linearly with time. The instantaneous radian frequency is found from

$$\omega = \frac{d\theta}{dt} = \omega_c + m_p \omega_m \cos \omega_m t. \qquad (10.10\text{-}7)$$

Comparison of this relation with Eq. (10.10-2) shows that the maximum frequency deviation produced by phase modulation is $(f_d)_p = m_p f_m$. The quantity $(f_d)_p$ is proportional both to the amplitude and to the frequency of the modulating signal. When the phase of a sine wave is modulated, the frequency is also modulated. The instantaneous frequency change is proportional to the derivative of the phase-modulating signal. Therefore, one method of obtaining an FM signal with frequency of

$$\omega = \omega_c + A(t)$$

is to use $\int A(t)dt$ as the modulating signal of a phase modulator. For example, if the desired FM signal frequency is

$$\omega = \omega_c + 2\pi k_f M \cos \omega_m t, \qquad (10.10\text{-}8)$$

then the phase-modulating signal should be

$$\frac{2\pi k_f M}{\omega_m} \sin \omega_m t.$$

The system of Fig. 10.10-3 can be converted from a phase modulator to a frequency modulator simply, by inserting an integrator between the modulating-signal generator and the amplitude modulator. Instead of applying the carrier signal to the system input, a subharmonic might be applied. The modulation can then be performed to produce a low-level, comparatively low-frequency FM signal. A harmonic of this signal is then amplified and limited by the class-C output amplifier. Greater indexes of modulation can be obtained by cascading phase modulators.

One method of obtaining an FM waveform involves heterodyning in connection with multipliers. This can be a good method to control the index of modulation. The operation called heterodyning or mixing occurs when a signal is multiplied by an auxiliary sinusoidal signal to form side-bands. To illustrate this method, the following example is given. Assume that we want to generate an FM signal having a carrier frequency of 200 MHz with a maximum frequency deviation of 25 kHz. The modulating-signal frequency varies from 50 to 5000 Hz. The system is based on a phase modulator that has a modulation index $m_p = 0.25$ at 50 Hz and operates at a frequency of 200 kHz. At the lowest modulating frequency $f_m = 50$ Hz, the phase modulator will produce a frequency deviation of

$$f_d = m_p f_m = 0.25 \times 50 \text{ Hz} = 12.5 \text{ Hz}.$$

To lead to the maximum deviation of 25 kHz requires a frequency multiplication of

$$\frac{25 \times 10^3}{12.5} = 2000.$$

Multiplying the frequency deviation of 12.5 Hz by 2000 leads to the required

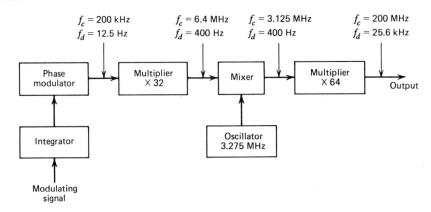

$f_c = 200$ kHz $f_c = 6.4$ MHz $f_c = 3.125$ MHz $f_c = 200$ MHz
$f_d = 12.5$ Hz $f_d = 400$ Hz $f_d = 400$ Hz $f_d = 25.6$ kHz

Figure 10.10-4   An FM waveform generator using heterodyning in connection with multipliers.

25 kHz deviation, but the carrier frequency would then be

$$2000 \times 200 \text{ kHz} = 400 \text{ MHz}.$$

Hence, it is necessary to heterodyne this signal down to a 200 MHz and preserve the correct deviation. It is more convenient to heterodyne the signal earlier in the multiplication chain, as indicated in the system of Fig. 10.10-4. In this system, the actual modulation takes place at a low frequency.

VCO-Type Frequency Modulator

The VCO, to be discussed in Chapter 11, can also be used to obtain an FM waveform. Generally, the signal will be produced at a lower frequency and will be a frequency-modulated rectangular wave if the modulating signal is applied to the control voltage input. When the signal is multiplied and mixed to produce the correct carrier frequency and deviation, the tuned circuit stages select the proper harmonic converting the rectangular wave to a higher-frequency sinusoid.

10.11   BASIC RF SWEEP GENERATOR

Requirements and Characteristics

The sweep frequency (or sweep) generator is a frequency-modulated generator used to plot automatically frequency response on the screen of an oscilloscope. The sweep signal from the generator is a sinusoidal voltage of constant amplitude in the rf range, whose frequency is smoothly and continuously varied over an entire frequency band, usually at a 60-Hz or another low audio rate. In an rf sweep generator, the center frequency of the FM waveform deviates periodically between an upper- and lower-frequency limit. Note that the frequency deviation is a sweep width. The deviation which must be used depends on the bandwidth employed by the circuit under test. Consider the following example. The FM detector (demodulator) in the sound section of a television receiver has a center frequency of 4.5 MHz and employs a comparatively narrow-band circuit. The TV rf tuner in a VHF channel has a much higher center frequency and employs a comparatively large bandwidth. Thus, if we wish to display the frequency-response curves of these two sections, we will set the center frequency of the sweep generator to 4.5 MHz with a sweep width of 75 or 100 kHz for the receiver sound section and may set the center frequency of the generator to 214 MHz with a sweep width of 10 or 15 MHz for the rf tuner.

A typical sweep generator used for TV–FM service has a range of center frequencies from 40 kHz to 214 MHz in addition to a continuously tunable band from 88 to 108 MHz. Most rf sweep generators have a maximum output within a range from 0.1 to 0.7 V_{rms}. In order to minimize the distortion of the displayed response curve, the sweep oscillator must provide essentially flat output that has good waveform with a minimum harmonic content.

Development of an rf Response Curve

Figure 10.11-1a shows a basic arrangement for display of a frequency-response curve on a scope screen. An FM signal from the generator is applied to the tuned circuit under test. The source resistance R_s and the LC circuit form a voltage divider, and the maxmum signal voltage is applied to the diode detector at the resonant frequency of L and C. The vertical amplifier of the scope is then energized by a varying voltage that is determined by the frequency-response characteristic of the tuned circuit. Note that the sweep-signal deviation is produced by a 60-Hz sine-wave voltage, and the scope is horizontally deflected by the 60-Hz sine-wave voltage supplied by

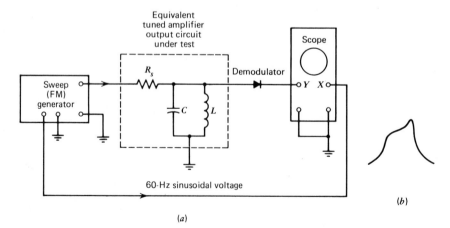

(a)

(b)

Figure 10.11-1 (a) Arrangement of test instruments for display of a frequency-response curve on an oscilloscope. (b) A response curve displayed on a 60-Hz sine-wave horizontal deflection.

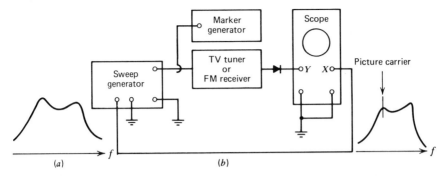

Figure 10.11-2   (a)An rf response curve. (b) Display for a frequency-response curve with a beat marker.

the sweep generator. Thus, the resulting pattern is automatically synchronized.

Since we employ the 60-Hz sine-wave voltage for horizontal deflection, the tuned circuit under test is swept twice each $\frac{1}{60}$ s (see Figs. 10.10-1b and 10.10-1c). The sweep signal proceeds from its low-frequency limit to its high-frequency limit, and then back to its low-frequency limit. However, because of the reversal of the horizontal-deflection voltage each $\frac{1}{120}$ s, the two curves, due to the twice-swept effect, are superimposed as shown in Fig. 10.11-1b. Although this form of display is quite usable, it is customary to blank the return trace and thereby develop the response curve as shown in Fig. 10.11-2a, where the baseline indicates the zero-volt level in the pattern.

In visual-alignment procedures, the beat markers (called pips or birdies) at the picture carrier frequencies are the most common form of frequency indication. The marker is typically produced by placing the output leads from the marker generator near the output leads from the sweep generator. If an ordinary signal generator has been precisely calibrated, it can be used as a marker generator; calibration is usually provided by quartz-crystal frequency standards built into the signal generator. The display for a frequency-response curve with a beat marker is shown in Fig. 10.11-2b.

10.12   OPERATION PRINCIPLE OF RF SWEEP GENERATOR

Oscillator Frequency Modulated by Means of a Varactor Diode

From Section 10.4 we know that the frequency of an LC oscillator can be changed by varying C or L. Thus a sweep frequency generator can be formed by connecting a reverse-biased varactor (Varicap) diode across one of the tank capacitors of a Colpitts oscillator, as shown in Fig. 10.12-1. The oscillation frequency can be changed as long as a triangle wave or a sinusoidal voltage at a low frequency is applied to the diode. The rf choke is used to

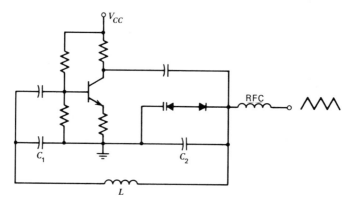

Figure 10.12-1 Oscillator frequency modulated by means of a reverse-biased varactor diode.

prevent the rf current from flowing into the power supply. The sweep frequency is produced by the varactor diode. The capacitors C_1 and C_2 can be made variable for fine frequency adjustment.

Oscillator Frequency Modulated by Means of a Saturable Reactor

A sweep-frequency generator can also be formed by using a current-controlled inductor (saturable reactor) as the tank coil of an LC oscillator. As shown in Fig. 10.12-2, the tank coils are wound on a ferrite core, and the permeability of the core is varied by a 60-Hz magnetic field from a control winding. Since a B-H curve has its maximum linearity at medium flux densities, a dc bias current is passed through the control winding, in addition to the 60-Hz ac current; thus, linear frequency deviation is obtained.

Frequency-Channel Conversion

If we want to decrease the sweep frequency, we may use a different inductance L by the way of L-channel conversion. If the L-channel conversion

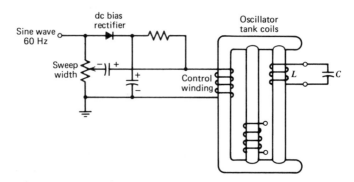

Figure 10.12-2 Resonant frequency modulated by means of saturable reactor (current-controlled inductor).

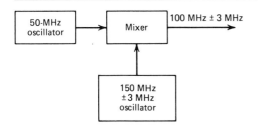

Figure 10.12-3 Arrangement for frequency-channel conversion.

is employed as a coarse frequency adjustment, the sweep-frequency range will be relatively reduced. For example, the sweep-frequency range may be from 0 to 6 MHz at a center frequency of 150 MHz. If the L-channel conversion is used, then the sweep frequencies will be only from 0 to 1 MHz at 50 MHz. In order to avoid the reduction of the sweep-frequency range as the center frequency is decreased, we may use a frequency conversion arrangement as in Fig. 10.12-3 for the coarse frequency adjustment. Suppose we mix an oscillator frequency of 50 MHz with a sweep signal of 150 MHz $\pm$ 3 MHz, then we may obtain the four different signals: 50 MHz, 150 MHz $\pm$ 3 MHz, 100 MHz $\pm$ 3 MHz, and 200 MHz $\pm$ 3 MHz. Since we need 100 MHz $\pm$ 3 MHz, we filter out the three other signals by a proper filter circuit. Thus, we complete the frequency-channel conversion.

10.13   WIDE-RANGE PLL-SYNTHESIZED RF SIGNAL GENERATORS

A PLL Frequency Synthesizer

The PLL frequency synthesizer is a new type of wide-range rf signal generator, which uses phase-locked-loop (PLL) circuits to generate the output frequencies. A block diagram of PLL frequency synthesizer, shown in Fig. 10.13-1, consists of a voltage-controlled oscillator (VCO), divide-by-N counter, phase detector, crystal reference oscillator, low-pass filter, dc amplifier, and a buffer amplifier.

The rf VCO operates at frequency f_0 that is a function of the dc control voltage V. The phase detector makes a comparison between the output frequency of the divide-by-N counter and a low-frequency reference oscillator signal so that the control voltage V is created. The divide-by-N counter divides the input frequency f_0 by an integer N. The integer division ratio is set by the binary word applied to the program-N inputs of the counter. The output frequency of the counter is equal to f_0/N. The frequency f_0/N and a low-frequency reference f_r (in the 5–10-kHz range) are applied to opposite inputs of the phase detector circuit. As long as f_0/N is equal to f_r, the VCO will oscillate at the frequency determined by the dc amplifier offset control and N-code applied to the counter. But if the VCO should shift, then the two frequencies applied to the phase detector are no longer equal, hence the dc error voltage at the output of the phase detector varies in a direction

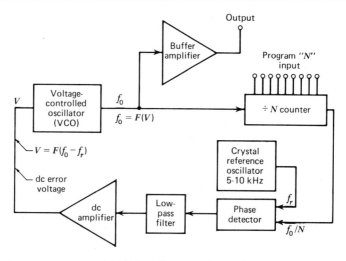

Figure 10.13-1   Simplified block diagram of a PLL frequency synthesizer.

and amount needed to pull the VCO back on frequency. Thus, the PLL circuit constantly corrects changes in the VCO frequency. Therefore, the stability and precision of this PLL synthesizer are essentially those of the crystal reference oscillator.

A Typical PLL-Synthesizer rf Signal Generator

The Philips PM5326 rf signal generator (Fig. 10.13-2) is an accurate instrument. Its block diagram is shown in Fig. 10.13-3. Based on the phase-locked-

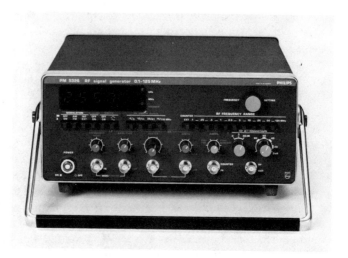

Figure 10.13-2   Philips PM5326 RF signal generator. (Courtesy of Philips.)

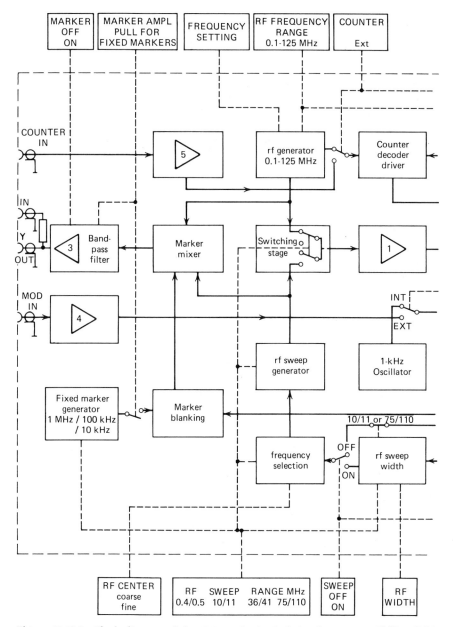

Figure 10.13-3 Block diagram of the PLL-synthesized rf signal generator, Philips PM5326 (Courtesy of Philips.)

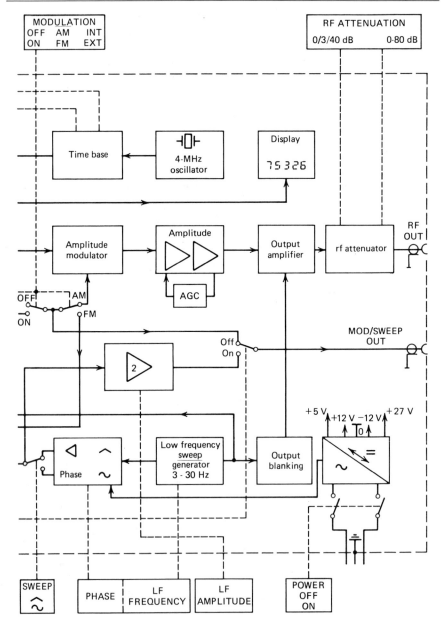

Figure 10.13-3 (*Continued*)

loop (PLL) principle the rf carrier is very stable and easy to set. The frequency is electronically counted (MOS technique) with crystal reference and five-digit LED display; so faults in setting and reading that are characteristic for common generators with circular or linear scales are eliminated. The amplitude of the carrier frequency able to be modulated is precise and stabilized throughout the whole frequency ranges. Radiation of the internal generated frequencies possibly escaping the case or the mains cable is reduced by the separate cast rf housing within the instrument, so that reliable use of the high attenuation can be made. The instrument includes a separate sweep-frequency oscillator modulator, the ranges of which are optionally suited for the common AM/IF, FM/IF, TV/IF, and VHF range as well. The modulator is also used as frequency modulator for the FM/IF and VHF ranges. Together with the modulating ranges of the center frequency, the width and marker distance of the spectrum are automatically switched over. When a modulating range is chosen, the crystal-controlled adjustable reference carrier frequency is used as "traveling marker" for pointing out the response in the band pass. The modulated or frequency-modulated signal is then amplitude stabilized, defined, attenuated, and fed to the rf output.

Rf Circuitry, Amplitude Modulation.   The rf generator produces high frequencies. The RF FREQUENCY RANGE push-button array allows selecting the desired range, while the continuous control FREQUENCY SETTING determines the exact frequency. Via switching stage and amplifier 1 the rf frequency is fed to the amplitude modulator passing either unmodulated or amplitude modulated by the internal 1-kHz oscillator or by external MOD-IN low frequency due to the selected push-buttons at MODULATION. The amplitude of the rf frequency is stabilized in the control circuit amplitude with automatic gain control AGC in the feedback path. The output amplifier can be interacted by the output blanking stage: when sweeping with internal triangle, the rf signal is blanked during flyback. The rf attenuation sets the output continuously from 0 to -80 dB, in addition, the rf attenuator has two fixed stages of -3 dB and -4 dB.

Sweep Section and Frequency Modulation.   The rf sweep generator produces and modulates high frequencies for the sweep ranges 0.4/0.5, 10/11, 36/41, 75/110 MHz, selected by push-buttons at RF SWEEP RANGE. The carrier for the frequency modulation and the center frequency for the modulating ranges, respectively, are set in frequency selection, activated by RF SWEEP RANGE just mentioned and coarse and fine adjustments are by the double continuous control rf center. Pressed button AM/FM activates frequency modulation of the ranges 10/11 or 75/110 MHz with 1 kHz internal or with external signal via MOD-IN input socket. If one of the RF SWEEP RANGE buttons is pressed, the frequency-modulated high frequency is fed via the switching stage to the main rf output path. Owing to the different frequency ranges, the maximum sweep width is adapted by the RF SWEEP RANGE. The control RF WIDTH reduces the width. The sweeping signal and the signal for X deflection of an indicator or oscilloscope at the MOD/ SWEEP OUT socket can be switched off by the push-button SWEEP OFF/ ON. In this case the modulating signal is available at the output.

Frequency Marker.   The marker mixer superimposes the frequency of the rf generator on the sweep frequency of the rf sweep generator. The low-frequency beat is filtered in the band-pass filter amplifier 3 and fed to the OUT-Y-IN socket for the Y channel of an indicator (oscilloscope). Each frequency of the rf generator can be used as a frequency marker, that is, "traveling marker." Pulling the button Marker AMPL generates fixed markers with many harmonics. The fundamental wave of the fixed marker is selected by RF SWEEP RANGE, so a marker spectrum with the correct distance due to the sweep range is generated. When using the triangular sweep mode the square-wave output of the LF sweep generator blanks the frequency markers in the marker blanking due to flyback.

Display Circuitry. The frequency of the rf generator is fed to the counter decoder driver, which is controlled by the time base. The time base, that is, one measuring period for the counter, is changed by the chosen frequency range. The stage of the counter at the end of one measuring period represents the frequency that is multiplex-displayed on the five-digit display. Pushbutton COUNTER EXT enables the display circuitry to work as normal counter. Amplifier 5 feeds the signal at the COUNTER IN socket directly to the counter, decorder, and driver.

The power supply provides the stabilized dc voltages of $+5$ V, $+12$ V, -12 V, and $+27$ V.

An IC 565 PLL Frequency Synthesizer

The 565 integrated circuit shown in Fig. 10.13-4 is a popular PLL unit; it contains a phase detector, amplifier, and voltage-controlled oscillator (VCO). This 565 PLL will work as an FM demodulator if R_1, C_1, and C_2 are externally connected and the VCO output is connected back to the phase

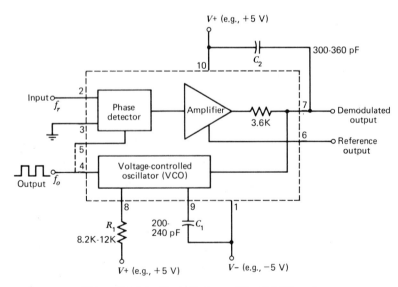

Figure 10.13-4 Block diagram of the 565 PLL unit.

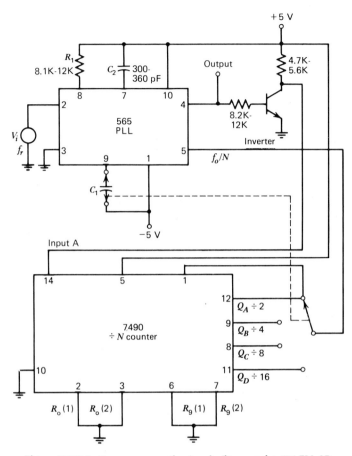

Figure 10.13-5 Frequency synthesizer built around a 565 PLL IC.

detector. In this case, R_1 and C_1 are used to set the free-running or center frequency (f_0) of the VCO, so that the frequency is given by

$$f_0 \simeq \frac{0.3}{R_1 C_1} \qquad (10.13\text{-}1)$$

with the limitation that $2K \leq R_1 \leq 20K$. There are two important frequency bands specified for a PLL: The capture range (f_c) of a PLL is the frequency range centered about the VCO free-running frequency, f_0, over which the loop can acquire lock with the input signal. Once the PLL has achieved capture, it can maintain lock with the input signal over a somewhat wider frequency range termed the lock range (f_L). The lock range and capture range of the 565 PLL in Fig. 10.13-4 are given by

$$f_L \simeq \pm \frac{8 f_0}{V+} \qquad (10.13\text{-}2)$$

and

$$f_c \simeq \pm \frac{1}{2\pi} \sqrt{\frac{2\pi f_L}{(3.6 \times 10^3)C_2}} \, . \tag{10.13-3}$$

The external capacitor C_2 is used to set the pass band of the low-pass filter. The 565 PLL can be used for various applications, including frequency synthesis, frequency demodulation, frequency-shift-keyed (FSK) decoders, etc.

The 565 PLL used in frequency synthesis is shown in Fig. 10.13-5. This frequency synthesizer employs a 565 PLL as frequency multiplier, a 7490 IC as a frequency divider, and an inverter inserted between them. As indicated in Fig. 10.13-1, the frequency divider ($\div N$ counter) is inserted between the VCO output and the phase comparator (detector) so that the loop signal to the comparator is at frequency f_0/N while the VCO output is f_0. This output is N times the input frequency (f_r) as long as the loop is in lock. The input signal can be crystal stabilized at f_r with the resulting VCO output at f_0 if the loop is set up to lock at the fundamental frequency (when f_0/N = f_r). The input V_i at frequency f_r is compared to the input (frequency f_0/N) at pin 5 of the PLL. An output of VCO at f_0 is connected through an inverter circuit to provide an input at pin 14 of the 7490 IC, which varies between 0 and $+5$ V. Using the output at pin 12, which is divided by two from that at the input to the 7490 IC, the signal at pin 4 of the PLL is two times the input frequency (f_r) as long as the loop remains in lock. The VCO can only vary over a limited range from its center frequency and so it may be necessary to change the VCO frequency whenever the divider value is changed. As long as the PLL circuit is in lock, the VCO frequency f_0 will equal Nf_r. When f_0 is readjusted to be in the capture-and-lock range, the closed loop resulting in the VCO output becomes exactly Nf_r at lock.

BIBLIOGRAPHY

Bell, D. A., Electronic Instrumentation and Measurements, Reston Publishing. Reston, Virginia, 1983, Chap. 15.

Carr, J. J., *Elements of Electronic Instrumentation and Measurement.* Reston Publishing, Reston, Virginia, 1979, Chap. 9.

Comer, D. J., *Electronic Design with Integrated Circuits.* Addison-Wesley, Reading, Massachusetts, 1981, Chap. 8.

Copper, W. D., *Electronic Instrumentation and Measurement Techniques,* 2nd ed. Prentice-Hall, Englewood Cliffs, New Jersey, 1978, Chap. 11.

Hnatek, E. R., *Applications of Linear Integrated Circuits.* Wiley, New York, 1975, Chap. 4.

Instruction Manual of Philips PM5326 RF Signal Generator, 1978.

Linear Data Book. p. 9-38 to p. 9-42 (PLL 565), National Semiconductor Corporation, 1980.

Millman, J., *Microelectronics.* McGraw-Hill, New York, 1979, Chap. 17.

RCA Phase-Locked Loop COS/MOS CD4046A 60 μW at 10 kHz, *RCA COS/MOS Integrated Circuit Data Book*. RCA Solid State Corp., [Box 3200] Somerville, New Jersey, 1976, pp. 614–617.

Question

10-1   What is the difference between an audio generator and an audio oscillator.

10-2   Give the Barkhausen conditions required for a sinusoidal oscillation to be sustained.

10-3   The Wien bridge in the oscillator circuit rejects feedback frequencies other than _____.

10-4   (a) Sketch a simple circuit of the Wien-bridge oscillator. (b) Which elements determine the frequency of oscillation? (c) Which elements determine the amplitude of oscillation?

10-5   Explain the FET amplitude stabilization used in the op-amp Wien-bridge oscillator of Fig. 10.2-2.

10-6   Which elements determine the frequency of oscillation and which components determine the amplitude of oscillation in the circuit of Fig. 10.2-3?

10-7   Explain the operation of the typical wide-band signal generator shown in Fig. 10.3-1.

10.8   What is the distinction between a Colpitts and a Hartley oscillator?

10-9   (a) Draw the equivalent circuit of a quartz crystal. (b) Sketch the reactance versus frequency function. (c) Over what portion of the reactance curve do we desire oscillations to take place when a modified Pierce crystal-controlled oscillator is formed? Explain.

10-10   Describe two methods of producing amplitude modulation.

10-11   Describe the operation principle of a variable transconductance multiplier.

10-12   What is an FM sweep generator?

10-13   Describe the operation principle of the PLL synthesizer.

Problems

10.1-1   A standard audio generator is matched with its load ($R_L = 600 \, \Omega$), Calculate the rms voltage across the load if the signal generator is producing $+3$ VU.

Answer.   1.095 V.

10.1-2   Show that the general solution of the differential equation $y'' + \omega_0^2 y = 0$ is given by $y = \cos \omega_0 t + A \sin \omega_0 t$.

10.2-1   Suppose that in Fig. 10.2-1 $R_1 = R_3 = 2.7K$, $C_2 = C_3 = 0.5$ μF, $R_4 = 360$ Ω, and a thermistor is used as the bridge element R_2 with a temperature coefficient of $-4\%/°C$. (a) Calculate the oscillation frequency and the value of R_2 at resonance (balance). (b) Assume that the operating temperature is 2.5°C higher than room temperature. Estimate the rating of the thermistor needed.

Answer.   (a) $f_0 \simeq 120$ Hz, $R_2 = 720$ Ω at resonance. (b) 800 Ω.

10.2-2   The modified Wien-bridge oscillator is shown in Fig. P10.2-2. Find (a) the oscillation frequency and (b) the minimum ratio R_2/R_4

Answer.   (a) $f_0 = \frac{1}{2} \pi \sqrt{LC}$; (b) $(R_2/R_4)_{min} = R/R_3$.

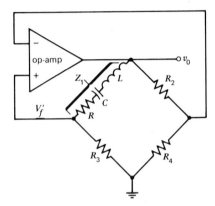

Figure P10.2-2

10.2-3   (a) An op-amp oscillator uses the network shown in Fig. P10.2-3. Prove that

$$\frac{V_f'}{V_0} = \frac{1}{3 + j(\omega RC - 1/\omega RC)}.$$

(b) Show that the oscillation frequency is $f = 1/2\pi RC$ and the gain must exceed 3. (c) Draw the op-amp oscillator circuit and compare its gain and amplitude stabilization with those of the Wien-bridge oscillator.

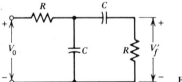

Figure P10.2-3

10.2-4   Calculate the oscillation frequency of the oscillator circuit shown in Fig. 10.2-3.

Answer.   1 kHz.

10.3-1   Figure P10.3-1 shows the op-amp Wien-bridge oscillator with Zener-diode amplitude stabilization. (a) Calculate the loop gain when both Zener diodes are initially nonconducting. (b) Explain why the Zener diodes in the circuit can be used for automatically controlling the gain of the oscillator and, thus, stabilizing the amplitude of the sine wave.

Answer. $-\beta A = 1.04$.

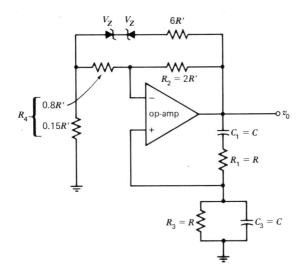

Figure P10.3-1

10.4-1   A standard rf signal generator is connected to a 50-Ω load. Find the voltage produced across the load if the generator attenuator is set to -10 dBm.

Answer.   0.0707 V.

10.4-2   (a) Consider a Colpitts oscillator, using the basic configuration of Fig. 10.4-1*a* and taking into account the resistance r_3 in series with inductor L_3. Show that the oscillation frequency is given by

$$\omega^2 = (1/L_3)[1/C_1 + (1/C_2)(1 + r_3/R_0)].$$

(b) If $r_3/R_0 \ll 1$, show that the minimum amplifier gain required for oscillation is

$$A_v = (C_1/C_2) + [(C_2 + C_1)/L_3]R_0 r_3.$$

[*Hint:* Use Eq. (10.4-2) and set $-\beta A \geq 1$.]

10.4-3   (a) Figure P10.4-3*a* shows a FET oscillator circuit with bias and power supplies omitted for simplicity. If the resistances of the inductors are r_1 and r_2, respectively, find the oscillation frequency.

(b) Find the circuit output resistance R_0, using Fig. P10.4-3b. (c) Find the value of R_s for which the value of the loop gain will just equal unity.

Answer. (a) $\omega^2 = \dfrac{1}{C_3(L_1 + L_2)} \cdot \dfrac{1 + r_2/R_0}{1 + \dfrac{(r_1 L_2 + r_2 L_1)(\mu + 1)}{R_0(L_1 + L_2)}}$.

If $r_1 = r_2 = 0$, $\omega^2 C_3 (L_1 + L_2) = 1$, which checks with Eq. (10.4-4); or $\omega^2 C_3 L_{eq} = 1$ with $L_{eq} = L_1 + L_2 + 2M$ for series aiding. (b) $R_0 = r_d + (\mu + 1)R_s$.

(c) $R_s = \dfrac{1}{\mu + 1}\left(\dfrac{(\mu L_1 - L_2)L_2}{(r_1 + r_2)C_3(L_1 + L_2)} - r_d\right)$

for small r_1 and r_2.

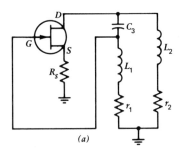
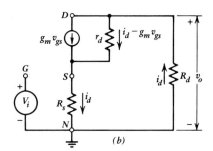

(a)   (b)

Figure P10.4-3

10.6-1 A carrier of 50 W is amplitude modulated under conditions of 100% modulation. (a) Find the total power of the modulated wave. (b) Find the energy contents of the upper and lower side-bands. (c) Find the additional percentage of the total current.

Answer. (a) 75 W; (b) 12.5 W, each; (c) 22.5%.

10.6-2 (a) The carrier strength of an AM signal is 1000 W. When the modulation is 30%, calculate the average side-band power. (b) Repeat part a for a modulation of 45%.

Answer. (a) 45 W; (b) 101.25 W.

10.7-1 In a practical difference amplifier, the difference-mode gain is given by

$$A_d = \tfrac{1}{2}\frac{h_{fe}R_c}{R_s + h_{ie}}.$$

Assuming the signal source resistance $R_s \ll h_{ie}$ and $r_{bb'} \ll r_{b'e}$, verify that

$$A_d = \tfrac{1}{2}g_m R_c \quad \text{and} \quad g_{md} = \frac{I_0}{4V_T}.$$

10.8-1   (a) Verify Eq. (10.8-1) for the reactance of a quartz crystal. (b) Prove that the ratio of the parallel-to series-resonant frequencies is given approximately by $1 + \frac{1}{2} C_m/C_0$. (c) If $C_m = 0.04$ pF and $C_0 = 2.0$ pF, by what percentage is the parallel-resonant frequency greater than the series-resonant frequency?

Answer.   (c) 1%.

10.9-1   The crystal in the FET circuit of Fig. 10.9-1*a* operates in its inductive region. (a) Draw the equivalent oscillator circuit. (b) Find the input impedance (Z_{gd}) and the negative resistance from the equivalent circuit where the crystal impedance ($R_3 + j\omega L_3$) is removed. (c) Draw the equivalent circuit for Z_{in} and Z_3. (d) Find the oscillation condition and frequency ω_0.

Answer.   (b) $Z_{in} = Z_{gd} = -\dfrac{g_m}{\omega^2 C_1 C_2} - j\dfrac{1}{\omega}\left(\dfrac{1}{C_1} + \dfrac{1}{C_2}\right).$

The first term is the negative resistance.

(d) $g_m/\omega^2 C_1 C_2 > R_3$; $\omega_0^2 = \dfrac{C_1 + C_2}{L_3 C_1 C_2}.$

10.9-2   Assume that the quartz crystal used in Fig. 10.9-5 has a motional resistance of negligible value compared to the reactance of its motional inductance. Show that the oscillation frequency is related to the timing resistance R by

$$\omega_0 = \frac{1}{\sqrt{LC}} \cdot \frac{1}{\sqrt{1 - R^2 C/L}}.$$

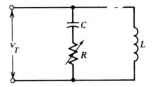

Figure P10.9-2

10.13-1   A 10–15-MHz PLL operates from a 5-kHz reference. Calculate the N code required to produce a VCO output frequency of 10 MHz when a divide-by-100 predivider is used ahead of the divide-by-N counter.

Answer.   20.

11

Comparators, Function, and Pulse Generators

11.1 SCHMITT TRIGGERS

Basic Schmitt Trigger

The Schmitt trigger is an emitter-coupled bistable multivibrator. It is level sensitive, and switches the output state at two distinct triggering levels, one called a lower-trigger level (LTL) and the other an upper-trigger level (UTL). Since the output of a Schmitt trigger is always a rectangular waveform, a distorted pulse at the input will be reshaped into a rectangular pulse at the output. The main applications of the circuit are as a voltage comparator and a pulse shaper. Figure 11.1-1a shows a basic circuit and an example of Schmitt trigger. Two silicon npn transistors are used. Voltage $V_{BE(\text{sat})} = 0.8$ V, $V_{BE(\text{cutin})} = V_\gamma = 0.5$ V, $V_{BE(\text{active})} = 0.7$ V, and $V_{CE(\text{sat})} = 0.2$ V. The operation of the circuit is approximately illustrated as follows.

With input $v_i = 0$, Q_1 is off and Q_2 is saturated; hence, output $v_o \simeq 10/5K = 2$ V. Since the base of Q_1 is reverse biased by $V_E = 2$ V, any voltage less than 2.5 V ($= V_E + V_\gamma$) applied to the input will not produce the base current I_{B_1}. When $v_i = 2.5$ V, the current I_{B_1} begins to flow, and so the collector voltage V_{C_1} is decreased. This reduced voltage coupled to the base of Q_2 decreases the collector current I_{C_2} from saturation, so that the emitter voltage V_E is reduced. This regeneration continues until the UTL [$= V_E + V_{BE(\text{sat})}$] of 2.8 V is reached. When v_i is at the UTL, Q_2 is off and Q_1 is saturated; thus $v_o = 10$ V, and $V_E = 10/10K = 1$ V. The switching occurs very fast, so that output v_o rapidly changes from about 2 to 10 V. In order to reduce v_o from 10 to about 2 V, input v_i must be decreased. As v_i is decreased to the LTL ($= V_E + V_{BE(\text{sat})} = 1.8$ V), current I_{C_1} reduces from

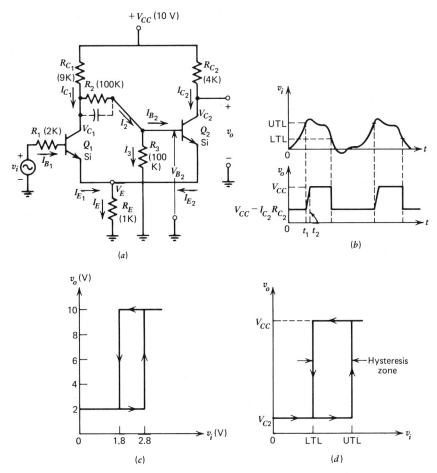

Figure 11.1-1 Schmitt trigger: (a) basic circuit; (b) input and output waveforms; (c) an approximate plot of output voltage versus input voltage for the illustrative example; (d) the general hysteresis zone.

saturation and I_{C_2} rises from zero, resulting in an increase of V_E. The rapid regeneration causes Q_1 to cut off and Q_2 to saturate, so that $v_o \approx 2$ V is finally reached. The relationship of output v_o and input v_i is shown as the plot of Fig. 11.1-1c. Output v_o is either about 2 V at lower state or 10 V at upper state. When v_o is at the lower state, v_i must be increased to the UTL (= 2.8 V) and then v_o can be switched to the upper state. Once the output is at the upper state, v_o maintains 10 V until v_i is reduced to the LTL (= 1.8 V), at which v_o rapidly returns to about 2 V.

Circuit Analysis

In Fig. 11.1-1a, transistors Q_1 and Q_2 share a common emitter resistor, R_E. For either transistor to be in a conducting state, the base voltage (V_B) is the sum of the emitter voltage (V_E) and the base–emitter voltage (V_{BE}), or

$$V_B = V_E + V_{BE}. \tag{11.1-1}$$

With input $v_i = 0$, Q_1 is off and Q_2 is on (saturated). The base current I_{B_2} is given by

$$I_{B_2} = I_2 - I_3. \tag{11.1-2}$$

The emitter voltage is given by

$$V_E = I_{E_2} R_E, \tag{11.1-3}$$

where

$$I_{E_2} = (h_{FE} + 1) I_{B_2} \tag{11.1-4}$$

if Q_2 is in the active region of operation. If Q_2 is in saturation,

$$I_{E_{2(\text{max})}} = \frac{V_{CC} - V_{CE(\text{sat})}}{R_{C_2} + R_E}. \tag{11.1-5}$$

The equivalent circuits for the condition $v_i = 0$ are illustrated in Fig. 11.1-2. Voltage V_{TH} is the Thevenin equivalent voltage at the base of Q_2. From Fig. 11.1-2a

$$V_{TH} = \frac{R_3 V_{CC}}{R_{C_1} + R_2 + R_3}. \tag{11.1-6}$$

The Thevenin resistance is given by

$$R_{TH} = \frac{(R_{C_1} + R_2) R_3}{R_{C_1} + R_2 + R_3}. \tag{11.1-7}$$

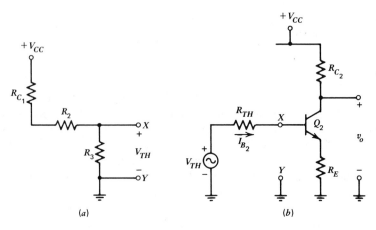

Figure 11.1-2 Equivalent circuit for the Schmitt trigger when input $v_i = 0$. (a) Determining the Thevenin voltage at the base of Q_2. (b) The Thevenin equivalent circuit.

From Fig. 11.1-2*b*,

$$V_{TH} - I_{B_2}R_{TH} - V_{BE_2} - (h_{FE} + 1) I_{B_2}R_E = 0.$$

Hence, the base current of Q_2 is

$$I_{B_2} = \frac{V_{TH} - V_{BE(sat)}}{R_{TH} + (h_{FE} + 1) R_E}. \tag{11.1-8}$$

Substitution of Eqs. (11.1-4) and (11.1-8) into Eq. (11.1-3) yields

$$V_E = (h_{FE} + 1) \frac{(V_{TH} - V_{BE(sat)}) R_E}{R_{TH} + (h_{FE} + 1) R_E}, \tag{11.1-9}$$

for Q_2 in the active region, or

$$V_E = I_{E2(max)} R_E = \frac{R_E}{R_{C_2} + R_E} (V_{CC} - V_{CE(sat)}), \tag{11.1-10}$$

for Q_2 in the saturation region.

For Q_1 to turn on (saturate), the required minimum input voltage is the so-called upper-trigger level (UTL):

$$UTL = V_E + V_{BE(sat)}, \tag{11.1-11}$$

where $V_{BE(sat)} \simeq 0.8$ V (Si). When Q_1 begins to turn on, voltage V_{C_1} begins to fall. This results in a decrease in I_{B_2}, causing voltage V_{C_2} to rise. During these changes, current $I_E = I_{E_1} + I_{E_2} =$ constant. Finally,, when Q_1 saturates, Q_2 turns off. Now voltage $V_{CE(sat)} \simeq +0.2$ V (Si) is not sufficient to turn on Q_2. The transition time is very short and is indicated as the interval between t_1 and t_2 in Fig. 11.1-1*b*. The output level remains at V_{CC} as long as Q_2 is off.

The input potential level required to turn off Q_1 and turn on Q_2 is the so called lower-trigger level (LTL). When input v_i drops to a point where Q_1 begins to turn off, voltage V_{C_1} increases until it is high enough to supply current I_{B_2}. Then Q_2 turns on. Again, a transition period exists, with both transistors conducting until Q_1 turns off. The LTL is calculated with the aid of the equivalent circuit of Fig. 11.1-3. From Fig. 11.1-3*a*, the Thevenin voltage and resistance at the collector of Q_1 are

$$V_{TH} = \frac{(R_2 + R_3) V_{CC}}{R_{C_1} + R_2 + R_3} \tag{11.1-12}$$

and

$$R_{TH} = \frac{R_{C_1} (R_2 + R_3)}{R_{C_1} + R_2 + R_3}. \tag{11.1-13}$$

For Q_2 to turn on,

$$V_{B_2} = V_E + V_{BE(sat)},$$

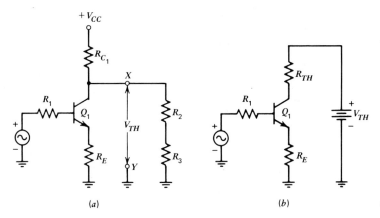

Figure 11.1-3 Determining the LTL of the Schmitt trigger. (a) Equivalent circuit. (b) Simplified circuit.

where

$$V_E = \frac{R_3 V_{C_1}}{R_2 + R_3} - V_{BE(\text{sat})} \tag{11.1-14}$$

and

$$V_{C_1} = V_{TH} - R_{TH}I_{C_1} \tag{11.1-15}$$

when Q_1 is at the end of the transition period. Substitution of Eqs. (11.1-12) and (11.1-13) into (11.1-15) yields

$$V_{C_1} = \frac{(V_{CC} - I_{C_1}R_{C_1})(R_2 + R_3)}{R_{C_1} + R_2 + R_3}. \tag{11.1-16}$$

Thus,

$$V_E = \frac{(V_{CC} - I_{C_1}R_{C_1})R_3}{R_{C_1} + R_2 + R_3} - V_{BE(\text{sat})}. \tag{11.1-17}$$

Substituting V_E/R_E for I_{C_1} and solving for V_E, we obtain

$$V_E = \frac{V_{CC} R_E R_3 - V_{BE(\text{sat})}R_E (R_{C_1} + R_2 + R_3)}{R_E (R_{C_1} + R_2 + R_3) + R_{C_1} R_3}. \tag{11.1-18}$$

The lower-trigger level (LTL) is

$$\text{LTL} = V_E + V_{BE(\text{sat})}, \tag{11.1-19}$$

where the value of V_E is less than that in the UTL expression [Eq. (11.1-11)].

Examples of input and output waveforms for a Schmitt trigger are shown in Fig. 11.1-1b. A significant difference between the Schmitt trigger and a flip-flop is that the Schmitt trigger has no memory. The difference between

UTL and LTL defines a hysteresis zone, as shown in the plot of Fig. 11.1-1d. Arrows in the plot indicate the switching sequence. The hysteresis zone may be reduced to zero by adjusting the loop gain to unity ($|\beta A| = 1$). A loop gain $|\beta A| \geq 1$ must be maintained for switching to occur. To ensure $|\beta A| \geq 1$, the following relationships must hold:

$$R_{C_2} \leq R_{C_1} \geq \frac{R_2}{h_{FE} - 1} \simeq \frac{R_2}{h_{FE}} \qquad (11.1\text{-}20)$$

and

$$R_1 \ll h_{FE} R_E. \qquad (11.1\text{-}21)$$

The value of R_1 equals the sum of the source and external base resistances. In practice, a loop gain greater than unity is maintained.

EXAMPLE 11.1-1   The Schmitt trigger of Fig. 11.1-1a uses *npn* silicon transistors with values of $V_{CE(sat)} = 0.2$ V, $V_{BE(sat)} = 0.8$ V, and $h_{FE_1} = h_{FE_2} = 100$. The circuit parameters are $V_{CC} = 10$ V, $R_{C_1} = 1.3$ K, $R_{C_2} = R_1 = 1$K, $R_2 = 6.8$K, $R_3 = 8.2$K, and $R_E = 240$ Ω. Determine (a) if switching occurs, (b) UTL, (c) LTL, and (d) the output levels.

Solution.   (a) For switching to occur, the loop gain must be greater than unity. By Eq. (11.1-20),

$$\frac{R_2}{h_{FE} - 1} = \frac{6.8\text{K}}{100 - 1} \simeq 68 \ \Omega \ll R_{C_1} = 1.3\text{K}.$$

By Eq. (11.1-21),

$$h_{FE} R_E = 100 \times 240 = 24\text{K} \gg R_1 = 1\text{K}.$$

Hence, we conclude that switching will occur.
(b) By Eqs. (11.1-6) and (11.1-7),

$$V_{TH} = \frac{8.2 \times 10}{1.3 + 6.8 + 8.2} = 5.03 \text{ V}$$

and

$$R_{TH} = \frac{(1.3 + 6.8)\,8.2}{1.3 + 6.8 + 8.2} \text{ K} = 4.07\text{K}.$$

By Eq. (11.1-8),

$$I_{B_2} = \frac{V_{TH} - V_{BE_2}}{R_{TH} + (h_{FE} + 1)\,R_E} = \frac{5.03 - 0.8}{4.07 + (100 + 1)\,0.24} = 149 \ \mu\text{A}.$$

For saturation,

$$I_{B2(min)} = \frac{I_{C2(max)}}{h_{FE}} = \frac{V_{CC} - V_{CE(sat)}}{(R_{C_2} + R_E)\,h_{FE}} = \frac{10 - 0.2}{(1 + 0.24)\,100} = 80 \ \mu\text{A}.$$

Since $I_{B2} > I_{B2(min)}$, Q_2 is in saturation. Hence, $V_{BE2} = V_{BE(sat)} = 0.8$ V. The value of V_E is

$$V_E = \left(\frac{V_{CC} - V_{CE(sat)}}{R_{C_2} + R_E}\right) R_E = \left(\frac{10 - 0.2}{1 + 0.24}\right) 0.24 = 2 \text{ V}.$$

Thus

$$\text{UTL} = V_E + V_{BE(sat)} = 2 + 0.8 = 2.8 \text{ V}.$$

When v_i reaches UTL, Q_1 begins conducting and goes into saturation. The emitter voltage is

$$V_{E_1} = I_{E1(max)} R_E = \frac{(10 - 0.2) \times 0.24}{1.3 + 0.24} = 1.53 \text{ V}.$$

Assuming that Q_1 ultimately becomes saturated,

$$V_{B_2} = \frac{(1.53 + 0.2) \times 8.2}{6.8 + 8.2} = 0.95 \text{ V}.$$

Since $V_{B_2} < V_{E_1}$, Q_2 is indeed off.
(c) By Eq. (11.1-18),

$$V_E = \frac{10\,(0.24)(8.2) - 0.8\,(0.24)(1.3 + 6.8 + 8.2)}{0.24\,(1.3 + 6.8 + 8.2) + 1.3\,(8.2)} = 1.14 \text{ V}$$

By Eq. (11.1-19),

$$\text{LTL} = V_E + V_{BE(sat)} = 1.14 + 0.8 = 1.94 \text{ V}.$$

(d) When Q_1 saturates, $v_o = V_{CC} = 10$ V. When Q_2 saturates,

$$v_o = V_{CC} - I_{C2(max)} R_{C_2} = 10 - (100 \times 80 \text{ }\mu\text{A})(1\text{K}) = 2 \text{ V}.$$

Schmitt trigger integrated circuits are commercially available. For example, the TI-13, TI-14, and TI-132 chips behave as positive NAND gates with totem-pole outputs and with a hysteresis (V_H) of 0.8 V. In addition, the μA710 op-amp is usually used as a Schmitt trigger.

11.2 AN INTEGRATED-CIRCUIT OPERATIONAL-AMPLIFIER COMPARATOR

The input section of an integrated-circuit operational amplifier is a difference-amplifier stage. A differential-voltage comparator is a high-gain, differential-input, single-ended output amplifier. The function of this device is to compare a signal voltage v_s on one input with a reference voltage V_R on the other, and produce a logic "1" or "0" at the output when one input is higher than the other. In other words, the comparator provides an output

v_o which is at one fixed dc voltage level or at another dc level depending on whether v_s is larger or smaller than V_R. Specifically, if the outputs are V_{o_1} and V_{o_2}, then $v_o = V_{o_1}$ when $v_s < V_R$ and $v_o = V_{o_2}$ when $v_s > V_R$. When v_s passes through V_R, the output makes an abrupt transition from V_{o_1} to V_{o_2} or vice versa, depending on the direction of v_s as it passes through V_R. In this comparator type, the common-mode gain between the two inputs is ideally zero. It is only the difference in the two inputs that controls the output.

The circuit of the Fairchild μA710 op-amp–type comparator is shown in Fig. 11.2-1. The input stage is a difference amplifier ($Q_1 - Q_2$) with a constant-current source (Q_3). Two inputs are available, one for reference voltage V_R and one for input signal v_s. If v_s is applied as V_1 and V_R as V_2, output v_o will be out of phase with v_s; while if v_s is applied as V_2 and V_R as V_1, output v_o will be in phase with v_s. The circuit involving Q_3 serves to provide a large incremental (ac) resistance to minimize the common-mode voltage gain, while reasonable dc emitter currents (I_{E_1} and I_{E_2}) are still allowed. Using typical transistor parameters the incremental resistance seen looking into the collector of Q_3 is found to be of the order of 250K. Since

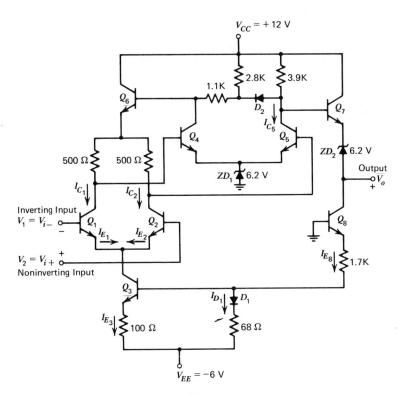

Figure 11.2-1 Circuit diagram of Fairchild μA710 op-amp comparator. (Courtesy Fairchild Camera and Instrument Corp.)

this incremental resistance is so high, the common-mode rejection ratio (CMRR is the full differential voltage gain divided by common-mode voltage gain) of the input stage is also high. In any event, CMRR $\simeq 10^4$.

The amplified signal from Q_2 is applied to Q_5, which provides additional gain. Q_7 is an emitter-follower output stage. Zener diodes ZD_1 and ZD_2 provide the necessary dc voltages to allow the dc coupling needed and to arrange proper output voltage levels.

Q_4 and Q_6 provide positive feedback to increase the gain from input to collector of Q_2. The increase of this gain will minimize the range over which the input must swing to carry the output from one limiting voltage to the other. To see this, let V_R be applied as V_2 and v_s as V_1. Then if v_s increases, the collector voltage of Q_2 will also increase. The collector voltage of Q_1 will decrease, as will the base voltage of Q_4. Therefore, the collector of Q_4 and the base of Q_6 will rise. Finally, the emitter voltage of Q_6 will increase and the collector of Q_2 will rise farther. Hence, when an input signal drives the collector of Q_2 in either direction, the feedback serves to push the collector of Q_2 further in the same direction. Thus, the feedback is positive, which increases the gain from input to collector of Q_2.

The Fairchild μA710 op-amp is usually used to perform the voltage-comparator function of a Schmitt trigger. The functional diagram of the circuit is shown in Fig. 11.2-2a. The reference voltage is connected to the noniverting terminal of the op-amp to establish the triggering level. Input signals above the reference level cause the output to switch, and signals below the level force the output to return to its original state. The hysteresis zone is a few millivolts. The circuit of Fig. 11.2-2b, however, has a controlled hysteresis zone V_H. The zone is a difference of UTL and LTL, and is a function of the feedback resistors R_1 and R_2, as expressed in the following:

$$V_H = \text{UTL} - \text{LTL} = \frac{2R_1 V_o}{R_1 + R_2} = \frac{\Delta v_o R_1}{R_1 + R_2}, \qquad (11.2\text{-}1)$$

where

$$V_o \equiv V_Z + V_D, \qquad (11.2\text{-}2)$$

$$\text{UTL} = V_R + \frac{R_1}{R_1 + R_2}(V_o - V_R) = v_2 \quad \text{for } v_s < v_2, \qquad (11.2\text{-}3)$$

and

$$\text{LTL} = V_R - \frac{R_1}{R_1 + R_2}(V_o + V_R) = v_2 \quad \text{for } v_s > \text{UTL}. \qquad (11.2\text{-}4)$$

If v_s is less than v_2 and is now increased, then v_o remains constant at $+V_o$, and $v_2 = \text{UTL} = \text{constant}$ until $v_s = \text{UTL}$. At this triggering potential, the output regeneratively switches to $v_o = -V_o$ and remains at this value as long as $v_s > \text{UTL}$. This transfer characteristic is indicated in Fig. 11.2-2c. If v_s is now decreased, then the output remains at $-V_o$ until $v_s = \text{LTL}$. At this triggering potential, a regenerative transition takes place, and the output returns to $+V_o$ almost instantaneously. For the parameter values given in

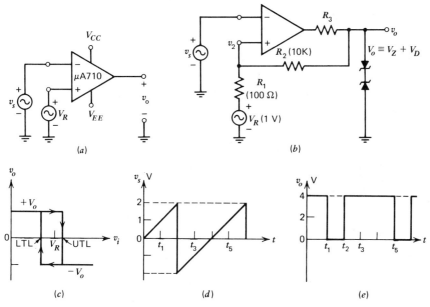

Figure 11.2-2 The op-amp comparator used as an inverting Schmitt trigger: (a) functional diagram; (b) and (c) practical Schmitt-trigger circuit and its hysteresis zone; (d) and (e) input and output waveforms of the circuit of Fig. 11.2-2a when the reference voltage $V_R = 1V$.

Fig. 11.2-2b and with $v_o = \pm 7$ V, UTL $= 1 + (0.1 \times 6)/10.1 = 1.059$ V, and LTL $= 1 - (0.1 \times 8)/10.1 = 0.921$ V. Hence, $V_H = $ UTL $-$ LTL $= 0.138$ V. For μA710, one of the output is $\Delta v_o = 4$ V $- 0$ V.

Now, consider the operation of the circuit shown in Fig. 11.2-2a when the sawtooth wave of Fig. 11.2-2d is applied to the inverting input, the reference voltage $V_R = 1$ V is applied to the noninverting input, and the maximum output is 4 V. For $0 < t < t_1$, the input rises from 0 to 1 V. Since the reference is constant at 1 V, the input is at a lower level than the reference. The op-amp provides an output of 4 V owing to its very high gain. A small difference in voltage between the inverting and noninverting input terminals causes the op-amp to saturate and deliver the maximum output of 4 V. For $t_1 < t < t_2$, the inverting input is more positive than the reference voltage, and so the output is at 0 V. For $t_2 < t < t_5$, the input is again less than the reference, and so the output is at 4 V. The output waveform is shown in Fig. 11.2-2e.

11.3 SQUARE-WAVE GENERATORS

Astable Op-amp Multivibrator

Intergrating the output of the Schmitt comparator of Fig. 11.2-2b by means of an RC low-pass network and applying the capacitor voltage to the inverting terminal in place of the external signal result in a square-wave generator, as shown in Fig. 11.3-1*a*. This circuit is called an astable multivibrator. The operating frequency can be changed by varying resistor R_f. Frequency stability is held very high by the used Zener diodes and capacitor C. The Magnitude and symmetry of the waveform depend on the matching of the two Zener diodes.

Suppose capacitor C has been initially charged by voltage V_o and is now exponentially charged toward V_{sq} $(= V_z + V_D)$. Then the voltage across C is given by

$$v_c(t) = V_o + (-V_o + V_{sq})(1 - e^{-t/RC}). \qquad (11.3\text{-}1)$$

Consider the circuit of Fig. 11.3-1*a*. When the positive output voltage holds itself across the series resistors R_1 and R_2, it will charge capacitor C through R_f. When the charging causes the voltage across C to be higher than the voltage across R_2 (or $V_C > V_{R2}$), the output voltage v_{sq} is immediately converted from positive to negative peak, so that capacitor C discharges. When V_C is less than V_{R2}, v_{sq} will be immediately converted to positive. Therefore, oscillations will be sustained.

In the multivibrator circuit, the feedback factor is $\beta = R_2/(R_1 + R_2)$. From the waveforms shown in Fig. 11.3-1*b*, we see that the initially charged voltage is $V_0 = -\beta V_{sq} = -\beta(V_z + V_D)$, where V_{sq} is the square-wave peak voltage. Thus, referring to Eq. (11.3-1), the voltage across capacitor C can also be expressed as

$$v_C(t) = -\beta V_{sq} + (\beta V_{sq} + V_{sq})(1 - e^{-t/RC}). \qquad (11.3\text{-}2)$$

The waveforms in Fig. 11.3-1*b* indicate that capacitor C requires the time $T/2$ to be charged up to $+\beta V_{sq}$; as the capacitor voltage reaches $+\beta V_{sq}$, the output state inverts immediately. Thus

$$v_C(T/2) = +\beta V_{sq} = -\beta V_{sq} + (\beta V_{sq} + V_{sq})(1 - e^{-T/2RC}), \qquad (11.3\text{-}3)$$

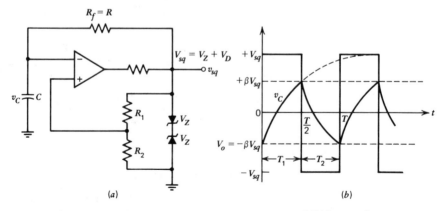

(a) (b)

Figure 11.3-1 (a) Op-amp square-wave generator and (b) its waveforms.

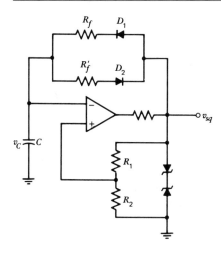

Figure 11.3-2 The astable op-amp multivibrator producing asymmetrical square waves.

or

$$v_C(T/2) = V_{sq} - V_{sq}(1 + \beta)e^{-T/2RC} = \beta V_{sq}.$$

Solving for T:

$$(1 + \beta)e^{-T/2RC} = 1 - \beta; \quad \frac{1 + \beta}{1 - \beta} = e^{T/2RC};$$

hence,

$$T = 2R_f C \ln \frac{1 + \beta}{1 - \beta}. \tag{11.3-4}$$

If $\beta = 0.475$ is chosen, then $\ln(1.475/0.525) = \ln 2.81 = 1$; hence, the period is

$$T = 2R_f C, \tag{11.3-5}$$

and the oscillation frequency is

$$f = \frac{1}{2R_f C}. \tag{11.3-6}$$

The frequency range of 10 Hz to 10 kHz is usually chosen for proper operation of the square-wave generator.

If it is desired to obtain an asymmetrical square wave, the feedback resistor R_f is replaced by the two diode-resistor combination, as in the circuit of Fig. 11.3-2. During the interval when the output is positive, diode D_1 conducts but D_2 is off. Therefore, the circuit reduces to that in Fig. 11.3-1a except that output V_{sq} is reduced by the diode drop. Because the period is independent of V_{sq} then T_1 is given by $T/2$ in Eq. (11.3-4). During the interval when the output is negative, D_1 is off and D_2 conducts. Therefore, the discharge-time constant is now $R_f' C$, and, consequently, T_2 is given by $T/2$ in Eq. (11.3-4) with R_f replaced by R_f'. If $R_f' = 2R_f$, then $T_2 = 2T_1$.

Astable Transistor Multivibrator

The astable transistor multivibrator consists of two RC-coupled common-emitter stages. In this circuit one stage conducts while the other is cut off until a point is reached at which the conditions of the stages are reversed. In the free-running collector-coupled multivibrator of Fig. 11.3-3, the output of Q_1 is coupled to the base of Q_2 through the feedback capacitor C_2', and the output of Q_2 is coupled to the base of Q_1 through the feedback capacitor C_1'. An increase in the collector current of Q_1 causes a decrease in the collector voltage which, when coupled through C_2' to the base of Q_2, causes a decrease in the collector current of Q_2. The resultant rising voltage at the collector of Q_2, when coupled through C_1' to the base of Q_1, drives Q_1 further into conduction. This regenerative process occurs rapidly, driving Q_1 into heavy saturation and Q_2 into cutoff. Q_2 is maintained in a cutoff condition by C_2' (which was previously charged to the supply voltage through R_{C_1}) until C_2' discharges through R_2 toward the collector-supply potential. When the junction of C_2' and R_2 reaches a slight positive voltage, however, Q_2 begins to conduct and the regenerative process reverses. Q_2 then reaches a saturation condition, Q_1 is cut off by the reverse bias applied to its base through C_1', and the junction of C_1' and R_1 begins charging toward the collector-supply voltage. The two transistors are therefore alternately cut off. The multivibrator has two quasistable states and oscillates continuously between them. The result is an approximate square-wave output.

The waveforms at the collectors and bases for the multivibrator of Fig. 11.3-3 are shown in Fig. 11.3-4. At $t = 0-$ (immediately before $t = 0$), Q_2 is in saturation while Q_1 is at cutoff. For $t < 0$, v_{B_1} is below $V_{BE(\text{sat})} = V_\sigma$, $v_{C_1} = V_{CC}$, $v_{B_2} = V_\sigma$ and $v_{C_2} = V_{CE(\text{sat})}$; then C_1' charges through R_1, and v_{B_1} rises exponentially toward V_{CC}. At $t = 0+$ (immediately after $t = 0$),

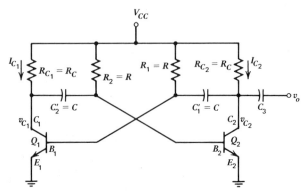

Figure 11.3-3   The free-running collector-coupled astable multivibrator.

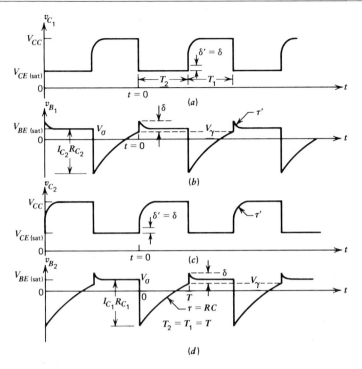

Figure 11.3-4 Waveforms of the collector-coupled astable multivibrator.

v_{B_1} reaches V_γ ($= V_{BE(\text{cutin})}$), and Q_1 conducts. As Q_1 goes to saturation, v_{C_1} falls by $I_{C_1}R_{C_1}$ to $V_{CE(\text{sat})}$ (Fig. 11.3-4a). This fall in v_{C_1} causes an equal fall of $I_{C_1}R_{C_1}$ in v_{B_2}. The fall in v_{B_2} cuts off Q_2, and its collector increases toward V_{CC}. This rise in v_{C_2} coupled through C_1' to the base of Q_1, causing overshoot δ in v_{B_1} (Fig. 11.4b), and the abrupt rise by δ in v_{C_2}. v_{B_1} and v_{C_2} change exponentially with $\tau' = (R_{C_2} + r_{bb}') C_1'$ to the levels V_σ and V_{CC}, respectively. v_{B_2} is $V_{BE(\text{sat})} - I_{C_1}R_{C_1}$ at $t = 0+$, and increases exponentially with $\tau_2 = R_2 C_2$ toward V_{CC}. At $t = T_2$, v_{B_2} reaches V_γ and a reverse transition takes place. Waveshapes in Q_1 during T_1 are the same as the waveforms in Q_2 during T_2.

During the duration T_2, Q_2 is off, and the change in v_{B_2} may be calculated from the simplified circuit of Fig. 11.3-5. As described above, $v_{B_2} = V_\sigma$ for $t < 0$. Since the collector of Q_1 and the base of Q_2 are coupled by capacitor C_1', an abrupt change in v_{C_1} must result in the same discontinuous change at the base of Q_2. At $t = 0+$, voltage v_{C_1} drops by $I_{C_1}R_{C_1}$, and so $v_{B_2} = V_\sigma - I_{C_1} R_{C_1} = V_I =$ initial value. For $t > 0$, v_{B_2} will rise exponentially toward V_{CC} with $\tau = (R + R_0)C \simeq RC$. At $t = \infty$, $v_{B_2} = V_{CC} = V_F =$ final value. Hence,

$$v_{B_2} = V_{CC} - (V_{CC} - V_\sigma + I_{C_1} R_{C_1})e^{-t/\tau}. \qquad (11.3\text{-}7)$$

This equation is derived from the general solution for a single-time-constant (RC) circuit:

$$v_o = V_F + (V_I - V_F)e^{-t/RC}. \qquad (11.3\text{-}8)$$

When $v_{B_2} = V_\gamma$, $t = T_2$. Then Eq. (11.3-7) becomes

$$\frac{V_{CC} - V_\gamma}{V_{CC} - V_\sigma + I_{C_1} R_{C_1}} = e^{-T_2/\tau}.$$

Hence,

$$T_2 = \tau \ln \frac{V_{CC} + I_{C_1} R_{C_1} - V_\sigma}{V_{CC} - V_\gamma}. \qquad (11.3\text{-}9)$$

Since Q_1 is driven into saturation, $I_{C_1} R_{C_1} = V_{CC} - V_{CE(\text{sat})}$, then

$$T_2 = \tau \ln \frac{2V_{CC} - V_{CE(\text{sat})} - V_{BE(\text{sat})}}{V_{CC} - V_\gamma},$$

or

$$T_2 = \tau \ln 2 + \tau \ln \frac{V_{CC} - \frac{1}{2}(V_{CE(\text{sat})} + V_{BE(\text{sat})})}{V_{CC} - V_\gamma}. \qquad (11.3\text{-}10)$$

At 25°C, $V_{CE(\text{sat})} + V_{BE(\text{sat})} \simeq 2V_\gamma$; therefore, Eq. (11.3-10) becomes

$$T_2 \simeq \tau \ln 2 + 0.69(R + R_0)C \simeq 0.69RC, \qquad (11.3\text{-}11)$$

where $R \gg R_0$, the resistance of a transistor in saturation. For a symmetrical circuit with $R_1 = R_2 = R$ and $C_1' = C_2' = C$, the duration T_2 is $T_2 = T_1 = T_t/2$, and the period is

$$T_t \simeq 1.38RC. \qquad (11.3\text{-}12)$$

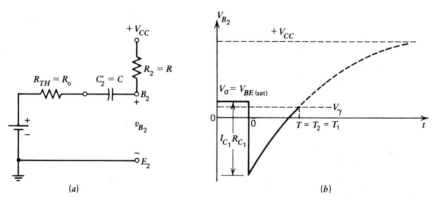

Figure 11.3-5 (a) Simplified circuit for calculating the base voltage of Q_2. (b) Voltage variation at the base of Q_2 during the time when Q_2 is off.

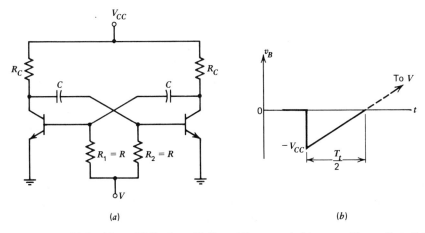

Figure 11.3-6 (a) Astable multivibrator with R_1 and R_2 connected to an auxiliary voltage V. (b) Approximate waveform of v_B in the circuit of (a) if $V_{BE_1} = V_{BE_2} = 0$.

The period T_t is changed by connecting R_1 and R_2 to an auxiliary voltage V (the collector supply remains V_{CC}), as shown in Fig. 11.3-6a. If V is varied, then T_t changes according to

$$T_t = \frac{1}{f} = 2RC \ln \left(1 + \frac{V_{CC}}{V} \right) ,\qquad (11.3\text{-}13)$$

provided that V is large compared with the junction voltages. Thus, this circuit becomes a voltage-to-frequency converter. In order to derive Eq. (11.3-13), the forward-biased junction voltages are assumed to be zero. The approximate waveform of v_B is shown in Fig. 11.3-6b. Applying Eq. (11.3-8) to this waveform, we have

$$v_B = V - (V_{CC} + V)e^{-t/RC}.$$

At $t = T_t/2$, $v_B = 0$,

$$e^{T_t/2RC} = \frac{V + V_{CC}}{V} = 1 + \frac{V_{CC}}{V} .$$

Thus, we obtain Eq. (11.3-13).

11.4   BASIC FUNCTION GENERATORS

Introduction

Almost all function generators produce square waves, triangular waves, and sine waves. Some function generators also produce sawtooth, pulse, and asymmetrical square waves. These generators commonly cover the frequency range from 0.01 Hz to 11 MHz. This type of instrument is typically rated to have a 600-Ω output impedance, and has a coarse, uncalibrated, output level control.

In most function generators the signal is originated from either a square-wave multivibrator or a triangular-wave generator, but not from sine-wave oscillators, because most have problems with amplitude and frequency stability unless properly compensated by the designer.

A Basic Function Generator with Signals Originated from a Square-Wave Multivibrator

A basic function generator is shown in Fig. 11.4-1. Its principal components are an op-amp comparator (A_1), op-amp integrator (A_2), low-pass filter, sine-wave amplifier (A_3), and an output amplifier (A_4). Feedback resistor R_f and capacitor C are connected between noninverting input and output terminals of the comparator and integrator, respectively. Both the loop involving R_f and that involving C_f provide positive feedbacks. Note that the reference voltage of 0 V (ground potential) is applied to the $-$input for comparator A_1. This, $v_{i-} = v_{ref} = 0$. The input signal v_i is the feedback signal v_f, which is applied to the $+$input: $v_{i+} = v_i = v_f$. Hence, the difference voltage is

$$v_d = v_{i+} - v_{i-} = v_i - 0 = v_i = v_f. \qquad (11.4\text{-}1)$$

Whenever v_f is slightly greater (about 70 μV) than 0 V, v_d will be positive and output v_{o1} will go to $+V_{o1}$ ($= + V_{sat} =$ maximum output voltage for

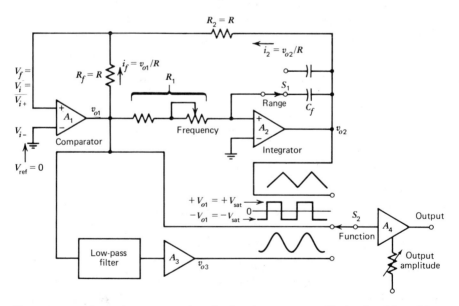

Figure 11.4-1 A basic sine–square–triangular function generator with signals originated from a square-wave multivibrator.

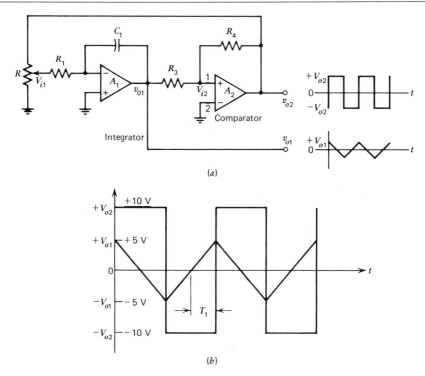

Figure 11.4-2 A basic function generator with signals originating from an intergrator. (a) Circuit. (b) Waveforms.

an op-amp, called the saturation voltage). On the other hand, if v_{fb} is slightly less than 0 V, v_d will be negative and output v_{o1} will go to $-V_{o1}$ $(= -V_{sat})$.

When the circuit in Fig. 11.4-1 is first turned on, the output voltage of comparator A_1 will snap to a peak value V_{01}, thus producing current $i_f = v_{o1}/R$. Voltage V_{o1} will also cause the feedback capacitor C_f in the integrator to start charging, thus causing its output v_{o2} to start rising. Voltage v_{o2} produces current $i_2 = v_{o2}/R$. When v_{o2} rises to the point where i_2 equals i_f, then v_{o1} snaps back to zero, also making i_f zero. But v_{o2} is still at the same level, so now i_2 is larger than i_f. This makes voltage v_{o1} pass through zero to become high in the negative direction. With $v_{o1} = -V_{o1}$ now, capacitor C_f will start to discharge, so voltage v_{o2} is reduced, toward zero.

The square wave is a composite of a fundamental sine wave and an infinite number of odd harmonics. When the square wave is applied to the low-pass filter, the harmonics will be removed, leaving only the sine-wave fundamental. Since the filtering causes a substantial loss of amplitude, the amplifier stage A_3 must follow the filter section.

A Basic Function Generator Signals Originated from an Integrator

The integrator–comparator combination shown in Fig. 11.4-2a is another basic function generator whose signals originate from the integrator. In the comparator circuit A_2, the input voltage and feedback voltage are both applied to the noninverting terminal 1, so that the output state is the same as the input state. Suppose the output voltage $v_{o2} = +10$ V, resistors $R_3 = R_4$, and the output impedance of op-amp A_1 is almost zero. Then the voltage v_{i2} across R_3 is $+5$ V; this is applied to A_2, causing its output voltage to be positive. This positive output is fed back to the inverting integrator A_1, causing it to start integrating, and its output voltage v_{o1} is then increased in the negative direction. When this negative voltage exceeds -5 V, the A_2 action immediately occurs, so that its output voltage snaps from $+10$ V to -10 V. This negative voltage applied to A_1 starts to integrate and the voltage v_{o1} gradually rises from -5 V. When v_{o1} exceeds $+5$ V, the A_2 action again occurs immediately, so that its output snaps from -10 V to $+10$ V. The circuit will continuously operate in such a manner, and the output v_{o1} will be a triangular wave and v_{o2} will be a square wave, as shown in Fig. 11.4-2b.

The integrator (A_1) output is given by

$$v_{o1} = -\frac{V_{i1}}{R_1 C_1} \int_0^{T_1} dt = -\frac{V_{i1}}{R_1 C_1} T_1. \tag{11.4-2}$$

From Fig. 11.4-2b we see that the period of a complete cycle is

$$T = 4T_1$$

and the frequency is

$$f = \frac{1}{4T_1}.$$

Since $T_1 = (V_{o1}/V_{i1}) R_1 C_1$, the frequency becomes

$$f = \frac{V_{i1}}{4R_1 C_1 V_{o1}}. \tag{11.4-3}$$

Usually, values of R_1 and C_1 are chosen by a band switch for the frequency change. The potentiometer R in Fig. 11.4-2a is used for a variable-frequency band.

11.5 VCO-TYPE FUNCTION GENERATORS

The VCO Composed of Multiplier, Integrator, and Comparator

Equation (11.4-3) indicates that the oscillation frequency may also be changed by varying the input voltage V_{i1}. One way to do this is to form a voltage-controlled oscillator (VCO) by adding a multiplier at the input terminal of the integrator in the circuit of Fig. 11.4-2a.

The multiplier is a differential amplifier ($Q_1 - Q_2$) with a constant-current stage (Q_3), as shown in Fig. 10.7-1. The output voltage is given by

$$v_{o2} \simeq 10I_0R_c(v_{b1} - v_{b2}). \qquad (10.7\text{-}8)$$

If signal A is used to control the magnitude of I_0 and signal B represents the voltage ($v_{b1} - v_{b2}$), then the output will be proportional to product of the two signals, or

$$V_o = AB.$$

Hence, the frequency is changed by varying voltage V_o instead of V_{i1} in Eq. (11.4-3).

Figure 11.5-1 shows a wide-range voltage-controlled oscillator. This circuit is similar to the basic function generator except that a multiplier and one more op-amp have been added. The main features of this circuit are that the triangular output voltage may reach a peak-to-peak value and the stability is high due to the action of the diodes.

The waveshaper is usually a combination of different-level clippers, as shown in Fig. 11.5-2. It is used to convert the triangular wave into a sine wave. The basic circuit of Fig. 11.5-2a operates as follows. When the input is within 1 V, the output level is the same as the input. When the input exceeds 1 V, diode D_1 conducts, and the output is decreased owing to R_1–R_2 division. When the input exceeds 1.2 V, both D_1 and D_2 conduct, then R_2 and R_3 are in parallel, and so the output is again decreased. When the input exceeds 1.5 V, diodes D_1, D_2, and D_3 conduct, then R_2, R_3, and R_4 are in parallel, and so the output is further decreased. Thus the output will eventually become a sine wave, as shown in Fig. 11.5-2b. The basic circuit

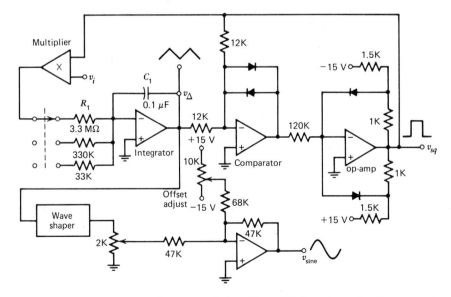

Figure 11.5-1   A wide-range voltage-controlled oscillator (VCO) used as a function generator.

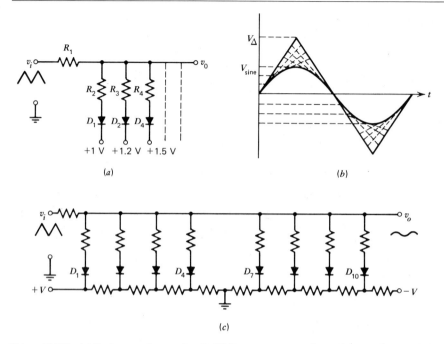

Figure 11.5-2 (a) Basic waveshaper circuit. (b) Sine-wave output from the waveshaper of Figs. 11.5-2a or 11.5-2c with a triangular wave input. (c) Practical waveshaper circuit.

of Fig. 11.5-2a is replaced by the practical circuit of Fig. 11.5-2c, where a group of resistors is used to divide the voltages so that the required different levels are obtained.

The VCO Composed of Comparator, CMOS Switch, and Integrator

A CMOS digitally controlled switch (double-pole double-throw, DPDT) has been added between the Schmitt comparator and the op-amp integrator in the voltage-controlled oscillator of Fig. 11.5-3a. The CMOS switch consists of a CMOS inverter feeding an op-amp follower, which drives R_1 from a very low output resistance. The oscillator frequency depends linearly upon a modulation voltage v_m.

Suppose that the Schmitt comparator output is $v_{sq} = V_{sq}$, where V_{sq} exceeds the modulation voltage $+v_m$. Then, in the CMOS digitally controlled switch, Q_1 is on and Q_2 is off. The input v_1 to the integrator is the output $-v_m$ from the voltage follower. Thereby, the output $v_\Delta(t)$ from the integrator increases linearly with a sweep speed of v_m/R_1C_1 V/s until v_Δ reaches the comparator threshold level $\beta V_{sq} = V_{sq}R_4/(R_3 + R_4)$. Then the Schmitt output changes state to $v_{sq} = -V_{sq}$, as indicated in Fig. 11.5-3b. Now Q_1 is off, Q_2 is on, and the CMOS switch output becomes $+v_m$, resulting in a linear negative ramp $v_\Delta = -v_mt/R_1C_1$ until the negative threshold $-\beta V_{sq}$ is reached. Since the two half-cycles are identical, the peak-to-peak voltage of the triangular wave is given by

$$\frac{v_m}{R_1 C_1} \frac{T}{2} = \beta V_{sq} - (-\beta V_{sq}) = 2 \frac{R_4}{R_3 + R_4} V_{sq}. \qquad (11.5\text{-}1)$$

The oscillator frequency is given by

$$f = \frac{1}{T} = \frac{R_3 + R_4}{4 R_1 C_1 R_4} \frac{v_m}{V_{sq}}. \qquad (11.5\text{-}2)$$

This equation indicates that the oscillation frequency varies linearly with the modulation voltage v_m. The linearity may extend from below 2 mV to above 2 V. The voltage-controlled oscillator of Fig. 11.5-3a is a system of a frequency-modulated square or triangular waveform.

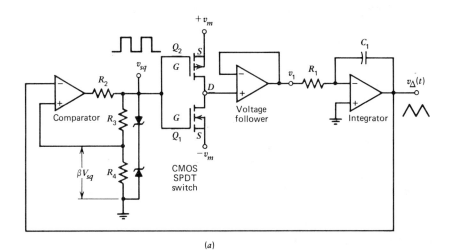

(a)

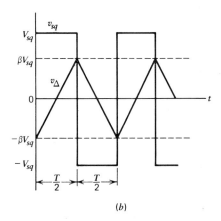

(b)

Figure 11.5-3 (a) A voltage-controlled oscillator whose frequency varies linearly with the modulating voltage v_m. (b) The square wave v_{sq} and the triangle wave v_Δ.

11.6 PULSE GENERATORS USING TRANSISTORS OR OP-AMPS

Monostable Transistor Multivibrator

The one-shot or monostable multivibrator is used to generate a rectangular waveform (pulse) having a width determined essentially by the values of a few circuit components. It has only one permanently stable state and one quasistable (unstable) state. On being triggered the circuit switches to its quasistable state where it remains for a predetermined duration (T_W) before returning to its original stable state.

The basic monostable multivibrator and its waveforms are shown in Fig. 11.6-1. The stable state exists when Q_2 is on and Q_1 is off. A negative trigger applied to the base of Q_2 at time $t = 0$ causes the transistor to turn off. The voltage (v_{c_2}) at the collector of Q_2 increases, forcing Q_1 to turn on. As Q_1 turns on and saturates, its collector voltage v_{c_1} drops from V_{CC} to $V_{CE(sat)}$. Since the voltage across a capacitor (C) cannot change instantaneously, the voltage (v_{B_2}) at the base of Q_2 experiences a drop of the same magnitude.

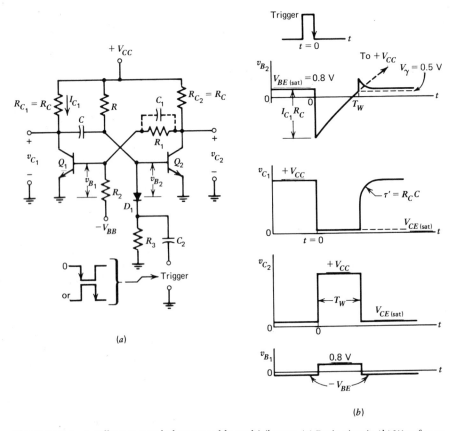

Figure 11.6-1 A collector-coupled monostable multivibrator. (a) Basic circuit. (b) Waveforms.

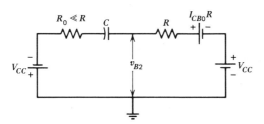

Figure 11.6-2 Simplified equivalent circuit used for computing the pulse width T_W affected by I_{CBO}.

Thus at $t = 0^+$, the negative value of v_{B_2} is given by

$$v_{B_2} = V_{BE(\text{sat})} - I_{c_1}R_c = 0.8 - I_{c_1}R_c \text{ (for Si)},$$

and Q_2 is kept off. The simplified equivalent circuit for computing the voltage v_{B_2} and the delay time T_W is the same as that of Fig. 11.3-5a. Thus, from Eq. (11.3-11), the pulse width of v_{c_2} is approximately given by

$$T_W \simeq 0.69RC. \tag{11.6-1}$$

The value of R is chosen to be less than $h_{FE(\text{min})}R_{C_2}$ in order to make Q_2 saturate in its stable state. The values of R_1 and R_2 are chosen so that Q_1 is off in the stable state and is saturated in the quasistable state. The value of R_c $(=R_{c_1}=R_{c_2})$ should be much less than R in order to avoid switching delays. If these conditions are satisfied, then the multivibrator will operate properly.

Consider the effect of the reverse saturation current I_{CBO} on the pulse width T_W. During the interval, when Q_2 is off, a nominally constant current I_{CBO} flows out of the base of Q_2. Assume now that capacitor C is disconnected from the junction of resistor R with the base of Q_2. Then the voltage at the base of Q_2 with C disconnected would be not V_{CC} but $V_{CC} + I_{CBO}R$. It therefore appears that capacitor C, in effect, charges through R from a source $V_{CC} + I_{CBO}R$. Hence, since I_{CBO} increases with temperature, the time T_W will decrease. The initial voltage v_{c_1} in the stable state is $V_{CC} - I_{CBO}R_c$, where now I_{CBO} is the collector current of Q_1 when it is off. When the multivibrator is triggered, $v_{c_1} = V_{CC} - (I_{c_1} + I_{CBO})R_c$; if $I_{CBO} = 0$, $v_{c_1} = V_{CC} - I_{c_1}R_c$. Thus the drop in v_{c_1} when the circuit is triggered is smaller by $I_{CBO}R_c$ than it would otherwise be. This is a second way in which I_{CBO} affects T_W. Since, however, $R_c \ll R$, this second effect is negligible in comparison with the first. To take account of the effect of I_{CBO} assume that $V_{CE(\text{sat})} = V_{BE(\text{sat})} = V_\gamma = 0$. During the interval T_W, Q_2 is off, and the change in v_{B_2} can be calculated from the simplified equivalent circuit shown in Fig. 11.6-2. At $t = 0$, v_{B_2} is zero and plunges to $-V_{CC}$. Using superposition method, we find

$$v_{B_2} \simeq (V_{CC} + I_{CBO}R)(1 - e^{-t/\tau}) - V_{CC}e^{-t/\tau}$$

where $\tau = (R + R_0)C \simeq RC$. At $t = T_W$, v_{B_2} is again zero. Then

$$0 = (V_{CC} + RI_{CBO}) - (2V_{CC} + RI_{CBO})e^{-T_W/\tau},$$

from which we obtain

$$T_W \simeq \tau \ln 2 - \tau \ln \frac{1 + I_{CBO}R/V_{CC}}{1 + I_{CBO}R/2V_{CC}} = \tau \ln 2 - \tau \ln \frac{1 + \phi}{1 + \phi/2}. \qquad (11.6\text{-}2)$$

where $\phi \equiv I_{CBO}R/V_{CC}$. From Eq. (11.6-2) we see that the delay time T_W decreases as the temperature increases.

One method of temperature compensating a monostable multivibrator is to connect R not to $+V_{CC}$ but to a voltage source V whose value decreases as the temperature increases. The voltage V may be obtained from a voltage divider $r_1 - r_2$ across V_{CC} in which the value of r_1 increases with temperature as it is connected to the $+V_{CC}$ terminal. In this case, V is provided from the tap between r_1 and r_2.

The monostable circuit is the basis of electronic timers. Integrated-circuit timers are commercially available.

Op-amp Monostable Multivibrator

The square-wave generator of Fig. 11.3-1 is modified in Fig. 11.6-3 to operate as a monostable multivibrator by adding a diode D_1 clamp across capacitor C and by introducing a negative trigger pulse through diode D_2 to the non-inverting terminal. The circuit is in its stable state with the output at $v_o = +V_o$ and with the capacitor clamped at diode D_1 on voltage $V_D \simeq 0.7$ V. The feedback voltage is

$$\beta v_o = \frac{R_2 v_o}{R_1 + R_2}, \qquad (11.6\text{-}3)$$

and $\beta V_o > V_D$. If the magnitude of the trigger amplitude is greater than $\beta V_o - V_D$, it will cause the comparator to switch to an output $v_o = -V_o$. As indicated in Fig. 11.6-3b, capacitor C will now charge exponentially with a time constant $\tau = RC$ through R toward $-V_o$ because D_1 becomes reverse biased. When the capacitor voltage v_c becomes more negative than $-\beta V_o$, the comparator output swings back to $+V_o$. The capacitor now begins charging toward $+V_o$ through R until v_c reaches V_D, and C again becomes clamped at $v_c = V_D$. Following Eq. (11.3-8), the capacitor voltage for $t > 0$ is expressed as

$$v_c = -V_o + (V_D + V_o)e^{-t/\tau}. \qquad (11.6\text{-}4)$$

When $v_c = -\beta V_o$, $t = T_W$, then Eq. (11.6-4) becomes

$$T_W = RC \ln \frac{1 + V_D/V_o}{1 - \beta}. \qquad (11.6\text{-}5)$$

If $V_o \gg V_D$ and $R_1 = R_2$ so that $\beta = \frac{1}{2}$, then $T_W = 0.69RC$.

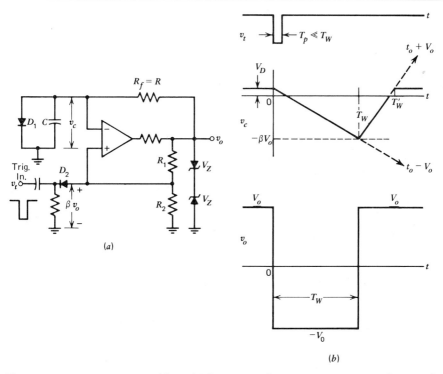

Figure 11.6-3 Op-amp monostable multivibrator or pulse generator. (a) Circuit diagram. (b) Waveforms (including triggering pulse v_t, capacitor waveform v_c, and output pulse v_0).

The duration T_p of the trigger pulse v_t should be much less than the delay time or width T_W of the generated pulse. The diode D_2 serves to avoid malfunctioning if any positive noise spikes are present in the triggering line. The capacitor voltage v_c does not reach its quiescent value $v_c = V_D$ until time $T'_W > T_W$. Consequently, there is a recovery time $T'_W - T_W$ during which the circuit may not be triggered again.

A Three-Transistor Pulse Generator

The pulse generator circuit of Fig. 11.6-4a consists of a triggering stage Q_1 and a monostable multivibrator Q_2–Q_3. Transistor Q_1 is saturated for less than half a cycle when a sinusoidal input v_i is applied, and is off for the remainder of the period. The output of Q_1 is a square wave, which applied to the RC differentiator, becomes spikes. The positive spike passes through diode D and is applied to the base of Q_2 for triggering. On being triggered the one shot switches to its quasistable state where it remains for a predetermined delay time ($T_W \approx 0.69R_TC_T$) before returning to its original stable state. The waveforms are shown in Fig. 11.6-4b.

Duty Cycle and Pulse Characteristics

The basic difference between a pulse generator and a square-wave generator concerns the duty cycle. The duty cycle is the ratio of the pulsewidth and the period (or pulse repetition time),

$$\text{Duty cycle} = \frac{\text{Pulsewidth}}{\text{Period}}.$$

The duty cycle is also the ratio of the resolving time to the mean spacing between pulses. The resolving time of a pulse (or a differentiator) equals the ratio of the pulse area to height.

The duty cycle of a square-wave generator remains at 50% as the frequency of oscillation is varied. The duty cycle of a pulse generator may vary; very-short-duration pulses give a low duty cycle. Short-duration pulses reduce the power dissipation in the component under test.

The pulse characteristics are shown in Fig. 11.6-5. The time required for the pulse to increase from 10 to 90% of its normal amplitude is called the rise time (t_r). The time required for the pulse to decrease from 90 to 10% of its maximum amplitude is called the fall time (t_f). When the initial am-

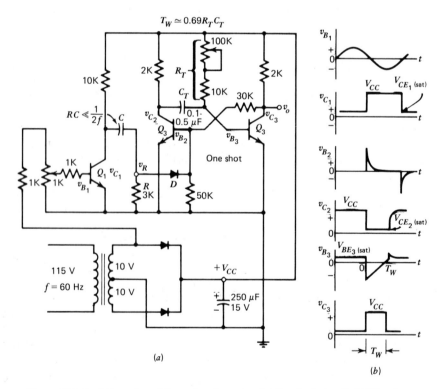

Figure 11.6-4 A three-transistor pulse generator. (a) Practical circuit. (b) Waveforms.

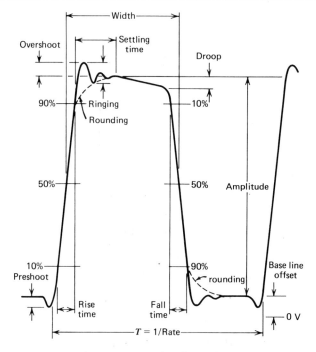

Figure 11.6-5 Pulse characteristics.

plitude rise exceeds the correct value, overshoot occurs. The overshoot may be visible as a single pip, or ringing (i.e., nearly undamped oscillation) may occur. When the maximum amplitude of the pulse is not constant but decreases slowly, the pulse is said to droop or tilt.

11.7 PULSE GENERATORS USING 555 TIMERS

Basic 555 Timer

The 555 monolithic IC timer is a very versatile unit. NS/SE 555 is a Signetics Corporation designation. The same unit is available as MC 1555 from Motorola. Although the 555 timer is considered to be an analog circuit, it can also function as a monostable (one-shot) multivibrator to provide accurate time delays, or as an astable multivibrator to generate clock pulses. Moreover, it is capable of supplying up to 200 mA of load current and can be operated over a wide voltage range, from 4 to 18 V. A block diagram of this integrated circuit is shown in Fig. 11.7-1. To use the 555 timer as a one-shot or an astable, the operation of the basic 555 circuit must be understood.

The trigger comparator compares a voltage of $V_{CC}/3$ to the voltage at the trigger input. When the trigger voltage drops to a value of $V_{CC}/3$ or less, the

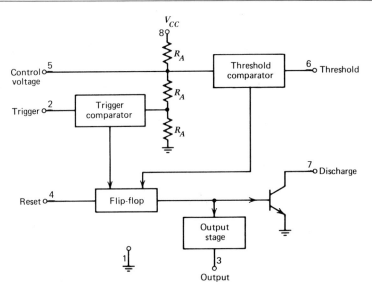

Figure 11.7-1   Block diagram of the 555 timer.

comparator produces an output that sets the flip-flop to the low-output state. Then the flip-flop cuts the discharge transistor off and drives the output positive through the output inverting buffer.

The threshold comparator continually compares a voltage of $2V_{CC}/3$ to the voltage at the threshold input. When the threshold voltage equals or exceeds $2V_{CC}/3$, the comparator produces a voltage that sets the flip-flop output to the high state. The flip-flop then drives the base of the discharge transistor so that the transistor can saturate when the collector circuit is completed. The output pin is now low. A low voltage on the reset input can also drive the flip-flop to this state.

A 555 timer that has been wired to act as a timed switch is shown in Fig. 11.7-2. The voltages as a function of time at trigger pin 2, threshold pin 6, and output pin 3 are shown in Fig. 11.7-3. When switch S is closed (pin 2 grounded), the input voltage at pin 2 goes from high ($\approx V_{CC}$) to low (0 V) (Fig. 11.7-3a), the output voltage at pin 3 goes from low to high (Fig. 11.7-3c), and the internal switch of pin 7 opens to let capacitor C charge through resistor R (Fig. 11.7-3b). When the threshold input at pin 6 reaches $2V_{CC}/3$, the flip-flop output is set to high state, so that the output at pin 3 goes low (Fig. 11.7-3c), and the internal switch of pin 7 closes to let capacitor C discharge (Fig. 11.7-3b). The period of high-output voltage at pin 3 is

$$T = 1.1RC, \tag{11.7-1}$$

which will be derived in the next paragraph.

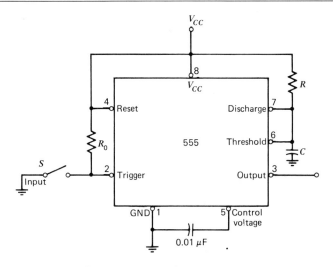

Figure 11.7-2 The 555 timed switch.

The 555 Timer as a One-Shot

To use the 555 timer as a one-shot, the configuration of Fig. 11.7-4 can be employed. It is triggered electronically by applying a suitable external voltage to pin 2, whereas the timed switch in Fig. 11.7-2 is activated by using an input switch S. In Fig. 11.7-4, the 0.01 μF capacitor is connected to the control-voltage output to filter out the noise supply. With the trigger input held at a voltage greater than $V_{CC}/3$ and the threshold input at V_{CC}, the

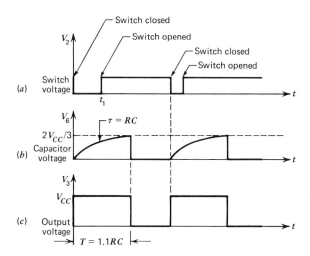

Figure 11.7-3 The voltages as a function of time at the trigger pin 2, threshold pin 6, and output pin 3 for the 555 timed switch of Fig. 11.7-2.

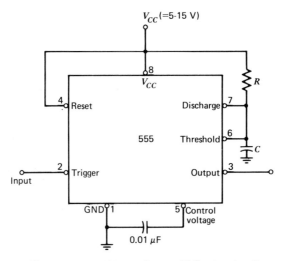

Figure 11.7-4   555 one-shot multivibrator circuit.

output is at the low level of 0 V. The discharge transistor is saturated, and the capacitor C is at the 0-V level. When the trigger input is dropped below $V_{CC}/3$, the flip-flop changes states, shutting off the discharge transistor and driving the output positive to approximately V_{CC}. As the capacitor charges toward V_{CC}, the threshold input reaches the voltage $2V_{CC}/3$ and the flip-flop changes back to the original state, dropping the output to 0 V and again discharging the capacitor. The period T is the time taken for the capacitor to charge from 0 V to a voltage of $2V_{CC}/3$ with a final (target) voltage of V_{CC}. This time can be found by writing the general equation for capacitor voltage, plugging in the correct initial and final values, and solving for T:

$$v_C = v_i + (v_f - v_i)(1 - e^{-t/\tau}), \qquad (11.7\text{-}2)$$

where v_i is the initial voltage, v_f is the final voltage, and τ is the time constant RC. For this case $v_i = 0$ and $v_f = V_{CC}$,

$$v_C = 2V_{CC}/3 = V_{CC}(1 - e^{-T/RC}). \qquad (11.7\text{-}3)$$

Solving for T we find

$$T = RC \ln 3 = 1.1RC. \qquad (11.7\text{-}1)$$

The trigger pulse must be shorter than T for Eq. (11.7-1) to be valid. If the trigger input remains low after the period is over, the output remains high. When the input trigger pulse is longer than T, an $R'C'$ differentiator should be used to ensure that the pulse reaching the trigger terminal is short enough. Resistor R' connects from V_{CC} to pin 2, and capacitor C' is inserted between the input line and pin 2. The trigger input is then held positive until the leading edge of the trigger pulse arrives. This transition is coupled through the capacitor to initiate the period. Pin 2 then charges toward V_{CC} with a time constant largely determined by the $R'C'$ product of the trigger circuit. This value must be smaller than that of the one-shot period T. The range of T is from 10 μs to several hours by varying the elements R and C. Examples:

$$T = 1.1RC = 1.1\,(10\text{K})(0.001\ \mu\text{F}) = 11\ \mu\text{s};$$

$$T = 1.1\,(10\ \text{M}\Omega)(100\ \mu\text{F}) = 1100\ \text{s}.$$

The output of the 555 monostable multivibrator has only one (mono-)

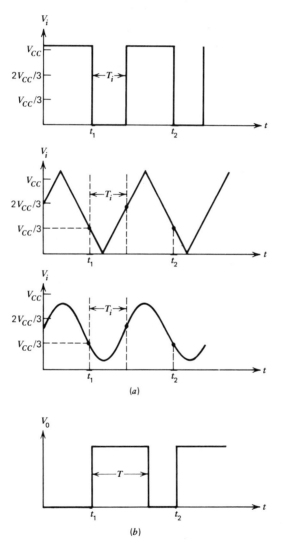

(a)

(b)

Figure 11.7-5   Input signals $(T_i < T)$ and output waveform of the one-shot multivibrator in Fig. 11.7-4. For proper triggering, the input signal must go from high (more than $2V_{CC}/3$) to low (less than $V_{CC}/3$) and then back to more than $2V_{CC}/3$ in a time less than the output pulse width. (a) Input voltages, $T_i < T$. (b) Output voltage.

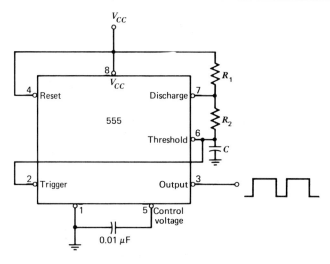

Figure 11.7-6   The 555 timer used as an astable multivibrator.

stable value, 0 V. A trigger signal applied to pin 2 will produce a fixed-width pulse. A significant application of a one-shot multivibrator is waveshaping. As shown in Fig. 11.7-5, a wide variety of input signals can be fed into the 555 one-shot, but the output will always have the same pulse height and width. Thus, the one-shot often serves as an interface between an input transducer whose output may vary and subsequent circuits that require a standard input pulse for their operation.

The 555 Timer as an Astable Multivibrator

Figure 11.7-6 shows a 555 timer wired to make an astable multivibrator, composed of a capacitor and two resistors. Note that the capacitor voltage (v_C), pin 6, is connected directly to the input trigger, pin 2. The voltage v_C can swing between $V_{CC}/3$ and $2V_{CC}/3$. It rises toward $2V_{CC}/3$ with a final voltage of V_{CC} and a time constant of $(R_1 + R_2)C$. When it reaches $2V_{CC}/3$, the threshold comparator causes the flip-flop to change state (see Fig. 11.7-1). The discharge transistor saturates, and the capacitor voltage heads toward ground with a time constant of R_2C. When the voltage drops to $V_{CC}/3$,

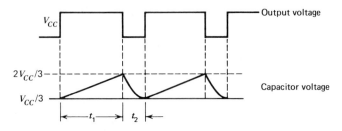

Figure 11.7-7   Typical waveforms for a 555 astable multivibrator.

the trigger comparator changes the state of the flip-flop and shuts off the discharge transistor. The capacitor again charges toward V_{CC}. The output and capacitor voltage waveforms of the 555 are shown in Fig. 11.7-7.

The duration of the positive portion of the waveform is t_1 and is calculated from Eq. (11.7-2), with $v_i = V_{CC}/3$, $v_f = V_{CC}$, and $\tau = (R_1 + R_2)C$. This gives

$$v_C = 2V_{CC}/3 = V_{CC}/3 + (V_{CC} - V_{CC}/3)(1 - e^{-t_1/\tau}).$$

Solving for t_1 we get

$$t_1 = (R_1 + R_2)C \ln 2 = 0.7(R_1 + R_2)C. \qquad (11.7\text{-}4)$$

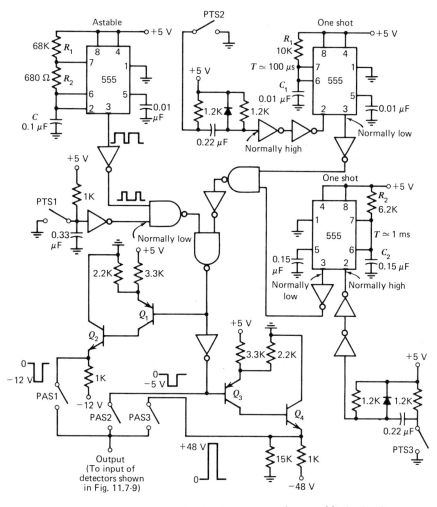

Figure 11.7-8 Practical 555 pulse generators with control logic circuit.

Similarly, the portion of t_2 can be found to be

$$t_2 = 0.7R_2C. \tag{11.7-5}$$

The duty cycle of the astable multivibrator is the ratio of the duration of the positive portion of the period to the total period, or

$$\text{Duty cycle} \equiv \frac{t_1}{t_1 + t_2} = \frac{R_1 + R_2}{R_1 + 2R_2}. \tag{11.7-6}$$

If $R_1 \gg R_2$, the duty cycle will be a maximum value of 100%. If $R_2 \gg R_1$, the duty cycle will approach a minimum value of 50%.

The ratio of the time the output transistor is on (low voltage) to the total period is called the ON duty cycle:

$$\text{ON duty cycle} = \frac{t_2}{t_1 + t_2} = \frac{R_2}{R_1 + 2R_2}. \tag{11.7-7}$$

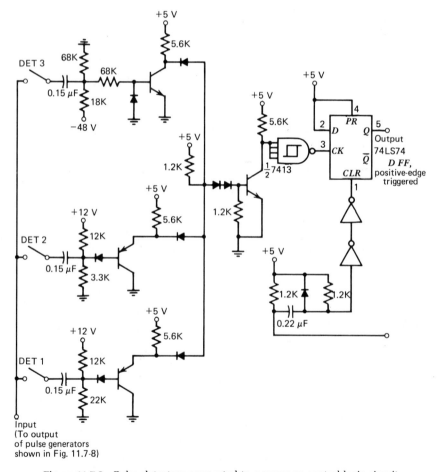

Figure 11.7-9 Pulse detectors connected to a common control logic circuit.

If the resistor R_2 in Fig. 11.7-6 is shunted by a diode with its anode connected to pin 7 and cathode to pin 6, then the capacitor can be charged and discharged through the separate resistors; in this case, R_1 and R_2 can be varied separately to give nearly a full range of output duty cycle from 0 to 100%.

Practical 555 Pulse Generators

The 555 pulse generators shown in Fig. 11.7-8 are used as the signal sources to trace the printed-circuit boards for trouble shooting by the way of the pulse detector shown in Fig. 11.7-9. In Fig. 11.7-8, the astable multivibrator produces a pulse train. The pulse width is $0.7(R_1 + R_2)C$ or 4.8 ms. The time between pulses is $0.7R_2C$ or 47.6 μs. The pulse widths of the outputs from the two monostable multivibrators are $1.1R_1C_1 \simeq 100$ μs and $1.1R_2C_2 \simeq 1$ ms. The outputs of these generators are controlled by the logic circuit. The pulse signals are selected by switches PTS1, PTS2, and PTS3; only one signal can be taken each time. The chosen signal can be amplified through one of the two complementary symmetry circuits (Q_1–Q_2 and Q_3–Q_4), resulting in the -12 V or $+48$ V amplitude. The different pulse amplitudes are selected by closing switch PAS1, PAS2, or PAS3, which can be automatically controlled with computer programming.

The output pulse from the generator system is applied to the proper detector. Each detector in Fig. 11.7-9 consists of two diodes and one transistor stage with the common control logic circuit at the end. The chosen pulse from the detector output is sent to the PC board for trouble shooting.

11.8   A TYPICAL HIGH-POWER PULSE GENERATOR

Block Diagram Description

The front end of the pulse generator Model 114A (Systron-Donner) is primarily composed of stages of multivibrators, pulse differentiators, and a pick-off amplifier. To provide sufficient drive power, the final stage of this front end is connected to the input of an output amplifier, as shown in the block diagram of Fig. 11.8-1.

The time reference for the instrument is provided by the repetition rate oscillator (or by an external trigger source, not shown in the diagram). The repetition rate oscillator is an emitter-coupled astable multivibrator. The pulses from the repetition rate oscillator or the single-cycle pushbutton are shaped into negative-going spikes by the trigger multivibrator. Each pulse from the trigger multivibrator shifts the delay one-shot to its quasistable (unstable) state. The duration that the delay one-shot remains in its quasistable state is determined by the size of the timing capacitor selected by the delay/advance control. The rectangular pulse from the delay one-shot is differentiated by the pulse differentiator and coupled to the synchronous output amplifier and to the pulse mode switch for the selection of

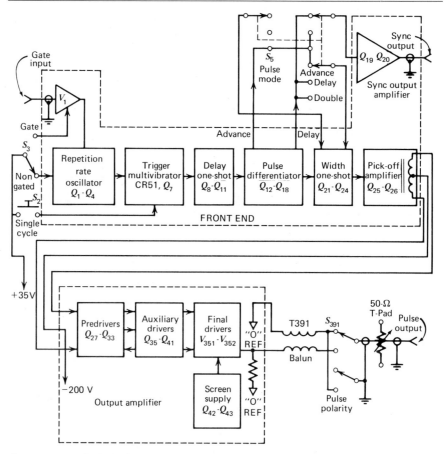

Figure 11.8-1   Block diagram of pulse generator Model 114A (Systron-Donner), showing internal trigger, gated or nongated mode. (Courtesy of Systron-Donner Corp.)

pulse/advance or double pulse mode. At the advance position, the output pulse occurs ahead of the sync pulse by the time interval determined by the delay/advance controls. At delay position, the output pulse occurs after the sync pulse by the time interval determined by the delay/advance controls. At the double position, two pulses per repetition rate period are generated at the output. The separation of the two pulses is determined by the delay/advance controls. The width one-shot is shifted to its quasistable state by the positive spike of the differentiated pulse from the pulse differentiator. The duration that the width one-shot is in the quasistable state is determined by the setting of the width controls. The output pulse from the width one-shot passes through the pick-off amplifier and is then applied to the predrivers. The predrivers and auxiliary drivers supply voltage and current drive for the leading and trailing edge triggers that generate the output pulse by turning the final drivers on and off. The output amplifier is isolated from the front end circuit by two transformers and isolated from the output by a balun, which allows for the selection of either a positive or negative output pulse.

The simplified circuits for the main sections of Model 114A pulse generator are shown in Figs. 11.9-1–11.9-3, 11.9-5, 11.9-6, and 11.10-7. These circuits are analyzed in Sec. 11.9 and 11.10.

Principal Performance Characteristics

The Model 114A is a high-performance laboratory and production-line instrument capable of 200-W peak pulse power at repetition rates up to 1 MHz. The output pulse is polarity selectable, and the amplitude is variable up to 100 V into 50 Ω. Pulse delay, advance, and width are variable up to 10 ms. Rise and fall times are fixed at less than 13 ns for amplitudes up to 50 V and 17 ns on the 100 V range. The instrument shown in Fig. 11.8-1 provides single or double pulse output and can be gated by an external source.

The repetition rate oscillator operates in five ranges from 10 Hz to 1 MHz. One pulse is generated at the output each time the single cycle pushbutton is depressed. For gated operation, the input gate must be about 8–40 V to obtain gated pulse bursts at main and sync output; input impedance is 250K or more. Minimum pulse spacing (delay) in the double pulse mode is 1 μs. The maximum duty cycle is 50% for amplitude of 10 V or less, 25% for amplitude from 10 to 20 V, or 10% for amplitudes from 20 to 100 V. Overshoot is 5% or less. Preshoot is 2% or less. Droop is 6% or less. The irregular random variation or jitter in repetition rate period is less than 0.5% of repetition rate period + 1 ns. Jitter in delay/advance is less than 0.05% of delay/advance setting + 1 ns. Jitter in pulse width is less than 0.05% of width setting + 1 ns. Output impedance is 50 Ω for amplitudes up to 50 V or 1500 Ω for amplitudes from 50 to 100 V. Output is protected against short and open circuits. An overload light illuminates when the duty cycle is exceeded. The amplitude, width, rise time, and fall time of the sync output pulse are 25 V into 2K, approximately 400 ns, less than 50 ns, and less than 300 ns, respectively. In the gated mode, output pulses occur only for time interval of an externally applied gate pulse. The power required is approximately 250 W, 50–60 Hz, at 100–125 Vac. Operating temperature is from 0 to 50°C.

11.9 ANALYSIS OF THE PULSE GENERATOR FRONT-END CIRCUIT

Repetition Rate Oscillator

The repetition rate oscillator of Fig. 11.9-1 is an emitter-coupled astable multivibrator. Transistors Q_1 and Q_2 comprise the basic oscillator. The frequency of the oscillator is determined by the value of capacitor C_T and the setting of potentiometer R_T. Assume that Q_1 has just turned on. The collector of Q_1, clamped at approximately 5 V by Zener diode, ZD_1, steps negatively.

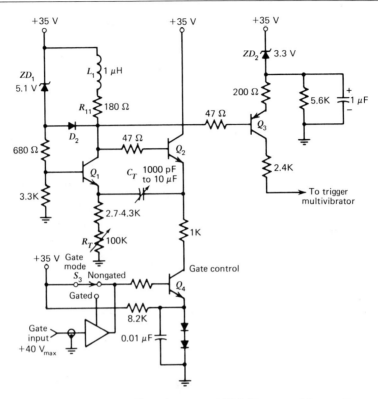

Figure 11.9-1   Repetition rate oscillator (10 Hz to 1 MHz). (Courtesy of Systron-Donner Corp.)

The negative step at the base of Q_2 produces a negative transient at its emitter, which is coupled through C_T to the emitter of Q_1. This regenerative feedback increases the conduction through Q_1. With Q_1 conducting, C_T charges through R_{11} and Q_1. The charging of C_T causes the voltage at the emitter of Q_2 to go negative. When the emitter voltage becomes negative enough to overcome the bias, Q_2 conducts, coupling a positive transient through C_T to the emitter of Q_1, and Q_1 is cutoff. While Q_1 is off, capacitor C_T discharges through timing resistor R_T and the voltage on the emitter of Q_1 starts to go negative. When the voltage of the emitter of Q_1 becomes negative with respect to fixed voltage on its base, Q_1 conducts and the cycle repeats itself.

The negative-going rectangular pulses developed at the collector of Q_1 are inverted by Q_3 and coupled to the trigger multivibrator $(D_{51}–Q_7)$.

Trigger Multivibrator and Delay One-Shot

The trigger multivibrator in Fig. 11.9-2 consists of transistor Q_7 and tunnel diode D_{51}. This circuit develops the negative-going spikes that trigger the delay one-shot. The spikes are developed by differentiating the pulses from the repetition rate oscillator and removing the positive spikes by means of a clamp diode. The leading edge of the pulse from the repetition rate oscillator or single-cycle pushbutton shifts tunnel diode D_{51} to its high-voltage state. The positive-going step produced by the tunnel diode turns off Q_7, causing its collector to step negatively. With Q_7 in a nonconducting state, capacitor C_{53} charges through resistor R_{54}. When the charge on C_{53} becomes large enough to overcome the reverse bias, Q_7 turns on. Capacitor C_{53} then discharges through Q_7 and the voltage on its collector rises toward 0 V. The trailing edge of the input pulse, which occurs after C_{53} is discharged, shifts tunnel diode D_{51} to its low-voltage state and the collector of Q_7 is prevented from going positive by clamp diode D_{61}.

The delay one-shot $(Q_8–Q_{11})$ is a monostable multivibrator. Each incoming pulse from the trigger multivibrator shifts the one-shot to its quasistable state. The duration that it remains in the quasistable state is determined by the selected timing capacitor (C_T) and the setting of the delay vernier (R_T).

In the stable condition, Q_8 is on, since its base is at approximately 0 V and the base of Q_9 is approximately -0.7 V. The negative-going spike from the trigger multivibrator turns Q_8 off, causing the base of Q_{10} to go positive. The positive transient, produced by Q_8 turning off, is coupled through C_T to the base of Q_9, causing Q_9 to conduct. Capacitor C_T then charges through

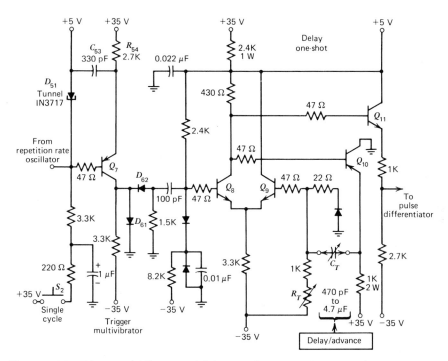

Figure 11.9-2 Trigger multivibrator and delay one-shot circuit. (Courtesy of Systron-Donner Corp.)

R_T and the base of Q_9 begins to go negative. When the base of Q_9 reaches approximately -0.7 V, Q_9 turns off, returning the circuit to the stable state. The rectangular pulses developed at the collector of Q_8 are coupled to the pulse differentiator by emitter-follower Q_{11}.

Pulse Differentiator

The pulse differentiator in Fig. 11.9-3 is composed of two differential amplifiers (Q_{12}–Q_{13} and Q_{17}–Q_{18}) and a one-shot multivibrator (Q_{14}–Q_{16}). The leading edge of the rectangular pulse from the delay one-shot turns Q_{12} on and Q_{13} off, and the trailing edge reverses the conducting states. The pulse produced at the collector of Q_{13} is differentiated by resistor R_{85} and inductor L_{81}. The differentiated pulse is coupled to the advance position on the pulse mode switch. Diode D_{81} removes the negative half of the differentiated pulse, and the positive half is coupled to the base of Q_{14}. The one-shot multivibrator portion of the pulse differentiator functions essentially the same as the delay one-shot described earlier. The positive half of the differentiated pulse turns Q_{14} on and Q_{15} off. Both transistors remain in this state until the charge on C_{83} is positive enough to turn Q_{15} on again. The resulting negative-going pulse at the collector of Q_{14} is coupled to the base of Q_{17} by emitter-follower Q_{16}. Transistor Q_{17} is turned off by the leading edge of the pulse and is turned on by the trailing edge of the pulse. The pulse at the collector of Q_{17} is differentiated by resistor R_{99} and inductor L_{82}, then coupled to the double and delay positions of pulse mode switch.

Figure 11.9-4 shows the simplified equivalent output circuit and the input and output waveforms of the Q_{13} and Q_{17} stages. Since the instantaneous voltage $L di/dt$ across an inductor cannot be infinite, the current through the inductor cannot change discontinuously. Hence, an inductor acts as an open circuit at the time of an abrupt change in voltage. Therefore, at the instant of the leading or trailing edge of the applied rectangular pulse, a spike will be produced owing to the parallel combination of R_{85}–L_{81} or R_{99}–L_{82} at the collector.

Synchronous Output Amplifier

The sync output amplifier in Fig. 11.9-5 consists of transitors Q_{19} and Q_{20}. transformer T_{111}, and switch S_6. Positive feedback for the amplifier is taken through capacitor C_{113}. Sync polarity switch S_6 reverses the polarity of the sync output by grounding either side of the coupling transformer. Resistor R_{120} and inductor L_{111} are used to decrease the circuit Q of the pulse transformer so that the overshoot is reduced.

Response of the Pulse Transformer

The equivalent circuit of a practical pulse transformer is shown in Fig. 11.9-6a, where L_M represents the magnetizing inductance portion of the primary inductance (L_p), and L_L represents the leakage inductance. L_L can be meas-

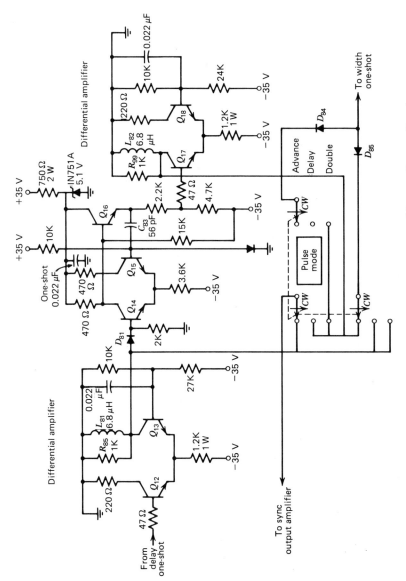

Figure 11.9-3 Pulse differentiator circuit. (Courtesy of Systron-Donner Corp.)

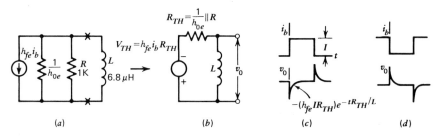

Figure 11.9-4 (a)–(b) Output equivalent circuit of Q_{13} and Q_{17}. (c) Waveforms of Q_{13}. (d) Waveforms of Q_{17}.

ured with an impedance bridge provided that the secondary is shorted. If this experiment is repeated with the secondary open circuited, then L_M $(\approx L_p)$ will be measured. Usually, L_L is much less than L_M. Typically, L_L is in the range from 0 to 100 μH, while L_M is from 10 to 100 mH. Resistors R_p and R_s represent the ohmic resistances of the primary and secondary windings, respectively. Capacitor C is used to take into account the effects of capacitance, which exists between the primary and secondary windings; it is placed in parallel with the primary winding of the ideal transformer. The pulse transformers normally use a turns ratio of less than 10 in order to minimize the undesired effects of C. The value of C is usually in the range

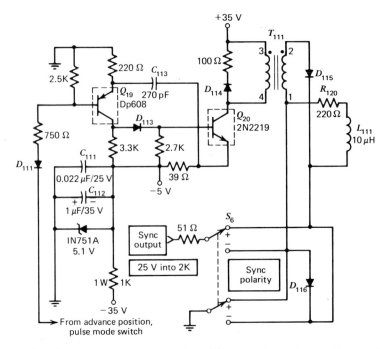

Figure 11.9-5 Synchronous output (pulse) amplifier. (Courtesy of Systron-Donner Corp.)

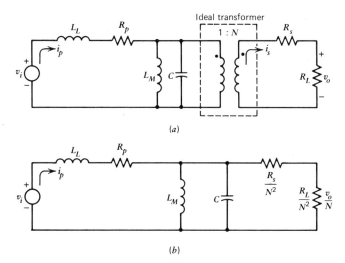

Figure 11.9-6 (a) Practical pulse transformer equivalent circuit. (b) The same circuit secondary resistances reflected back to primary.

up to 100 pF. The circuit of Fig. 11.9-6a can be simplied somewhat by reflecting the secondary resistances R_s and R_L back to the primary, as shown in Fig. 11.9-6b.

Several things concerning the circuit of Fig. 11.9-6b are described as follows. For a step or pulse input the transitions (rising or falling edges) contain the high-frequency components, while the flat, constant portions contain the low-frequency components. The pulse transformer, then, must respond to a wide range of frequencies. In Fig. 11.9-6b it appears that as far as the output is concerned the circuit acts as a band-pass circuit. That is, at very high frequencies the impedance of L_L will be high so that i_p will be diminished. The impedance of C will become very small so that most of i_p will be shunted away from the reflected load. Thus at high frequencies the output will drop. On the other hand, at very low frequencies it is the magnetizing inductance (L_M) that causes the output to drop, since the impedance of L_M becomes very small at low frequencies and shunts current away from the reflected load. Thus, the output of the transformer will generally be a distorted version of the input.

A procedure to be followed in analyzing this circuit utilizes a high-frequency analysis and a low-frequency analysis, which are both concerned with the circuit's response to the rising edge of the step and the response to the constant portion of the step, respectively. By combining the two responses, a good approximation of the complete response can be made.

Consider the pulse transformer equivalent circuit with its parameters given in Fig. 11.9-7a. The input is a 10-V step with rise time $t_r = 0.1$ μs. The turns ratio in this case is $N = 1$, so the voltage across the reflected load will be the same as the actual secondary load voltage. The secondary

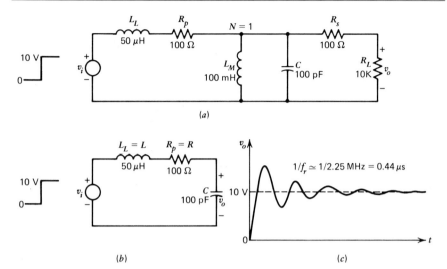

Figure 11.9-7 (a) Pulse transformer equivalent circuit with a turns ratio of $N = 1$. (b) Circuit simpliied for high-frequency analysis. (c) Response to rising edge of input step.

winding resistance (R_s) is much less than R_L. Since $t_r = 0.1$ µs, frequencies as high as $f_H = \frac{1}{2}t_r = 5$ MHz must be considered. Then,

$$X_M = 2\pi f L_m = (6.28) \times (5 \times 10^6) \times 10^{-1} = 3.14 \text{ M}\Omega$$

and

$$X_C = \frac{1}{2}\pi f C = 318 \ \Omega.$$

Since L_M and R_L are shunted by C, the circuit is simplified for the high-frequency analysis, as shown in Fig. 11.9-7b. This simplified circuit is a series RLC circuit, which is analyzed as follows.

Applying KVL to the circuit of Fig. 11.9-7b at the instant of the input step rising edge, we obtain

$$V = R_i + L\frac{di}{dt} + \frac{1}{C}\int idt. \tag{11.9-1}$$

Differentiating, we get the homogeneous equation:

$$0 = L\frac{d^2i}{dt^2} + R\frac{di}{dt} + \frac{i}{C}. \tag{11.9-2}$$

Assume a trial solution $i = Ke^{st}$. Substituting in Eq. (11.9-2), we find $Ke^{st}(LS^2 + RS + 1/C) = 0$, or the characteristic equation:

$$S^2 + \frac{R}{L}S + \frac{1}{LC} = 0. \tag{11.9-3}$$

The roots of this quadratic occur at

$$S = S_1, S_2 = -\frac{R}{2L} \pm \sqrt{\left(\frac{R}{2L}\right)^2 - \frac{1}{LC}} . \tag{11.9-4}$$

The borderline case occurs when

$$\left(\frac{R}{2L}\right)^2 = \frac{1}{LC} \tag{11.9-5}$$

Let us call the value of resistance that satisfies this equation the critical resistance, R_K. Substituting in Eq. (11.9-5), we get

$$R_K = 2 \sqrt{\frac{L}{C}} . \tag{11.9-6}$$

Then, we can write Eq. (11.9-4) as

$$S_1, S_2 = -\frac{R_K}{2L}\frac{R}{R_K} \pm \sqrt{\left(\frac{R_K}{2L}\right)^2 \left(\frac{R}{R_K}\right)^2 - \frac{1}{LC}}. \tag{11.9-7}$$

Let

$$\omega_0 = 2\pi f_0 = 1/\sqrt{LC}. \tag{11.9-8}$$

Substituting Eqs. (11.9-6) and (11.9-8) into Eq. (11.9-7), we obtain

$$S_1, S_2 = -\omega_0 \frac{R}{R_K} \pm \omega_0 \sqrt{\left(\frac{R}{R_K}\right)^2 - 1}. \tag{11.9-9}$$

The damping factor is defined as

$$d \equiv \frac{R}{R_K} . \tag{11.9-10}$$

Then Eq. (11.9-9) becomes

$$S_1, S_2 = -\omega_0 d \pm \omega_0 \sqrt{d^2 - 1}. \tag{11.9-11}$$

A series RLC circuit has a Q given by

$$Q = \frac{2\pi f_0 L}{R} = \frac{1}{R} \sqrt{\frac{L}{C}} . \tag{11.9-12}$$

Then Eq. (11.9-10) becomes

$$d = \frac{R}{2\sqrt{L/C}} = \frac{1}{2Q} . \tag{11.9-13}$$

Hence, Eq. (11.9-11) becomes

$$S_1, S_2 = -\omega_0 \left(\frac{1}{2Q}\right) \pm \omega_0 \sqrt{\frac{1}{4Q^2} - 1} . \tag{11.9-14}$$

If $d > 1$ ($Q < \frac{1}{2}$) or $R^2/4 > L/C$, then S_1 and S_2 is two real roots, and the circuit is overdamped. If $d < 1$ ($Q > \frac{1}{2}$) or $R^2/4 < L/C$, then two complex roots result, and the circuit is underdamped. If $d = 1$ ($Q = \frac{1}{2}$) or $R^2/4 = L/C$, then there is a double real root to the characteristic equation, and the circuit is critically damped.

From the parameters of the pulse transformer,

$$\frac{R^2}{4} = 2500 < L/C = 500,000.$$

This indicates that the circuit of Fig. 11.9-7b is underdamped, and the output will contain ringing oscillations in response to the rising edge of the step. If this circuit were valid for all frequencies, output v_o would be expected to settle at 10 V. The initial portion of v_o is shown in Fig. 11.9-7c. The circuit Q is

$$Q = \frac{1}{R} \sqrt{\frac{L}{C}} = 7.07,$$

and the ringing frequency is

$$f_r \simeq \frac{1}{2\pi\sqrt{LC}} \simeq 2.25 \text{ MHz.}$$

It is desirable to try to reduce the overshoot as indicated at the v_o waveform. This can be done by decreasing the circuit Q. Since the decrease in Q also causes a slower rise time in v_o, a compromise usually has to be made; a Q value of about 0.8 gives an overshoot of 9% and a rise time equal to approximately one-fourth of the ringing period.

The pulse transformer equivalent circuit of Fig. 11.9-7a can be simplified at low frequencies by neglecting the effect of L_L, which will have a very low impedance compared to R_p. The effects of C and R_L can be neglected also, since they are both shunted by L_M. The resulting simplified circuit is shown in Fig. 11.9-8a. The output v_o will be exponentially decaying from 10 V to 0, as shown in Fig. 11.9-8b. The time constant is $\tau = L_M/R_p = 1$ ms, thus v_o will essentially reach steady state in 5τ or 5 ms.

The complete v_0 waveform shown in Fig. 11.9-8c is a combination of the results of Fig. 11.9-7c and Fig. 11.9-8b. The results of the above analysis can be easily extended to pulse inputs. The response to a pulse of width $t_p = 200$ μs is shown in Fig. 11.9-9. The output waveform has ringing oscillations on both the rising and falling edges owing to high-frequency response of the transformer; the tilt in the output pulse is due to the low-frequency response of the transformer. In this case, $\tau = L_m/R_p = 100$ mH/100 Ω = 1 ms, and $t_p = 200$ μs = 0.2 ms = 0.2 τ, hence the exponential portion goes through 0.2τ during the pulse width interval. This results in a tilt of

$$1 - e^{-t_p/\tau} = 1 - e^{-0.2} \simeq 18\%$$

from 10 V down to 8.2 V.

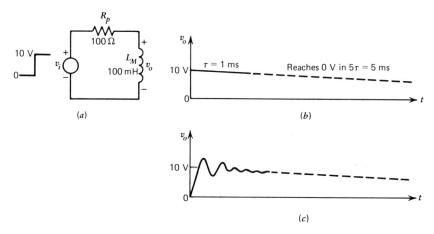

Figure 11.9-8   (a) Pulse transformer circuit simplified for low-frequency analysis. (b) Output response to flat portion of input. (c) Composite total output waveform.

Width One-Shot and Pick-Off Amplifier (Fig. 11.9-10)

The operation of the width one-shot is the same as the delay one-shot described earlier. When the pulse mode switch is in the advance position, the width one-shot is triggered by the negative half of the differentiated pulse from Q_{17} (Fig. 11.9-3), and in the delay position the width one-shot is triggered by the negative half of the differentiated pulse from Q_{13}. However, in the double pulse mode, the width one-shot is triggered by the negative half of the differentiated pulses from both Q_{13} and Q_{17}. The output of width one-shot is coupled to the pick-off amplifier by emitter-follower Q_{24}.

The leading and trailing edges of the rectangular pulse from the width one-shot are converted to sharp negative spikes by the pick-off amplifier. Transistor Q_{25}, the left half of the differential amplifier, is normally off and is turned on by the leading edge of the pulse from the width one-shot, thereby

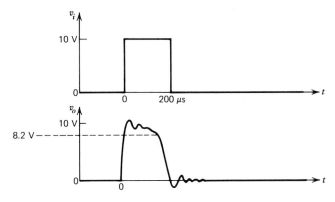

Figure 11.9-9   Output response to pulse input.

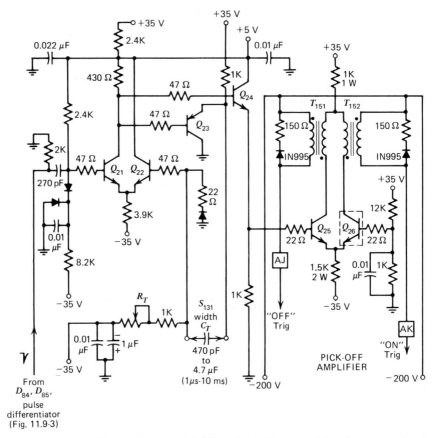

Figure 11.9-10 Width one-shot and pick-off amplifier. (Courtesy of Systron-Donner Corp.)

producing a sharp negative spike at the collector of Q_{25} and a sharp positive spike at the collector of Q_{26}. The trailing edge of the pulse from the one-shot reverses the conduction state of Q_{25} and Q_{26} and the polarity of the spikes at the collector. The phase of the spikes is inverted by transformers T_{151} and T_{152}. Only the negative spikes are utilized to trigger the predrivers.

11.10 BLOCKING OSCILLATORS AND PULSE GENERATOR DRIVERS

Astable Blocking Oscillator

An astable blocking oscillator conducts for a short period of time and is blocked out (cut off) for a much longer period. A basic blocking oscillator is shown in Fig. 11.10-1. The base coil (winding 1–2) and collector coil (winding 3–4) must be connected for regenerative feedback. This feedback through winding 1–2 and capacitor C_1 causes current through the transistor to rise rapidly until saturation is reached. The transistor is then cut off until discharged through resistor R_1. The output waveform is a pulse, the width of which is primarily determined by winding 1–2. The time between pulses (blocking time) is determined by the time constant R_1C_1.

The stray winding capacitance determines a resonant frequency. When the oscillator is cut off, $v_C = V_{CC}$. When base (C_1) discharges to the level of $V_{BE(cutin)} + V_{EE}$, current i_C starts to flow, inducing a voltage on the base that aids current flow. The circuit shuts off at the peak value of the cycle. A pulse-repetition rate of 400 means that there are 400 on and off conditions per second.

The waveforms are shown in Fig. 11.10-2. The transistor is on during intervals ab and cd, and off during bc and de. From a to b, energy is stored up in the magnetic field. When cut off at b, this energy must be dissipated and produces a high "back voltage." This release of stored energy in the transformer causes an oscillation within the base coil which dies out rapidly. The negative loop on the base-coil voltage waveform adds to the waveform of v_B and causes the rounding effect. In the waveform of v_{fb}, the area under the positive loop equals that under the negative loop. The natural resonant frequency determines the time intervals between a and b, and between c and d.

Monostable Blocking Oscillator (Emitter Timing)

A triggered blocking oscillator (emitter-timing) and the equivalent circuit are shown in Fig. 11.10-3. For simplifying the circuit, let us assume that a triggering signal is momentarily applied to the collector to lower its voltage. By transformer action and with the indicated winding polarities the base will rise in potential.

Applying Kirchhoff's voltage law to the outside loop (Fig. 11.10-3b), including both the collector and base meashes, gives

$$V = \frac{V_{CC}}{n + 1}, \qquad (11.10\text{-}1)$$

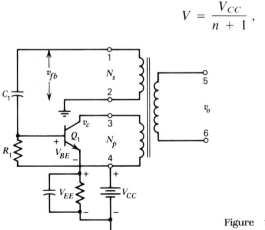

Figure 11.10-1 Basic blocking oscillator. (Typically $N_s/N_p = 3$.)

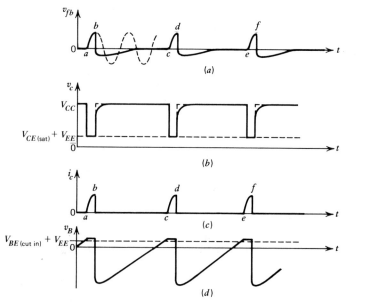

Figure 11.10-2 Waveforms of the base-coil voltage (v_{fb}), collector voltage, and base voltage in the basic blocking oscillator.

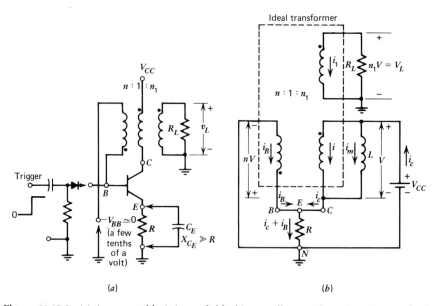

Figure 11.10-3 (a) A monostable (triggered) blocking oscillator with emitter timing; (b) the equivalent circuit from which to calculate the current and voltage waveforms.

where V is the voltage drop across the collector winding during the pulse. Since the voltage drop across R is

$$V_{EN} = nV = (i_C + i_B)R,$$

the emitter current, which is constant, is given by

$$-i_E = i_C + i_B = \frac{nV}{R} = \frac{n}{n+1}\frac{V_{CC}}{R}.$$ (11.10-2)

since the sum of the ampere-turns in the ideal transformer is zero,

$$i - ni_B + N_1 i_1 = 0.$$ (11.10-3)

The current in the load circuit is

$$i_1 = -\frac{n_1 V}{R_L}.$$ (11.10-4)

Since V is a constant, the magnetizing current is given by

$$i_m = \frac{Vt}{L}.$$ (11.10-5)

From Kirchhoff's current law at the collector mode, we have

$$i = i_C - i_m = i_C - \frac{Vt}{L}.$$ (11.10-6)

Substituting from Eqs. (11.10-4) and (11.10-6) into Eq. (11.10-3), we have

$$i_C - \frac{Vt}{L} - ni_B - \frac{n_1^2 V}{R_L} = 0.$$ (11.10-7)

Solving Eqs. (11.10-2) and (11.10-7) and using Eq. (11.10-1), we obtain

$$i_B = \frac{V_{CC}}{(n+1)^2}\left(\frac{n}{R} - \frac{n_1^2}{R_L} - \frac{t}{L}\right)$$ (11.10-8)

and

$$i_C = \frac{V_{CC}}{(n+1)^2}\left(\frac{n^2}{R} + \frac{n_1^2}{R_L} + \frac{t}{L}\right).$$ (11.10-9)

Notice that the collector-current waveform is trapezoidal with a positive slope, the base current is also trapezoidal with a negative slope, and the emitter current is constant during the pulse. These current waveforms and the voltage waveforms are pictured in Fig. 11.10-4. If the damping is inadequate, the backswing may oscillate, as indicated by the dashed curve in Fig. 11.10-4e, and regeneration will start again at the point marked X.

At $t = 0^+$, $i_C < h_{FE}i_B$, the operating point on the collector characteristics of Fig. 11.10-5 is at point P and the transistor is in saturation. As time passes, i_C increases and the operating point moves up the saturation line in Fig.

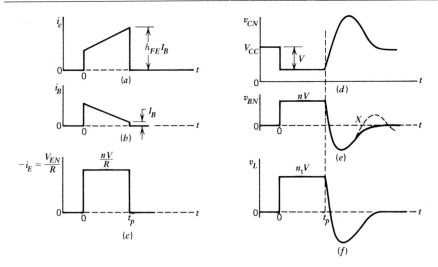

Figure 11.10-4 The current and voltage waveforms in a monostable blocking oscillator with emitter timing (Fig. 11.10-3).

11.10-5. While i_C grows with time, i_B is decreasing and eventually point P' is reached at $t = t_p$, where $i_B = I_B$ and

$$i_C = h_{FE}I_B. \tag{11.10-10}$$

At this point P' the transistor comes out of saturation and enters its active region. Because the loop gain exceeds unity in the active region, the transistor is quickly driven to cutoff by regenerative action, and the pulse ends. Since the regeneration which terminates the pulse starts when the transistor

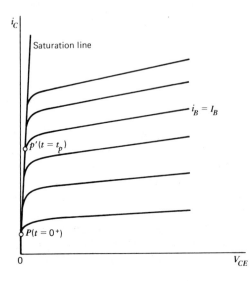

Figure 11.10-5 Collector characteristics. The path of the collector current is along the saturation line from P to P'. The pulse ends at P', where the transistor comes out of saturation.

comes out of saturation, the pulse width t_p is determined by the condition given by Eq. (11.10-10).

Applying Eq. (11.10-10) to Eqs. (11.10-8) and (11.10-9) with $t = t_p$, we find

$$h_{FE}i_B = \frac{h_{FE}V_{CC}}{(n + 1)^2}\left(\frac{n}{R} - \frac{n_1^2}{R_L} - \frac{t_p}{L}\right)$$

$$= \frac{V_{CC}}{(n + 1)^2}\left(\frac{n^2}{R} + \frac{n_1^2}{R_L} + \frac{t_p}{L}\right),$$

$$(h_{FE} + 1)\frac{t_p}{L} = \frac{h_{FE}n}{R} - \frac{n^2}{R} - (h_{FE} + 1)\frac{n_1^2}{R_L}.$$

Hence,

$$t_p = \frac{nL}{R}\frac{h_{FE} - n}{h_{FE} + 1} - \frac{n_1^2 L}{R_L}. \qquad (11.10\text{-}11)$$

Usually, $\frac{1}{5} \leq n \leq 1$, and $h_{FE} \gg n$. Then Eq. (11.10-11) becomes

$$t_p \simeq \frac{nL}{R} - \frac{n_1^2 L}{R_L}. \qquad (11.10\text{-}12)$$

Therefore, the pulse width t_p is independent of h_{FE} and depends only on passive elements n, L, R, etc. We may conclude that the blocking oscillator of Fig. 11.10-3 is a simple circuit which yields a pulse of very stable duration.

A positive trigger pulse (wider than the blocking oscillator pulse) may be applied through a diode to the base, as shown in Fig. 11.10-3a. A small capacitor may be used across the emitter resistor R ($\ll X_{C_E}$) to improve the rise time of the oscillator pulse.

Predrivers of the Typical Pulse Generator

The on-predriver and off-predriver of the model 114 A pulse generator are the two nearly identical blocking oscillators, as shown in Fig. 11.10-6. These two oscillators develop the voltage drive for the pulses that trigger the final drivers on and off to develop the main output pulse. The on-predriver produces negative pulses, while the off-predriver produces positive pulses. However, the operations of both the predrivers are the same. The on-predriver is explained as follows.

The sharp negative-going spikes from the pick-off amplifier are inverted by Q_{27} and coupled through emitter follower, Q_{28}, to the base of Q_{29}. The increase in the collector current of Q_{29} applies regenerative feedback to the base of Q_{29} through the transformer action of T_{321}. When the collector current in Q_{29} reaches a steady state, the regenerative feedback ceases and Q_{29} cuts off. The sharp negative spike developed at the collector of Q_{29} is coupled through the third winding of T_{301} to the base of emitter-follower Q_{30}. The output pulse of the on-predriver is taken at the emitter of Q_{30}, and applied to the on driver.

On/Off Drivers, Latch, and Final Drivers

The on/off drivers, latch and final drivers are shown in Fig. 11.10-7. The on and off drivers develop the current drive for the pulses that trigger the final drivers and the on/off latch keeps the final drivers on for the time interval of the output pulse.

During the time period between output pulses, Q_{41} is off since its emitter–base junction is biased slightly below the turn-on by forward-biased germanium diode D_{333} and silicon diode D_{331}. Pulses from the on-predriver turn on parallel transistors Q_{35}, Q_{36}, and Q_{37}. Current flow through the parallel

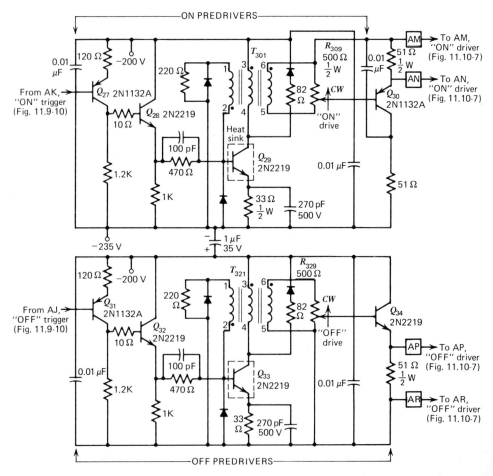

Figure 11.10-6 On-predriver and off-predriver of Model 114A pulse generator. (Courtesy of Systron-Donner Corp.)

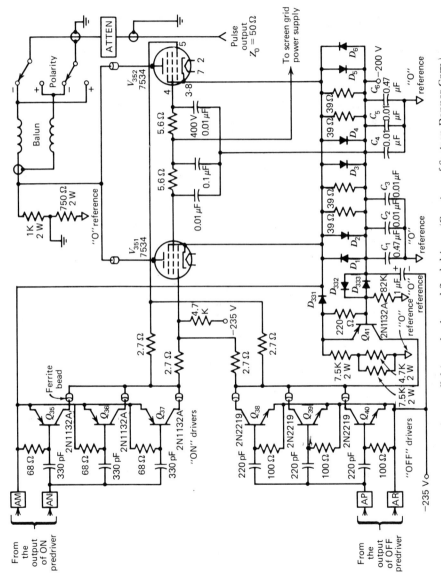

Figure 11.10-7 The on/off drivers, latch, and final drivers. (Courtesy of Systron-Donner Corp.)

355

transistors causes final drivers V_{351} and V_{352} to conduct. The conduction of the final drivers forward biases diodes D_2 through D_6, thus increasing the voltage on the emitter of Q_{41} by approximately 0.7 V. With the increase in emitter voltage Q_{41} conducts and its collector current maintains the voltage drop across R_{355} that keeps the final drivers conducting after the incoming on-driver pulse has subsided. Pulses from the off predriver turn on parallel transistors Q_{38} through Q_{40}, which bias V_{351} and V_{352} off.

The final drivers are current sources consisting of two parallel-operated pentodes. Output pulses are developed when V_{351} and V_{352} are biased on and off by the "on" and "off" drivers. The pulse-amplitude control is provided by the adjustment of the screen power supply. Diodes D_2 through D_5 clamp the cathodes of V_{351} and V_{352} to -200 V and minimize switching aberrations. Capacitors C_1 through C_6 are filter capacitors.

11.11   A HIGH-RATE PULSE GENERATOR

Introduction

Multipurpose high-rate (up to 10 MHz) pulse generators are commercially available. The HP-8002A pulse generator shown in Fig. 11.11-1 is a typical example. Its output impedance is 50 Ω for amplitudes from 0.02 to 5 V in seven ranges. Pulse widths are from 30 ns to 3 s in five ranges.

As shown in the block diagram of Fig. 11.11-1, the repetition rate is controlled by the internal or external trigger source. The instrument output pulse is a result of the original signal processed through a delay circuit, a width generator, a pulse shaper, a rise-time and fall-time arrangement, a positive or negative output amplifier, and an attenuator.

Repetition Rate Circuits

Internal Triggering   The internal repetition circuit is shown in Fig. 11.11-2. While transistors Q_7 and Q_8 are initially nonconducting, a selected ramp capacitor C at the emitter of Q_8 is charged in the negative direction by current source Q_9. When the voltage across C is more negative than 0.7 V, Q_8 begins to conduct. Then the voltage across diodes D_2 and D_3 and resistor R_{22} appears at the base of Q_7, causing it to conduct also. The voltage across resistor R_{24} causes Q_8 to conduct more. This regeneration continues until both transistors are in saturation. Diodes D_2 and D_3 will increase the loop gain if the current through Q_8 is small. Capacitor C_{17} will increase the loop gain at the high frequencies.

Now, capacitor C discharges through Q_8, R_{22}, D_2, and D_3, so that the voltage at base of Q_7 goes up slowly. When the current of Q_7 is not enough to maintain the transistor in saturation, Q_7 will drive Q_8 to tend to cutoff, and both transistors will finally reach cutoff owing to regeneration processing. Thus, the result will be a negative pulse produced from the emitter

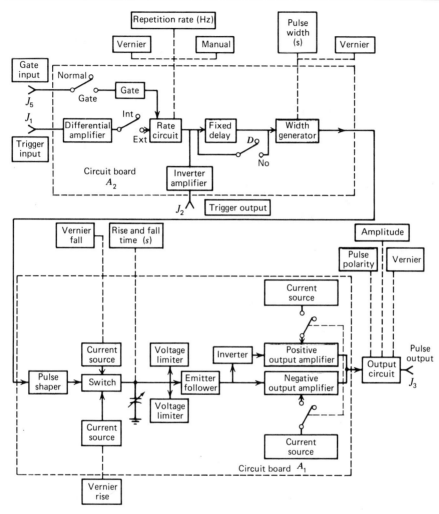

Figure 11.11-1   Block diagram of HP-8002A pulse generator. (Author's revision of 8002A circuit.)

of Q_7. The repetition rate depends on the values of capacitor C, its charge current, and the constant current from Q_9. The rate can be adjusted by vernier R_1.

During the internal triggering period, diodes D_{16} and D_{17} clamp the collectors of Q_{10} and Q_{11} at a voltage level (about 11 V), which is determined by Zener diode D_{18}. If during this time an external signal is connected to the trigger input J_1, then the collector voltages will not be raised, and so no signal will be applied to the Schmitt trigger Q_{12}–Q_{13}.

External Triggering   The pulse generator is triggered by negative-going slope of an external negative signal when the repetition rate switch S_1 is in position "EXT $-$", or by positive-going slope of an external positive pulse

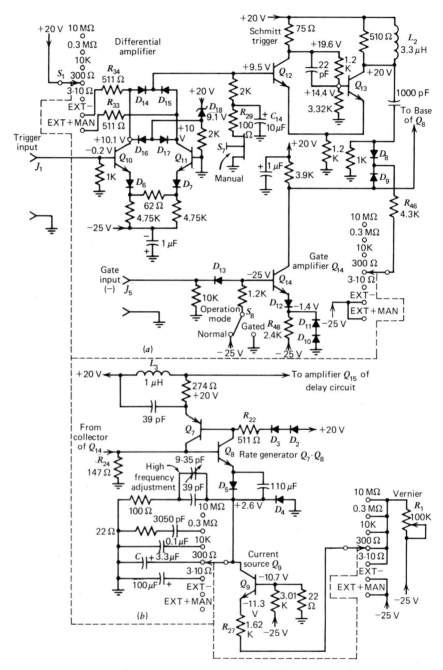

Figure 11.11-2 (a) External triggering circuit and gate amplifier. (b) Rate generator with current source. (Courtesy of Hewlett Packard.)

when S_1 is in position "EXT +/MAN". The external trigger signal (through J_1) is applied to the difference amplifier Q_{10}–Q_{11}. Diodes D_6 and D_7 are used to prevent Q_{10} and Q_{11} from reverse-bias breakdown.

In position "EXT −", resistor R_{34} is connected to +20 V, so that diode D_{16} is reverse biased. When the applied negative trigger is large enough, transistor Q_{10} is cut off. Its increased collector voltage is then applied to the base of Q_{12} through the D_{14} so that Schmitt trigger Q_{12}–Q_{13} operates, which produces a positive-going spike across inductor L_2.

In position "EXT +", R_{33} is connected to +20 V, so that D_{17} is reverse biased. When no external positive trigger is applied, both transistors Q_{10} and Q_{11} are in conduction, whereas Q_{12} is at cutoff. When a sufficient positive trigger is applied, Q_{10} in conduction causes Q_{11} to cut off. The increased voltage at the collector of Q_{11} is applied to the base of Q_{12} through the diode D_{15}, so that Schmitt trigger Q_{12}–Q_{13} operates, which still produces a positive spike.

In any external-triggering position, R_{46} is connected to −25 V, so that the base of Q_8 is at about −1 V, forcing Q_8 into cutoff. Diode D_4 conducts through constant current Q_9, resistor R_{27}, and then returns to ground, causing Q_7 and Q_8 to cut off. When the Schmitt trigger produces a spike, diode D_8 conducts, forcing D_9 into cutoff. At this time the voltage at the base of Q_8 increases, causing Q_8 and Q_7 to conduct. Therefore, as long as a proper trigger is applied, the emitter of Q_7 will produce a negative pulse.

When switch S_1 is in position "EXT +/MAN" and manual pushbutton S_7 is pressed down, capacitor C_{14} charges, so that a positive voltage appears at the base of Q_{12}, and a pulse is produced. When S_7 is released, the voltage across C_{14} discharges through R_{29}.

Gatting Mode

When the gated switch S_8 is in normal position, the base of Q_{14} is negative, and so this transistor is nonconducting. Diodes D_{10} and D_{11} serve to avoid the voltage applied to the base of Q_{14} more negative than −1.4 V. When S_8 is in the "gated" position, the base of Q_{14} goes up to 0 V, and the current of Q_{14} flows through D_{12} and R_{48} and returns to ground. When operated with external triggering, the base of Q_8 is negative, forcing the transistor into cutoff. Then, the constant current from Q_9 flows through D_4 and returns to ground. Now if a negative signal is applied to the gate input (J_5), Q_{14} is cut off and Q_8 conducts since its base voltage is going up. When operated with internal triggering, a pulse train will be produced until the gate input signal (−) is removed. The leading edge of the output pulse is synchronous with that of the input gate signal. Gating can be operated with either of the external triggering modes. In this case, the base of Q_8 is negative. The output pulse is produced only when both the input trigger and gate signal are applied.

Trigger Output and Delay Circuit

As shown in Fig. 11.11-3, the negative pulse from the rate generator is applied to the base of inverter Q_{16}, and the positive trigger output is obtained from

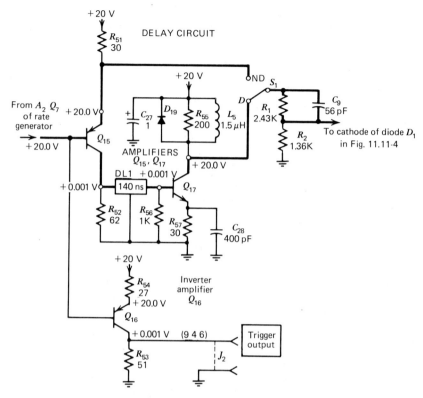

Figure 11.11-3 Trigger output and delay circuit. All resistances are in ohms and capacitances in microfarads unless otherwise indicated. (Courtesy of Hewlett Packard.)

its collector. The same negative pulse is also sent to the base of Q_{15}, and the positive pulse from the collector passing through delay line DL_1 (140 ns) is amplified by Q_{17} and differentiated by inductor L_5. The positive spike is removed by diode D_{19} and the negative spike is sent to width generator Q_1–Q_6 shown in Fig. 11.11-4.

Width Generator

The pulse width is determined by the selector S_2 and the adjustment C_8 in the width generator of Fig. 11.11-4. The signal from the delay circuit controls the Schmitt trigger Q_1–Q_2. Initially, Q_1 is off and Q_2 is conducting, so that a voltage drop developed across R_8 forces Q_3 into conduction. In this case, the constant current from Q_4 through Q_3 returns to ground, forcing Q_5 into cutoff and Q_6 into conduction in the second Schmitt trigger.

When a negative pulse from the delay circuit forces Q_1 into conduction and Q_2 into cutoff, the chosen width capacitor is linearly charged to +4 V by the constant-current source Q_4, so that the second Schmitt trigger (Q_5–

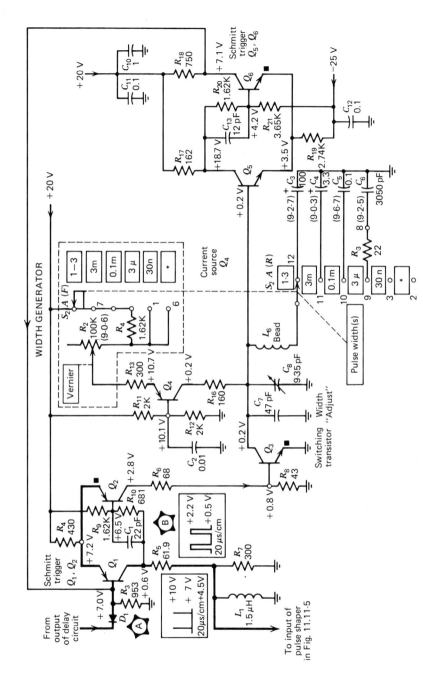

Figure 11.11-4 Pulse width generator. All resistances are in ohms and capacitances in microfarads unless otherwise indicated. ■ indicates conducting transistor between pulses. (Courtesy of Hewlett Packard.)

361

Q_6) changes its stable state. Then the voltage at the collector of Q_6 becomes positive and feeds back to Q_1, forcing Q_1 into cutoff. Hence, this circuit recovers its original state. The output pulse from the collector of Q_1 through differentiation becomes a positive and a negative spikes which are sent to the pulse shaper shown in Fig. 11.11-5.

Pulse Shaping Circuit

The pulse shaping circuit consists of a Schmitt trigger $Q_1–Q_2$, a transistor switch Q_3, and an emitter follower Q_4. The positive and negative spikes from the width generator force the Schmitt trigger into action, so that a square wave is produced at the collector of Q_2. R_{10} and C_3 are connected in series with Q_3 to reduce ringing oscillations at the pulse shaper output for the rise and fall times (t_r and t_f) other than 10 ns. A fast pulse with $t_r = t_f \leq 10$ ns will become slower if it appears across the $R_{10}–C_3–Q_3$ path. The circuit output is matched with the emitter-follower Q_4 for a low-output impedance. Two Zener diodes D_1 and D_2 are used to obtain two output pulses with different dc levels.

Integrator Circuit

The integrator circuit, shown in Figs. 11.11-6a and 11.11-6b, determines the pulse rising and falling characteristics. The rise and fall times are inversely proportional to the charge and discharge currents of the ramp capacitor. The pulse from the pulse shaping circuit is sent to Q_6 and Q_7. When Q_6 conducts, Q_7 cuts off, or vice versa. The constant currents from Q_{11} and Q_{16} flow through Q_6 and Q_7 toward the ramp capacitor, respectively.

When Q_6 is off, the current from Q_{11} flows through D_3 toward the voltage source $Q_{12}–Q_{14}$. On the other hand, when Q_7 is off, the current from Q_{16} flows through D_4 toward the voltage source $Q_{17}–Q_{18}$. These voltage sources limit the voltages at the emitters of the switching transistors Q_8 and Q_9. The emitter-followers Q_{10} and Q_{15} serve to provide the stable voltages at the bases of Q_{11} and Q_{16}.

The simplified integrator circuit is shown in Fig. 11.11-7. During the time interval t_1, transistor Q_6 cuts off and diode D_3 conducts. Meanwhile, Q_7 conducts and D_4 cuts off. Thus the current through Q_7 linearly charges the ramp capacitor in the negative direction until a value equal to the clamping voltage of -8.4 V is reached. During the interval t_2, Q_7 cuts off and D_4 conducts. Hence the current from Q_{16} flows through D_4. Meanwhile, Q_6 conducts and D_3 cuts off. Hence the current from Q_{11} passes through Q_6, and the ramp capacitor linearly discharges toward the positive direction, until its voltage is $+0.7$ V clamped by D_9. The discharge time determines the fall time and the base-line voltage of the pulse. The switching transistors Q_8 and Q_9 connected to the ramp capacitor will reduce the lead inductance.

Emitter-Follower and Inverter Circuits

The emitter-follower and inverter circuits are shown in Figs. 11.11-8a and 11.11-8b. The emitter-followers Q_{58} and Q_{21} are used to obtain a low-output

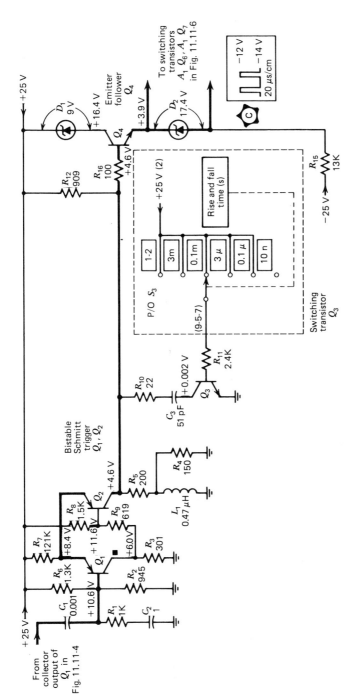

Figure 11.11-5 Pulse shaping circuit. All resistances are in ohms and capacitances in microfarads unless otherwise indicated. ■ indicates conducting transistor between pulses. (Courtesy of Hewlett Packard.)

363

impedance. The emitter-followers Q_{28}–Q_{29} and Q_{22}–Q_{23} are used to isolate the positive and negative output amplifiers. Zener diodes D_{10}, D_{11}, D_{12}, D_{13}, and D_{14} serve to decrease the collector voltage so that power dissipation is decreased. D_{15} is used to improve the dc level of the negative output signal.

Transistors Q_{24} and Q_{30} comprise a positive output inverting amplifier. The constant current source Q_{25} serves to improve the linear operation and common-mode rejection. When a negative pulse is applied to Q_{24}, most of the current flows through Q_{30}. The voltage source Q_{26}–Q_{27} provides the

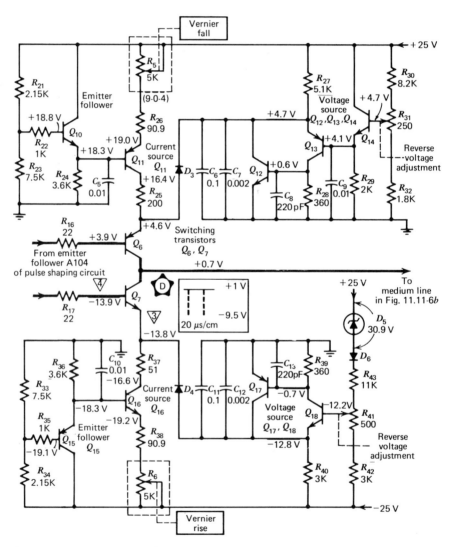

Figure 11.11-6a First section of integrator circuit. All resistances are in ohms and capacitances in microfarads unless otherwise indicated. (Courtesy of Hewlett Packard.)

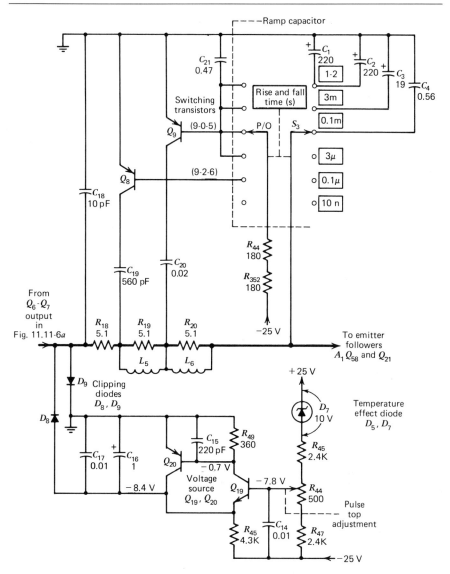

Figure 11.11-6b Second section of integrator circuit. All resistances are in ohms and capacitances in microfarads unless otherwise indicated. (Courtesy of Hewlett Packard.)

voltage at the collector of Q_{24}. Potentiometer R_{72} is used to adjust the baseline voltage of the positive pulse.

Output Circuits

The negative and positive output amplifiers are shown in Figs. 11.11-9 and 11.11-10, respectively. Each amplifier provides a 10-V open-circuit voltage and a 50-Ω output impedance. Both amplifiers are similarly designed.

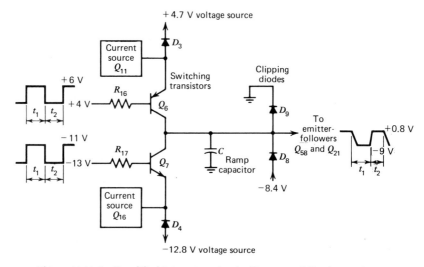

Figure 11.11-7 Simplified integrator circuit. (Courtesy of Hewlett Packard.)

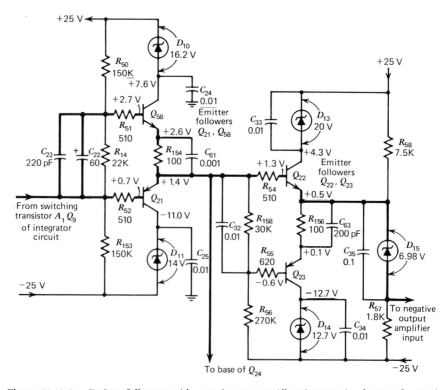

Figure 11.11-8a Emitter followers with negative output. All resistances in ohms and capacitances in microfarads unless otherwise indicated (Courtesy of Hewlett Packard.)

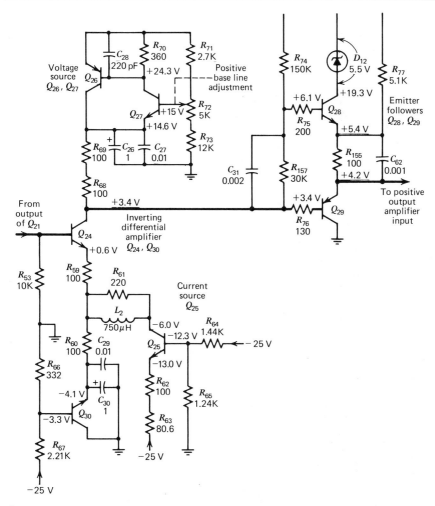

Figure 11.11-8b Inverting differential amplifier and emitter followers with positive output. All resistances in ohms and capacitances in microfarads unless otherwise indicated. (Courtesy of Hewlett Packard.)

The signal applied to the bases of transistors Q_{31}–Q_{34} controls the division of the current passing through this transistor group and the other group Q_{35}–Q_{37}. All the current flowing into transistors Q_{31}–Q_{34} results in the base-line voltage of the negative output pulse. All the current flowing into Q_{35}–Q_{37} results in the maximum voltage of the negative output pulse. Potentiometer R_{91} is used to change the voltage at bases of Q_{35}–Q_{37}, thus adjusting the base-line voltage of the negative pulse. Resistors R_{102} or R_{103} or R_{104} serve as the internal load of the negative output.

In the positive output amplifier of Fig. 11.11-10, the two transistor groups

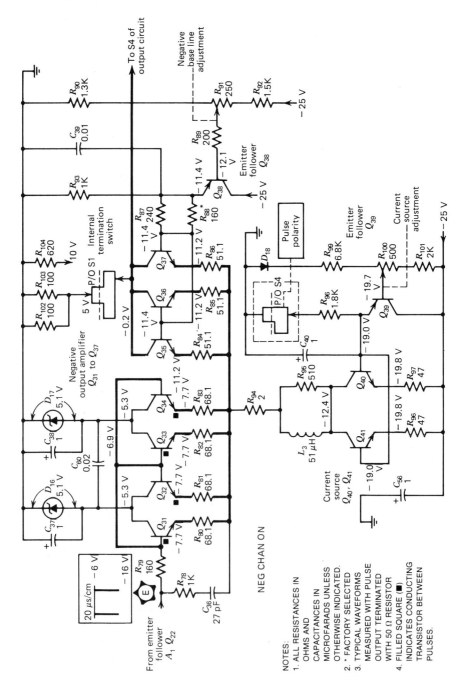

Figure 11.11-9 Negative output amplifier with current source and emitter followers. (Courtesy of Hewlett Packard.)

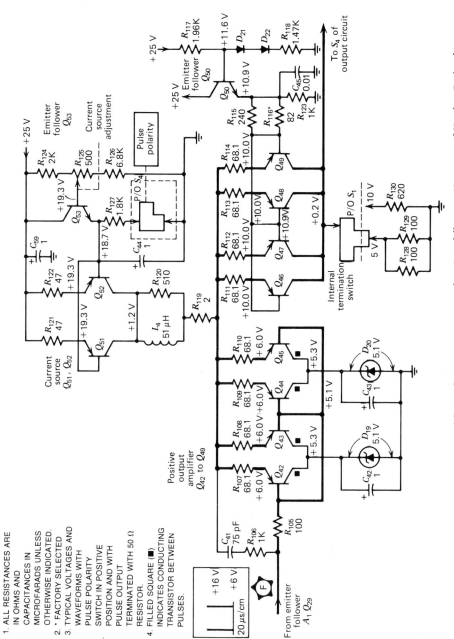

Figure 11.11-10 Positive output amplifier with current source and emitter followers. (Courtesy of Hewlett Packard.)

Q_{42}–Q_{45} and Q_{46}–Q_{49} determine the base-line and maximum voltages of the positive output pulse, respectively.

Only one output amplifier is permitted to be operated at a time; either negative or positive output can be chosen by switch S_4, which is connected to an attenuator assembly through the output terminal.

BIBLIOGRAPHY

Bell, D. A., *Electronic Instrumentation and Measurements*. Reston Publishing, Reston, Virginia, 1983, Chap. 15.

Carr, J. J., *Elements of Electronic Instrumentation and Measurement*. Reston Publishing, Reston, Virginia, 1979, Chap. 9.

Comer, D. J., *Electronic Design with Integrated Circuits*. Addison-Wesley, Reading, Massachusetts, 1981, Chap. 3.

Instruction Manual of HP-8002A Pulse Generator. Hewlett Packard.

μA *Linear Data Book*. Fairchild Camera and Instrument Corporation, 1982, p. 9-3 to 9-14 (555 and 556 Timer).

McWane, J., *Introduction to Electronics and Instrumentation*. Breton Publishers, North Scituate, Massachusetts, 1981, Chap. 11.

Millman, J., *Microelectronics*. McGraw-Hill, New York, 1979, Chap. 17.

Millman, J., and H. Taub, *Pulse, Digital, and Switching Waveforms*. McGraw-Hill, New York, 1965, Chaps. 11 and 16.

Operation and Maintenance Handbook of Model 114A Pulse Generator. Systron-Donner Corporation, Datapulse Division, 10150W. Jefferson Blvd., Culver City, California.

Taub, E., and D. Schilling, *Digital Integrated Electronics*, McGraw-Hill, New York, 1977, Chap. 2.

Questions

11-1 Sketch the hystersis zone of a transistor Schmitt trigger. List the principal characteristics and uses of this basic circuit.

11-2 Draw a practical circuit for an IC op-amp comparator used as a Schmitt trigger. Sketch its hystersis zone and describe its operation.

11-3 Draw the system of a square-wave generator using one IC comparator. What parameters determine the loop gain? What parameters determine the oscillation frequency?

11-4 Draw the circuit of an astable transistor multivibrator used as a voltage-to-frequency converter. What parameters determine the frequency?

11-5 Draw the system of a basic function generator and describe its operation.

11-6 Draw the system of a VCO consisting of a comparator, a CMOS switch, a voltage follower, and an intergrator. Describe the operation of the system.

11-7   Draw the circuit of a monostable transistor multivibrator and describe its main characteristics.

11-8   Draw the configuration of an op-amp monostable multivibrator and explain its operation by referring to the capacitor and output waveforms.

11-9   Explain the operation of a basic blocking oscillator.

11-10   Explain the operation of the practical 555 pulse generators shown in Fig. 11.7-8.

11-11   List the principal performance characteristics of the typical high-power pulse generator.

11-12   Describe the operation of the width one-shot and the pick-off amplifier in Fig. 11.9-10.

11-13   Describe the operation of the on/off, latch, and final drivers in Fig. 11.10-7.

11-14   List the main performance characteristics of the typical high-rate pulse generator.

11-15   Describe the operation of the internal repetition rate circuit in Fig. 11.11-2.

11-16   Explain the charge and discharge operation of the ramp capacitor in the integrator circuit of Fig. 11.11-7.

Problems

11.1-1 In the Schmitt trigger of Fig. 11.1-1a, $V_{CE(\text{sat})} = 0.2$ V, $V_{BE(\text{sat})} = 0.8$ V, $h_{FE_1} = H_{FE_2} = 100$, and the circuit parameters are given as indicated. Determine: (a) If switching occurs, (b) UTL, (c) LTL, and (d) the output levels.
        Answer.   (b) UTL = 2.76 V; (c) LTL = 1.55 V; (d) 10 V and 2.16 V.

11.2-1 In Fig. 11.2-1, Q_3 is intended as a constant-current source. Use a small-signal circuit representation to compute the resistance seen looking into the collector of Q_3. Use $h_{ie} = 1$K, $h_{FE} = 50$, and $1/h_{oe} = 100$K.
        Answer.   528K.

11.3-1 (a) Consider the square-wave generator of Fig. P11.3-1, where non-identical Zener diodes V_{Z1} and V_{Z2} are used. If output v_0 is either $+V_{o1}$ or $-V_{o2}$, where $V_{o1} = V_{Z1} + V_D$ and $V_{o2} = V_{Z2} + V_D$, verify that the duration of the positive section is given by

$$T_1 = RC \ln \frac{1 + \beta V_{o2}/V_{o1}}{1 - \beta}.$$

        (b) Verify that T_2 (the duration of the negative section) is given by the same equation with V_{o1} and V_{o2} interchanged. (c) If $V_{o1} > V_{o2}$, is T_1 greater or less than T_2? Explain.

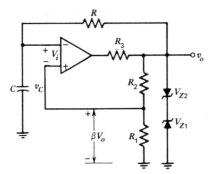

Figure P11.3-1

11.3-2 Consider a symmetrical collector-coupled astable multivibrator using *npn* silicon transistors. The circuit and device parameters are $V_{CC} = 6$ V, $R_c = 560$ Ω, $R = 5.6$K, $C = 50$ pF, $h_{FE} = 40$, $V_{CE(sat)} = 0.2$ V, $V_{BE(sat)} = 0.8$ V, $V_\gamma = 0.5$ V and $r_{bb'} = 200$ Ω. Calculate (a) the waveforms at the base and collector of one transistor and plot to scale, (b) the frequency of oscillation, and (c) the recovery time (the 10–90% time). [*Hint:* (a) Draw the collector C_1 to base B_2 equivalent circuit for calculating the overshoot at B_2 at $t = T^+$, where $T = T_1 = T_2$. In this equivalent circuit, $V'_{BE} = I'_B r_{bb'} + V_{BE(sat)}$ and $V_{C1} = V_{CC} - I'_B R_C$, where I'_B is the loop current. From Fig. 11.3-4, the jumps in voltage at B_1 and C_1 are, respectively, $\delta = I'_B R_{bb'} + V_{BE(sat)} - V_\gamma$ and $\delta' = V_{CC} - I'_B R_C - V_{CE(sat)}$. From the relationship $\delta = \delta'$ the expression for I'_B can be obtained.]

Answer. (a) $\delta = \delta' = 0.68$ V; (b) $f_0 \simeq 2.59$ MHz; (c) $t_r = 83.6$ ns.

11.3-3 Design a complementary-pair astable multivibrator circuit for minimal consumption of power.

11.4-1 A constant voltage of 2 V is applied to the input of the integrator in Fig. 11.4-2 for 3 s. Find the value of C_1, if $R_1 = 1$ MΩ and $V_{o1} = -12$ V.

Answer. 0.5 μF.

11.5-1 A free-running sweep generator is shown in Fig. P11.5-1, where $-V$ is a constant negative voltage and where $R' \ll R$. (a) Explain the operation, write the expressions for $v(t)|_{max}$ and $v(t)|_{min}$, and sketch the waveform $v(t)$. (b) Find the reference voltage V_R and sweep voltage V_s so that the sweep extends from 0 to V_s. (c) Find the sweep time T_s. (d) Explain how the sweep voltage can be frequency modulated and find f in terms of the modulating voltage v_m. (*Hint:*

Write the expression for input v^+ by using superposition theorem. Note that the noninverting comparator input impedance is much higher than R_1 or R_2).

Answer. (a) $v(t)|_{max} = \dfrac{-V_R}{R_1}(R_1 + R_2) + \dfrac{R_2}{R_1}V_o;$

$$v(t)|_{min} = \dfrac{-V_R}{R_1}(R_1 + R_2) - \dfrac{R_2}{R_1}V_o.$$

(b) $V_R = \dfrac{-R_2 V_o}{R_1 + R_2}$; $V_s = V_{max} - V_{min} = \dfrac{2R_2}{R_1}V_o.$

(c) $T_s = T_1 + T_2 = \dfrac{V_s RC}{V}$;

(d) $f = \dfrac{V}{V_s RC}$.

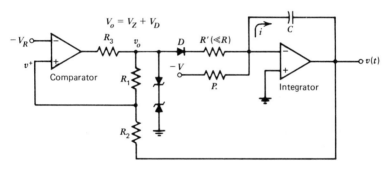

Figure P11.5-1

11.6-1 In the monostable multivibrator of Fig. 11.6-1a, the pulse width T_W is a function of I_{CBO}. (a) By what percentage change is T_W decreased as the temperature is increased from a value for which I_{CBO} is negligible to the temperature for which $I_{CBO} = 100$ μA? Assume $V_{CC} = 22.5$ V and $R = 120K$. (b) Indicate a method of temperature compensation using a resistor whose resistance increases with temperature.

Answer. T_W decreases 28%.

11.6-2 (1) A TTL monostable multivibrator SN74121 is shown in Fig. P11.6-2a. Its pulse width is given by $T_W = 0.7RC$. Its trigger pulse is shown in Fig. P11.6-2b. Draw the Q output waveform. (2) Assume that the A input is not grounded, but is connected to a 100-kHz clock, and B input is a logical 1 instead of the trigger pulse. Draw this new circuit and its input and $\bar{Q}$ output waveforms.

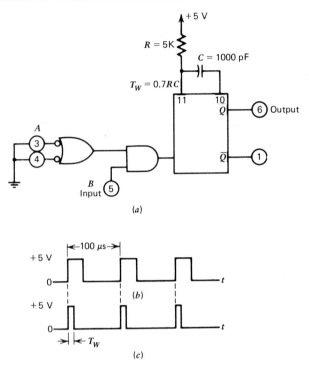

Figure P11.6-2   (a) TTL monostable multivibrator SN74121. (b) Input trigger. (c) Output pulse.

11.7-1   Chose the values of R and C for the 555 monostable circuit of Fig. 11.7-4 to generate a 1-ms output pulse. If $C = 0.1 \ \mu F$, $R = ?$ If $R = 50K$, $C = ?$

11.7-2   Calculate the pulse width delivered by the monostable circuit of Fig. 11.7-4 if $R = 22K$ and $C = 0.005 \ \mu F$.

11.7-3   The 555 astable circuit of Fig. 11.7-6 has the following component values: $R_1 = 1.5K$, $R_2 = 8.6K$, and $C = 0.1 \ \mu F$. Calculate (a) pulse width, (b) frequency of oscillation, and (c) ON duty cycle.
  Answer.   (a) 0.7 ms; (b) 770 Hz; (c) 46%.

11.10-1   (a) Compute the loop gain for the blocking oscillator of Fig. 11.10-3a by proceeding as follows: Open the transformer lead connected to the base and apply a 1-V signal to the base–ground circuit. Place the impedance R' $[=(h_{FE} + 1)R]$ across the base winding to simulate the base input impedance of the transistor. Since the input to the base is 1 V, the amount of voltage induced in the base winding equals the loop gain ($\beta A = nV = 1$). Assume $V_{BN} = V_{EN} = 1$ V. (b) Verify that the following inequality must be satisfied in order that the loop gain exceed unity:

$$R_L > \frac{n_1^2 R(1 + h_{FE})}{n h_{FE} - n^2} .$$

[*Hint:* (a) Draw the circuit shown in Fig. P11.10-1 for calculating the loop gain.]

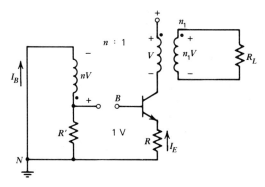

Figure P11.10-1

11.10-2 Design a pulse generator consisting of a blocking oscillator with its emitter resistor directly coupled to the output inverter.

12

Telemetering
Transmitters
And Receivers

12.1 ELECTRONIC TELEMETERING SYSTEMS

Introduction to Telemetering Systems

Telemetry is a process that is used to measure data at a location remote from the data source. Electronic telemetering systems acquire data at relatively inaccessible locations, then process, transmit, recover, and display the data at the location where it is to be used. Six subsystems are combined to form a complete telemetering system as in Fig. 12.1-1: the transducer subsystem (1) senses and converts physical and electrical parameters into suitable electrical quantities for application to the multiplexer–programmer subsystem (2), which electrically times and encodes the sensed data from the transducer subsystem into a form that can be transmitted; the radio transmitter subsystem (3) conveys the encoded information to the radio receiver subsystem (4), which acquires and processes the data; within the demultiplexing–decoding subsystem (5), the originally encoded data are separated and decoded so that the information originally sensed at the source can be visually presented by the display subsystem (6).

A multiplexer is actually a signal selector gating circuit that is the logic equivalent of a multiposition selector switch. A number of input signals are fed into the multiplexer, and by using control inputs, one of the signals is selected and made available at the output. Multiplexers can be used in both digital and analog applications. The multiplexer–programmer subsystem in Fig. 12.1-1 is made of a number of individual assemblies, such as a timer

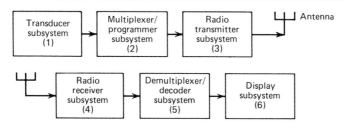

Figure 12.1-1 An electronic telemetering system.

assembly, a multiplexer assembly, an analog-to-digital converter, and a memory. The multiplexer assembly is composed of one digital multiplexer subassembly and one analog multiplexer subassembly. An analog multiplexer subassembly consists of numerous integrated gate packages interconnected to perform the multiplexing of a number of analog signals. A basic transmitter and a basic receiver used for telemetry operate by the same principles as for communication.

Block Diagrams of Radio Transmitters

The radio transmitter is essentially a device for producing radio-frequency (rf) energy that is controlled by the information to be transmitted. A block diagram of a basic amplitude-modulated (AM) transmitter is shown in Fig. 12.1-2. Here the desired frequency is generated at a low-power level by a stable oscillator, usually a crystal oscillator. This is followed by a buffer, which simultaneously increases the power to drive the final rf power amplifier. The modulating signal is applied to the buffer to modulate the amplitude of the rf carrier.

Frequency-modulated (FM) transmitters usually operate at frequencies above 30 MHz. One of the commonly used FM types is the reactance modulation. An example of this type is the oscillator frequency modulated by a varactor-diode modulator. A block diagram of the basic FM transmitter is shown in Fig. 12.1-3. In this case the modulated wave is generated at a low-power level, and a chain of amplifiers and frequency multipliers is used to develop the required transmitter power and frequency.

Block Diagrams of Radio Receivers

The block of a superheterodyne receiver for AM is shown in Fig. 12.1-4. The rf amplifier is used to amplify the antenna signal, provide tuning for the antenna in order to enhance the selectivity of the receiver, and provide a better signal-to-noise ratio than would be obtained if no rf amplifier were provided. The mixer heterodynes the signal of frequency f_s from the rf amplifier or antenna with the signal of frequency f_0 from the local oscillator to accomplish the frequency conversion from the antenna frequency to the intermediate frequency, $IF = f_0 - f_s$. The mixer output is an AM waveform whose frequency is IF, typically 455 kHz for AM. The IF amplifier is a

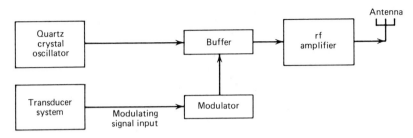

Figure 12.1-2   Block diagram of basic AM radio transmitter.

narrow-band amplifier with a large gain that provides the main amplification of the signal prior to demodulation (detection). The output of the IF amplifier is applied to the demodulator (detector). Typically, the AM detector is a half-wave rectifier that recovers the modulating or audio signal from the AM waveform by removing either the top or the bottom portion of the AM envelope. The detector output after filtering is applied to the audio voltage amplifier, which increases the signal level to that suitable for driving the audio power amplifier, which furnishes power to the speaker. For a basic telemetering use, the output from the audio voltage amplifier drives the recorder or other display system. The automatic-volume-control (AVC) network controls the gain of all or part of the stages that precede the demodulator, in order to maintain a nearly constant signal level at the demodulator.

It is instructive to observe the stage gains in the receiver of Fig. 12.1-4. The signal voltage is typically stepped up 10 times (at 600 kHz) in the rf amplifier, 25 times (at 600 kHz) in the mixer, 85 times (at 455 kHz) in the IF amplifier, 80 times (at 400 Hz) in the LF (audio) voltage amplifier, and 15 times (at 400 Hz) in the audio power amplifier. Hence, the total voltage gain of the receiver is equal to the product of the individual stage gains, or 25,500,000 times. Most of the gain prior to demodulation is provided by the IF amplifier. Most of the selectivity is also provided by the IF amplifier.

A disadvantage of the AM transmitting and receiving systems is that they are sensitive to noise. This disadvantage is largely overcome in FM systems by causing the carrier frequency to vary or deviate over a small range in

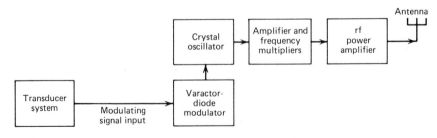

Figure 12.1-3   Block diagram of basic FM radio transmitter.

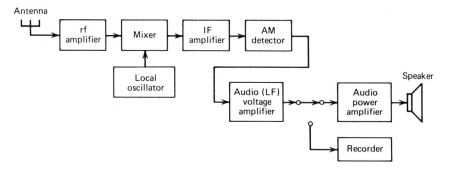

Figure 12.1-4   Block diagram of basic AM radio receiver.

accordance with the modulating signal (audio) waveform (see Fig. 10.10-1). The noise immunity of FM transmission systems results primarily from the fact that FM detectors (demodulators) respond to variation in carrier frequency rather than to variations in carrier amplitude. By incorporating amplitude limiters into the IF amplifier of an FM receiver, most of the amplitude variations that may have been produced by undesired electrical noise can be removed. Moreover, FM detectors can be designed to be largely immune to amplitude variations. The FM broadcast band ranges from 88 to 108 MHz, while the AM broadcast band ranges from 540 to 1600 KHz. The FM stations are located in the band at 200-kHz intervals, while the AM station frequencies are assigned so that the minimum separation between carrier frequencies of adjacent stations is 10 kHz. The FCC permits FM (AM) transmission of audio frequencies ranging from 50 to 15,000 Hz (up to 5000 Hz). The fidelity of FM transmissions is higher as compared with AM.

The block diagram of a superheterodyne receiver for FM is shown in Fig. 12.1-5. The standard IF for FM is 10.7 MHz. The IF amplifier stages have

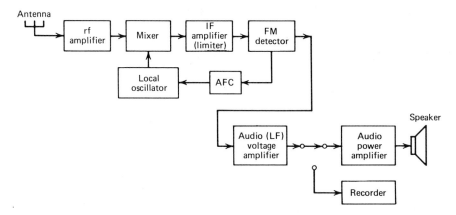

Figure 12.1-5   Block diagram of basic FM radio receiver.

amplitude limiting in order to provide noise immunity. The most-popular FM demodulator is the ratio detector, which has the advantage of inherent limiting action. A limiter will always precede a Foster–Seeley discriminator, if it is used as an FM demodulator. An FM detector is designed to discriminate between deviations above and below the center frequency and to develop an output voltage with an amplitude proportional to these deviations. An automatic frequency control (AFC) is used to control the oscillator frequency (f_0). This control eliminates the effect of frequency f_0 varying with changes in temperature or other factors. The frequency of antenna signal is

$$f_s = f_0 - \text{IF}.$$

12.2   ILLUSTRATION OF MULTIPLEXED TELEMETRY SYSTEMS

In order to accommodate more than one telemetry signal modulated on the same carrier, an information processor called an encoder/decoder arrangement is usually employed in the multiplexed telemetry system, as shown in Fig. 12.2-1. As an illustration of the encoding process, encoders 1, 2, and 3 may modulate the subcarriers of 1.3, 1.7, and 2.3 KHz, respectively. Then, decoders 1, 2, and 3 would include the band-pass fitters with the center frequencies of 1.3, 1.7, and 2.3 KHz, respectively. Subcarriers and rf carrier may be frequency modulated or amplitude modulated. If the subcarriers are amplitude modulated and the rf carrier is frequency modulated, an AM–FM telemetry system is formed. Similarly, the FM–AM telemetry system and FM–FM system may be utilized.

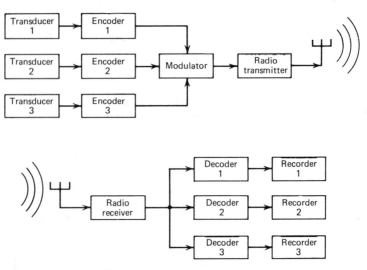

Figure 12.2-1   A three-signal telemetry system. It may be one of AM–FM, FM–AM, and FM–FM systems.

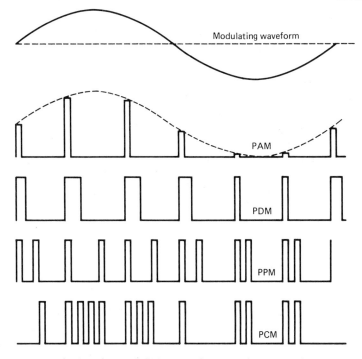

Figure 12.2-2   Four basic pulse-modulation waveforms used in many telemetry systems. PAM = pulse-amplitude modulation; PDM = pulse duration modulation; PPM = pulse-position modulation; PCM = pulse-code modulation.

Telemetry systems may also employ the four basic pulse-modulation waveforms to accomplish the pulse-amplitude modulation (PAM), pulse-duration modulation (PDM), pulse-position modulation (PPM), and pulse-code modulation (PCM), as shown in Fig. 12.2-2.

1.   PAM means that the modulating waveform is sampled by the pulse train, and that the heights of successive pulses correspond to the instantaneous amplitudes of the modulating waveform. In turn, the demodulated pulse waveform at the receiver with an *RC* low-pass filter reproduces the modulating sinusoidal waveform.

2.   PDM uses a varying pulse width at constant amplitude. PDM pulses follow one another at uniform intervals. The width of each pulse is proportional to the instantaneous amplitude of the modulating waveform.

3.   PPM uses pulses that follow one another at nonuniform intervals. The pulses have constant amplitude and constant width. However, they occur at a time rate that is proportional to the instantaneous amplitude of the modulating waveform.

4.   PCM means that the modulating waveform is sampled, with each sample assuming a nearest whole value, such as 0.1, 0.2, 0.3, etc., to the

instantaneous amplitude of the modulating waveform. In turn, these values are coded in terms of pulses and spaces. In binary PCM, these pulses and spaces correspond to binary bits.

The advantages of pulse modulation are a good signal-to-noise ratio and the operation in time-sharing multiplexing. This mode of operation means that the blank spaces between pulses can be used to transmit another pulse signal on the same carrier. Comparatively large bandwidths are required for all forms of pulse modulation. When constant-amplitude pulses are used, a limiter is contained at the receiver to improve the signal-to-noise ratio. Pulse modulation is extensively used in telemetry systems and outer-space communication because there is large spectrum space available in the microwave region.

12.3   A RADIATION TELEMETRY TRANSMITTER

Introduction

The modulating signal is a pulse from a radiation detector (transducer). The circuit shown in Fig. 12.3-1 is a special AM transmitter, similar to a code-modulation type. The only difference between these is that the modulating pulse waveform requires appropriate pulse shaping, while the telegraph code does not. The information conveyed by this circuit is radiation intensity, expressed in count rate (counts per second) or exposure rate (milliroentgens per hour; a roentgen is defined as 2.58×10^{-4} coulombs per kilogram of dry air). The main characteristics of this circuit are the following: operating frequency is 27.1611 MHz, within the assigned scientific channel (26.96–27.28 MHz); power output is about 0.5 W; pass band of the tuned circuits is 10 kHz; antenna is of resonant half-wave type, matched to a two-wire transmission line; reliable (line-of-sight) communication distance up to nearly 0.5 km.

Circuit Analysis

The circuit in Fig. 12.3-1 consists of quartz-crystal oscillator Q_1, buffer Q_2, keyer Q_3, rf power amplifier Q_4, and modulator section Q_5–Q_7. Q_1–Q_4 make up the rf section. A 20 Vdc power supply is provided for both the sections. At the power line connected to the rf section, there are two rf chokes and four capacitors returned to ground, comprising decoupling networks. These networks are used to prevent from parasitic oscillation occurring due to the signal current flowing through the power-source resistance. At the power line connected to the modulator section, there is a π-type filter (1000 μF–100 Ω–1000 μF) used to stabilize the power-line current when Q_5-Q_7 stages are switching. When trace the signal path, the dc voltage-source line may be regarded as ac ground.

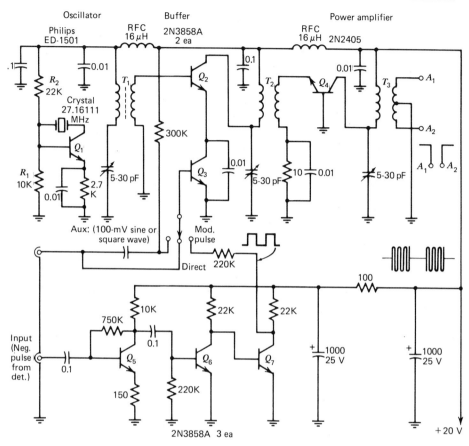

Notes: Unless otherwise specified, all resistors are of ½ W, ± 10%, and all capacitors are of μF.
Transformers T_1: winding diameter D = 9 mm; wire diameter d = 0.5 mm; primary turns N_p
            = 11T, secondary turns N_s = 3T;
      T_2: D = 9 mm; d = 0.8 mm; N_p = 9T; N_s = 2T;
      T_3: D = 12 mm; d = 1 mm; N_p = 6T; N_s = 6T.

Figure 12.3-1 Transmitting circuit for radiation telemetry. [Chiang, H. H. and J. K. Chang,
Nuclear Science Journal, **13** (4) 218 (1976).]

   In the rf section, Q_1 is in clase A operation, while Q_2 and Q_4 are in class
B operation. In the Q_1 oscillator stage, the *BE* junction forward-bias and *BC*
junction reverse-bias sources are provided by the voltages across R_1 and
R_2, respectively. The quartz crystal is the feedback element connected
across the collector and base. A quartz crystal has piezoelectric effect. Its
equivalent circuit contains a motional inductance, a motional capacitance,
and a motional resistance connected in a series form, across which an electrostatic capacitance shunts. In this case, the crystal must be operated in the
inductive region, so that the crystal inductive reactance combines the base

capacitance and the collector capacitance, and thus a basic LC oscillation circuit is formed. Under this condition, the collector tank must be tuned at a frequency slightly higher than resonant point. Then, continuous oscillations are maintained owing to a proper regenerative feedback. Since Q_2 and Q_4 are in class B operation, no forward bias is provided to their BE junctions. Since I_{CEO} is much larger than I_{CBO} ($\approx I_{CEO}/\beta$), we prefer CB configuration rather than CE form for the power amplifier Q_4; its emitter coil is returned to ground via a 10-Ω resistor, which is used for temperature compensation and shunted by a by-passed capacitor. With keyer Q_3 turned on, the buffer and the power amplifier are tuned to the resonant frequency, causing the modulated wave to be transmitted. A small 0.5-W lamp can be used as a dummy load instead of the antenna when we tune the transmitter.

The modulator (Q_5–Q_7) is a switching circuit. Normally Q_5 conducts, and Q_7 is saturated, while Q_6 is at cutoff; hence, the magnitude of the negative pulse input must be high enough to trigger Q_5 into cutoff. Since the modulator output is a positive square wave with almost constant amplitude, the system counting will result in high accuracy provided that the transmitting and receiving circuits maintain their normal performances. The modulator may be operated within frequencies up to 50 kHz. The required magnitude of the input pulse is in the range of 0.15–2 V. When the telemetry transmission system provides a pass band of 10 kHz, the count rate up to 5000 counts/s can be obtained.

12.4   A RADIATION TELEMETRY RECEIVER

Figure 12.4-1 shows the receiving circuit for radiation telemetry. It is of the AM superheterodyne type. The oscillator frequency f_0 = 27.53888 MHz. The antenna input frequency f_s = 27.16111 MHz. Hence IF = $f_0 - f_s$ = 377.77 kHz (standard IF = 455 kHz for AM). The pass band is 10 kHz. The circuit can be divided into an rf section and a pulse section. At the dc power line connected to the rf section, there are two capacitors (470 μF, 0.1 μF) and two resistors (100 Ω, 470 Ω) making up decoupling networks. At the dc power line connected to the pulse section, a decoupling network contains a 100-Ω resistor and a 0.1-μF nonpolarized-capacitor (C_1) shunted by a 200-μF polarized-capacitor (C_2). When an interference impulse with improper polarity is applied across C_2, it may be deficient for a little while, during which time the decoupling effect can be maintained by C_1. In the rf section (Q_1–Q_5), all stages are in class A operation, and each emitter resistor is connected with a by-passed capacitor to ground except for the mixer Q_3. The primaries of IF transformers T_4–T_6 are tapped for impedance matching.

The local oscillator (Q_2) is of quartz crystal type operating by the same principle as described in Section 12.3. The oscillator signal is applied to the emitter of mixer Q_3 and the amplified antenna signal is applied to the base, resulting in an IF signal produced at the collector resonant circuit. The IF

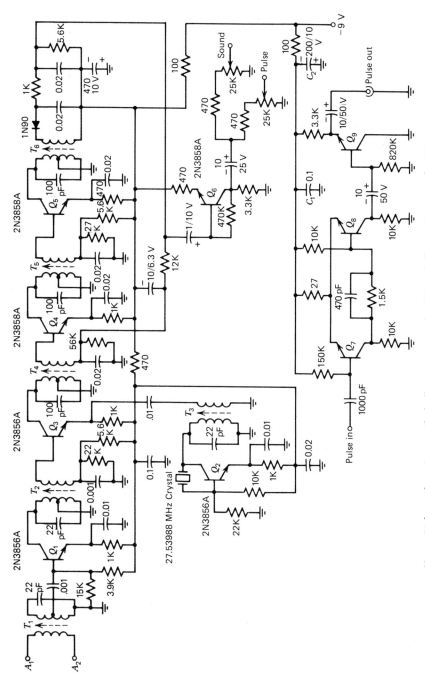

Notes: Unless otherwise specified, all resistors are of ±10% and all capacitors are of μF. Transformers (wire diam. = 0.2 mm).

T_1: $N_{p1} = 1T$, $N_{p2} = 1T$, $N_{s1} = 1T$, $N_{s2} = 6T$, $N_{s2} = 1T$; T_2: $N_p = 7T$, $N_s = 7T$; T_3: $N_p = 7T$, $N_s = 1T$.
Others are of usual IFT for BC radio set.

Figure 12.4-1 Receiving circuit for radiation telemetry. [Chiang, H. H. and J. K. Chang, *Nuclear Science Journal*, **13** (4) 219, (1976).]

signal is then amplified by the two successive IF stages (Q_5–Q_6). From the detector to the final stage is the pulse section. The diode detector is a half-wave rectifier. There are two paths for the rectified and filtered output: one is the AVC (automatic volume control) filter network, consisting of a 12K resistor and a 10-μF capacitor. Through this network, a negative AVC voltage is applied to the base of Q_4, thus reducing its forward voltage V_{BE}. The other path for the demodulated signal is the 1-μF coupling capacitor to the base of Q_6. The amplified signal from Q_6 is applied to the Schmitt trigger (Q_7–Q_8) for pulse shaping. Through the emitter follower (Q_9), the output pulse is counted by a scaler (electronic counter).

12.5   AN FM TRANSMITTER

Circuit Analysis

An FM transmitting circuit is shown in Fig. 12.5-1. It consists of a frequency-modulated oscillator (Q_1), a frequency doubler (Q_2), a class AB amplifier (Q_3), a frequency tripler (Q_4), a driver stage (Q_5), and a power amplifier ($Q_6 \parallel Q_7$). The crystal frequency is 20 MHz; the output frequency is 120 MHz. The output power is 3 or 4 W. The antenna impedance is 50 Ω. All transistors are operated in common-base configuration except the oscillator, which is in common-emitter form. The oscillator is of a Miller quartz-crystal control type. In order to obtain continuous oscillations, the quartz crystal must be operated in the inductive region, and the collector tank must be tuned to a frequency slightly less than the resonant point, so that the tank becomes an inductive load. In this case, a basic LC oscillating circuit is formed. The interelectrode capacitance C_{CB} is the feedback element. Oscillations are started and maintained due to the crystal piezoelectric effect and the proper feedback.

The oscillator tank is shunted by a capacitance branch, which includes a capacitor ($C_1 = 5$ pF), a varactor diode, a 1.8K resistor, and a secondary coil, connected in series. A -15 V source is connected to a voltage divider, which consists of a 75K and a 25K resistors; at the junction tap, an isolation resistor (50K) is connected to the anode of the varactor diode, so that the diode is reverse biased. When the modulating signal v_m from the input transformer T_1 is increased in a positive direction, the reverse bias across the diode is also increased, and therefore, the diode capacitance is decreased, causing the total capacitance of the tank to decrease; hence, the oscillator frequency is increased. [Check using the formula $f_{\mathrm{MHz}} \simeq 159/\sqrt{L_{\mu\mathrm{H}} C_{\mathrm{pF}}}$.] On the other hand, when the modulating signal is negative, the oscillator frequency is decreased. The FM wave is shown in Fig. 10.10-1.

At the doubler and tripler tanks, the resonant frequencies are set to the second and third harmonics, respectively. The tank coils of Q_3–Q_5 stages

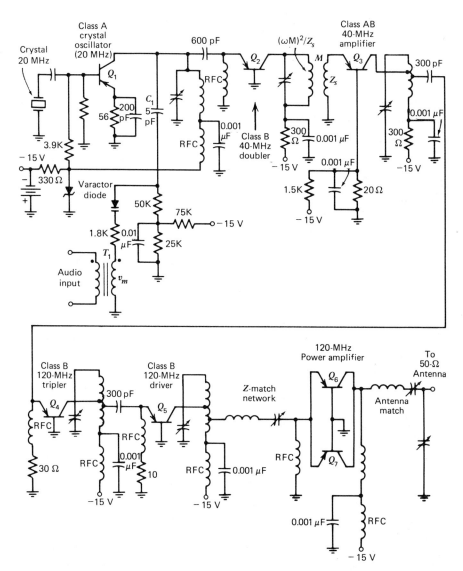

Figure 12.5-1 FM transmitter circuit. Note: In a class A amplifier, the collector current flows for the full ac cycle (360°). In a class AB amplifier, collector current flows for more than half the ac cycle but less than a full ac cycle. In a class B amplifier, the collector current flows for exactly 180° of the full ac cycle. In a class C amplifier, the collector current flows for less than 180° of the full ac cycle.

are tapped in order to match the collector impedance. Each emitter choke at Q_4–Q_5 stages is connected to ground via a small damping resistor. The final amplifier stage has one tuned LC series network at the input and another at the output for matching impedance. The V_{CC} source line is connected to each stage via a decoupling RFC or resistor with a decoupling capacitor bypassed to ground. In order to obtain stable oscillations, a regulated voltage is provided for the oscillator bias.

Tuning for Resonance

In the circuit of Fig. 12.5-1, all tanks are of parallel-resonant LC circuit. At resonance, the line current is minimum, but the circular current is maximum. ($I_{cir} = QI_{line}$; $Z_{tank} = QX_L$ = max.) The tank output at each class B stage is a sinusoidal waveform owing to the flywheel effect. For tuning the transmitter, a 3- or 4-W lamp may be used as a dummy load instead of the antenna. The V_{CC} source line may be connected to a stage via a dc milliammeter, which indicates a dip when the tank of the stage is tuned to resonance. If all resonant circuits of the transmitter are adjusted to resonant points, the small lamp at the output terminal will light.

We start to tune the transmitter from the first stage Q_1, and adjust the final power amplifier last. When we tune the final stage, the preceding stage may be slightly detuned owing to the Miller effect. In this case, retuning is needed. In the transformer-coupled stage, the impedance reflected from secondary to primary is $(\omega M)^2/Z_s$, where Z_s is the secondary impedance.

Preemphasis and Deemphasis

In order to reduce the noise contribution of the audio high frequencies (HF) the volume of the HF is increased for FM broadcast transmission and is reduced within the audio section of the FM receiver. The boosting of the audio HF at the transmitter is termed preemphasis, and the reduction of the audio HF at the receiver is termed the deemphasis. For realistic reproduction, the amount of deemphasis at the receiver must equal the preemphasis at the transmitter. Simple networks such as RL high-pass filter and RC low-pass filter are employed to accomplish preemphasis and deemphasis, respectively. The FCC has specified that the time constant for these networks is to be 75 μs. In FM receivers, the deemphasis network follows the FM detector.

12.6   AN FM RECEIVER

General Description

Figure 12.6-1 shows a superheterodyne FM receiving circuit. It is divided into an FM tuner and an audio section. The FM tuner consists of tuned rf amplifier Q_1, mixer (including local oscillator) Q_2, IF (= 10.7 MHz) amplifier

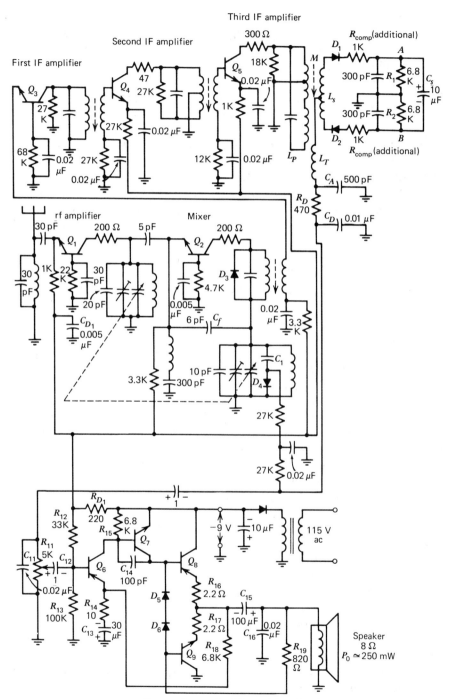

Figure 12.6-1 An FM receiving circuit.

389

stages Q_3–Q_5, and demodulator D_1–D_2. The demodulator is a ratio detector. The two diodes at both ends of the final IF transformer secondary are connected in opposite polarity and in series with two load resistors for half-wave rectification. The ratio of the voltages across the two load resistors varies with the change of the input frequency. The audio section is a direct-coupled, transformerless (*TL*) amplifier, including four transistors (Q_6–Q_9: *pnp -npn-pnp-npn*), connected as complementary pairs. The final stage (Q_8–Q_9) is a complementary symmetry push–pull class AB amplifier. Other stages in the receiver are in class A operation. In the FM tuner, Q_1–Q_3 are operated in common-base mode, while Q_4 and Q_5 are in the common-emitter mode. At the -9 V power line, there is a decoupling network consisting of R_{D_1} ($=220\ \Omega$) and C_{D_1} ($=0.005\mu$F). Except for the base circuits of Q_1 and Q_2, all the base and emitter biasing resistors in the FM tuner are shunted by the by-passed capacitors. The -9 V source line can be considered as ac ground.

Broadcast FM Tuners

The incoming FM signal has a bandwidth of 2×75 or 150 kHz (15 times the AM channel width), and all of the FM tuned circuits must have reasonably uniform response over a 150-kHz channel. The 150-kHz channel equals a swing of 75 kHz above and below the carrier frequency. The sum of a 150-kHz channel and a 50-kHz guard band is a 200-kHz interval. The FM stations are located in the band at 200-kHz intervals. Most broadcast FM tuners are designed for high-fidelity response; this response requires tuned circuits with 150-kHz bandwidth. As in an AM radio practice, the bandwidth of an FM circuit is defined as the frequency span between the 70.7% of maximum amplitude points on the response curves. Since the rf and mixer stages process frequencies in the 100-MHz region, much smaller coils and capacitors are utilized than in AM receivers that process frequencies in the 1-MHz region. A transistor that operates satisfactorily in an AM 1-MHz circuit may have no response in an FM 100-MHz circuit. However, a transistor will operate satisfactorily at a frequency lower than rated. All the circuitry up to the IF amplifier is called the FM front end, which contains an rf amplifier, a mixer, and an oscillator.

At the front end of the FM tuner, the antenna coil is shunted by 30-pF capacitor to filter out the HF interference. The collector series resistors at the Q_1, Q_2, Q_4, and Q_5 stages are used to suppress parasitic oscillations. All tuned circuits of the IF stages are shunted by damping resistors (27K, 27K, 18K) in order to obtain a proper bandwidth.

The rf amplifier Q_1 is followed by the frequency converter or mixer Q_2, which is a self-oscillating arrangement, called an autodyne converter. Two basic conditions are required to obtain frequency conversion: A local oscillator signal must be provided, and a nonlinear resistance must be ultilized for mixing. A transistor has a suitable nonlinear characteristic when appropriately biased. The major advantage of the autodyne converter is its simplicity and economy. Note that the tuning capacitor at the rf stage Q_1 is ganged with the oscillator tuning capacitor; the oscillator section is designed so that the frequency of oscillation is always 10.7 MHz higher than the frequency to which the Q_1 collector coil is tuned. Hence, transistor Q_2 combines mixer and oscillator functions. The adjustable capacitors across the Q_1 collector coil and the oscillator coil are used as trimmers. The input section of the autodyne converter (Q_2) tunes over the FM broadcast band from 88 to 108 MHz, and the output section supplies a 10.7-MHz signal, no matter to what station the receiver is tuned.

The collector coil at the mixer Q_2 is shunted by a diode D_3, and the oscillator coil is shunted by a diode D_4 in series with capacitor C_1. D_3 is used to emphasize the feedback signal. D_4 is used as a variable-voltage capacitor diode for automatic frequency control (AFC). The control voltage is supplied by a portion of the ratio-detector output. The oscillator action can be explained as follows: When the power switch is turned on, the Q_2 collector current flows through the oscillator coil; the feedback voltage from this coil is applied between the base and emitter through the feedback capacitor C_f ($=6$ pF). Then the forward voltage V_{BE} is increased owing to the feedback, causing the collector current to increase. The increase of this current further increases the voltage V_{BE}. This regeneration causes the circuit to oscillate. At the emitter of Q_2, there is a RFC in series with a 300-pF capacitor returned to ground. This LC trap is used to minimize interference. The AFC and ratio detector are discussed in Section 12.7.

Complementary-Type Audio Amplifier

The audio signal from the output of the ratio detector is applied to the audio input stage (Q_6) via a potentiometer (R_{11}), which operates as a volume control. The capacitor C_{11} connected across R_{11} is used for tone compensation. The R_{12}–R_{13} voltage divider furnishes the reverse bias and forward bias to transistor Q_6. The voltage across R_{13} supplies the V_{BE} bias to Q_6 through the speaker, R_{19}, Q_9, R_{17}, and R_{18}. A negative-feedback voltage is developed across the R_{14}–C_{13} combination. The dc voltage across R_{15} furnishes the forward-bias V_{BE} to Q_7. The reverse bias for the CB junction of Q_7 is provided from the positive polarity (ground) of the 9 Vdc source through the speaker, R_{19}, and diodes, D_6 and D_5. The forward-voltage across D_5 and D_6 furnishes the forward bias to the bases of Q_8 and Q_9. The 2.2-Ω resistors in the emitter circuit not only limit the final stage operated in class AB mode, but are also employed for temperature compensation. When a positive signal is applied to the final stage, the current in Q_8 is decreased while the current in Q_9 is increased, so that the emitter output positive signal drives the speaker. If the input signal is negative, the current in Q_9 is decreased while in Q_8 it is increased so that the emitter output negative signal drives the speaker. Therefore, Q_8 and Q_9 are in push–pull operation.

C_{14} at Q_7 stage is a feedback capacitor used to stabilize the signal. C_{16} is a tone-compensated capacitor. C_{12} and C_{15} are ac coupling (dc blocking) capacitors. The advantages of the complementary–symmetry output stage are that the transformerless operation saves on weight and cost and balanced push–pull input signals are not required. The direct-coupled amplifier stages improve the low-frequency response.

12.7 RATIO DETECTOR AND AFC CIRCUITS

Operation Principle of the Ratio Detector

A basic ratio detector circuit is shown in Fig. 12.7-1. The transformer includes a tertiary winding L_T connected to the secondary center tap. Two diodes, D_1 and D_2, at both ends of the secondary are connected in opposite polarity and in series with load resistors R_1 and R_2 for half-wave rectification. Capacitors C_1 and C_2 complete the circuit for the 10.7-MHz voltage developed across the tertiary winding so that the voltage at the low end of the tertiary winding is applied to the C_1–C_2 sides of diodes D_1 and D_2. A large capacitor C_s is connected across the R_1–R_2 load. C_s is used for stabilization. It is largely responsible for the AM-limiting properties of a ratio detector. The output is taken from the tertiary winding. Audio signal is developed across capacitor C_A. R_D and C_D form a deemphasis network to reduce the ratio detector response at high frequencies. Both the primary and secondary are tuned to the FM IF center frequency (f_c) of 10.7 MHz. The primary voltage is applied to the secondary center tap by L_T, the tertiary winding, which consists of a few turns of wire coupled tightly to the low end of I_p. Since L_T is tightly coupled to L_p, no phase shift occurs from L_p to L_T; the tertiary voltage is in phase with the primary voltage. Thus, a voltage applied to the secondary center is in phase with the primary voltage. Let the tertiary voltage be V_p', which is in phase with the primary voltage V_p. Let the primary current, secondary input voltage, and current be I_p, V_i, and I_s, respectively.

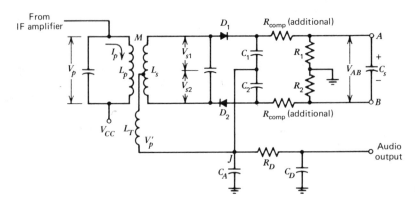

Figure 12.7-1 Basic ratio detector circuit. When its operation is explained, the compensation resistors (R comp) can be neglected.

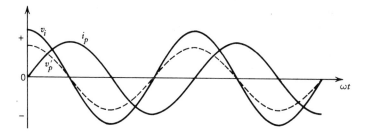

Figure 12.7-2 Waveforms of i_p, v_p', and v_i.

The waveforms of i_p, v_p', and v_i are shown in Fig. 12.7-2, where

$$i_p = I_m \sin \omega t,$$

$$v_i = M\frac{di_p}{dt} = \omega M I_m \cos \omega t = V_m \cos \omega t,$$

and

$$V_p' = V_{pm}' \cos \omega t.$$

At resonance or the center frequency f_c, the primary line current is in phase with V_p (or V_p'), while the circular current I_p lags V_p (or V_p') by 90°. Let V_{s1} and V_{s2} be the two equal halves of the secondary induced voltage due to I_s, and let V_1 and V_2 be the forward voltages applied to D_1 and D_2, respectively. V_1 is the phasor sum of V_p' and V_{s1}; V_2 is the phasor sum of V_p' and V_{s2}.

Phasor diagrams for the detector circuit are shown in Fig. 12.7-3. When input frequency f_{in} equals f_c, equal voltages ($|V_1| = |V_2|$) at 10.7 MHz are applied to the diodes, as shown in Fig. 12.7-3a. When $f_{in} > f_c$, I_s lags V_i by θ_L, and so $|V_1| > |V_2|$, as shown in Fig. 12.7-3b. When $f_{in} < f_c$, I_s leads V_i by θ_c, and so $|V_1| < |V_2|$, as shown in Fig. 12.7-3c.

When $f_{in} = f_c$, $|V_1| = |V_2|$, diodes D_1 and D_2 conduct equally. Current flow to the right through D_1 is equal to the flow to the left through D_2. Therefore, no current flows in the tertiary winding, and C_A remains uncharged. Current only flows around the loop $L_s–D_1–R_1–R_2–D_2$, with equal voltage drops V_1 and V_2 across R_1 and R_2, respectively. The voltage between

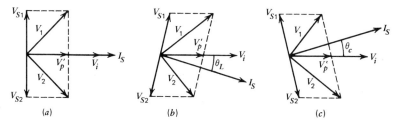

Figure 12.7-3 Phasor diagrams for a ratio detector: (a) $f_{in} = f_c$; (b) $f_{in} > f_c$; (c) $f_{in} < f_c$.

points A and B is $V_{AB} = V_1 + V_2$. Thus the stabilizing capacitor C_s (a few microfarads) charges to voltage V_{AB}. Since C_s charges fast and discharges slowly, it can hold V_{AB} constant.

When $f_{in} > f_c$, $|V_1| > |V_2|$, or $r_{d1} < r_{d2}$; hence, D_1 conducts more than D_2. This indicates a positive output taken from point J. A current equal to the difference in D_1 and D_2 currents flows via the tertiary-winding terminal J. Capacitor C_A is charged positively, and the voltage developed on C_A represents the output of the detector.

When $f_{in} < f_c$, $|V_1| < |V_2|$, or $r_{d1} > r_{d2}$; hence, D_2 conducts more than D_1. This indicates a negative output taken from point J or C_A charged negatively.

When an input FM signal is applied, the diode rectification causes voltages V_1 and V_2 to develop across R_1 and R_2, respectively. The polarity of V_1 and V_2 is such that they add to give V_{AB}. The average amplitude of the input signal determines the value of V_{AB}.

Now the sum of V_1 and V_2 equal to V_{AB} is held constant by the large capacitor C_s, so that as f_{in} changes to increase V_1, V_2 must necessarily decrease. When f_{in} varies, the ratio of V_1 to V_2 varies accordingly. The long time constant of R_1–R_2–C_s circuit makes the ratio detector unresponsive to amplitude modulation.

Ratio detector circuits can sometimes become "overstabilized" so that when an increase in amplitude suddenly occurs, the output of the ratio detector actually drops momentarily. This can be avoided by connecting compensation resistors (R_{comp}) as shown in Fig. 12.7-1. These limit maximum diode current and reduce AM sensitivity at high signal levels.

Ratio-Detector Response

Figure 12.7-4 shows a ratio-detector response curve or S curve. The central, linear region is the active or useful portion of the curve. The output is zero when the input frequency equals the center frequency (f_c) or 10.7 MHz. When the input frequency increases, the output becomes positive and increases with frequency to the positive peak of the S curve. This peak occurs at the upper limit of the IF amplifier bandwidth. If the frequency becomes higher, the input amplitude falls off quite sharply. This produces the downward turn of the response, forming the upper portion of the S curve. A similar response is obtained as the input frequency falls below the center frequency. The output becomes increasingly negative when the frequency becomes lower. The negative peak occurs at the lower limit of the IF amplifier bandwidth.

The S curve is obtained when the primary and secondary windings of the tuned transformer are resonated precisely at 10.7 MHz each. For high-fidelity reproduction, the bandwidth from A to B is made 400 kHz, approximately. The response curve of a discriminator is, for all practical purposes, the same as that of a ratio detector.

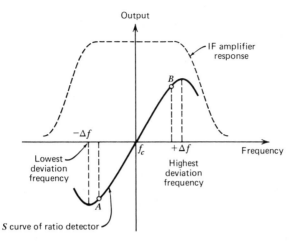

Figure 12.7-4 A typical plot of ratio-detector output as a function of frequency. Dotted-line curve indicates IF amplifier response.

Automatic Frequency Control (AFC)

Referring to the practical circuit in Fig. 12.6-1, a simplified AFC-circuit is drawn as in Fig. 12.7-5. The oscillator coil is shunted by a reverse-biased diode. The dc component from the detector output is applied to the diode anode through the AFC filter. A diode has junction capacitance, and it is equivalent to a capacitor when it is reverse biased. If the value of reverse bias is variable, the diode is equivalent to a variable capacitor. Therefore, the oscillator frequency can be varied by varying the value of reverse bias on the AFC diode. If the local-oscillator frequency happens to drift slightly off its correct value, the AFC circuit reacts and brings the oscillator back to practically its normal frequency. This control action has definite limits.

As discussed previously, the output voltage from a ratio detector will have a positive or a negative polarity, depending on whether the incoming frequency is above or below the center frequency of the tuned transformer.

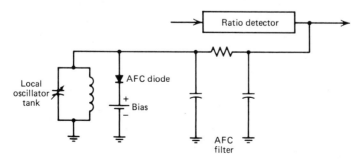

Figure 12.7-5 Simplified AFC circuit.

If the local oscillator drifts to a higher frequency, the IF frequency drifts to a higher frequency, and so the ratio-detector output voltage drifts in a positive direction, on the average. This average positive dc component is separated from the AF component by the AFC filter, and is applied to the AFC diode. Since part of the fixed reverse bias on the diode is cancelled, an increase of junction capacitance results. This increased junction capacitance reduces the oscillating frequency, thus returning the IF center frequency to practically its correct value.

The reverse-biased diode in Fig. 12.7-5 can be replaced by a zero-biased diode used in an AFC system, as in the practical circuit of Fig. 12.6-1.

When tuning an FM receiver the AFC should be switched off, otherwise a weak station may be skipped due to the holding action of the AFC. The AFC would hold to a strong station, then jump to the next strong station.

12.8   A MICROPOWER TEMPERATURE BIOTELEMETER

Introduction

The micropower temperature biotelemeter is used to measure temperature and transmit the result by an FM transmitter to a distant commercial-type FM receiver. The two-channel telemeter circuit is shown in Fig. 12.8-1. It consists of two multivibrators, and transmitting oscillator, and a mixing network. The temperature conditioning system incorporates a free-running multivibrator in which the output "off" period is related to sensor (thermistor) resistance and where the "on" period is fixed. The period, and thus frequency, is sensor resistance dependent. Small thermistors exhibiting large resistance changes in response to temperature variations (approximately $-5\%/°C$) allow the monitoring of relatively small changes in body temperature in the range of $37 \pm 2°C$ in mammals. The circuit is connected to a mixing network and a tunable VHF (90–105 MHz) oscillator, whose frequency is modulated by the summed subcarrier oscillator (SCO) outputs.

A commercially available FM receiver is adequate in providing reception at ranges to at least 3 m. Received signals may be either counted directly with a frequency counter or demodulated to an analog voltage. Multiple channel situations can be monitored in the same manner with prior separation of the subcarrier frequencies by band-pass filtering.

The design makes use of small printed circuit modules for the carrier oscillator and subcarrier oscillators, which are fabricated and assembled in small "sandwich" fashion.

Subcarrier Oscillator Operation

Each temperature subcarrier oscillator in Fig. 12.8-1 is an astable multivibrator whose design provides reliable operation at minimum current drain. The 120K collector resistances of high-gain transistors Q_1–Q_4 (Amperex

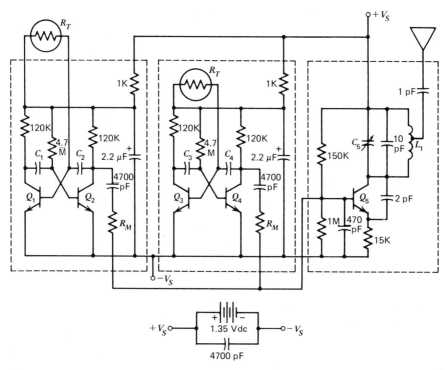

Figure 12.8-1 Circuit diagram of a two-channel telemeter. [Jeutter, D. C. and E. Fromm, *IEEE Transaction on Biomedical Engineering*, **BME-27** 244 (1980).]

A143) are selected for 10 μA collector current when each is in saturation. The thermistor (R_t) must exhibit the resistance required by the timing circuit to provide operation at the desired SCO center frequency at a nominal body temperature of 37°C. The thermistor chosen for the circuit has a resistance of 10 MΩ at 25°C (Fenwall CA71L1) and may be located within the telemeter or remotely in a sealed probe. The fixed timing resistances at 4.7 MΩ to match the thermistor resistance at 37°C. Timing capacitors C_1 and C_2 (C_3 and C_4) are selected following assembly to provide the desired operating frequency with a symmetical waveshape of the multivibrator (i.e., 50% duty cycle for the output signal) when in combination with the timing resistor and the thermistor.

Single output is coupled to the VHF oscillator through a simple network providing the required signal mixing for multichannel system, SCO isolation, and the gain (attenuation) necessary for establishing FM preemphasis and VHF oscillator FM bandwidth. The 4700-pF capacitors couple SCO outputs to the mixing network, and the 1-V output of the SCO is appropriately attenuated by the mixing resistor of R_M associated with each channel and the input resistance (360K) of the carrier oscillator. The 1000Ω resistors in conjunction with the 2.2-μF tantalum capacitors form decoupling networks that effectively isolate the SCO circuits from one another, thereby minimizing crosstalk through the battery supply. The resistor is selected to drop little voltage (10 mV), but to provide desired isolation.

VHF Transmitter Operation

The VHF transmitter is a tunable Colpitts-type oscillator that is frequency modulated by the summed subcarrier oscillator signal applied to its base. The active device chosen for the oscillator is the Amperex LDA407, which has sufficiently high gain at the frequencies and low oscillator currents encounted in this design. The transistor is available as a leadless inverted device composed of a transistor chip mounted in a miniature rectangular ceramic frame with gold pads at the comers for electrical contact. The inductor L_1 is constructed to fit around the VHF oscillator module. The 10-pF fixed capacitor and the 4-pF adjustable capacitor (C_5) in combination with L_1 determine the operating frequency, while the 2-pF capacitor is primarily responsible for the feedback necessary oscillation. The bias network formed by the 150-kΩ and 1-MΩ resistors and the 15-kΩ emitter resistor set the dc operating point of the oscillator, while the 470-pF bypass capacitor limits the bandwidth of the transmitter.

Adjustment of the feedback necessary for oscillation requires consideration of the intended operating environment of the oscillator. Feedback can be adjusted by varying the ratio of the feedback capacitor (2 pF) to the tank circuit capacitance, and is dictated by how heavily the circuit is loaded by its surroundings. Loading is increased slightly by encapsulation and somewhat more when immersed in physiologic saline or implanted in animal tissue. In practice, the oscillator is loosely capacitively coupled (by a 1pF capacitor) to one thermistor probe lead for improved signal radiation.

Implantable Temperature Probe

The thermistor probe construction is shown in Fig. 12.8-2. A small bead thermistor is installed in a 1-cm-long polyethylene tubule, 1.4 mm ID, after attachment of Teflon-insulated stainless-steel stranded lead wires. This end of the sensor is then filled with a polyethylene–paraffin mixture drawn into the lumen while in its melted state. The result is a hermetically sealed remote temperature sensor that will operate for 6 months without exhibiting effects of moisture penetration. To provide a flexible but sealed interface with the telemeter, the proximal ends of the lead wires are enclosed in the lumen of a similar 1.5-cm-long polyethylene tubule, followed by filling of the lumen with the polyethylene–paraffin mixture. These tubules are subsequently attached securely to the telemeter mainframe board and the temperature probe leads are soldered to the appropriate connections.

Calibration and Performance

The timing capacitors in the subcarrier oscillators are selected to provide the desired operating center frequency and waveshape at 37°C. The trimmer

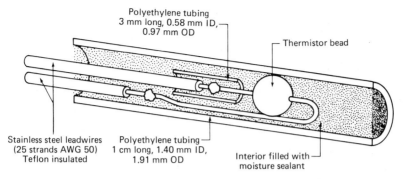

Figure 12.8-2 Thermister probe construction. [Jeuller, D. C. and E. Fromm, *IEEE Transactions on Biomedical Engineering*, **BME-27** (5), 245 (1980).]

capacitor C_5 (Johanson 9401) has a capacitance range of 0.3–4 pF, which enables the VHF oscillator to be set within a 15-MHz range of operating frequencies. The oscillator frequency is set 2-MHz higher than the intended operating frequency to allow for encapsulation and implantation loading effects.

The modulation index of the carrier oscillator was measured at the 1.7- and 3.9-kHz modulation frequencies with a News–Clark model 1672 receiver. The modulation index measured was 3.5, suggesting a satisfactory signal-to-noise ratio.

The temperature–frequency characteristics of a two-channel telemeter are illustrated in Fig. 12.8-3. The slopes of the responses for each channel

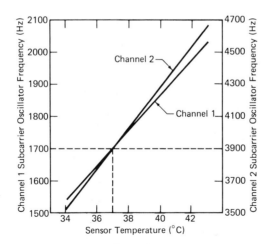

Figure 12.8-3 Frequency–temperature relationships in a two-channel temperature telemeter. The slopes of the straight lines are identical when expressed as 3.3% frequency change per degree centigrade. [Jeutter, D. C. and E. Fromm, *IEEE Transactions on Biomedical Engineering*, **BME-27** (5) 246 (1980).]

are equivalent when expressed as a percentage change in frequency per degree centigrade from its set point at 37°C. Temperature sensitivity is seen to be a 3.3% change in frequency per degree centigrade. Temperature non-linearity of this curve in the 30–42°C range is 0.15%. Supply voltage versus SCO frequency tests for the system show a linear dependence of 0.92% change in frequency per 0.1 V supply change in the range of 1.35 ± 0.25 V.

12.9 TRANSMITTER FOR RADIOTELEMETRY OF INTRAARTERIAL PRESSURE IN HUMANS

Transducer Amplifier and Subcarrier Oscillator

Figure 12.9-1 shows the circuit diagram of a transducer amplifier and a temperature-compensated subcarrier oscillator with stabilized power supplies. The pressure signal from a Statham P37 pressure transducer connected to an intraarterial cannula is amplified and filtered by a second-order low-pass filter with a nominal gain of 350 and 3 dB cutoff frequency of 80 Hz. The output of this filter modulates, linearly, the frequency of a subcarrier oscillator over the range 750–1250 Hz, these frequencies approximately corresponding to 0 and 250 mmHg, respectively. Any effects on the subcarrier frequency of changes in battery voltages or temperature have been minimized by incorporating two 3-V stabilized power supplies and a temperature-compensation circuit that adjusts the voltage on the catching diodes D_1 and D_2 according to the temperature sensed by thermistor Th_1. The subcarrier oscillator is a voltage-to-frequency converter whose frequency (= 1/period) is given by Eq. (11.3-13).

Crystal-Controlled FM Transmitter

Figure 12.9-2 shows the crystal-controlled FM transmitter circuit. The smoothed square-wave output from the subcarrier oscillator modulates the frequency of a voltage-controlled crystal oscillator about a center frequency of 17.06 MHz. Subsequent frequency multiplication of this signal produces an aerial output at the frequency of 102.360 MHz with a frequency modulation of ±4 kHz. The half-wave aerial is sewn into the harness worn by the patients. With this arrangement, the aerial characteristics are modified by the proximity of the patient's body, which also absorbs a large proportion of the signal fed to the aerial. However, the convenience of this aerial arrangement readily justifies the necessary increase in transmitter output power. Ranges of more than 70 m were achieved in steel-framed hospital buildings with loss of signal less than 5% of the time. The transmitter is powered by two 5.4-V mercury batteries (Malloy "Duracell" TR-134R) and a 1.35-V mercury cell (Malloy "Duracell" Rm-1). These provide sufficient power for approximately 30 h of continuous operation or up to 300 h in the intermittent mode.

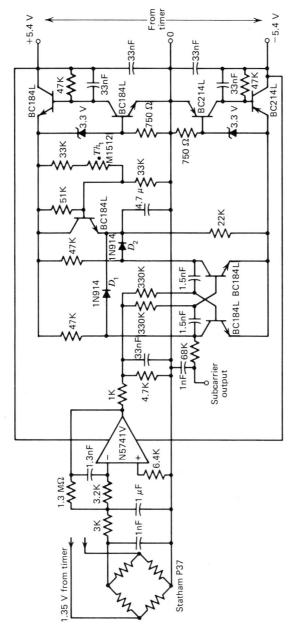

Figure 12.9-1 Transducer amplifier and temperature-compensated subcarrier oscillator with sta-bilized power supplies. [Figures 12.9-1, 12.9-2, and 12.9-3: From Brash, H. M., J. B. Irving, B. J. Kirby, and K. W. Donald, *Medical and Biological Engineering*, **14** (1), 61–63 (1976).]

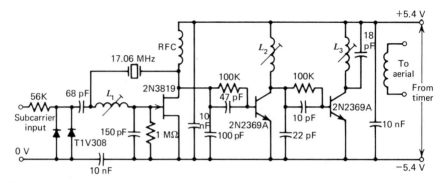

Figure 12.9-2   Crystal-controlled FM transmitter.

Circuit of the Timer for Intermittent Operation

For many purposes, data are only required at 5-min periods, thus battery power can be conserved by turning the circuit on only 30 s during each 5-minute period. Figure 12.9-3 shows the circuit of the 4.5-min and 30-s timers used. The 30-s timer (lower half of Fig. 12.9-3) is based on a conventional unijunction transistor circuit. At the end of this timing period, the TIS43 unijunction transistor discharges the 10-μF tantalum capacitor, turning on the BC184L transistor and applying a negative 2-ms pulse to the bistable reed relays R_2, R_3, and R_4, thereby switching off the transmitter and transducer and removing the -5.4 V supply from the 30-s timer. The 47-μF timing

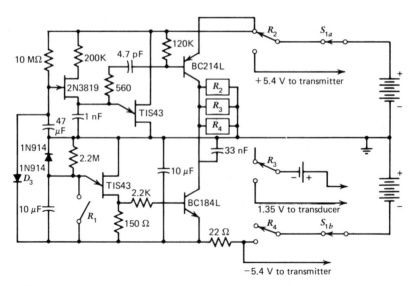

Figure 12.9-3   Circuit of the timer for intermittent operation of the transmitter.

capacitor of the 4.5-min timer is held at -5 V by diode D_3 until the end of the 30-s interval, after which the 47-μF tantalum capacitor begins to charge up through the 10-MΩ resistor. When the voltage on this capacitor is approximately zero, the 2N3819 field-effect transistor triggers the second TIS43 unijunction transistor causing the BC214L transistor to apply a positive 2-ms pulse to the bistable reed relays. This switches on the transmitter and transducer, starts the 30-s timer, and resets the 4.5-min timer, completing the cycle. Continuous operation of the transmitter during calibration and testing is achieved by placing a permanent magnet externally on the transmitter case adjacent to the reed switch R_1 to inhibit the 30-s timer.

Signal Reception and Waveform Recording

Radio-frequency signals from the patient are received and demodulated by a crystal-controlled receiver with an intermediate frequency of 10.7 MHz and a 3 dB bandwidth of 20 kHz. The resulting subcarrier output is demodulated by a pulse-counting discriminator and then filtered by a 70-Hz fourth-order Butterworth active low-pass filter, the output of this filter being a fascimile of the original pressure waveform. When the transmitter is turned off, the filter output in the frequency range 20–70 Hz increases markedly in the presence of receiver noise, as the pressure waveform has a very small, but significant, frequency component in this band. By means of a high-pass filter and detector this increase in noise output provides a very sensitive indication of "transmitter on." Initially, the pressure waveform was recorded with an ultraviolet recorder, in which the paper drive was started by this "transmitter-on" signal. The blood pressure and heart rate were computed by manual analysis of the calibrated tracings, but the custom-built analog system now in use computes the average values of systolic and diastolic pressure, and of heart rate, from 12 beats during each 30-s measuring interval. These average values are punched, along with the time, onto paper tape for off-line computer analysis and are also recorded on an industrial point-chart recorder at a paper speed of 25 mm/h, thereby providing a compact record. In addition, the pressure waveform for the 12 heart beats analyzed is recorded on magnetic tape along with the time of measurement, offering a subsequent check on the quality of the basic analog data.

Advantages of the Apparatus

The apparatus has successfully been used for monitoring for periods of 3–36 h. It has three main advantages over other techniques for prolonged blood-pressure monitoring. First, it is unobtrusive, so that patients rapidly become unaware of it; second, activity is completely unrestricted, and third, the effects of an activity can be observed while it is being carried out, without the knowledge of the subject. At constant temperature and battery voltage the zero drift over a 24-h period is typically 1 mm Hg or less.

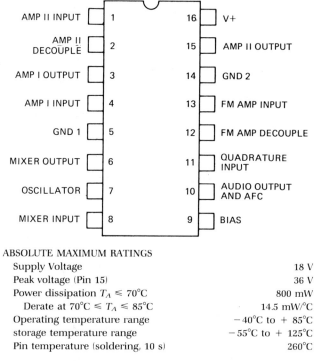

AMP II INPUT	1	16	V+
AMP II DECOUPLE	2	15	AMP II OUTPUT
AMP I OUTPUT	3	14	GND 2
AMP I INPUT	4	13	FM AMP INPUT
GND 1	5	12	FM AMP DECOUPLE
MIXER OUTPUT	6	11	QUADRATURE INPUT
OSCILLATOR	7	10	AUDIO OUTPUT AND AFC
MIXER INPUT	8	9	BIAS

ABSOLUTE MAXIMUM RATINGS

Supply Voltage	18 V
Peak voltage (Pin 15)	36 V
Power dissipation $T_A \leq 70°C$	800 mW
Derate at $70°C \leq T_A \leq 85°C$	14.5 mW/°C
Operating temperature range	$-40°C$ to $+85°C$
storage temperature range	$-55°C$ to $+125°C$
Pin temperature (soldering, 10 s)	260°C

Figure 12.10-1 μA721 connection diagram and main ratings. (Courtesy Fairchild Camera and Instrument Corp.)

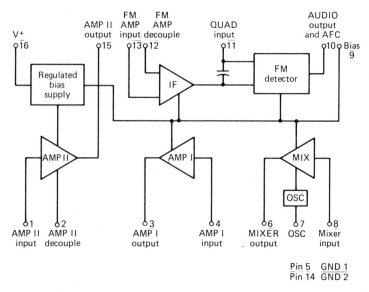

Pin 5   GND 1
Pin 14   GND 2

Figure 12.10-2 Functional blocks of Fairchild μA721. (Courtesy Fairchild Camera and Instrument Corp.)

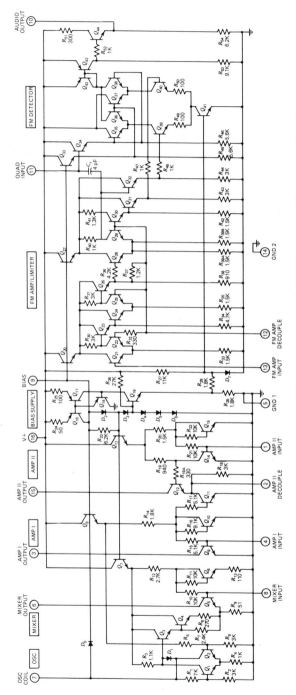

Figure 12.10-3 Equivalent circuit of μA721. (Courtesy Fairchild Camera and Instrument Corp.)

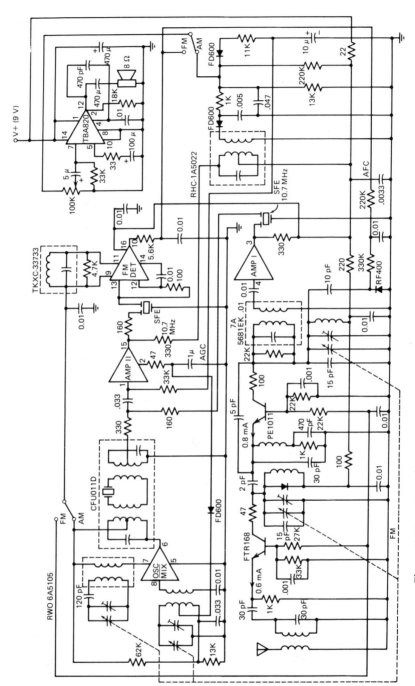

Figure 12.10-4 Typical AM/FM receiver with μA721. Two 10.7 mHz ceramic filters (SFE) are used in the FM IF amplifiers. (Courtesy Fairchild Camera and Instrument Corp.)

12.10 A MONOLITHIC AM/FM RADIO RECEIVER SYSTEM

Introduction to the Radio Integrated Circuit

The Fairchild μA721 is a monolithic AM/FM radio-receiver system manufactured using the planar epitaxial process. Several gain blocks are available to perform AM conversion, rf, and IF or FM IF amplification. The FM limiter and quadrature detector are also included. The μA721 is designed to operate over a very wide voltage range (3.5–16 V) making it useful in a variety of applications including portable, home, and automobile AM/FM radios, and industrial communications systems. The connection diagram of μA721 is shown in Fig. 12.10-1.

The independent functional blocks of the μA721 shown in Fig. 12.10-2 can be arranged for a variety of AM and FM receiver applications. Amplifier I, amplifier II, and the AM oscillator mixer have open-collector output circuits that give complete flexibility for selecting load impedances. The FM IF amplifier limiter and detector section needs only a few external components. It can be designed to be free of alignment, if ceramic filters are used for selectivity and the conventional quadrature coil is replaced by ceramic resonator. With appropriate external circuit arrangements, these four functional blocks can be combined to cover the FM IF amplification, FM detection, and all AM functions in most AM/FM receivers.

Two ground systems are used in the μA721 to minimize the chance of high-frequency instability owing to common-ground impedances. However, for any application, care must be taken in the selection and placement of external components and in the layout of the printed circuit board to ensure stable system operation. The equivalent circuit of μA721 is shown in Fig. 12.10-3. A typical AM/FM radio receiver with μA721 is shown in Fig. 12.10-4.

12.11 FM QUADRATURE DETECTOR

A quadrature detector is a phase-shift type of detector. Its operation depends on the shifting phase of a resonant circuit voltage relative to a loosely coupled driving voltage. The voltage induced in each half of the secondary at resonance is 90° out of phase with the primary driving voltage, as shown in Fig. 12.11-1. The resonant-circuit voltage is said to be "in quadrature" with the driving voltage. Moreover, the phase difference varies as the driving frequency varies above and below the natural frequency of the resonant circuit.

Figure 12.11-2 illustrates the general scheme and the action of the quadrature detector when the input f_{in} is equal to, is less than, and is greater than the resonant circuit frequency f_c. The driving and quadrature signals drive series-connected gates, as shown in Fig. 12.11-2a. The gates indicated

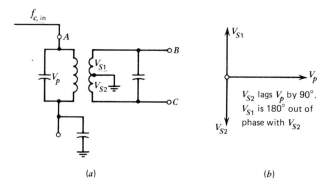

(a)   (b)

Figure 12.11-1 (a) Tuned transformer with the input signal of the same frequency as that to which the primary and secondary are tuned. The circuits are loosely coupled. (b) Phasors indicating the phase relationships of voltages, V_p, V_{s1}, and V_{s2}.

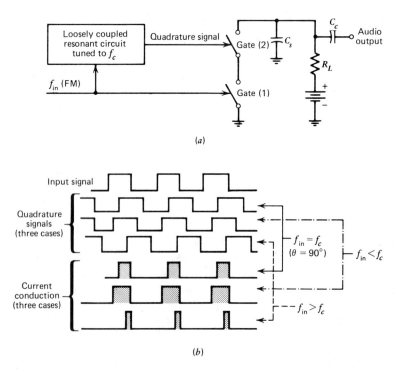

(a)

(b)

Figure 12.11-2 (a) General scheme of a quadrature detector. (b) Quadrature signal varies in phase relative to the input signal.

as switches are capable of being cut completely on or off by a small positive or negative voltage, respectively. Both gates must be on for conduction through R_L to occur. In Fig. 12.11-2b, square waves are shown, owing to the off–on controlling characteristics of the gates. When f_{in} equals f_c, the input and quadrature signals are 90° out of phase ($\theta = 90°$), and the resulting conduction is indicated. A change in f_{in} causes the phase of the quadrature signal to shift, varying the conduction as shown. Hence, conduction occurs as a series of pulses whose width varies according to the relationship of f_{in} to f_c. The pulses occur at the rate of the IF frequency, and the pulses themselves bear little resemblance to the audio signal of the FM. Capacitor C_s smooths the pulses so that the audio signal is derived directly from the pulses.

The integrated-circuit quadrature detectors are characterized by an FM IF input to the chip, an external quadrature resonant circuit, and an audio output directly from the chip.

12.12   REVIEW OF PROPERTIES OF RESONANT CIRCUITS

Now, let us review the properties of resonant circuits, since they are extremely important to the operation of the tuned amplifiers in radio transmitters and receivers. The resonant frequency, voltage, and current magnification factors, bandwidth, etc. will be discussed as follows.

Series *RLC* Resonant Circuits

Resonant Frequency.   At the resonant frequency (f_r), the power factor (PF = cos θ) is unity, implying that the circuit of Fig. 12.12-1 is purely resistive. Hence, the net reactance (X_T, Ω) or the difference of inductive reactance ($X_L = 2\pi f L$, Ω) and capacitive reactance ($X_C = \frac{1}{2\pi f C}$, Ω) equals zero. The total circuit impedance (Z, Ω) is equal to resistance $R(\Omega)$, or

$$Z = R + j(X_L - X_C) = R = \text{min.}$$

Thus

$$2\pi f_r L = \frac{1}{2\pi f_r C,}$$

or

$$f_r = \frac{1}{2\pi\sqrt{LC}} , \qquad (12.12\text{-}1)$$

where the resonant frequency f_r is expressed in Hz, inductance L in henrys (H) and capacitance C in farads (F).

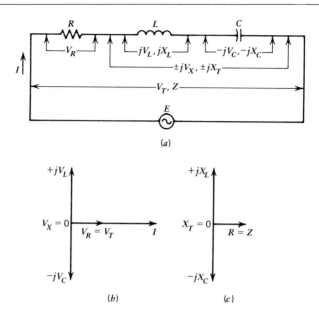

Figure 12.12-1　Series *RLC* circuit at resonance: (*a*) circuit; (*b*) phasor diagram; (*c*) impedance diagram.

Voltage Magnification Factor.　Q is defined as the ratio of stored energy to dissipated energy. In a series circuit, Q equals the ratio of reactance to resistance. Since the energy in the resonant circuit is stored alternately in the inductance and the capacitance, Q becomes the ratio of either reactance (X_{Lr} or X_{Cr}) to the resistance (R) or

$$X_{Lr} = X_{Cr} = Q_r R = Q_r Z. \qquad (12.12\text{-}2)$$

In voltage form,

$$Q_r = \frac{IX_{Lr}}{IR} = \frac{IX_{Cr}}{IR} = \frac{V_{Lr}}{V_T} = \frac{V_{Cr}}{V_T}$$

or

$$V_{Lr} = V_{Cr} = Q_r V_T. \qquad (12.12\text{-}3)$$

In practice, Q_r is usually in the range of about 10 to 100. In a series *RLC* circuit at resonance, the quality factor Q_r is called the voltage magnification factor.

Bandwidth.　Bandwidth (BW) in a series resonant circuit is defined as the frequency difference between the two conditions at which the net reactance equals resistance. At the below-resonance bandwidth frequency f_1, the capacitive reactance X_{C1} will be greater than the inductive reactance X_{L1}:

$$X_{T1} = X_{C1} - X_{L1} = R. \qquad (12.12\text{-}4)$$

At the above-resonance bandwidth frequency f_2, the inductive reactance X_{L2} will be greater than the capacitive reactance X_{C2}:

$$X_{T2} = X_{L2} - X_{C2} = R. \qquad (12.12\text{-}5)$$

It follows, then, that

$$X_{T1} = X_{T2} = R. \qquad (12.12\text{-}6)$$

The impedance triangles referring to the lower frequency f_1 and the higher frequency f_2 are shown in Figs. 12.12-2a and 12.12-2b, respectively. At f_1 or f_2,

$$X_T = R,$$

$$Z = \sqrt{R^2 + X_T^2} = \sqrt{2}R,$$

$$\phi = \arctan \frac{X_T}{R} = \arctan 1 = 45° = \pi/4 \text{ rad}.$$

Hence, at the bandwidth frequencies (f_1 and f_2), the impedance is $\sqrt{2}$ times the resistance, and the phase angle ϕ is 45° ($\pi/4$ rad).

At zero frequency, X_C is infinitely high and X_L equals zero. Hence, the circuit in effect consists of an infinitely high capacitive reactance in series with a resistance. The impedance is therefore infinite, and the phase angle is $+90°$ ($+\pi/2$ rad).

If the frequency were to become infinitely high, X_L would become infinite and X_C would be zero. The effective circuit would be that of an infinite inductive reactance in series with resistance. The impedance would be infinite, and the phase angle would be $-90°$ ($-\pi/2$ rad).

Between zero frequency and resonance, and between resonance and infinite frequency, the transition of impedance and phase angle values is gradual, the rate of change depending on the specific circuit values.

Figure 12.12-3 shows how the current varies with respect to frequency in a series circuit with the values $L = 100$ μH, $C = 250$ PF, and $R = 100$ Ω. If the voltage across the circuit is maintained constant at 100 mV while the frequency increases from zero, the circuit current ($I = V_T/Z$) will in-

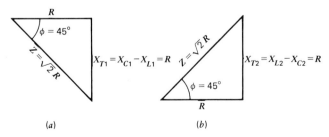

(a)   (b)

Figure 12.12-2 Series RLC circuit impedance triangles for the bandwidth frequencies: (a) f_1 (below resonance); (b) f_2 (above resonance).

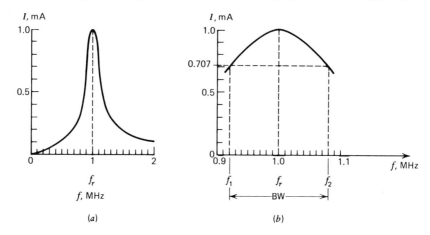

Figure 12.12-3   Series *RLC* circuit current–frequency curves.

crease from zero to a maximum value of $I = V_T/R = 1$ mA at resonance. The current will then decrease toward zero at frequencies above resonance. The current at the bandwidth frequencies f_1 and f_2 will be $I = V_T/\sqrt{2}R = 0.707$ mA, as shown in Fig. 12.12-3b, which is a magnified view of the current variations at frequencies close to resonance.

Bandwidth is a measure of a circuit selectivity. High selectivity is narrow bandwidth; low selectivity is broad bandwidth. The bandwidth of a circuit is the frequency difference $(f_2 - f_1)$ between points at which the power response is one-half that of the maximum at resonant frequency. The relationship between the powers at the bandwidth frequencies (P_1 at f_1 and P_2 at f_2) and the power at resonance (P_r at f_r) is

$$P_1 = P_2 = \frac{P_r}{2}. \qquad (12.12\text{-}7)$$

This can be written in the form

$$I_1^2 R = I_2^2 R = \frac{I_r^2 R}{2},$$

$$I_1^2 = I_2^2 = \frac{I_r^2}{2},$$

or

$$I_1 = I_2 = \frac{I_r}{\sqrt{2}} = 0.707 I_r. \qquad (12.12\text{-}8)$$

Equation (12.12-8) leads to another definition of bandwidth: The bandwidth of a series *RLC* circuit is the frequency difference $(f_2 - f_1)$ between conditions where the magnitudes of the currents are $1/\sqrt{2}$, or 0.707, of the magnitude at resonant frequency. At the lower frequency f_1 (Hz) and the higher frequency f_2 (Hz), the bandwidth BW (Hz) is

$$BW = f_2 - f_1. \qquad (12.12\text{-}9)$$

Figure 12.12-4 gives curves for X_C and X_L, expanded to show the region close to resonance. At frequency f_2.

$$X_{T2} = X_{L2} - X_{C2} = R. \qquad (12.12\text{-}10)$$

But from Fig. 12.12-4, it can be seen that $X_{C2} \simeq X_{L1}$. Hence,

$$X_{L2} - X_{L1} \simeq R,$$

$$2\pi L(f_2 - f_1) \simeq R,$$

$$\frac{R}{2\pi L} = f_2 - f_1 = BW,$$

$$\frac{R}{2\pi f_r L} = \frac{f_2 - f_1}{f_r} = \frac{BW}{f_r},$$

or

$$BW = \frac{f_r}{O_r}. \qquad (12.12\text{-}11)$$

Parallel *RLC* Resonant Circuit

In the parallel *RLC* circuit of Fig. 12.12-5, the total admittance Y (siemens, S) is the phasor sum of the element admittances

$$Y = G + j(B_C - B_L),$$

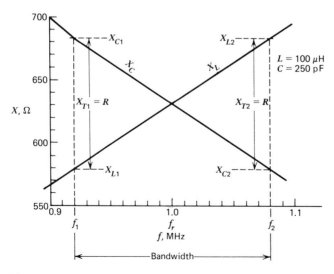

Figure 12.12-4 Series resonant circuit reactance–frequency curves.

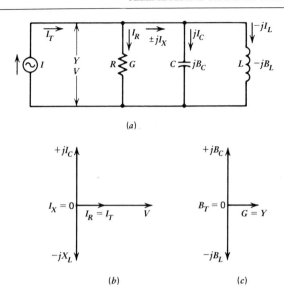

Figure 12.12-5   Parallel *RLC* circuit at resonance: (a) circuit; (b) phasor diagram; (c) admittance diagram.

where the conductance G $(= 1/R)$ is expressed in S, the capacitive and inductive susceptances B_C $(= 1/X_C)$ and B_L $(= 1/X_L)$ in S.

Resonant Frequency.   At resonant frequency f_r (Hz), the power factor is unity, the circuit is purely resistive, the admittance consists only of the circuit conductance, the net susceptance (B_T) is zero, and the phase angle (ϕ) is zero:

$$B_C = B_L; \quad Y = G + j(B_C - B_L) = G = 1/R = \text{min.}$$

$$\omega_r C = 1/\omega_r L. \tag{12.12-12}$$

Hence,

$$f_r = \frac{1}{2\pi\sqrt{LC}}. \tag{12.12-13}$$

Current Magnification Factor.   At resonance, the current in parallel *RLC* circuit is determined solely by the resistance. The capacitive and inductive currents can be many times greater than the total circuit current. The quality factor Q of a parallel circuit is the ratio of either susceptance to conductance:

$$Q_r = \frac{B_{Cr}}{G} = \frac{B_{Lr}}{G}$$

or

$$B_{Cr} = B_{Lr} = Q_r G = Q_r Y. \tag{12.12-14}$$

In current form.

$$Q_r = \frac{VB_{Cr}}{VG} = \frac{VB_{Lr}}{VG} = \frac{I_{Cr}}{I_R} = \frac{I_{Lr}}{I_R}$$

or

$$I_{Cr} = I_{Lr} = Q_r I_R = Q_r I_T, \qquad (12.12\text{-}15)$$

where I_{Cr}, I_{Lr}, and $I_T (=I_R)$ are the capacitive current (A), inductive current (A), and total circuit current (A), respectively. In practice, Q_r is greater than unity, and so I_{Cr} and I_{Lr} are each greater than I_T. In a parallel RLC circuit, the quality factor is called the current magnification factor.

Bandwidth. At two frequencies, f_1 below rersonance and f_2 above resonance, the net susceptance (B_T, S) equals the conductance ($G = 1/R$, S). f_1 and f_2 are the lower and higher bandwidth frequencies, respectively:
At f_1,

$$B_{T1} = B_{L1} - B_{C1} = G. \qquad (12.12\text{-}16)$$

At f_2,

$$B_{T2} = B_{C2} - B_{L2} = G. \qquad (12.12\text{-}17)$$

$$B_{T1} = B_{T2} = G,$$

The admittance triangles for f_1 and f_2 are shown in Fig. 12.12-6. For either bandwidth frequency,

$$B_T = G,$$

$$Y = \sqrt{G^2 + B_T^2} = \sqrt{2}G,$$

and

$$\phi = \arctan \frac{B_T}{G} = \arctan 1 = 45° = \pi/4 \text{ rad.}$$

At zero frequency, B_L is infinitely high and B_C equals zero. Hence, at zero frequency the circuit consists of an infinitely high inductive susceptance

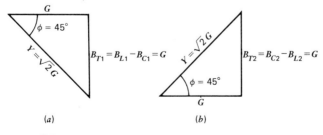

Figure 12.12-6 Parallel RLC circuit admittance triangles for the bandwidth frequencies: (a) f_1 (below resonance); (b) f_2 (above resonance).

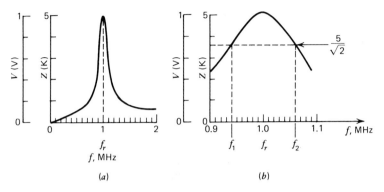

Figure 12.12-7 Parallel *RLC* circuit curves of impedance–frequency and voltage–frequency. Subsidiary scales to the left of (*a*) and (*b*) give the voltage values with a constant current of 0.2 mA.

in parallel with a conductance. The admittance is infinite, and the phase angle is $-90°$ or $-\pi/2$ rad.

If the frequency could become infinitely high, B_C would become infinite and B_L zero. In this case the circuit would be in effect an infinite capacitive susceptance in parallel with a conductance. The admittance would be infinite, and the phase angle would be $+90°$ or $+\pi/2$ rad.

Between zero frequency and resonance, and between resonance and infinite frequency, the change in admittance and phase angle is gradual, the rate of change depending on the circuit values involved.

Figure 12.12-7 shows how the impedance and voltage magnitudes of a parallel-resonant circuit vary with changing frequency. In a parallel resonant circuit supplied with constant current, maximum power is dissipated at the resonant frequency. The relationship between the powers at the bandwidth frequencies f_1 and f_2 and the power P_r at resonance is

$$P_1 = P_2 = \frac{P_r}{2}$$

or

$$V_1^2 G = V_2^2 G = \frac{V_r^2 G}{2},$$

$$V_1^2 = V_2^2 = V_r^2/2.$$

Hence,

$$V_1 = V_2 = V_r\sqrt{2} = 0.707V_r,$$

where V_1, V_2, and V_r are the circuit voltages at f_1, f_2, and f_r, respectively. The bandwidth is

$$BW = f_2 - f_1. \tag{12.12-18}$$

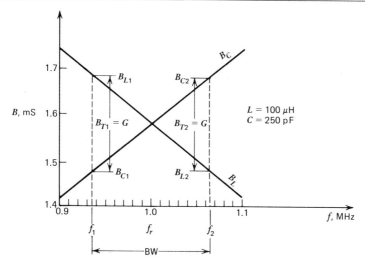

Figure 12.12-8 Parallel resonant circuit susceptance–frequency curves.

Curves for B_C and B_L, expanded to show the region close to resonance, are given in Fig. 12.12-8. These curves indicate that $B_{L2} = B_{C1}$. At frequency f_2, $B_{T2} = B_{C2} - B_{L2} = G$. Then

$$B_{C2} - B_{C1} = G; \quad 2\pi C(f_2 - f_1) = G;$$

$$BW = f_2 - f_1 = \frac{G}{2\pi C}; \quad \frac{BW}{f_r} = \frac{G}{2\pi f_r C}.$$

Hence,

$$BW = \frac{f_r}{Q_r}. \tag{12.12-19}$$

Practical Parallel Resonant Circuit

In practice, a parallel resonant circuit consists of a capacitor and an inductor connected in parallel. In addition to its inductance, a practical inductor also exhibits a resistance owing to the resistance of the wire used to wind the inductor. The resistance is not necessarily small, and it is important in determining the sharpness of the resonance of the circuit of which the inductor is a part. At high frequencies current tends to flow only on the surface of the conductor (skin effect), causing the effective resistance to be many times the low-frequency value. Figure 12.12-9 illustrates the variation of impedance as a function of frequency. Note that the resistance of the inductor, R_s, is taken into account and that it affects the maximum impedance. As R_s increases, the maximum impedance decreases. The maximum impedance also depends on the L/C ratio. Circuits with large values of L and small

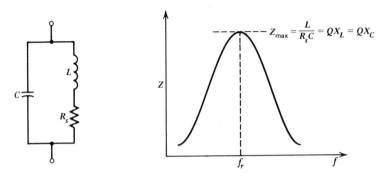

Figure 12.12-9 Variation of impedance with frequency for a practical parallel resonant circuit.

values of C have higher resonant impedances. This ratio cannot be made arbitrarily large, however, because of distributed capacitance effects of practical inductors. Note further that the resonant impedance equals the reactance of the inductor or capacitor at the resonant frequency multiplied by the Q of the circuit.

A practical parallel resonant circuit is frequently shunted by a resistor R_p (as shown in Fig. 12.12-10) to lower the Q of the circuit to a new value we denote as Q_p:

$$Q_p = R_T/X_L, \qquad\qquad (12.12\text{-}20)$$

where $R_T = R_p \parallel Q^2 R_s$; $Q \geq 10$ or $X_L \geq 10R_s$ at f_r.

Tuned Transformers

Figure 12.12-11 illustrates a circuit in which two resonant circuits tuned to the same frequency are magnetically coupled to form a tuned transformer. The frequency response depends on the Q of the individual tuned circuits and very strongly on the degree of coupling between the inductors. When the inductors are well separated (loose coupling), a frequency response is

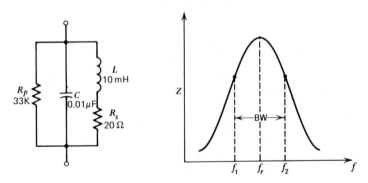

Figure 12.12-10 An example illustrating a practical parallel resonant circuit shunted by a resistor. $f_r = 15{,}915$ Hz; $X_L = X_C \simeq 1000\ \Omega$; $Q = 50$; $Q_p \simeq 20$.

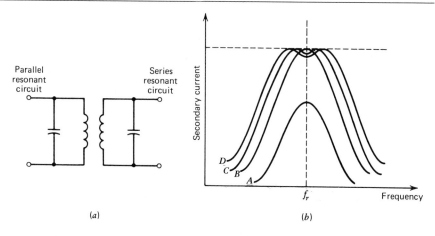

Figure 12.12-11 (a) Tuned transformer and (b) its frequency response.

obtained as shown by curve *A* in Fig. 12.12-11*b*. As the inductors are moved closer together, the peak of the frequency-response curve reaches a maximum at the point of critical coupling, illustrated by curve *B*. A further increase in coupling causes a broadening of the response with no increase in the level of the peak. Finally, with degree of coupling well beyong critical coupling, a double-humped response curve (curve *D*) develops, which is a characteristic of overcoupled tuned circuits.

BIBLIOGRAPHY

"Transmitter for Radiotelemetry of Intra-Arterial Pressure in Man," *Medical and Biological Engineering,* **14**(1), 61–63 (1976).

Carr, J. J., *Elements of Electronic Instrumentation and Measurement.* Reston Publishing, Reston, Virginia, 1979, Chaps. 21 and 22.

Green, C. R. and R. M. Bourque, *The Theory and Servicing of AM, FM, and FM Stereo Receivers.* Prentice-Hall, Englewood Cliffs, New Jersey, 1980.

Gregg, W. D., *Analog and Digital Communication.* Wiley, New York, 1978.

Jeutter, D. C. and E. Fromm, "A Modular Expandable Implantable Temperature Biotelemeter," *IEEE Transactions on Biomedical Engineering,* **BME-27**(5), 242–247 (1980).

Questions

12-1 Draw a block diagram indicating a basic telemetry system.

12-2 What are the advantages of pulse modulation?

12-3 What are the differences between class A and class *B* operations? Why do the buffer and rf power stages use class *B* operation rather than class A in the circuit of Fig. 12.3-1?

12-4   Why must the quartz crystal of the oscillator in the circuit of Fig. 12.3-1 be operated in the inductive region?

12-5   How is the transmitter (Fig. 12.3-1) tuned?

12-6   What is the purpose of the keyer transistor (Q_3, Fig. 12.3-1)?

12-7   Describe the operation of mixer Q_3 in Fig. 12.4-1.

12-8   What is AM detection?

12-9   What is the purpose of the varactor diode in the circuit of Fig. 12.5-1?

12-10   Describe the operation principle of the Miller crystal oscillator in the circuit of Fig. 12.5-1.

12-11   How is the FM transmitter (Fig. 12.5-1) tuned?

12-12   Define the bandwidth of an FM circuit.

12-13   Describe the operation principle of the autodyne converter (Fig. 12.6-1)?

12-14   What are the purposes of diodes D_1–D_6 used in the circuit of Fig. 12.6-1?

12-15   Describe the push–pull operation of Q_8 and Q_9 (Fig. 12.6-1).

12-16   What is the function of an FM detector?

12-17   What is the purpose of AFC in an FM receiver?

12-18   Why should the AFC be switched off when tuning an FM receiver?

12-19   What is the purpose of the tertiary winding of a ratio detector?

12-20   What is the purpose of the stabilizing capacitor of a ratio detector?

12-21   What is the purpose of the compensating resistors of a ratio detector?

12-22   How does the combination of preemphasis and deemphasis reduce the noise at audio HF in an FM system?

12-23   Briefly describe the principle of operation of the subcarrier oscillator in Fig. 12.8-1.

12-24   How is the feedback adjusted for the transmitter (Fig. 12.8-1)?

12-25   Briefly describe the principle of operation of the timer in Fig. 12.9-3.

12-26   Briefly describe the principle of operation of a quadrature detector.

Problems

12.12-1   A 500-pF capacitor and a 50-μH inductor (with an effective resistance of 20 Ω) are connected in series. Find (a) the resonant frequency f_r and (b) bandwidth BW.

    Answer.   (a) 1.007 MHz; (b) $f_r/Q_r = 63.65$ kHz.

12.12-2   A capacitance of 500 pF, an inductance of 50 μE, and a resistance of 5K are connected in parallel. (a) Find the resonant frequency

f_r and bandwidth BW. (b) If a current of 1 mA at the resonant frequency is supplied to the complete circuit, what are the currents in each element.

Answer.   (a) f_r = 1.007 MHz; BW = 63.65 kHz. (b) $I_R = I_T$ = 1 mA; $I_C = I_L$ = 15.82 mA.

12.12-3   A series resonant circuit consisting of a variable capacitor and a 300-μH inductor with an effective resistance of 10 Ω is connected between a receiving antenna and ground and is used as a wave trap to reject an undesired signal of 400 kHz. What value should the variable capacitor be set?

Answer.   527.7 pF.

12.12-4   A practical parallel resonant circuit is shown in Fig. P12.12-4. (a) Verify that the resonant frequency is

$$f_r = \frac{1}{2\pi\sqrt{LC}} \sqrt{1 - \frac{R_s^2 C}{L}},$$

(b) and the impedance at resonance is

$$Z_r \simeq L/CR_s, \text{ if } X_{Lr} \gg R_s.$$

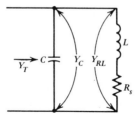

Figure P12.12-4

12.12-5   A 0.1-μF capacitor is connected in parallel with an inductor, whose series characteristics measured at 1 kHz are 200-mH inductance and 500-Ω resistance. Find the resonant frequency of the circuit.

Answer.   1053 Hz.

12.12-6   In the amplifier resonant network of Fig. P12.12-6, the Q of the 25-μH inductor is 75 at a frequency of 3 MHz. The value of the parallel capacitor makes the inductor resonant at 3 MHz. The transistor supplies 1 mA to the network at resonance. Find: (a) The value of capacitor C; (b) the resulting bandwidth BW; (c) the resonant impedance Z_r of the network; (d) the magnitude of the circular current I_{cir} at resonance.

Answer.   (a) 112.6 pF; (b) f_r/Q_r = 40 kHz; (c) 35.34K; (d) $Q_r I_T$ = 75 mA.

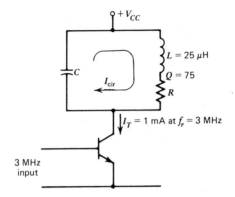

Figure P12.12-6

12.12.7   The circuit in Fig. P12.12-7 is an IF stage of a radio receiver that operates either on AM 450-kHz or FM 10.7 MHz signals. Draw the equivalent circuits at 450 kHz and 10.7 MHz to indicate that on AM stations the signal is delivered at output 1, while on FM stations the signal leaves the circuit at output 2.

Answer.   At 455 kHz, Z_{r1} = 71.48K and Z_2 = 62.90 Ω. (b) At 10.7 MHz, Z_{r2} = 29.58K and Z_1 = 76.01 Ω.

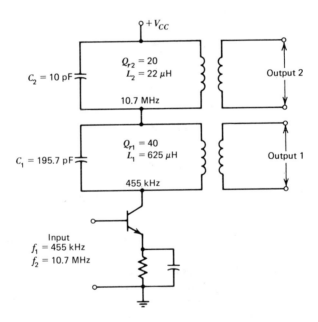

Figure P12.12-7

12.12-8   A resonant circuit (Fig. P12.12-8) in an AM receiver is required to tune over the band 525 kHz to 1.6 MHz. The variable tuning capacitor has a range of 30–300 pF. What values of inductance and trimming capacitance are needed?

[*Hint:* $C_{LF}/C_{HF} = (f_H/f_L)^2 = 9.288$.]

Answer. $L = 303.7 \ \mu\text{H}$. $C_{\text{trim}} = 2.58 \ \text{pF}$.

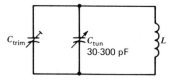

Figure P12.12-8

12.12-9 The broadcast AM receiver has an intermediate frequency (IF) of 455 kHz, which is a difference of the oscillator frequency (f_{osc}) and the signal frequency (f_{sig}). If the capacitor that tunes the oscillator is the same type (30–300 pF) as that which tunes the signal frequency circuit, what values of inductance and padding capacitance (C_p) are required for the oscillator circuit in Fig. P12.12-9? [*Hint:* The oscillator operates in the band (525 + 455) to (1600 + 455) kHz or 980 to 2055 kHz.]

Answer. $L = 232.95 \ \mu\text{H}$; $C_p = 181.9 \ \text{pF}$.

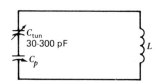

Figure P12.12-9

12.12-10 In the rf amplifier equivalent circuit (Fig. P12.12-10), $N_1 = 80$ turns, $L_1 = 600 \ \mu\text{H}$, coil $Q = 100$, resonant frequency $f_r = 1000$ kHz, and $R_i = 5\text{K}$. Find N_2.

Answer. $N_2 \simeq 9.2$ turns.

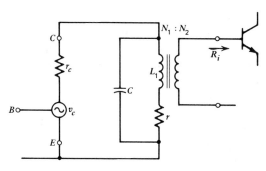

Figure P12.12-10

13

A Typical Triggered-Sweep Dual-Trace Oscilloscope

13.1 INTRODUCTION

The Tektronix 465B is selected as a typical dual-trace oscilloscope, and the circuit of its main sections are discussed in this chapter. These discussions may help the readers understand those circuits of similar oscilloscopes.

The 465B oscilloscope shown in Fig. 13.1-1 is a wide-band, dual-channel portable instrument, providing traces for two input channels, a trigger view from an external trigger input, and an add function. Calibrated deflection factors from 5 mV/div to 5 V/div are provided by the dc-to-100 MHz vertical system for the input channels and add function. Sweep trigger circuits are capable of stable triggering over the full bandwidth capabilities of the vertical deflection system. The horizontal deflection system provides calibrated sweep rates from 0.5 s/div to 0.02 μs/div along with delayed sweep features for accurate relative-time measurements. A × 10 magnifier extends the calibrated sweep rate to 2 ns/div. The instrument operates over a wide variation of line voltages and frequencies with maximum power consumption of approximately 100 W.

Increased measurement capabilities are achieved by the 465B when it is equipped with an optional Tektronix DM44 Digital Multimeter.

The following instrument specifications apply over an ambient temperature range of −15°C to +55°C unless otherwise specified.

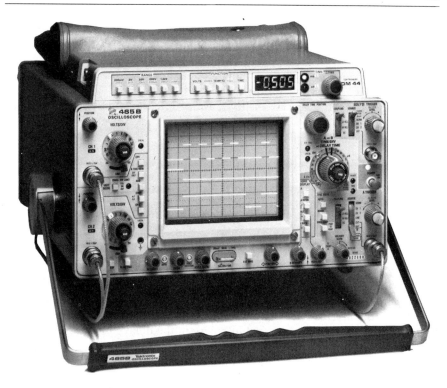

Figure 13.1-1   Tektronix 465B oscilloscope (equivalent to TEK 2235). (Courtesy Tektronix, Inc. © 1980.)

13.2   MAJOR SPECIFICATIONS

Vertical Deflection

Calibrated range is from 5 mV/div to 5 V/div. Accuracy is within 3%. At all deflection factors from 50-Ω terminated sources, the bandwidth measured at −3 dB is dc to 100 MHz from −15°C to +40°C or dc to 85 MHz from +40°C to +55°C. Rise time is 3.5 ns from +15°C to +40°C or 4.1 ns from +40°C to +55°C. The bandwidth may be limited to approximately 20 MHz by a bandwidth limit switch. In cascaded operation, channel 1 vertical signal output is connected to channel 2. Cascaded bandwidth is at least 50 MHz when signal output is terminated in 50 Ω. Lower −3 dB points with ac coupling 1× probe and 10× probe are 10 Hz or less and 1 Hz or less, respectively. Uncalibrated VAR control provides deflection factors continuously variable between the calibrated settings (steps) and extends deflection factor to at least 12.5 V/div. There are five display modes—namely, Ch 1,

Ch 2, ADD (normal and inverted), alternate, and chopped modes. The chopped-mode repetition rate is approximately 500 kHz. Common-mode rejection ratio is at least 20 dB at 20 MHz for common-mode signals of six divisions or less. Input resistance and capacitance is 1 MΩ ± 2% paralleled by about 20 pF. Dc-coupled maximum input voltage is 250 V (dc + peak ac) or 500 V (p-p ac at 1 kHz or less). Ac-coupled maximum input voltage is 250 V (dc + peak ac) or 500 V (p-p at 1 kHz or less). Delay Line permits viewing leading edge of displayed waveform.

Horizontal Deflection

Time Base *A* has a calibrated sweep range from 0.02 μs/div to 0.5 s/div; 10× MAG extends maximum sweep rate to 2 ns/div. Time Base *B* has a calibrated sweep range from 0.02 μs/div to 50 ms/div; 10× MAG extends maximum sweep rate to 2 ns/div. Unmagnified sweep accuracy is within ±3%. Magnified sweep accuracy is within ±4%. Time Base *A* provides continuously variable uncalibrated sweep rates between calibrated seetings. The Variable Range extends the slowest *A* sweep rate to at least 1.25 s/div. A trigger hold off increases *A* sweep holdoff time by at least a factor of 10. There are four horizontal display modes—namely, *A*, *A* intensified, *B* delayed, and *B* ends *A* for increased intensity in the delayed mode. Electronic switching is between intensified and delayed sweep. *A* sweep and *B* sweep may be viewed simultaneously. Calibrated sweep delay has a delay time range from 0.2 to 10× delay time/div settings of 200 ns to 0.5 s. Delay or differential time jitter is one part or less in 50,000 (0.002%) of 10× the *A* sweep time/div setting. The jitter is one part in 20,000 when operating from a 50-Hz line.

Triggering *A* and *B*

There are three *A* Trigger modes—namely, Normal (sweep runs when triggered), Automatic (sweep runs in the absence of a triggering signal and for signals below 30 Hz), and Single Sweep (sweep runs one time on the first triggering event after the rest selector is pressed). A Trigger Holdoff is an adjustable control that permits a stable presentation of repetitive complex waveforms. There are two *B* Trigger modes—namely, "*B* runs after delay time" (starts automatically at the end of the delay time) and "*B* triggerable after delay time" (runs when triggered). The *B* (delayed) sweep runs once, in each of these modes, following the *A* sweep delay time.

Sensitivities of triggering are different when different coupled types are used. Sensitivity for dc couping is 0.3 division internal or 50 mV external from dc to 10 MHz, increasing to 1.5 divisions internal or 150 mV external at 100 MHz. Requirements increase below 60 Hz for sensitivity with ac coupling. Requirements increase below 50 kHz for sensitivity with ac LF Reject coupling. Requirements increase below 60 Hz and above 50 kHz for sensitivity with ac HF Reject coupling. Trigger jitter is 0.5 ns or less at 100 MHz and 2 ns/div.

A Trigger View is an electronically switched trigger view circuit that displays the external signal used for *A* sweep triggering. This provides quick verification of the signal and time comparison between a vertical signal and the trigger signal, which can be displayed simultaneously. The deflection factor is approximately 100 mV/div (1 V/div with external ÷ 10).

Level adjustment is through at least ± 2 V in external and through at least ± 20 V in external ÷ 10. Internal level and slope control permits selection of triggering at any point on the positive or negative slope of the displayed waveform. There are six *A* sources—namely, Norm, Ch1, Ch2, line, external, and external ÷ 10. There are five *B* sources—namely, Start after delay, Norm, Ch1, Ch2, and External. External input resistance and capacitance are about 1 MΩ paralleled by about 20pF. Maximum external input is 250 V (dc + peak ac).

X-Y Operation

Full-sensitivity *X-Y* (Ch1 Horiz, Ch2 Vert) is 5 mV/div to 5 V/div, accurate ± 4%. Bandwidth is dc to at least 4 MHz. Phase difference between amplifiers is 3° or less from dc to 50 kHz.

Display

CRT display area is 8 cm × 10 cm. Accelerating voltage is 18 kV. Beam finder compresses trace to within graticule area for ease in locating an offscreen signal. A preset intensity level provides a constant brightness.

Z-Axis Input

Dc coupled, positive-going signal decreases intensity; 5-V p-p signal causes noticeable modulation at normal intensity; dc to 50 MHz.

13.3   BASIC BLOCK DIAGRAM OF THE 465B OSCILLOSCOPE

A basic block diagram of the 465B oscilloscope is shown in Fig. 13.3-1. Only the basic interconnections between the individual blocks are shown on this diagram. Each block represents a major circuit within the instrument. These circuits will be explained in the following sections.

Signals to be displayed on the CRT are applied to the CH1 or *X* input connector or the CH2 or *Y* input connector. These input signals are then amplified by the Preamp circuits (1, 2). Each Channel includes a separate vertical deflection factor, input coupling, balance, gain, and variable attenuation switches or controls. A trigger pickoff stage in each Vertical Preamp circuit supplies a sample of that channel's signal to the Trigger Generator circuit. A sample of the Channel 1 signal is also supplied to the CH 1 VERT SIGNAL OUT connector on the instrument near panel.

In the *X-Y* mode of operation the Channel 1 signal is connected to the input of the Horizontal Amplifier circuit to provide the *X*-axis deflection. The Channel 2 signal is amplified by the Vertical Output Amplifier circuit (5) to provide the *Y*-axis deflection. The Channel 2 Vertical Preamp circuit

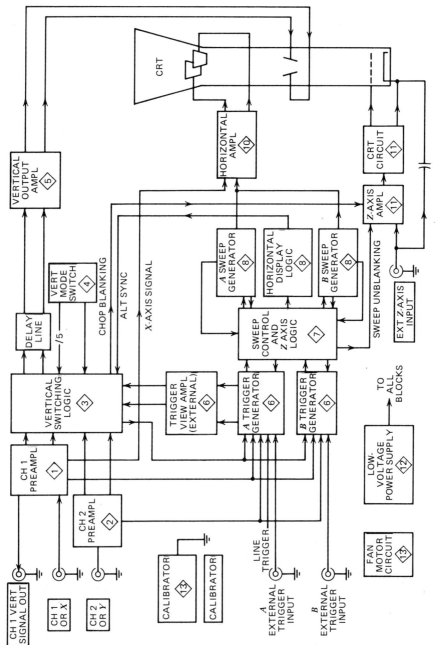

Figure 13.3-1 Basic block diagram of 465B oscilloscope. (Courtesy Tektronix, Inc. © 1980)

428

contains an invert feature to allow the operator to invert the Channel 2 signal displayed on the CRT.

The outputs of both Vertical Preamp circuits and the output of the *A* Trigger View Amplifier circuit (6) are connected to the Vertical Switching circuit (3). The Vertical Mode Switch circuit (4) uses the inputs from the VERT MODE switches and the *X-Y* position of the *A* TIME/DIV switch to set the Vertical Switching circuit into the chosen Vertical Mode of operation. The Vertical Switching circuit then will select the input(s) to be displayed on the CRT.

The chopped Blanking signal, used to blank the switching transients between channels when the chopped mode of operation is selected, is produced in the Vertical Switching circuit and fed to the Z-Axis Amplifier circuit (11). A Normal Trigger pickoff stage at the output of the Vertical Switching circuit provides a sample of the displayed signal(s) to the Trigger Generator circuits (6).

The output of the Vertical Switching circuit is connected to the Vertical Output Amplifier (5) through the Delay Line. The Vertical Output Amplifier circuit provides the final amplification for the signal before it is connected to the vertical deflection plates of the CRT. This circuit includes part of the BEAM FINDER switch, which, when activated, limits vertical deflection to within the graticule area so that location of off-screen displays is facilitated.

The *A* and *B* Trigger-Generator circuits (6) each create an output pulse which initiates the sweep produced by either the *A* or *B* Sweep Generator circuits (8). The input signal to the *A* and *B* Trigger Generator circuits can be individually selected from any of the following sources: Channel 1 signal, Channel 2 signal, signal(s) displayed on the CRT (Normal), signal(s) connected to the External Trigger Input connectors, or a line voltage sample (*A* Trigger only). Each trigger circuit contains level, slope, coupling, and source controls. The *A* External Trigger input is also fed to the *A* Trigger View Amplifier where it is amplified and made available to the Vertical Switching circuit for selection to be viewed on the CRT.

The *A* Sweep Generator circuit (8), when activated by the *A* Trigger Generator circuit, produces a linear sawtooth output signal, the slope of which is controlled by the *A* TIME/DIV switch. The TRIG MODE switches control the operating mode of the *A* Trigger Generator circuit. When AUTO is selected, the absence of an adequate trigger signal for about 100 ms after the end of holdoff causes an *A* sweep start gate to be generated. When NORM is selected, a horizontal sweep is presented only when triggered by an adequate trigger signal. Pushing the SINGL SWP pushbotton sets the Sweep Logic to initiate one sweep after a trigger pulse is received.

The Z-Axis Logic circuit (7) produces a gate signal to unblank the CRT so that the display can be presented. This gate signal is coincident with the sawtooth produced by the *A* Sweep Generator circuit. The *A* gate signal, which is also coincident with the sawtooth, is avialable at the *A* +GATE connector on the instrument rear panel. The Sweep Control Logic circuit (7) also produces a Horizontal Alternate Sync pulse. This pulse is fed to the Horizontal Display Logic circuit (8) to switch the display between A Intensified and B Delayed Sweeps when the ALT Horizontal Display mode is selected.

The B Sweep Generator circuit (8) is basically the same as the A Sweep Generator circuit. However, this circuit only produces a sawtooth output signal when a delay time period, determined by the DELAY TIME POSITION dial, has lapsed, or when a trigger pulse is received from the B Trigger Generator circuit. If the B TRIGGER SOURCE switch is set to the START AFTER DELAY position, the B Sweep Generator begins to produce the sweep immediately following the selected delay time. If the SOURCE switch is in one of the remaining positions, the B Sweep Generator circuit does not produce a sweep until it receives a trigger pulse from the B Trigger Generator circuit.

The output of either the A or B Sweep Generator is amplified by the Horizontal Amplifier circuit (10) to produce horizontal deflection for the CRT, except when the A TIME/DIV switch is in the fully counterclockwise (X-Y) position. The Horizontal Amplifier circuit contains a $\times 10$ magnifier that may be selected to increase the sweep rate 10 times in any A or B TIME/DIV switch position. Other deflection signals may be connected to the Horizontal Amplifier by using the X-Y mode of operation. When the TIME/DIV switch is set to X-Y, the X signal is connected to the Horizontal Amplifier circuit through the Channel 1 Vertical Preamp circuit.

The Z-Axis Amplifier circuit (11) determines the CRT intensity and blanking. The Z-Axis Amplifier circuit sums the current inputs from the INTENSITY control, Vertical Switching circuit (chopped blanking), Z-Axis Logic circuit (unblanking), and the external Z-Axis INPUT connector. The output level of the Z-Axis Amplifier circuit controls the trace intensity through the CRT circuit. The CRT circuit provides the voltages and contains the controls necessary for operation of the CRT.

The calibrator circuit produces a square-wave output (with accurate voltage and current amplitudes) that is useful for both checking the calibration of the instrument and the compensation of probes. The CALIBRATOR current loop provides an accurate current source for calibration of current measuring probe systems.

13.4 CHANNEL 1 AND CHANNEL 2 VERTICAL PREAMPLIFIERS

Introduction

The channel 1 preamplifier circuit is shown in Figure 13.4-1a–13.4-1c. Since the Channel 2 Preamplifier circuit is virtually the same as the Channel 1 Preamplifier circuit, the only discussion of the Channel 2 Preamplifier circuit will be on the differences in the operation between the two.

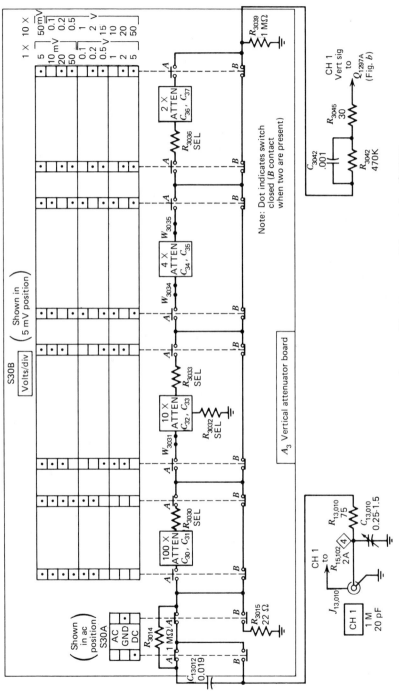

Figure 13.4-1 (a) Channel 1 Vertical Preamplifier. (Courtesy Tektronix, Inc. © 1980.)

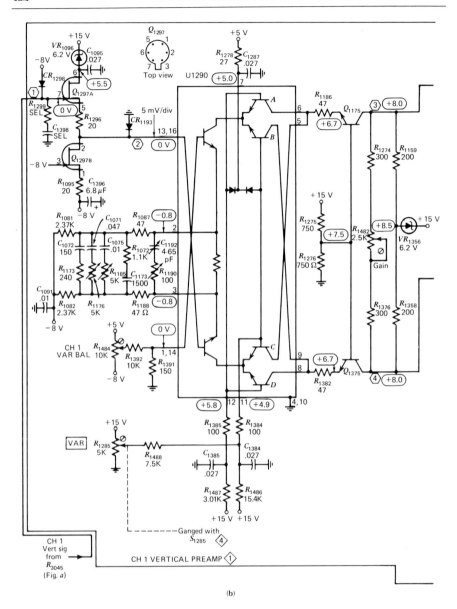

Figure 13.4-1   (*b*) (*Continued*)

The Channel 1 Preamplifier circuit provides control of input coupling, vertical deflection factor, gain, and dc balance. Input signals for vertical deflection on the CRT are connected to the CH 1 or *X* input connector. When the TIME/DIV switch is set to the *X-Y* mode, the input signal applied to the CH 1 or X connector provides the horizontal (*X*-axis) deflection.

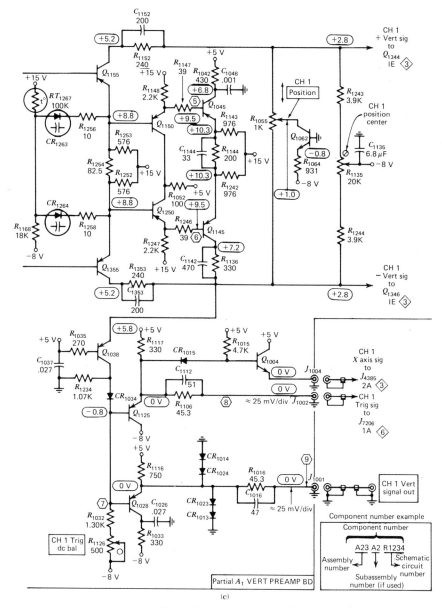

Figure 13.4-1 (c) (Continued)

Input Coupling

Signals applied to the input connector can be either ac coupled or dc coupled or, they can be internally disconnected from the input to the Vertical Input Amplifier stages. When Input Coupling Switch *S30A* is set for dc coupling, the input signal is coupled directly to the Input Attenuator stage. When ac coupled, the input signal passes through capacitor $C13012$ which prevents the dc component of the input signal from passing to the amplifier. In the GND position of $S30A$, the signal path is opened and the input of the amplifier is connected to ground through $R3015$. This provides a ground reference without the need to disconnect the applied signal from the input connector. $R3014$ is a high resistance connected across Input Coupling switch $S30A$, which allows $C13012$ to be precharged when the switch is in the GND position. Therefore, the trace remains within the viewing area of the CRT when the switch is placed in the AC position.

Input Attenuator

The effective overall deflection factor of each vertical channel of the 465B Oscilloscope is determined by the setting of the Channel VOLTS/DIV switch. The basic deflection factor of the Vertical Deflection System is 5 mV/div of CRT deflection. To achieve the deflection factor values marked on the front panel, precision attenuators are switched into the input to the Vertical Preamplifier circuit.

For VOLTS/DIV switch positions above 5 mV, frequency-compensated voltage dividers are switched into the circuit to produce the vertical deflection factors indicated on the front panel. Each channel has a $2\times$, $4\times$, $10\times$, and $100\times$ attenuator, which may be selected in various combinations. A constant attenuation is provided at all frequencies within the bandwidth range of the instrument. The Input Attenuators are designed to maintain the same input charateristics (1 MΩ and about 20 pF) for each setting of the VOLTS/DIV switch. Each attenuator contains an ajustable series capacitor to provide correct attenuation at high frequencies and an adjustable shunt capacitor to provide correct input capacitance.

Source Follower

The channel 1 signal from the input attenuator is connected to Source Follower $Q1297A$ through $R3042$, $C3042$, and $R3045$. Resistor $R3039$ provides the input resistance and $R3045$ functions as a damping resistor. $Q1297B$ is a constant current source for $Q1297A$. $Q1297A$ and B provide a high input impedance for the attenuators and the current drive needed for the first amplifier.

In the event that excessively high-amplitude signals are applied to Source Follower $Q1297A$, the signal will be limited by $CR1298$ and the gate-source junction of $Q1297A$. If the negative signal amplitude causes $CR1298$ to become forward biased, $Q1297A$ gate is champed to about -8.7 V. Excessive positive signal amplitude will forward bias the gate-source junction of $Q1297A$. As soon as gate current flows, the gate voltage will cease increasing. Gate current is limited to a safe value by the high resistance of R3042.

First Amplifier

The First Amplifier stage is an integrated emitter-coulped, push–pull, cascode amplifier $U1290$. The input signal on pins 13 and 16 is converted from a single-ended signal to a push–pull signal by a paraphase amplifier and then is fed to the common-base output stage to produce the current drive to $Q1175$ and $Q1375$. The CH 1 VAR VOLT/DIV control, which is connected to pin 11 of $U1290$, varies the gain of the First Cascode Amplifier stage. This control provides variable vertical deflection at each position of the VOLT/DIV switch. With the VAR control in its calibrated detent (wiper at ground), the A and D output transistors of $U1290$ are conducting. The B and C output transistors are biased off. Thus, the signal current available to the following amplifier stage is the collector current flowing in output transistors A and D.

When the VAR control is rotated out of its calibrated detent, the B and C output transistors of $U1290$ begin to conduct by an amount determined by the position of the VAR control. This causes two events to occur.

1. The signal current flowing in the A and D output transistors is reduced by the amount of signal current flowing in the B and C output transistors.

2. Output transistors A and C and output transistors B and D conduct current of opposite polarity. The output of transistor C is added to the output of transistor A to reduce the signal current available at pins 5 and 6, and the output current of transistor B is added to the output current of transistor D to reduce the signal current available at pins 8 and 9.

The component values selected for the variable function provide a variable attenuation ratio of approximately 2.5 to 1. Channel 1 Variable Balance adustment $R1484$ is adjusted so that no trace shift in the display occurs when rotating the VAR control. When the Channel 1 VAR control is out of its calibrated detent, Channel 1 UNCAL LED is illuminated. The components connected between pins 2 and 3 of $U1290$ provide frequency compensation for the stage.

$Q1175$ and $Q1375$ are common-base amplifiers that convert the output current signals from $U1290$ into voltage signals to be amplified in the Second Amplifier circuit. Gain adjust $R1482$ allows setting of the overall gain of the channel 1 Vertical Preamplifier by adjusting the signal voltage to the bases of $Q1155$ and $Q1355$.

Second Amplifier

$Q1155$ and $Q1355$, in conjunction with $Q1344$ and $Q1346$ in the Vertical Switching circuit, make up a push–pull cascode amplifier. $CR1263$, $CR1264$, and $R1267$ provide temperature compensation for the high-frequency gain amplifier to ensure constant gain in the presence of varying ambient temperature. As temperature increases, the resistance value of $RT1267$ decreases, and the reverse bias on both $CR1263$ and $CR1264$ decreases. $CR1263$ and $CR1264$ are voltage-control variable capacitors whose capacitance increases as reverse bias decreases. The increase in capacity at higher temperature provides additional high-frequency peaking to counteract the effects of increased temperature on the amplifier's gain.

The push–pull signals at the emitters of $Q1155$ and $Q1355$ are converted to a single-ended signal by $Q1150$, $Q1250$, $Q1045$, and $Q1145$. The current signal from $Q1145$ is converted to a voltage signal by common-base amplifier stage $Q1038$ and applied to the bases of $Q1125$ and $Q1028$. Q1028 provides the output signal to the CH 1 VERT SIGNAL OUT output connector located on the instrument rear panel. $CR1014$, $CR1024$, $CR1023$, $CR1013$ protect the emitter circuit of $Q1028$ in the event large signal levels are accidentally connected to the CH 1 VERT SIGNAL OUT connector. The output signal at the emitter of $Q1125$ is used as the trigger signal source in the CH 1 positions of the Trigger SOURCE switches and as the signal source for emitter-follower $Q1004$. $R1126$ adjusts the dc level of the CH 1 trigger source signal. When in the X-Y mode, $Q1004$ provides the X axis from the Channel 1 Preamplifier to the Horizontal Amplifier.

$R1055$ is the Channel 1 Vertical POSITION control. When set to its midposition, the constant current supplied by $Q1062$ flows equally through each side of $R1055$ into the collectors of $Q1155$ and $Q1355$. As the POSITION control is rotated off its midpoint, one side of the amplifier receives more current while the other side of the amplifier receives less current. This proportionally changes the amount of current flowing into the Delay Line Drivers, therefore causing the trace to be positioned vertically on the CRT. The midrange operating point of the POSITION control is set by adjusting $R1135$.

Channel 2 Preamplifier Differences

The only differences between the Channel 2 Preamplifier circuit and Channel 1 Preamplifier circuit are described as follows

The Source-Follower and First-Amplifier section of the Channel 2 Preamplifier circuit is shown in Figure 13.4-2. Input signals for vertical deflection on the CRT are connected to the CH2 or Y input connector. When the TIME/DIV switch is set to the X-Y mode, the channel 2 input signal provides the vertical (Y-axis) deflection.

Basically, the First-Amplifier stage in Channel 2 operates the same as the First-Amplifier stage in Channel 1. However, the Channel 2 circuit also contains the INVERT switching function. This allows the Channel 2 CRT display to be inverted. When pushed in, the INVERT switch changes the biasing on the output transistors of $U1790$ so that the normally inactive transistors (B and C) now carry the signal. Since their outputs are cross coupled from side to side, the output signal is of opposite polarity from the signal available when the INVERT switch is in the normal (button out) position. Channel 2 Invert Balance potentiometer $R1975$ allows the dc balance of the stage to be adjusted to eliminate base-line shift in the display when switching from a normal to an inverted display.

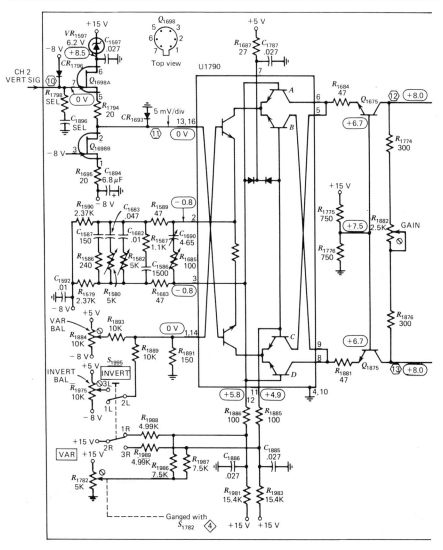

Figure 13.4-2 Source-Follower and First-Amplifier section of the Channel 2 Preamplifier circuit. (Courtesy Tektronix, Inc. © 1980.)

13.5 VERTICAL SWITCHING LOGIC CIRCUIT

Introduction

The Vertical Switching Logic circuit is shown in Fig. 13.5-1a–13.5-1c. This circuit determines the input signal or combination of input signals to be connected to the Vertical Output Amplifier. Possible input signal combinations that may be displayed are selected by a read-only-memory (ROM) integrated circuit that is controlled by the VERT MODE switches and the X-Y position of the TIME/DIV switch.

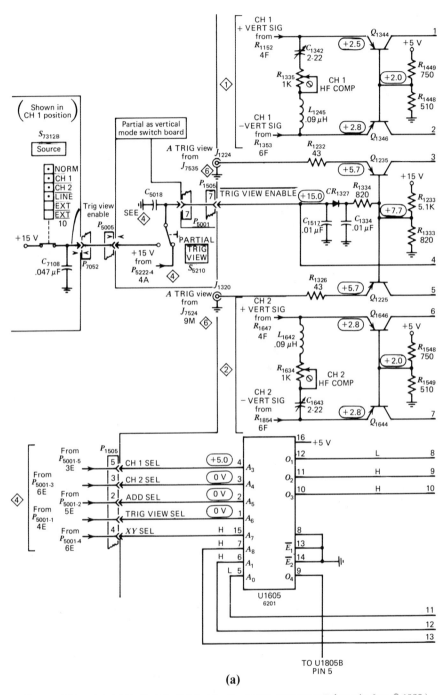

(a)

Figure 13.5-1 (a)–(c) Vertical-Switching Logic circuit. (Courtesy Tektronix, Inc. © 1980.)

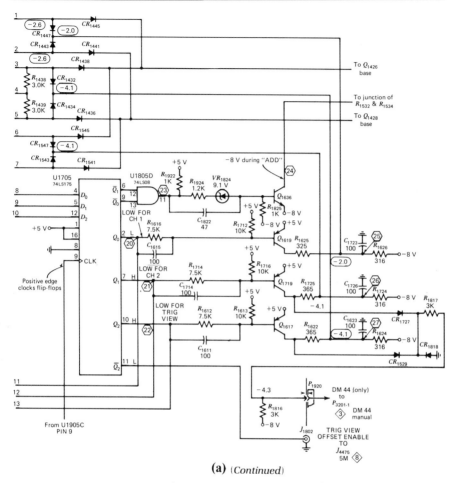

(a) (*Continued*)

Figure 13.5-1   (*Continued*)

Diode Gates

The Channel 1, Channel 2, and Trigger View Diode Gates, consisting of four diodes each, act as switches that are controlled by the Vertical Switching Logic circuit. Outputs Q_0, Q_1, and Q_2 of $U1705$ (pins 2, 7, and 10) control the switching transistors that switch the Diode Gates on or off. These output signals also are fed into the A_0, A_1, and A_2 inputs of ROM $U1605$ (pins 5, 6, and 7), to indicate the state of the switches. A LO indicates that a particular switch is on, and a HI indicates it is off. The ROM is programmed to use the state indicators from $U1705$ and the selected VERT MODE inputs to $U1605$ A_3 through A_7 (pins 4, 3, 2, 1, and 15) to turn on the correct Diode Gates for obtaining the selected signal, or combination of signals, to be displayed.

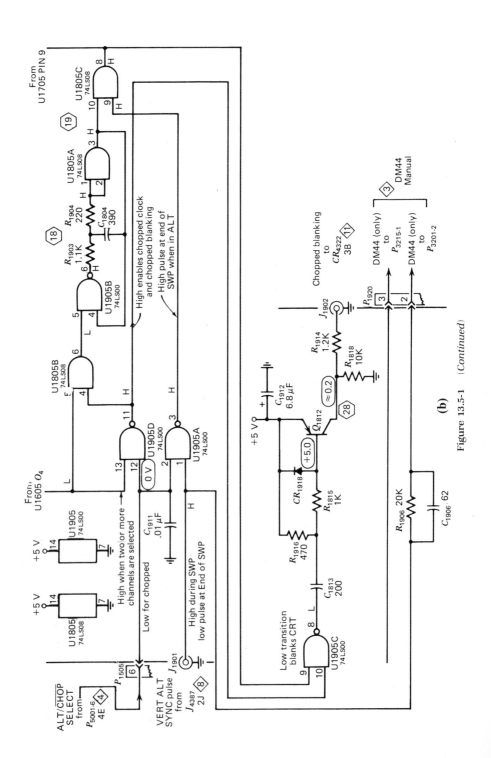

Figure 13.5-1 (Continued)

440

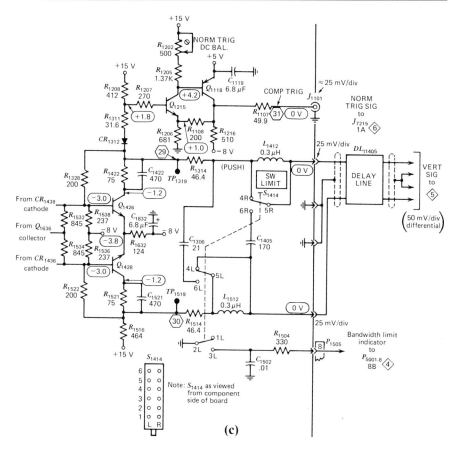

Figure 13.5-1

Channel 1 Display Only. When the CH 1 VERT MODE pushbutton is pressed, a HI is applied to the A_3 input of $U1605$. The A_4, A_5, and A_6 inputs will be LO and the A_7 input (controlled by the X-Y position of the A TIME/DIV switch) will be HI. This combination of inputs is decoded to switch transistor $Q1619$ on, which turns on the Channel 1 Diode Gate. The O_1 output of $U1605$ will be LO, the O_2 and O_3 outputs will be HI. The O_4 output, which controls the CHOP oscillator, will be low. A LO will disable the CHOP clock and $U1705$ will remain in a state that allows the Diode Gates to pass only the Channel 1 input signal to the Delay Line Driver.

With Channel 1 only selected, the Q_0 output of $U1705$ will be LO. The LO will turn on transistor $Q1619$ and the junction of $CR1447$ and $CR1443$ will be returned to the $+5$-V supply through $R1625$ and $Q1619$. This raises the junction voltage to reverse bias $CR1447$ and $CR1443$. Then $CR1445$ and $CR1441$ are forward biased and the Channel 1 input signal passes through $CR1441$ and $CR1445$ to the Delay-Line Driver.

CR1432 and CR1434 in the Trigger View Diode Gate and CR1547 and CR1543 in the Channel 2 Vertical Signal Diode Gate remain forward biased through the pull-down resistors to the -8-V power supply. Trigger View signals and the Channel 2 input signals are shunted to the -8-V supply and are blocked from the Delay Line Driver because CR1438 and CR1436 in the Trigger View Diode Gate and CR1545 and CR1541 in the Channel 2 Vertical Signal Diode Gate are reverse biased.

Channel 2 Display Only.   When CH 2 VERT MODE is selected, $Q1719$ turns on; $Q1619$ and $Q1617$ remain off. The center diodes of the Channel 1 Diode Gate and the center diodes of the Trigger View Diode Gate are forward biased, and the center diodes of the Channel 2 Diode Gate are reverse biased. Trigger View and Channel 1 signals are blocked from the Delay Line Driver, and Channel 2 input signal is connected to the Delay Line Driver.

A Trigger View (EXT Only).   When A TRIG VIEW is selected, the Channel 1 and Channel 2 Diode Gates are biased off, and the Trigger View Diode Gate is biased on to allow the External Trigger signal to be connected to the Delay Line Driver. In all single input selections (CH 1 only, CH 2 only, or A TRIG VIEW only) the clock is disabled and $U1705$ remains in a state to select only the single input chosen to be displayed.

Add Vertical Mode. $U1605$ is programmed to turn on transistor $Q1619$ (CH 1) and $Q1719$ (CH 2). The logic from $U1705$ will turn on $Q1636$ (ADD) at the same time, if ADD VERT MODE is selected. It is not necessary to select Channel 1 or Channel 2 to obtain the ADD display. With ADD selected, the Q_0 and Q_1 outputs of $U1705$ will be LO, and the $\overline{Q}_0$ and $\overline{Q}_1$ outputs will be HI. This will cause both Channel 1 and Channel 2 Diode Gates to be on. With $\overline{Q}_0$ and $\overline{Q}_1$ HI, $U1805D$ will be enabled to turn on $Q1636$. The junction of $R1532$ and $R1534$ will have -8 V applied to provide sufficient additional current to keep both Diode Gates turned on without altering the dc levels associated with the Delay Line Driver. By selecting additional VERT MODEs, it is possible to view the Channel 1 input, the Channel 2 input, the A External Trigger input, and ADD MODE on the CRT during one display cycle on four separate traces.

X-Y Mode.   When the A TIME/DIV switch is set to X-Y, a LO is applied to pin 15 of $U1605$. The ROM is programmed to produce outputs that turn on Channel 2 switching transistor $Q1719$ and disable the CHOP clock. This action causes the Channel 2 input to be connected to the Delay Line Driver for the Y-axis signal. It is not necessary to select Channel 2 VERT MODE, since the ROM will not respond to any other input while the X-Y mode is selected.

Chop Clock and Alternate Logic

$U1905B$, $U1805A$, $R1904$, $R1903$, and $C1804$ make up the Chop Clock Oscillator circuit. When the O_4 output of $U1605$ is HI and the CHOP MODE is selected, a HI is present on pin 5 of U1905B. U1805A will alternately put a HI and then a LO on pin 4 of U1905B. Assume an initial LO on pin 4 of U1905B. U1905B is a NAND gate, so its output will be HI. C1804 charges toward a HI and as soon as its charge reaches the threshold level of U1805A, U1805A will switch to a HI output. The HI output of U1805A pin 3 will assert a HI on U1905B pin 4. This HI is NANDed with the HI already present on pin 5 of U1905B to produce a LO at pin 6 of U1905B. Now C1804 has to discharge toward a LO. As soon as the charge on C1804 reaches the LO threshold of U1805A, U1805A will switch to a LO output and the cycle will repeat. The Chop Clock oscillator frequency is approximately 1 MHz and depends on the RC-time constant of R1903 and C1804, as well as the threshold level of U1805A.

When CHOP is selected, pin 2 of U1905A will be LO. Pin 3 of U1905A will be HI, which enables U1805C to pass the CHOP clock oscillator frequency to the pin 9 of U1705. U1705 changes state for every positive transition of the clock oscillator (once each cycle) effectively performing a divide-by-two. Therefore the CHOP frequency is approximately 500 kHz. The CHOP clock oscillator will not be enabled unless more than one input is selected. As stated previously, a single-input VERT MODE selection will cause U1705 to remain in a state will allow only the selected input to be passed on to the Delay Line Driver.

If multiple inputs are selected for display, U1705 will be clocked to select the appropriate inputs programmed for display. Transistors Q1619, Q1719, Q1617, and Q1636 (if ADD is selected) are being switched at the Chop clock frequency (CHOP MODE). As the displays are being incremented, the Q_0, Q_1, and Q_2 outputs of U1705 are being used as state indicators to the A_0, A_1, and A_2 inputs of ROM U1605 to indicate the next input to be selected for display. The order of priority of the switching is: CH 1, CH 2, ADD, then TRIG VIEW.

Alternate Trace Sync. With ALT Vertical Mode selected, a HI will be present at U1905D pin 12. If more than a single-input display is selected, a HI will also be present at pin 13. The CHOP clock will then be disabled. U1905A will be enabled to pass the Alternate Trace Sync pulse to U1805C pin 9, and U1805C will be enabled to pass the pulse to U1705 pin 9 (clock input). U1705 will now be switching between selected inputs at a rate determined by the Alternate Trace Sync pulse from the Horizontal Display Logic circuit.

If a single input is selected for display, U1805B pin 5 will be LO and the Chop clock will remain disabled. The Alternate Trace Sync pulse will be present at U1705 pin 9, but with only input selected by the VERT MODE switches, U1705 will not change state and the switching transistor for the selected input to be displayed will remain on.

When Alternate Horizontal Display is chosen, the Alternate Trace Sync input becomes a square wave with a period equal to twice the time between

Alternate Trace Sync pulses. This will allow the display of the *A* sweep and the *B* sweep before switching to the next vertical input signal to be displayed.

Chopped Blanking Amplifier

When CHOP Mode is selected, a LO on U1905D pin 12 holds U1905D pin 11 HI. This HI enables U1905C to pass the Chop pulse to C1813. C1813 and R1916 differentiate the Chop pulse to produce positive and negative spikes having sufficient fast rise times necessary for the Chopped Blanking. CR1918 limits the positive spike and R1815 limits the base current of Q1812. The positive portion of the waveform reverse biases Q1812, but when the waveform switches from the positive portion to the negative portion, Q1812 is driven rapidly into conduction. The blanking time is determined by the charging time of C1813 through R1916 and R1815. The positive-going output pulse, which is coincident with trace switching, is connected to the Z-Axis Amplifier through R1914.

Delay-Line Driver

The output from the Diode Gates is applied to the Delay Line Driver composed of Q1426 and Q1428. Transistors Q1426 and Q1428 are connected as feedback amplifiers, with R1328 and R1522 providing feedback from the collector to the base of their respective transistors. A sample of the signal in the collector circuit of Q1426 is used for triggering in the Normal mode of trigger operation. Bandwidth limit switch S1414 connects a π-type filter (composed of C1306, C1405, L1412, and L1512) between the output signal lines of the Delay Line Driver to reduce the upper -3 dB bandwidth limit of the Vertical Amplifer system to approximately 20 MHz. Resistors R1314 and R1514 provide reverse termination for any reflections in the delay line.

Normal Trigger Pickoff Amplifier

The trigger signal for Normal Trigger operation is obtained from the collector of Q1426. Normal Trigger DC Balance Adjustment R1202 sets the dc level of the normal trigger output signal such that the sweep is triggered at the zero level of the displayed signal whenever the trigger LEVEL control is set to zero. Q1215 and Q1118 are connected as a feedback amplifier, with the signal applied to the base of Q1215 and the feedback connected between the output and emitter of Q1215 through R1108.

13.6   VERTICAL MODE SWITCH

Introduction

The Vertical Mode Switch circuit shown in Fig. 13.6-1 produces the logic necessary for placing the Vertical Switching Logic circuit into the correct state for the Vertical Mode selected. The Scale-Factor Switching circuit

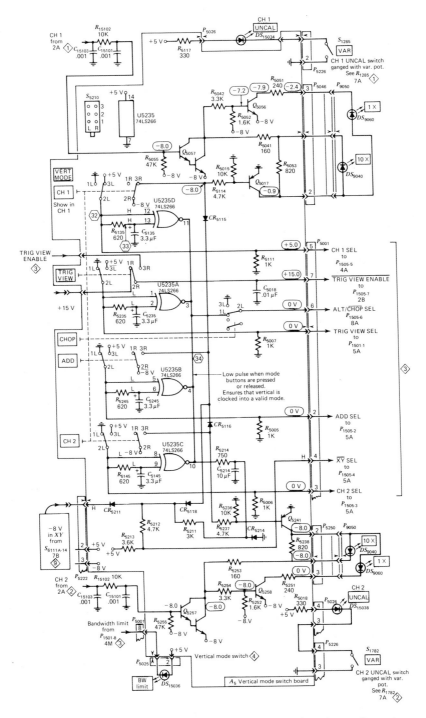

Figure 13.6-1 Vertical Mode Switch circuit. (Courtesy Tektronix, Inc. © 1980.)

selects the correct scale-factor LED to be illuminated with each Vertical Mode selection.

Exclusive-NOR Switch Circuit

The Exclusive-NOR Switching circuits produce a LO pulse on the ALT/CHOP select line whenever a VERT MODE pushbutton is either pressed or released. While the ALT Vertical Mode is selected, the LO pulse will enable the Chop oscillator, momentarily producing clock pulses which enable the Vertical Switching Logic circuit to select the correct Vertical Mode for the next display.

Assume that the CH 1 VERT MODE pushbutton has been pressed. The condition of Exclusive-NOR circuit $U5235D$ is as follows: A HI is on pins 12 and 13 of $U5235D$, and a HI is present at output pin 11; $C5135$ has charged to a HI level through $R5135$. When the pushbutton is released, a LO is immediately applied to pin 12, but $C5135$ is still charged to a HI level, so the HI remains on pin 13. These input conditions to the Exclusive-NOR circuit will produce a LO at output pin 11 to enable the CHOP oscillator. Then $C5135$ discharges toward a LO level through $R5135$, and when the LO threshold voltage is reached, both inputs to the Exclusive-NOR circuit will be LO. The output of $U5235D$ will switch back to HI, turning off the Chop oscillator.

When the X-Y Horizontal Display is selected, -8 V is applied through $R5212$ to the $\overline{XY}$ select input of the Vertical Switching Logic circuit (placing it in the X-Y Mode of operation) and through the series combination of $R5211$ and $R5214$ to the ALT/CHOP select to enable the Chop oscillator if the Alternate Vertical Mode is selected.

Scale-Factor Switch Circuit

The vertical deflection factor for each channel is indicated by back lighting the appropriate figures imprinted on the flange of each VOLT/DIV knob. Because the operations of the Channel 1 and Channel 2 Scale-Factor Switching circuits are similar, only the circuit action of the Channel 1 Scale-Factor Switching circuits is described.

When CH 1 or ADD Vertical Mode is selected, or when X-Y Horizontal Display Mode is selected, -8 V is applied to the $Q5017$ base biasing voltage divider network composed of $R5114$ and $R5015$. The base of $Q5017$ will be biased negative, saturating $Q5017$. When $Q5017$ is saturated, the $\times 10$ and $\times 1$ scale-factor LED $DS9040$ and $DS9060$ will have a return path to ground through the transistor and are enabled. ($10\times$ LED or $1\times$ LED is lighted depending on the type of probe or cable attached to the CH 1 OR X input connector.)

The X-Y position of A TIME/DIV switch $S611$ puts -8 V on both Channel 1 and Channel 2 scale-factor LED-enabling transistors $Q5017$ and $Q5241$ through blocking diodes $CR5115$ and $CR5116$. This allows the appropriate scale-factor LED to be illuminated on both CH 1 and CH 2 ($10\times$ or $1\times$).

With either a coaxal cable or $1 \times$ probe attached to the CH 1 OR X input connector, the probe-coding ring portion of the input connector is not contacted. $Q5057$ will be biased off by the -8 V through $R5055$. A voltage divider network composed of $R5053$, $R5041$, $R5042$, and $R5052$ between $Q5017$ and the -8-V supply will bias $Q5056$ into conduction; the $1 \times$ LED in series with $Q5056$ will be on. The $10 \times$ LED is in parallel with $R5053$, and the voltage drop across $R5053$ will not be sufficient to cause the $10 \times$ LED to light.

When a $10 \times$ probe equipped with a scale-factor-switching connector is attached to the CH 1 OR X input connector, the probe coding ring will be connected. The base of $Q5057$ will now be connected to ground through $R15102$ and an internal resistor located within the probe connector body. A bias voltage divider is formed, biasing $Q5057$ into saturation. The collector of $Q5057$ will drop to about -7.2 V. This voltage level is enough to light the $10 \times$ LED, and when it is applied to the $Q5056$ base bias voltage divider composed of $R5042$ and $R5052$, it is sufficient to bias $Q5056$ off and turn off the $1 \times$ LED.

13.7 VERTICAL OUTPUT AMPLIFIER

Introduction

The Vertical Output Amplifier circuit shown in Fig. 13.7-1 provides the final amplication for the vertical deflection signal. The circuit contains the Delay Line, input amplifier, part of the Beam Finder circuitry, part of the Trace Separation circuitry, and the output amplifier. Pushing the BEAM FIND button compresses an overscanned display to within the viewing area. The Trace Separation circuit provides vertical positioning of the B trace when the ALT Horizontal Display mode is selected.

Delay Line

Delay line $DL11405$ (Fig. 13.5-1) provides approximately 120-ns delay of the vertical signal to allow the Sweep Generator circuits sufficient time to initiate a sweep before the vertical signal reaches the deflection plates of the CRT. When using internal triggering, the instrument is allowed to display the leading edge of the signal originating the trigger pulse. Resistors $R2214$ and $R2218$ provide forward termination for the Delay Line.

Input Amplifier

The Input Amplifier consists of integrated circuit amplifier $U2225$ and the frequency compensation network connected across pins 2 and 3. Gain Adjust $R2025$ sets the gain of the amplifier. BEAM FIND switch $S4075$, when pressed, reduces the dynamic swing capabilities of the stage, thereby limiting the display to within the display area on the CRT.

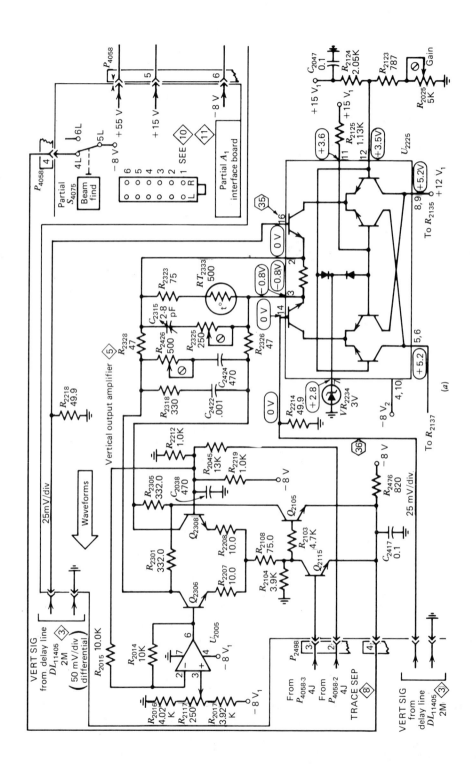

448

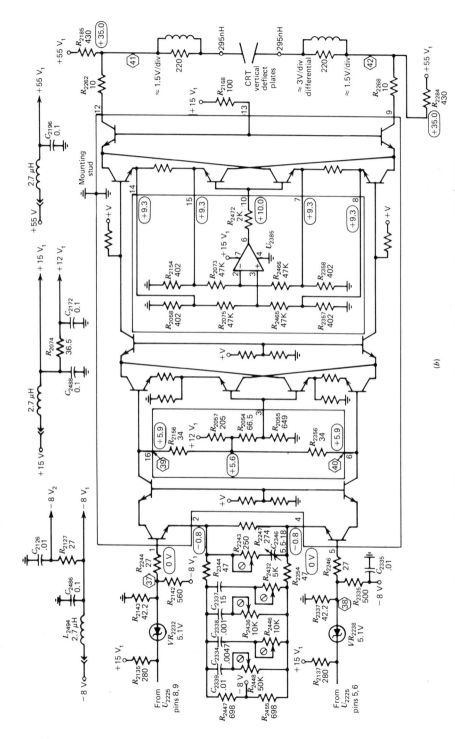

Figure 13.7-1 (a)–(b) Vertical Output Amplifier circuit. (Courtesy Tektronix, Inc. © 1980.)

449

Trace Separation

During *B* sweep the Trace Separation circuit (*Q*2115, *Q*2306, and *Q*2308) is enabled by the alternate pulse on the base of *Q*2115. This switches *Q*2115 on and *Q*2105 off. This switching action allows the *B* trace to be positioned vertically independently of the *A* trace. Normally, *Q*2105 provides a constant amount of current to *U*2225; however, when in ALT mode, turning *Q*2115 on supplies current through *Q*2306 and *Q*2308. The TRACE SEP control (*R*14498) supplies the variable voltage at the base of *Q*2308 determines the position of the *B* sweep display relative to the *A* sweep display.

The current supplied to *U*2225 must remain constant so that the *A* sweep display positioning is not affected. *U*2005 senses the voltage at the base of *Q*2308 and compensates by raising or lowering the voltage at the base of *Q*2306. Potentiometer *R*2117 centers the *B* trace around the *A* trace so that the *B* trace moves equally above and below the *A* trace.

Output Amplifier

Integrated circuit *U*2255 is a multistage cascode amplifier cell that provides the final amplication for the vertical signal. The input signal is applied push–pull between pins 1 and 5, and the output signal is taken from pin 9 and 12. Integrated-circuit amplifier *U*2385 monitors the emitter currents of the output transistors and automatically sets the dc levels of the output stage to obtain the maximum undistorted gain from *U*2255.

13.8   *A* AND *B* TRIGGER GENERATORS

Introduction

The Trigger Generator circuits shown in Fig. 13.8-1 produce trigger pulses to start the Sweep Generator circuits. These trigger pulses are derived either from the interal trigger signal (sampled from the vertical deflection system), an external signal connected to the external trigger connectors, or a sample of the line voltage applied to the instrument. Controls are provided in each circuit to select trigger level, slope, coupling, and source.

An *A* trigger View Amplifier is provided which amplifies the external *A* Trigger signal for application to the Trigger View Diode Gate where it may be selected for viewing. The trigger view display provides a method of making a quick and convenient check of the external trigger signal being used to trigger the *A* Sweep Generator. The external trigger input signal may be continually monitored by selecting the A TRIG VIEW Vertical Mode.

Since the *A* and *B* Trigger Generator circuits are virtually the same, only the *A* Trigger Generator circuit action and the differences between the *A* and *B* Trigger Generator circuits are described.

Trigger Source

Trigger SOURCE switch *S*7312 selects the source of the trigger signal. The sources available to the *A* Trigger Generator circuit are the signal(s) being displayed. (NORM), Channel 1 (CH 1), Channel 2 (CH 2), LINE, and EXT. The EXT/10 (*A* trigger circuit only) position attenuates the external trigger signal by a factor of 10. The *B* Trigger SOURCE switch does not have a LINE or an EXT/10 position, but has a STARTS AFTER DELAY position.

The STARTS AFTER DELAY position of the *B* Trigger SOURCE switch is used in conjunction with the DELAY TIME POSITION control. When

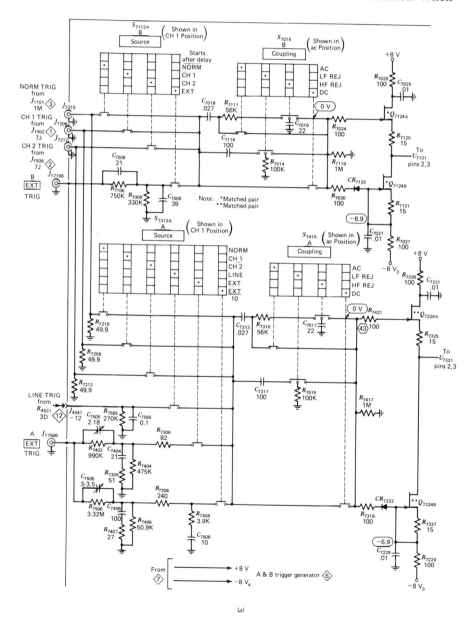

(*a*)

Figure 13.8-1   (*a*)–(*c*) *A* and *B* Trigger Generator circuits. (Courtesy Tektronix, Inc. © 1980.)

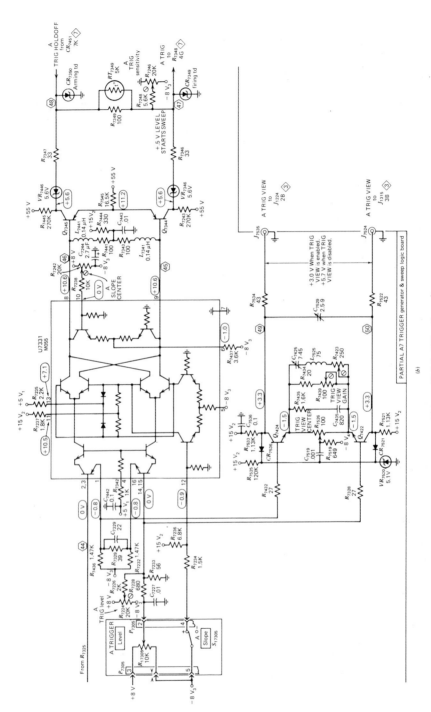

Figure 13.8-1 (Continued)

452

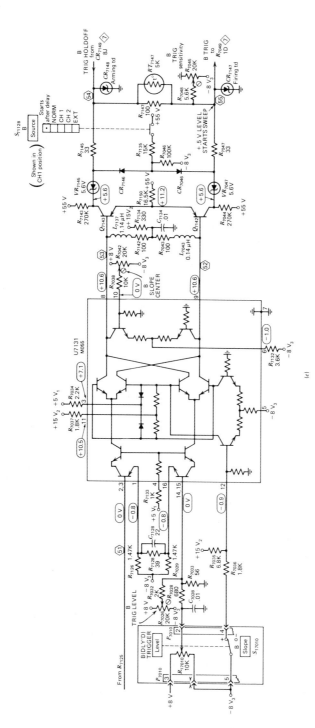

Figure 13.8-1 (Continued)

453

STARTS AFTER DELAY is selected as a trigger source for the *B* Sweep, the *B* Sweep will begin to run immediately after the delay time selected by the DELAY TIME POSITION control has elapsed.

In the LINE mode of triggering, a sample of the power-line frequency is obtained from the secondary of the power transformer in the LV power supply. To prevent unwanted attenuation of the trigger signal by the LF REJ circuit, the *A* Trigger COUPLING switch should not be in LF REJ when using line voltage as a trigger source.

Trigger Coupling

The Trigger COUPLING switches offer a means of accepting or rejecting certain components of the trigger signal. In the ac, LF REJ, and HF REJ mode of trigger signal, the dc component of the trigger signal is blocked by coupling capacitors *C*7317. Frequency components below 60 Hz are attenuated when using ac coupling, and components below about 50 kHz are attenuated when using LF REJ coupling. The higher-frequency components of the trigger signal are passed without attenuation. In the HF REJ mode of trigger coupling, the high-frequency components of the trigger signal (above about 50 kHz) and low-frequency components (below 60 Hz) are attenuated, while the remaining frequency components are passed with minimal attenuation. The DC mode of trigger coupling passes all signals from dc to at least 100 MHz without attenuation.

Input Source Follower

Transistor *Q*7324A is FET source follower. It provides both a high input impedance for the trigger signal and isolation between the Trigger Generator circuit and the trigger signal source. Diode *CR*7322 provides input protection for *Q*7324A if an excessively high-amplitude negative-going input signal is present. If a high-amplitude positive signal is applied, the source-gate junction of *Q*7324A becomes forward biased and clamps the voltage at that level. The second FET of the matched pair [*Q*7324B) is a high-pmpedance, constant-current source for *Q*7324A. Since both FETs are matched and mounted in the same heat sink, both will display equal temperature effects, and *Q*7324B will provide temperature compensation for *Q*7324A.

Paraphase Amplifier

*U*7331 is a paraphase amplifier stage that converts the single-ended input from Source Follower *Q*7324A into push–pull output applied to the Tunnel Diode Driver stage. Trigger Level Centering adjustment *R*7224 sets the level at pins 14 and 15 of *U*7331 that the display is correctly triggered when the LEVEL control is centered. LEVEL control *R*17305 varies the level at pins 14 and 15 of *U*7331 to select the point on a trigger signal where triggering occurs.

The slope of the input signal that triggers the Sweep Generator circuit is determined by the setting of SLOPE switch *S*17305. When the SLOPE switch is set to the "+" position, the output signal present at pin 8 of $U7331$ is in place with the input signal, and the output signal at pin 9 is inverted with respect to the input signal. When the SLOPE switch is set to the "-" position, the output signal at pin 8 is inverted with respect to the input signal, and the output signal at pin 9 is in phase with the input signal.

Tunnel Diode Driver

Transistors $Q7344$ and $Q7345$ are common-emitter amplifier stages that provide signal currents necessary to switch the Trigger Firing tunnel diode. $CR7350$ and $CR7349$ are approximately 4.7 mA tunnel diodes. Quiescently, $CR7349$ and $CR7350$ are biased into their low-voltage states. $Q7344$ cannot provide sufficient current to switch $CR7349$ to its high voltage state. However, $Q7345$ can provide sufficient current to bias $CR7350$ into its high voltage state. When $Q7345$ conducts triggering signal current, the anode of $CR7350$ steps positive to approximately $+0.5$ V. Since only approximately 1 mA of current is required to maintain $CR7350$ in its high-voltage state, this makes approximately 3 mA of additional current available with which to switch $CR7349$ to its high-voltage state. Thus, the next time $Q7344$ conducts signal current, $CR7349$ steps to its high-voltage state, sending a positive pulse to the logic circuit to initiate sweep action. The A Trigger Sensitivity adjustment, $R7246$, adjusts the tunnel diode bias to a level that will not allow $CR7349$ to be switched to its high-voltage state until $CR7350$ has been switched to its high-voltage state. At the end of the sweep time and during holdoff, a negative level is applied to the anode of $CR7350$, thereby resetting both $CR7349$ and $CR7350$ to their low-voltage states. The reset level remains during holdoff time to ensure that a sweep gating signal will not be generated until the sweep circuit has returned to its quiescent state.

Trigger View Amplifier

Transistors $Q7424$ and $Q7422$ make up half of a cascode, push–pull amplifier. In the Vertical Switching Logic circuit, $Q1235$ and $Q1225$ form the rest of the Trigger View Amplifier. The Trigger View Amplifier requires that the A Trigger SOURCE switch be set to EXT or EXT/10 and the A TRIG VIEW Vertical Mode be selected before the amplifier is enabled to pass the external trigger signal to the Diode Gate and on to the Vertical Output Amplifier. If the trigger view display is selected, the Vertical Switching Logic circuit will turn on the Trigger View Diode Gate during the proper time to pass the signal on to the Delay Line Driver.

A sample of the push–pull external trigger signal is taken from pins 1 and 16 of $U7331$ and amplified by $Q7424$ and $Q7422$. The Trigger View Centering control ($R7526$) is used to vertically position the Trigger View display. $R7439$ is adjusted to set the gain of the Trigger View Amplifier, and $C7425$, $L7525$, $C7529$, and $R7423$ provide HF compensation. Diodes $CR7520$, $CR7521$, and $CR7536$ are used to clamp the collectors of $Q7424$ and $Q7422$ to approximately $+5.7$ V whenever Trigger View is disabled. $Q1225$ and $Q1235$ will be reverse biased during this time.

13.9   SWEEP AND Z-AXIS LOGIC CIRCUIT

Introduction

The Sweep and Z-Axis Logic circuit, shown in Fig. 13.9-1, develops the logic levels necessary to control the sequence of events associated with sweep generation and CRT unblanking. The A and B +Gate signals are also generated in this circuit.

A Sweep Gate

The A Sweep Gate circuit consists of $Q7254$ and $Q7256$ (Fig. 13.9-1a). They form an emitter-coupled stage where only one transistor can be conducting at any time. The input signal to the stage is the positive-going trigger signal from the A Trigger Firing Tunnel diode in the A Trigger Generator circuit. The signal at the collector of $Q7254$ is connected to the A Sweep Z-Axis Gate circuit ($Q7181$ and $Q7182$) to control the CRT unblanking and to generate the A +Gate signal. The signal at the collector of $Q7256$ is connected to the emitter of Sweep Disconnect Amplifier $Q4497$ in the A Sweep Generator circuit to initiate A Sweep generation.

B Sweep Gate

The B Sweep Gate circuit consists of $Q7053$ and $Q7055$ (Fig. 13.9-1b). These transistors also form an emitter-coupled stage where only one transistor can be conducting at any time. The input signal to the stage is the positive-going trigger signal from the B Trigger Firing Tunnel diode in the B Trigger Generator circuit. The signal at the collector of $Q7053$ is connected to the B Sweep Z-Axis Gate circuit ($Q7093$ and $Q7095$) to control CRT unblanking and to generate the B +Gate signal. The signal at the collector of $Q7055$ is connected to the emitter of Sweep Disconnect Amplifier $Q4565$ in the B Sweep Generator circuit to initiate B Sweep generation.

Sweep Control Integrated Circuit

The Sweep Control integrated circuit is $U7535$ (Fig. 13.9-1a). Several functions are performed in this stage, depending on the mode of operation of the instrument sweep generators. The following is a brief explanation of the function associated with each pin of the IC.

Pin 1 is the Positive Auto Sense input. The signal connected here comes from the A Trigger Firing tunnel diode.

Pin 2 is a reference input to the Auto Sense circuit. A fixed dc level established by $R7277$ and $R7276$ is connected here.

Pin 3 is the +auto gate terminal. In the AUTO mode of operation, if no trigger signals are applied to pin 1 of $U7375$ during the approximately 100 ms following the end of holdoff, the gate level at pin 3 steps LO to turn $Q7256$ on which initiates a sweep.

Pin 4 is not used in this application.

Pin 5 is the input terminal for negative voltage supply.

Pin 6 is the auto gate timing terminal. $R7272$ and $C7273$ determine the amount of time between the end of holdoff and the generation of the auto gate.

Pin 7 output lights the TRIG LED when a triggered gate has occurred.

Pin 8 is the holdoff timing terminal. The RC network connected to this terminal (selected by the A TIME/DIV switch) determines the length of holdoff time.

Pin 9 is the ground terminal.

Pin 10 is the Holdoff output terminal. The gate level present here is LO during sweep holdoff time and HI otherwise.

Pin 11 output lights the READY LED when operating in the single sweep mode.

Pin 12 is the single sweep mode terminal. When +5 V is applied to this terminal, the sweep operates in the single sweep mode; when the terminal is left open or grounded, the sweep operates in the repetitive mode.

Pin 13 is not used in this application.

Pins 14 and 15 are the single sweep reset terminals. Pushing the SINGLE SWP button prepares the single sweep circuitry to respond to the next triggering event, and also causes the READY LED to come on.

Pin 16 is the holdoff start input terminal. The HI sweep reset gate pulse from the sweep generators is applied here to initiate sweep holdoff.

Pin 17 is the sweep disable output terminal. The gate level at this terminal is HI during holdoff and LO otherwise.

Pin 18 is the sweep lockout input; +5 V applied to this terminal disables all sweep action.

Pin 19 is the auto mode terminal. Grounding this terminal enables auto sweep operation.

Pin 20 is the input terminal for positive voltage supply.

Holdoff Timing

A resistor and capacitor network located in the A and B Timing Switch circuit to pin 8 of $U7375$ via pin 8 of $J4571$. Various resistor and capacitor combinations switch into the circuit depending on the setting of the A TIME/DIV switch. At sweep end pin 8 of $U7375$ is released, and the timing capacitors in the holdoff timing network start to charge. $Q7465$ is biased off

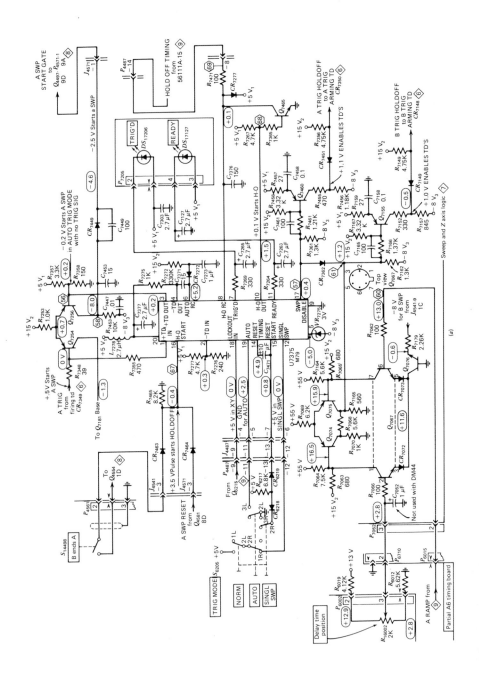

458

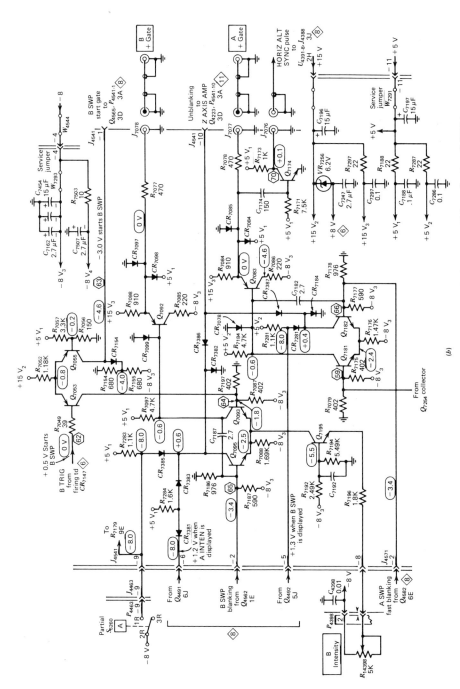

Figure 13.9-1 (a)–(b) Sweep and Z-Axis Logic circuit. (Courtesy Tektronix, Inc. © 1980.)

459

during the sweep holdoff time by a LO pin 10 of $U7375$, which allows the voltage at pin 8 to rise due to charging of the timing capacitors. When the charge on the capacitors rises to approximately $+4$ V at pin 8, pin 17 goes HI and pin 10 goes HI to turn on $Q7465$. The holdoff timing capacitors discharge rapidly through $Q7465$ as pin 8 goes LO. $C6121$, $C6122$, $C6112$, $R6115$, and $R6114$ in the A and B Timing Switch circuit compose the holdoff timing circuits which are switch selectable, and A Trigger HOLDOFF $R14498$ is a variable resistor that allows lengthening of the time constant to increase holdoff time.

A Sweep Holdoff Amplifier

The A Sweep Holdoff Amplifier is $Q7460$. The holdoff gate waveform from pin of $U7375$ is applied to the base of $Q7460$ through $R7461$ and $C7461$. When $Q7460$ is turned off (during holdoff time), its collector is LO and $CR7451$ is forward biased, thus resulting both the Arm and Firing Trigger tunnel diodes in the A Trigger Generator. When $Q7460$ is turned on (any time other than holdoff time), its collector level is HI and $CR7451$ is reverse biased. This allows the trigger tunnel diodes in the A Trigger Generator to respond to the next adequate triggering signal.

B Sweep Holdoff Amplifier

The B Sweep Holdoff Amplifier is $Q7155$. Its circuit action is identical to that described for the A Sweep Holdoff Amplifier with the exception that two gate signal sources control the state of the stage. The two sources are the holdoff gate from pin 17 of $U7375$ (through $CR7362$) and the collector of $Q7075$ in the Delay Pickoff Comparator. Both gate sources must be in their LO state for B Sweep to be triggerable. Either source in its HI state will disable the B Trigger Generator tunnel diodes.

A Sweep Z-Axis Gate

$Q7181$ and $Q7182$ (Fig. 13.9-1b) comprise the A Sweep Z-Axis Gate. They form an emitter-coupled stage where only one transistor can be conducting at any time. The controlling signals consist of inputs from the collector of $Q7254$ in the A Sweep Gate, the unblanking signal from $Q4582$ in the A Sweep Generator, and $Q4492$ in the Horizontal Display Logic circuit. The unblanking signal for use in the Z-Axis Amplifier is taken from the collector of $Q7182$ (through $CR7387$). The collector signal of $Q7181$ is applied to the A +Gate Emitter Follower ($Q7083$).

The Horizontal Display Logic circuit controls the bias voltage on $CR7281$. When the diode is reverse biased, as it is for all horizontal modes except for B DLY'D, -8 V is connected to the anode of $CR7281$ through $Q4492$. This allows the gate signal at the collector of $Q7182$ to pass through $CR7378$ creating the unblanking signal to the Z-Axis Amplifier. In the B DLY'D mode, $Q4492$ is turned off and $CR7281$ is forward biased through $R7281$ to the $+5$-V supply. $CR7378$ will now be reverse biased, and the A unblanking signal is blocked from reaching the Z-Axis Amplifier. In the ALT Horizontal Display mode, $CR7281$ will be reverse biased during the A sweep and forward biased during the B sweep.

B Sweep Z-Axis Gate

The B Sweep Z-Axis Gate is comprised of $Q7093$ and $Q7095$. These transistors form an emitter-coupled stage where only one transistor can be conducting at any time. The controlling signals come from the collector of $Q7053$ (B Sweep Gate), and the blanking signal from the collector of $Q4562$ in the B Sweep Generator. The emitter current in the gate transistors is supplied partly by $Q7195$ which is controlled by B INTENSITY control $R14398$. The B INTENSITY control sets the level of the B Sweep unblanking signal to control the B Sweep intensity separately from the overall display intensity. The collector of $Q7095$ supplies the unblanking signal to the Z-axis amplifier and the collector of $Q7093$ supplies the signal to the B + Gate Emitter Follower ($Q7092$).

When the A Horizontal display is selected, -8 V from $Q4491$ in the Horizontal Display Logic circuit is applied to the cathode of $CR7381$. This reverse biases $CR7383$ and allows the collector of $Q7095$ to be pulled positive through $CR7385$ and $R7283$ to the $+5$-V supply. $CR7386$ will be reverse biased and the B Sweep Z-Axis Gate ($Q7093$ and $Q7095$) will not affect CRT unblanking. When either A INTEN, ALT, or B DLY'D Horizontal Display is selected, -8 V is applied to the anode of $CR7385$ to reverse bias it and allow the Horizontal Display Logic circuit to control the B Sweep Z-Axis Gate.

In A Intensified Horizontal Display, $CR7381$ becomes reverse biased and $CR7383$ becomes forward biased. Diode $CR7386$ is still reverse biased, but when B sweep starts, the collector of $Q7095$ steps negative enough to forward bias $CR7386$ and add a slight amount of unblanking to the A Sweep unblanking already present. This provides further intensification for the B Sweep portion of an A Intensified display. In ALT Horizontal Display, the Horizontal Display Logic circuit controls the A Sweep Z-Axis Gate ($Q7181$ and $Q7182$) and the B Sweep Z-Axis Gate ($Q7093$ and $Q7095$). The B Sweep unblanking signal is added to the A Sweep unblanking signal during the A Intensified display; the A Sweep unblanking signal is blocked during the B DLY'D display. In B DLY'D Horizontal Display, the A Sweep Z-Axis Gate output diode $CR7387$ is held reverse biased, and the only unblanking signal presented to the Z-Axis amplifier input is the B Sweep unblanking signal.

A + Gate and B + Gate Emitter Followers

Emitter followers $Q7083$ and $Q7092$ provide the A + Gate and the B + Gate output signals available at the instrument rear panel. The output signals are positive-going rectangular waveforms, approximately $+5.5$ V in amplitude. The amplitude is set in the collectors of $Q7181$ and $Q7093$. For example, when $Q7181$ is conducting, the base of $Q7083$ can go more than approximately -0.7 V (limited by $CR7078$). When $Q7181$ is not conducting, the base of $Q7083$ rises to the $+5$-V power supply level through $R7184$. Diodes $CR7084$, $CR7085$, $CR7097$, and $CR7098$ provide protection against accidental application of damaging voltage levels to the A + Gate and B + Gate output connectors.

Horiz Alt Sync Pulse Amplifier

The pickoff amplifier for the Horiz Alt Sync pulse is $Q7174$. It is biased into saturation, so its quiescent output voltage is approximately zero. A sample of the A + Gate is coupled to the base of $Q7174$ by $C7174$ where the positive-going gate is integrated by the action of $C7174$ and $R7171$. The positive-going portion of the integrated signal cannot increase the collector current of $Q7174$ beyond its saturation level, so no signal output is obtained. When the A + Gate negative-going edge occurs, $C7174$ cannot change its charge instantaneously so the entire negative transition is left on the base of $Q7174$ across $R7171$. The negative peak of the signal is enough to cutoff $Q7174$, and the collector voltage rises in response to the base voltage decrease. The base voltage rapidly returns to a positive level, and the transistor again saturates, ending the Horiz Alt Sync pulse.

13.10 *A* AND *B* SWEEP GENERATORS

Introduction

The A and B Sweep Generator circuits shown in Fig. 13.10-1 produce sawtooth voltages that are amplified by the Horizontal Amplifier circuit to provide horizontal deflection on the CRT. These sawtooth voltages are produced on command (Sweep start gate) from the Sweep Logic circuits. The Sweep Generator circuits also produce gate waveforms that are used by the Z-Axis Logic circuit to unblank the CRT during sweep time and by the Sweep Logic circuit to terminate sweep generation. Since the B Sweep Generator circuit is very similar to the A Sweep Generator, the only discription of the B Sweep Generator will be on the differences in the operation between the two.

Disconnect Amplifier

After holdoff, but before the next sweep, Disconnect Amplifier $Q4497$ is biased on and conducts through $R4587$ and R_T back to $+V_T$. This sets the charge on C_T in presentation for the beginning of the next A Sweep and prevents current from the Miller integrator circuit from changing the charge on C_T. When the positive-going A Sweep Start Gate is applied to the emitter of $Q4497$, $CR4592$ becomes forward biased and turns off $Q4497$. Now the A Sweep starts, and the Miller Integrator circuit begins to change the charge on C_T. $Q4497$ will remain off until retrace is initiated and the A Start Sweep

Gate is removed. Then $Q4497$ will become forward biased again, and C_T will rapidly charge to its quiescent value for the start of the next A Sweep.

Sawtooth Generator

The Miller Integrator Circuit is composed of $Q4498$ and $Q4598$. It works on the principle that if the charging current to a capacitor can be held constant, then the charging curve will be linear rather than exponential. The action starts when Disconnect Amplifier $Q4497$ is turned off by the A Sweep Start Gate. The selected capacitor fot the chosen setting of the TIME/DIV switch (C_T) begins to charge through the R_T. This causes the junction of C_T and R_T to start positive in the direction of $+V_T$, thereby causing the gate of $Q4498$ to start positive. The $Q4498$ source then starts in a positive direction and increases the forward bias on $Q4598$ causing the collector voltage to move in a negative direction (less positive). This couples back through C_T and opposes the positive change at the gate of $Q4498$. Capacitor C_T is attempting to charge toward $+V_T$, but the action of $Q4498$ gate being held virtually constant, and the collector of $Q4598$ going more negative, results in the reduction of the charge on C_T (it discharges). The gate of $Q4498$ rises positive about 10 mV over the entire sweep generation time. Since the voltage at the gate of $Q4498$ remains relatively constant, both the voltage across R_T and the current through R_T (the current discharge from C_T) remain constant. The linear rate of discharge of C_T results in a linear ramp across it. The resultant output at the collector of $Q4598$ appears as a negative-going ramp, dropping from approximately $+13$ V to approximately $+2$ V.

When the ramp reaches $+2$ V, $Q4581$ sends a pulse to $U7375$ (Fig. 13.9-1a) initiating retrace. Transistor $Q4497$ turns on, and its collector goes more negative. This moves the gate of $Q4498$ in the negative direction, causing the voltage on the base of $Q4598$ to go more negative, thereby causing its collector voltage to go more positive. Now C_T charges rapidly through $Q4497$ to its quiescent state in preparation for the next A Sweep start gate.

Output Amplifier

The Output Amplifier ($Q4475$) is a common-base amplifier with the signal current driven into the emitter. $Q4475$ provides the output sawtooth current signal to the Horizontal Amplifier and furnishes a measure of isolation between the Sawtooth Generator and the Horizontal Amplifier. The Horizontal Display Switching circuit connects to this stage and controls the A sawtooth output in the various horizontal modes of operation. In the A and A INTEN modes of operation, the A sweep signal passes through $Q4475$ to the Horizontal Amplifier. In the ALT mode, $Q4475$ is enabled for the A sweep and turned off for the B sweep by the Horizontal Display Logic circuit. In the B DLY'D mode, $Q4475$ held off, and $Q4465$ in the B Sweep Generator is held on to pass the B sweep sawtooth to the Horizontal Amplifier.

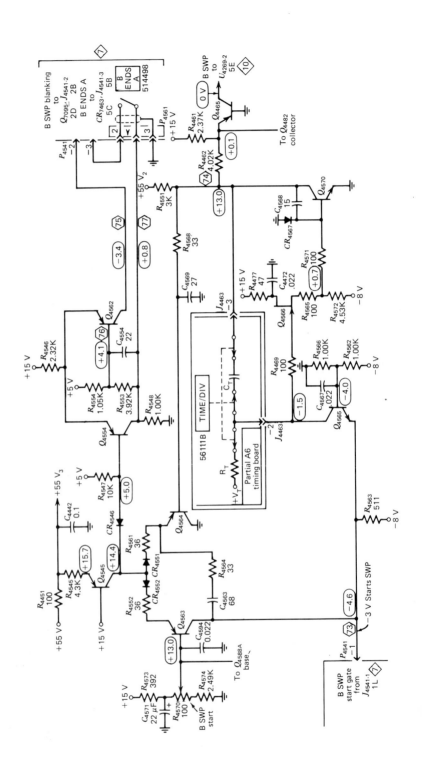

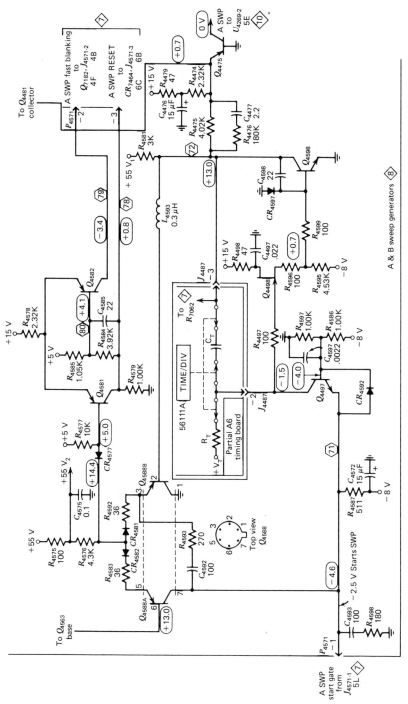

Figure 13.10-1 *A* and *B* Sweep Generator circuits. (Courtesy Tektronix, Inc. © 1980.)

465

A Sweep Start Comparator

Just before the sweep starts to run down, the levels at the bases of $Q4588A$ and $Q4588B$ are approximately equal. When the sweep starts to turn down, the base of $Q4588B$ goes negative, following the collector of $Q4598$. This increases the forward bias on $CR4581$, which in turn decreases the forward bias on $CR4582$. Very shortly after start of the sweep, $CR4582$ becomes reverse biased and interrupts the current through $Q4588A$. The circuit remains in this condition until after the sweep retrace is complete. When the circuit returns to quiescence, $Q4588A$ again begins to conduct through $R4587$. This sets the currents through $Q4497$, establishing the starting point for the sweep. Sweep start adjustment $R4570$ sets the base level of $Q4588A$. This level is also connected to the base of $Q4563$ in the B Sweep Generator to ensure that the B sweep starts at the same level as the A sweep.

A Sweep End of Sweep Comparator

The A Sweep End of Sweep Comparator is a switching circuit composed of $Q4581$ and $Q4582$. At quiescence, $Q4582$ is conducting and $Q4581$ is turned off. When the sweep starts to run, the negative-going ramp at the collector of $Q4598$ coupled through the base of $Q4588B$ and through $CR4581$ to the cathode of $CR4577$. When the collector of $Q4598$ reaches about $+2$ V, the cathode of $CR4577$ reaches about $+4$ V. $CR4577$ begins conducting and turns on $Q4581$, which then turns off $Q4582$. The resulting positive step at the collector of $Q4581$ is fed to pin 16 of $U7375$ where it is used to initiate retrace and holdoff. The negative-going pulse at the collector of $Q4582$ is applied to the A Sweep Z-Axis Logic Gate to blank the CRT as soon as a sweep-end command is generated.

B Sweep Generator Differences

There are two differences between the A Sweep Generator and B Sweep Generator. The first is that $Q4545$ is used as a constant current source in the B Sweep Start Comparator circuit. The second is that one of the outputs of the B Sweep Generator is controlled by the B ENDS A switch associated with the A Trigger HOLDOFF control. In the B ENDS A position, the end of the B sweep also ends the A display on the CRT.

Horizontal Display Switching

Switching transistors $Q4482$ and $Q4481$ are controlled by the Horizontal Display Logic circuit shown in Fig. 13.11-1. They are switched off and on as required to allow A sweep output transistor $Q4475$ and B sweep output transistor $Q4465$ to pass the A or B sawtooth to the Horizontal Amplifier.

When HORIZ DISPLAY is set to A, a LO on the base of $Q4481$ keeps the transistor biased off. The A sawtooth is allowed to pass the emitter of $Q4475$ to be amplified and sent to the Horizontal Amplifier. A HI on the base of $Q4882$ turns it on, and the B sawtooth passes to ground through the transistor, thereby preventing the B sawtooth from going to the Horizontal Amplifier. The same conditions exist when the HORIZ DISPLAY is set to A INTEN. Setting the HORIZ DISPLAY to ALT will cause the Horizontal Display Logic circuit to alternately turn $Q4481$ and $Q4482$ off and on to first pass the A sawtooth and then the B sawtooth to the Horizontal Amplifier. When the HORIZ DISPLAY is set to B DLY'D, $Q4482$ will be biased off and $Q4481$ will be on. The B sawtooth will go to the Horizontal Amplifier, and the A sawtooth will be shunted to ground.

13.11   HORIZONTAL DISPLAY LOGIC

Introduction

The Horizontal Display Logic circuit in Fig. 13.11-1 produces the signals that switch the A and B Sweep Generators and the A Sweep and B Sweep Z-Axis Gates. It also provides a Vertical Alt Sync pulse to the Vertical Switching circuit.

Vert Alt Sync Pulse

A gating circuit is formed by $U4391B$ and $U4391D$ to control the Vertical Alt Sync pulse. The pulse is used in the Vertical Switching circuit for clocking $U1705$ (Fig. 13.5-1a) whenever the ALT Vertical Mode is selected. In all Horizontal Display modes except ALT, a HI at pin 6 of $U4391B$ will put a LO at pin 3 of $U4391D$. Pin 2 of $U4391D$ has the positive-going Horiz Alt Sync pulse present. This pulse is inverted through $U4391D$ and fed to the Vertical Switching circuit. In ALT Horizontal Display mode, pin 6 of $U4391B$ will be LO, and the signal at pin 5 will control output pin 4. The signal present at pin 6 of $U4491A$, which changes state with every Horiz Alt Sync pulse, now controls $U4391D$, and the Vert Alt Sync pulse becomes a rectangular pulse having a period equal to twice the time between Horiz Alt Sync pulses.

A and B Sweep Switching

Flip-flop $U4491A$ controls Horizontal Display Switching transistors $Q4482$ and $Q4481$ in the A and B Sweep Generators. The HORIZ DISPLAY switch $S6260$ sets the flip-flop input to do one of the following:

1. Turn off $Q4481$ and turn on $Q4482$ to allow the A Sweep signal to go to the Horizontal Amplifier.

2. Turn off $Q4482$ and turn on $Q4481$ to allow the B Sweep signal to go to the Horizontal Amplifier.

3. Alternately turn the two transistors off and on to display both sweeps in ALT Horizontal Display mode.

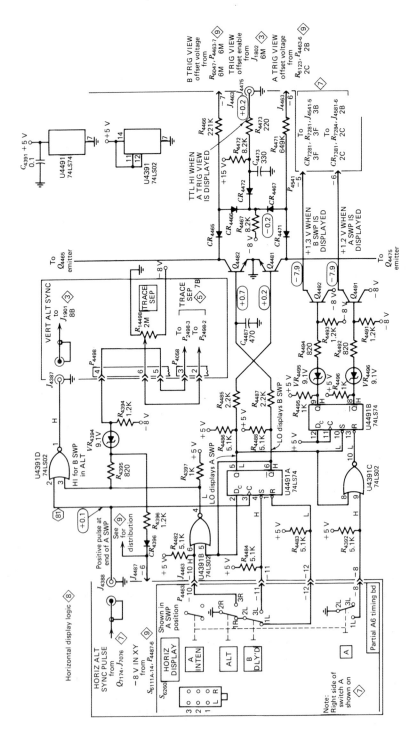

Figure 13.11-1 Horizontal Display Logic circuit. (Courtesy Tektronix, Inc. © 1980.)

U4491A is clocked by the Horiz Alt Sync when the Horizontal Display is set to ALT. When the Horizontal Display is set to A, A INTEN, or B DLY'D, the state of the outputs of U4491A is determined by the logic levels present pins 2 and 4. The HORIZ DISPLAY switches at these logic levels.

Flip-flop U4491B controls the A and B Sweep Z-Axis Gate switching transistors Q4492 and Q4491. These transistors switch the bias on CR7281 and CR7381 in the A and B Sweep Z-Axis Gates to either allow the unblanking gates to pass to the Z-Axis Amplifier or block them. U4491B is controlled by both the signal on pin 5 of U4491A and the A position of the HORIZ DISPLAY switch.

13.12   HORIZONTAL AMPLIFIER

Introduction

The Horizontal Amplifier circuit shown in Fig. 13.12-1 provides the output signals to the CRT horizontal deflection plates. The signal applied to the input of the Horizontal Amplifier is determined by the TIME/DIV switch and the HORIZ DISPLAY switch. This signal can come from either the sweep generators within the instrument of some external signal applied to the CH 1 OR X input connector (X-Y Horizontal Display). Horizontal positioning, $\times 10$ magnifier circuitry, and the horizontal portion of the beam finder circuitry are also contained in the Horizontal Amplifier.

X-Axis Amplifier

In all positions of the TIME/DIV switches except X-Y, the input signal to the base of U4269A will be sawtooth waveforms from the sweep generators. In the X-Y position, however, the sweeps are disabled, and the signal applied to U4269A comes from the Channel 1 Preamplifier via the X-Axis Amplifier. This stage includes Q4284, Q4285, and associated circuitry.

Transistor Q4284 is connected as a feedback amplifier, with R4285 as the feedback element. The input resistance is made up of R4384 and the gain-setting adjustment of R4381. When not operating in the X-T Horizontal Display, the base of Q4284 rises toward the $+15$-V supply, but is clamped at approximately $+5.7$ V by CR4287 and R4287. This reverse biases the base–emitter junction of Q4284. The base of Q4285 also rises to approximately $+5.7$ V and with junction of R4281–R4280 at approximately 0 V, Q4285 is biased off.

When the A TIME/DIV switch is set to X-Y position (fully counterclockwise), -8 V is applied to the junction of R4382 and R4383. In addition, $+5$ V is applied to emitter circuit of Q4285 through CR4285. This biases the X-Axis Amplifier into conduction. The $+5$ V is also applied to pin 18 of U7375 in the Sweep and Z-Axis Logic circuit to disable sweep generation.

Input Paraphrase Amplifier

The Input Paraphase Amplifier consists of $U4269A$ and $U4269B$ (part of a transistor array). This is an emitter-coupled amplifier stage that converts the single-ended input signal to a push–pull output signal. The signal at the collector of $U4269A$ is opposite in phase to the input signal, while the signal at the collector of $U4269B$ is in phase with the input signal. Thermistor $RT4373$ reduces its value with increases in ambient temperature to increase the gain of the stage. This compensates for slight changes in amplifier gain that occur as operating temperatures vary.

The Horizontal POSITION potentiometers $R14288A$ (Coarse) and $R14288B$ (Fine), are mounted on the same shaft in a mechanical arrangement

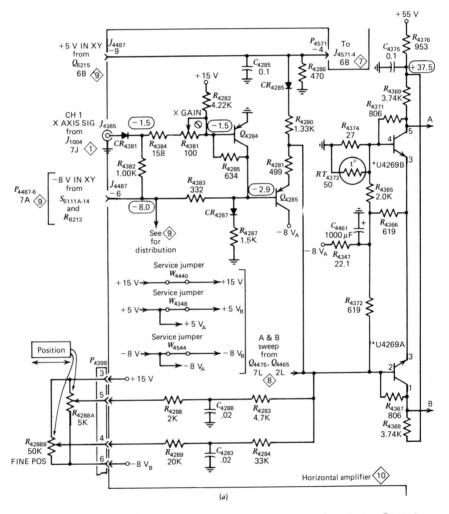

(a)

Figure 13.12-1 (a)–(c) Horizontal Amplifier circuit. (Courtesy Tektronix, Inc. © 1980.)

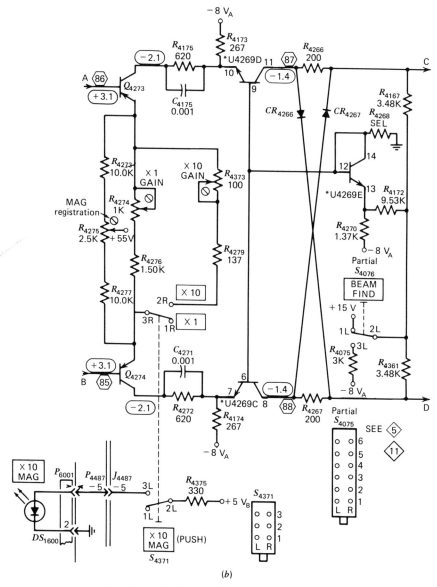

Figure 13.12-1 (*Continued*) * *Note*: transistor array.

that allows $R14288B$ to rotate about one-eighth turn in either direction before $R14288A$ moves. The Fine Potentiometer has approximately one-tenth the range of the Coarse potentiometer.

Gain Setting Cascode Amplifier

A cascode push–pull amplifier stage is composed of $Q4273$, $Q4274$, $U4269C$, and $U4269D$. The gain of the Horizontal Amplifier is controlled by adjusting

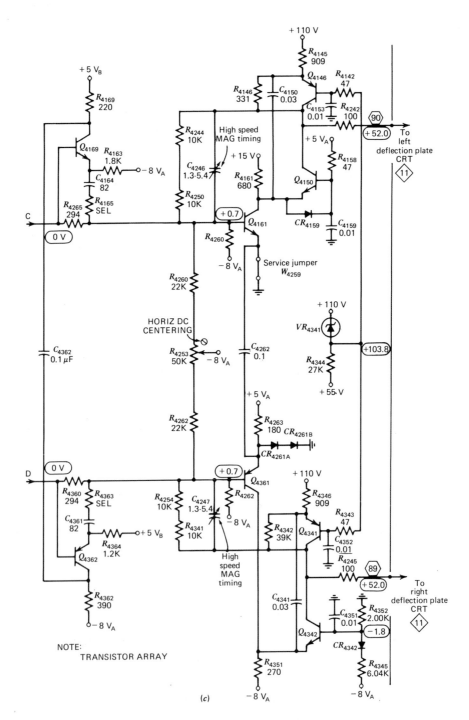

Figure 13.12-1 (Continued)

472

the resistance connected between the emitter of $Q4273$ and $Q4274$. The $\times 1$ Gain adjustment, $R4274$, is used to adjust the unmagnified horizontal gain and the $\times 10$ Gain adjustment, $R4373$, is used to adjust the magnified horizontal gain. Magnifier Registration adjustment, $R4275$, is used to balance the quiescent dc current in $Q4273$ and $Q4274$ so that a center screen display does not change positions when switching between magnified and unmagnified display.

When the BEAM FIND pushbutton is pressed, $+15$ V is removed from the junction of $R4167$ and $R4361$, and -8 V is applied to the junction through $R4075$. The following actions occur:

1. The base voltage of $U4269C$ and $U4269D$ is lowered to decrease the current through the two transistors. The reduced current flow decreases the voltage drop across $R4266$ and $R4267$ and the diode clamps ($CR4266$ and $CR4267$) clamp at much lower voltage. The action limits the horizontal deflection on the CRT.

2. $R4167$ and $R4361$ are now no longer connected to $+15$ V, therefore, less current flows through these resistors. This compensates for the decreased current flowing through $U4269C$ and $U4269D$ and maintains the output stage in a linear operating region.

Output Amplifier

The push–pull signal from the Gain Setting Cascode Amplifier is connected to the bases of $Q4169$ and $Q4362$, through $R4265$ and $R4360$ and on to the bases of $Q4161$ and $Q4361$. At lower sweep frequencies, the signal path is through $R4265$ and $R4360$ to the bases of $Q4161$ and $Q4361$. These transistors are inverting amplifiers whose collector signals drive the emitters of complementary amplifiers $Q4146$–$Q4150$ and $Q4341$–$Q4342$, respectively. Capacitor $C4262$ provides emitter peaking for fast ac signals. Capacitors $C4150$ and $C4341$ transfer part of the high-frequency signal to the emitters of $Q4146$ and $Q4341$ to maintain the gain of the output stage at high sweep speeds. Resistors $R4244$–$R4250$ and $R4254$–$R4341$ are the feedback elements, with $C4246$ and $C4247$ providing high-frequency compensation. As the frequency of the sweep signal increases, the reactance of $C4246$ and $C4247$ decreases and feedback current increases. To compensate for the increase in drive required to maintain the gain of the output stage, $Q4169$ and $Q4362$ (fast-path amplifiers) increase signal current to the bases of $Q4161$. High-frequency signal current is shunted around $R4265$ by $C4164$ and $R4165$, and $C4361$ and $B4363$ shunt high-frequency signal current around $R4360$. The Output Amplifiers are limited from being overdriven by $CR4267$ and $CR4266$. If the output signal from $U4269D$ or $U4269C$ becomes too large, the diodes become forward biased and prevent further increase in the signal level. These diodes operate mainly to clamp the signal whenever the $\times 10$ Magnification

Circuitry is operating. The signal level is limited to the forward drop across the diodes plus the drop across $R4266$ and $R4267$.

13.13   CRT CIRCUIT

Introduction

The CRT circuit shown in Fig. 13.13-1 provides the voltage levels and control circuitry necessary for operation of the cathode-ray-tube (CRT) circuit.

High-Voltage Oscillator

The high-voltage oscillator is made up of $Q14009$ and associated circuitry. It produces the drive for the high-voltage transformer $T4015$. When the instrument is turned on, transistor $Q4008$ is forward biased and conducts through the base circuit of $Q14009$ to forward bias $Q14009$. The increasing collector current of $Q14009$, through the primary winding of $T4015$, induces a voltage across the feedback winding. Because the feedback winding is connected to the base of $Q14009$ and the feedback is positive, the collector current increases rapidly toward saturation. Soon the rate of increase slows to a point where the voltage induced in the feedback winding starts to decrease. This decreases the current through $Q14009$, further decreasing the feedback voltage. The cycle continues until $Q14009$ turns off, and the magnetic field around the primary winding of $T4015$ starts to collapse. Transistor $Q14009$ is held off until the field collapses sufficiently to allow the base of $Q14009$ to become biased into conduction and the cycle repeated.

The voltage waveform at the collector of $Q14009$ is a sinusoidal wave at the resonant frequency of $T4015$. The amplitude of sustained oscillations depends on the average current delivered to the base of $Q14009$. Frequency of oscillation is approximately 50 kHz. Fuse $F4508$ protects the unregulated $+15$-V supply in the event the High-Voltage Oscillator stage becomes shorted. $C4006$ and $L4006$ decouple the unregulated $+15$-V supply to prevent current changes (present in the Hogh-Voltage Oscillator) from affecting the $+15$-V supply.

High-Voltage Regulator

Once the output voltage from the High-Voltage Oscillator has reached its stable level after the instrument is turned on, regulation occurs as follows. A sample of the -2450-V CRT cathode supply is applied to the base of $Q4228$ through $R4127D$ which, with the voltage supplied by the bias network composed of $R4332$, $R4127$, $C4327$, and $CR4329$, sets the forward bias on $Q4228$. Any change in the -2450 V changes the conduction level of $Q4228$ to produce a proportional dc change on its collector.

Assume that the -2450-V supply starts to go positive (less negative). The positive-going change is applied to the base of $Q4228$ and causes the collector

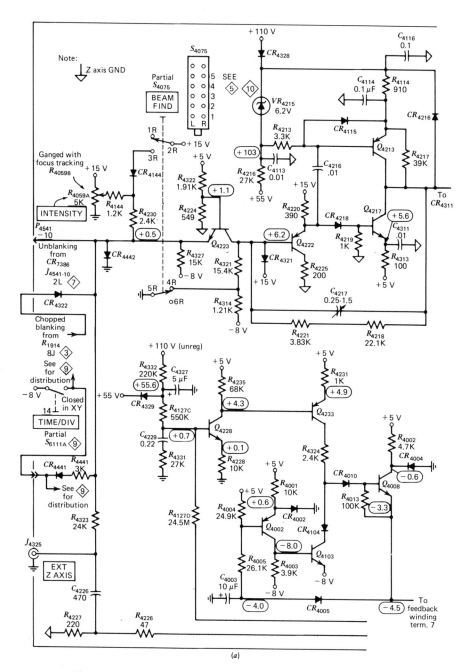

Figure 13.13-1 (a)–(c) CRT circuit. (Courtesy Tektronix, Inc. © 1980.)

current to increase, which, in turn, causes Q4233 and Q4008 to conduct harder. This results in greater bias current to the base of Q14009 through the feedback winding of T4015. Now Q14009 is biased closer to its conduction level, and it will conduct sooner in the oscillation cycle of T4015 to increase the average current delivered to the primary of T4015. This increases the amplitude of oscillation and induces a larger voltage into the high-voltage secondary of T4015 to correct the original positive-going

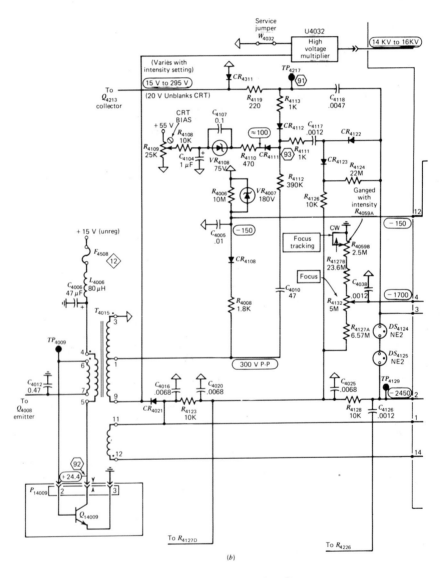

(b)

Figure 13.13-1 (Continued)

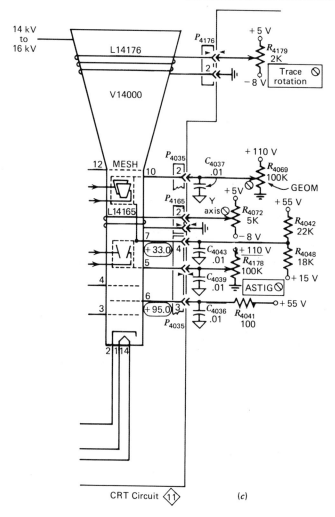

Figure 13.13-1 (Continued)

change. By sampling the output of the CRT cathode supply in this manner, the total output of the High-Voltage Supply is held relatively constant.

Overvoltage protection is provided by $Q4002$, $Q4103$, and associated circuitry. Normally $Q4002$ and $Q4103$ are biased off, but if the CRT cathode supply voltage approaches approximately -3000 V, the voltage level at the emitter of $Q4008$ will be approximately -6 V. At this point $Q4002$ will be biased into conduction, which, in turn, biases $Q4103$ into conduction to reduce the forward bias on $Q4008$. This reduces the base drive to $Q14009$ in order to limit the amplitude of oscillation in $T4015$, and the CRT cathode supply is prevented from going more negative than approximately -3000 V.

High-Voltage Rectifiers and Output

High-Voltage transformer $T4015$ has two secondary windings. One winding provides heater voltage for the CRT. The heater voltage is supplied from the High-Voltage Supply, since the CRT has a very low heater current drain, and this allows the cathode and heater of the CRT to be connected together to prevent cathode-to-heater breakdown. The high-voltage winding is the source for the negative CRT cathode potential and the source for High-Voltage Multiplier $U4032$. The CRT grid bias voltage is derived by a dc-restorer circuit that uses a sample of the signal in the high-voltage winding in conjunction with dc levels supplied by both the Z-axis Amplifier and the CRT negative cathode potential.

The positive accelerating potential is supplied by High-Voltage Multiplier $U4032$. Regulated output voltage is approximately $+15,500$ V. The negative cathode potential of -2450 V is supplied by half-wave rectifier $CR4021$. Voltage variations in this supply are monitored by the High-Voltage Regulator circuit to provide a regulated high-voltage output.

CRT Control Circuits

Focus of the CRT display is controlled by FOCUS control $R4132$. ASTIG adjustment $R4178$, which is used in conjunction with the FOCUS control to provide a well-defined display, varies the positive level on the astigmatism grid. Geometry adjustment $R4069$ varies the positive level on the horizontal deflection plate shields to the overall geometry of the display.

Two adjustments control the trace alignment by varying the magnetic field around the CRT. Y-axis adjustment $R4072$ controls current through $L14165$, which affects the CRT beam after vertical deflection but before horizontal deflection. Therefore, it affects only the vertical (Y) components of the display. TRACE ROTATION adjustment $R4179$ controls the current through $L14176$ and affects both vertical and horizontal rotation of the beam.

Z-Axis Amplifier

The Z-Axis Amplifier circuit controls the CRT intensity level from several inputs. The effect of these input signals is to either increase or decrease the trace intensity or completely blank portions of the display. Input transistor $Q4223$ is a current-driven, low-input-impedance amplifier. It provides termination for the input signals as well as isolation between the input signals and the following stages. Current signals from the various control sources are connected to the emitter of $Q4223$, and the algebraic sum of these signals determines the collector conduction level.

Transistors $Q4222$, $Q4217$, and $Q4213$ are configured in a feedback amplifier arrangement with $R4218$ and $R4221$ as feedback elements and with $C4217$ providing high-frequency compensation. Emitter-follower $Q4222$ provides the drive to complementary amplifier $Q4217$–$Q4213$. Diodes $CR4321$, $CR4218$, and $CR4115$ provide protection in the event of high-voltage arcing.

The Z-Axis portion of the Beam Find circuit acts on the input to the Z-

Axis Amplifier. When the BEAM FIND pushbutton is pressed, two events occur: First, $+15$ V is applied to the anode of $CR4144$, which lifts the emitter of $Q4223$ sufficiently positive to ensure nonconduction of the transistor. Second, $R4321$ becomes connected to -8 V through $R4314$ to establish a fixed and predetermined unblanking level at the output of the amplifier. Thus, the INTENSITY control and all of the input unblanking signals have no control over the intensity level of the CRT display whenever the BEAM FIND pushbutton is pressed, and a bright trace will be displayed.

DC Restorer

The DC Restorer circuit provides CRT control grid bias and couples both dc and low-frequency components of the Z-Axis Amplifier unblanking signal to the CRT control grid. This circuit allows the Z-Axis Amplifier output to control the intensity of the CRT display. The potential difference between the Z-Axis Amplifier output and the control grid (about 2465 V) prevents direct signal coupling.

The DC Restorer circuit's ac drive is taken from the center tap of $T4015$. Voltage on the center tap is approximately 300 V_{p-p} at 50 kHz. A sample of this sinusoidal voltage is fed through $C4010$ and $R4112$ to the junction of $CR4111$, $CR4112$, and $R4111$. The CRT Bias Adjustment ($R4109$) sets the voltage level on the cathode of $CR4111$ to approximately $+100$ Vdc. When the ac sample voltage rises to $+100$ V, $CR4111$ becomes forward biased and clamps the junction of $CR4111$ and $CR4112$ to approximately $+100$ V.

The Z-Axis Amplifier output voltage level is applied via $R4113$ to anode of $CR4112$. This voltage level varies between $+15$ and $+95$ V, depending on the setting of the INTENSITY control. The sample voltage will hold $CR4112$ reverse biased until the voltage falls below the Z-Axis Amplifier output level. At that point $CR4112$ becomes forward biased and clamps the junction of $CR4111$ and $CR4112$ to the Z-Axis Amplifier output level. Clamping the sample between $+100$ V and the positive voltage level set by the INTENSITY control produces an approximately square-wave signal with a positive dc offset level.

The DC Restorer circuit is referenced to the -2450 V, present on the CRT cathode, through $R4126$ and $CR4123$ to the junction of $C4117$ and $CR4122$. Initially, $C4117$ will charge to a level determined by the difference between the Z-Axis Amplifier output level and the -2450-V reference voltage. The charging path is from the -2450-V line, through $R4126$, $CR4123$, $C4117$, $R4111$, $CR4112$, and $R4113$ to the Z-Axis output.

Initially, $C4118$ will also be charged to approximately the same voltage as $C4117$ through $R4126$, $CR4123$, and $CR4122$ to the Z-Axis output.

When the sinusoidal sample voltage starts its positive transition from the lower clamped level ($+15$ to $+95$ V) toward the higher clamped level ($+100$ V), the charge on $C4117$ increases due to the rising voltage at the anode of $CR4111$. The additional charge acquired by $C4117$ is proportional to the amplitude of the positive transition of the clamped sample voltage.

When the sample voltage starts its negative transition from its upper clamped level toward its lower clamped level, the negative transition is coupled through C4117 to reverse bias CR4123 and forward bias CR4122. When CR4122 becomes forward biased, the charge on C4117 is transfered to C4118 as C4117 attempts to discharge to the Z-Axis output. The amount of charge that is transfered is proportional to the setting of the INTENSITY control, since the INTENSITY control sets the lower clamping level of the sample voltage from T4015.

If the INTENSITY control is set so the lower level of the sample voltage is clamped at +15 V, a voltage change of approximately 75 V is coupled through CR4122. The 75-V negative excursion is added to the charge already present on C4118. This causes the control grid to be sufficiently negative with respect to the CRT cathode to keep the CRT blanked. When the INTENSITY control is set to increase the display intensity, the lower clamping level of the sample voltage is moved toward the +100 V upper clamping level. This makes the swing of the negative transition less, therefore, less charge will be added to C4118. The voltage on the CRT control grid becomes less negative with respect to the cathode and allows more beam current to flow the CRT. The more positive the lower clamping level is made, the brighter the trace on the CRT.

During the period that C4117 is charging, the voltage on the control grid is held constant by the filter action of C4118 as it discharges through R4124 back to the −2450-V line. R4124 is a very high resistance, so the RC time constant of R4124 and C4118 is long with respect to the frequency of the sample voltage from T4015. Whatever charge is leaked off of C4118 during the positive transitions of the sample voltage will be replaced by C4117 when the sample voltage makes its negative transitions.

The fast rise and fall of the unblanking pulses from the Z-Axis Amplifier are coupled by C4118 to the grid to start the CRT beam current change. The DC Restorer output level then follows the Z-Axis output level to set the new bias level on the control grid.

In the event of a failure that causes a loss of potential on either the control grid or the cathode, protection against arcing is provided by DS4124 and DS4125.

13.14 CALIBRATOR

Multivibrator

The Calibrator shown in Fig. 13.14-1 consists of an astable multivibrator (Q4182 and Q4196) and an output amplifier (Q4291). The basic frequency of the multivibrator is approximately 1 kHz and is primarily determined by the resistance and capacitance of C4187, R4186, R4191, R4184, and R4185. Transistors Q4182 and Q4196 alternately conduct, producing a square-wave output signal. This output is taken from the collector of Q4196.

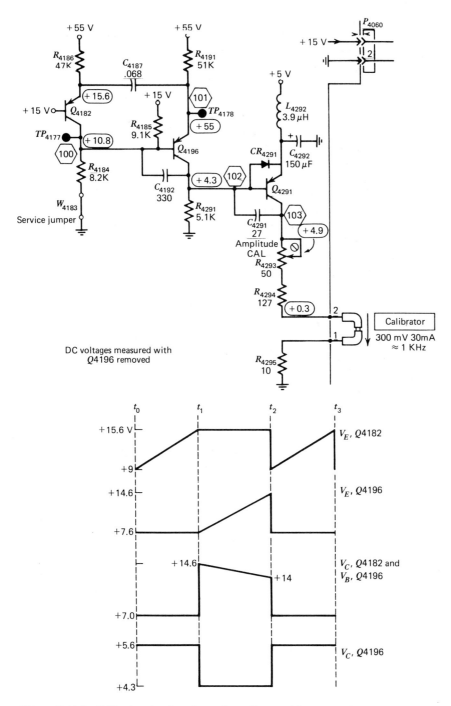

Figure 13.14-1 Calibrator circuit and waveform diagram. (Courtesy Tektronix, Inc. © 1980.)

481

Refer to the waveform diagram (Fig. 13.14-1) for the following discussion. At t_0, assume that the circuit is operating and has reached its normal operating conditions. Also assume $Q4182$ is off and $Q4196$ is on. At t_0, the emitter of $Q4182$ is at approximately $+9$ V, and the emitter of $Q4196$ is at $+7.6$ V with its base at approximately $+7.0$ V. From t_0 to t_1, $C4187$ charges toward the $+55$-V supply through $R4186$. When the emitter of $Q4182$ reaches $+15.6$ V, $Q4182$ becomes forward biased. At t_1, the collector of $Q4182$ rises to approximately $+14.6$ V, and since the base of $Q4196$ is directly connected to the collector of $Q4182$, $Q4196$ is cut off.

Now $C4187$ charges in the opposite direction. At t_1, as $C4187$ starts charging through $R4191$ and $Q4182$, the emitter of $Q4196$ rises from $+7.6$ V to approximately $+14.6$ V. When $+14.6$ V is reached at t_2, $Q4196$ begins to conduct, reducing both the charging current through $C4187$ and the collector current of $Q4182$. At t_2 the collector voltage of $Q4182$ drops in a negative direction and $Q4196$ conduction increases. The emitter of $Q4196$ drops from $+14.6$ to $+7.6$ V. This negative transition is coupled through $C4187$ to the emitter of $Q4182$ cut $Q4182$ off and the cycle repeats itself.

Output Amplifier

The output signal from the Multivibrator derives Output Amplifier $Q4291$ to produce a square-wave at the output. When the base of $Q4291$ goes positive, the transistor is cut off and its collector voltage drops to zero. When the base goes negative, $Q4291$ is biased into saturation, and the collector voltage rises in a positive direction to about $+5$ V. Amplitude adjustment $R4293$ is used to adjust the resistance between the collector of $Q4291$ and ground to control the amount of current allowed to flow. This in turn determines the voltage developed across $R4295$. The output voltage, at the calibrator current loop on the 465B oscilloscope, is 300 mV $\pm$ 1.0% and the output current is 30 mA $\pm$ 2.0%.

BIBLIOGRAPHY

Cameron, D., *Advanced Oscilloscope Handbook*. Reston Publishing, Reston, Virginia, 1977.
Instruction Manual of Tektronix 465B Oscilloscope. Tektronix, Inc., Beaverton, Oregon, 1980.

Questions

13-1   What are the specified bandwidth and rise time of the vertical amplifier in the Tektronix 465B oscilloscope?

13-2   What sweep rates are provided by the Time Base A?

13-3   What is the A Trigger Holdoff?

13-4   Why is the A Trigger View employed in the 465B?

13-5   What is the function of the Beam Finder?

13-6   A dc-coupled positive-going signal applied to the External Z-Axis input decreases the CRT intensity. Is it true? Trace and check the related circuit in Fig. 13.13-1 and then answer.

13-7   What controls are contained in the A and B Trigger Generator circuits?

13-8   What is the function of the Z-Axis Logic circuit (7)?

13-9   Do the B and A Sweep Generators produce sawtooth output signals at the same time?

13-10   What circuit determines the CRT intensity and blanking?

13-11   What are provided by the Input Attenuators in the Vertical Preamplifier?

13-12   What are provided by the FETs Q1297A and B in Fig. 13.4-1?

13-13   Describe the operation of the Chop Clock oscillator in the Vertical-Switching Logic circuit (Fig. 13.5-1).

13-14   Describe the operation of the Delay-Line Driver (Q1426–Q1428).

13-15   What is the function of the Delay Line DL11405?

13-16   How does the Exclusive NOR circuit U5235D (Fig. 13.6-1) work when the CH 1 VERT MODE pushbutton is first pressed and then released?

13-17   Explain the function of the Trace Separation circuit (Q2115, Q2306, and Q2308).

13-18   Explain the function of the input source follower Q7324A.

13-19   Describe the operation of the A Sweep Holdoff Amplifier Q7460.

13-20   Explain the operation of the Horiz Alt Sync Pulse amplifier Q7174.

13-21   What is the function of Disconnect Amplifier Q4497?

13-22   Explain the operating principle of the Miller Integrator circuit (Q4498–Q4598).

13-23   How does the A Sweep Start Comparator (Q4588A–Q4588B) work?

13-24   How does the A Sweep End of Sweep Comparator (Q4581–Q4582) work?

13-25   How does the gating circuit formed by U4391B and U4391D control the Vertical Alt Sync pulse?

13-26   Explain the operation of the output amplifier in the Horizontal Amplifier circuit (Fig. 13.12-1).

13-27   How does the High-Voltage Oscillator (Q14009) work?

13-28   Explain the overvoltage protection of the high-voltage regulator in the CRT circuit (Fig. 13.13-1).

Problems

13.5-1   Refer to the Vertical Switching Logic diagram in Fig. 13.5-1 for the following description. Input signals to U1605, a read-only memory (ROM), are as follows:

Input lines A_0 through A_2—Logic levels from the Q_0, Q_1, and Q_2 outputs of $U1705$ used to indicate the present state of the switching. (Q_0, Q_1, and Q_2 outputs are active when LO.)

Input lines A_3 through A_6—Logic levels selected by the VERT MODE switches (CH 1, CH 2, ADD, and A TRIG VIEW). A HI logic level present indicates that the Vertical Mode is selected.

Input lines A_7—Logic level controlled by the X-Y position of A TIME/DIV switch. A LO logic level is present when X-Y Horizontal Display is selected.

After $U1705$ is clocked, ROM $U1605$ uses the present data on its input lines (A_0 through A_7) to select the next output switching state to be presented to $U1705$. There are four output lines from $U1605$, O_1 through O_4. Output lines O_1 through O_3 carry the future data; the signal present on output line O_4 is the Chop Clock Oscillator enabling logic (Hi enables).

In the partial table shown in Fig. P13.5-1, no Vertical Modes

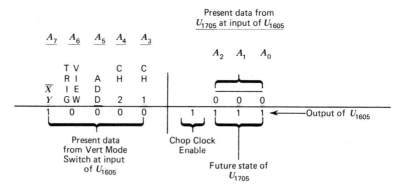

Figure P13.5-1   Partial table P13.5-1.

are selected, and the present data from $U1705$ is an undefinded column (i.e., $\overline{000}$ is not a Vertical Mode selection). In the example given, the Chop Clock Oscillator will be enabled, and the next clock pulse to $U1705$ will switch $U1705$ output to 111. Table P13.5-1 shows that column 111 is the idle state of $U1605$ (the state switched to when no Vertical Modes are selected). In the idle column, the future state of $U1705$ will remain 111, and the Chop Clock Oscillator will not be enabled. Each row across the table indicates the possible future states of $U1705$, while the column headings indicate the possible present state. The order of priority in switching (when multiple Vertical Modes are selected) is CH 1, CH 2, ADD, then A TRIG VIEW.

To use the table, determine the Vertical Mode(s) selected. Follow that row across the table. If the output of $U1705$ is at the present state indicated by a column heading, the data given in that column are the future state of $U1705$.

(a) Assume CH 1 and CH 2 Vertical Modes are selected. The present data from the Vertical Mode Switch is 10011. Move across that row to the 000 column. The data given indicate that if the present state $U1705$ is 000, the future state will be _____ and that the Chop Clock Oscillator will be enabled. Following across the row, each column, except 110 (CH 1), gives the same future state. If the present state is CH 1, the future state will be _____. From there, it will switch back to CH 1 for the chopped display of the Channel 1 and Channel 2 input signals.
Ans.   110 (CH 1); 101 (CH 2).

(b) Assume that X-Y Horizontal Display is selected. Go to any Vertical Mode selection in the bottom half of the table. All the columns indicate that regardless of the state of the $U1705$ output, the future state will be _____. When the output of $U1705$ goes to 101, the Chop Clock Oscillator will be turned off, and the Chop Clock to $U1705$ will cease.
Ans.   101 (CH 2).

(c) Assume ADD Vertical Mode is selected. In the Present Data from the Vertical Mode Switch column read 10100. Move across the row (adjacent to the data) to the 000 column and note that the future state of $U1605$ is _____. The output state required to switch to the ADD display is _____.
Answer.   (a) 110 (CH 1), 101 (CH 2); (b) 101 (CH 2); (c) 1100, 100.

13.12-1   A pulse generator with a t_r of 2 ns is utilized in determining the rise time of an amplifier. If a 100-MHz oscilloscope measured a t_r of 10 ns, what is the actual rise time of the amplifier? *Hint:* The actual rise time of device under test is

$$t_{r(\text{dev})} = \sqrt{t_{r(\text{means})}^2 - t_{r(\text{scope})}^2 - t_{r(\text{pG})}^2}$$

$$t_{r(\text{scope})} = \frac{0.35}{\text{BW}} .$$

Answer. $t_{r(\text{dev})} = 9.15$ ns.

13.12-2   The output pulse rise time of a video amplifier was measured as 0.5 μs. What is the approximate bandwidth of the amplifier?

13.12-3   To measure the relative phase of two signals at the same frequency, we similarly feed them into the two oscilloscope inputs and adjust the dual-trace oscilloscope's triggering controls so that the two signals begin on the screen at the same point in time. Triggering the signal on one input from the signal at the other generally pro-

Table P13.5-1 Read-Only-Memory $U1605$ Logic

$\dfrac{X}{Y}$	Vert Mode Present Data at $U1605$				Undefined			Trig View	ADD	CH 2	CH 1	Idle
	TRIG VIEW	ADD	CH 1	CH 2	$\overline{000}$	$\overline{001}$	$\overline{010}$	$\overline{011}$	$\overline{100}$	$\overline{101}$	$\overline{110}$	$\overline{111}$
1	0	0	0	0	1111	1111	1111	1111	1111	1111	1111	0111
1	0	0	0	1	1110	1110	1110	1110	1110	1110	0110	1110
1	0	0	1	0	1101	1101	1101	1101	1101	0101	1101	1101
1	0	0	1	1	1110	1110	1110	1110	1110	1110	1101	1110
1	0	1	0	0	1100	1100	1100	1100	0100	1100	1100	1100
1	0	1	0	1	1110	1110	1110	1110	1110	1110	1100	1110
1	0	1	1	0	1101	1101	1101	1101	1101	1100	1101	1101
1	0	1	1	1	1110	1110	1110	1110	1110	1100	1101	1110
1	1	0	0	0	1011	1011	1011	0011	1011	1011	1011	1011
1	1	0	0	1	1110	1110	1110	1110	1110	1110	1011	1110
1	1	0	1	0	1101	1101	1101	1101	1101	1011	1101	1101
1	1	0	1	1	1110	1110	1110	1110	1110	1011	1101	1110
1	1	1	0	0	1100	1100	1100	1100	1011	1100	1100	1100
1	1	1	0	1	1110	1110	1110	1110	1011	1110	1100	1110
1	1	1	1	0	1101	1101	1101	1101	1011	1100	1101	1101
1	1	1	1	1	1110	1110	1110	1110	1011	1100	1101	1110

duces this condition. The relative phase between the two signals we determine by measuring the time Δt between comparable zero-crossing points. (a) Write the expression for the phase difference ϕ. (b) If the period $T = 4.0$ ms and the difference between zero crossings in the same direction is $\Delta t = 1.0$ ms, what is the relative phase angle?

Answer.   (a) $\phi = \dfrac{\Delta t}{T} \times 360°$; (b) 90°.

14

Digital
Multimeter Design

14.1 INTRODUCTION

The analog-to-digital converter (ADC) is the main section of a digital voltmeter. The popular ADC systems are the counting ADC, successive-approximation ADC, parallel-comparator ADC, and the dual-ramp (slope) ADC. Among these the dual-ramp ADC is usually employed in the digital multimeters (DMMs). Motorola Semiconductor provides a three-chip dual-ramp ADC system: MC1505 analog subsystem, MC14435 digital subsystem, and MC14511 decoder/driver for display. Some manufacturers provide different types of single-chip ADC, as listed in Table 6-4. The dual-ramp digital voltmeter system has been discussed in Section 9.8 (Fig. 9.8-2).

The design of a $3\frac{1}{2}$ digit electronic multimeter is described in this chapter. This multimeter contains provisions for the measurement of voltage, resistance, capacitance, frequency, period, and time. This instrument includes the following seven sections:

1. Counter/display. This consists of the $3\frac{1}{2}$ digit counter, memory, decoder and the display devices. It has three inputs: count, transfer, and reset. It also gives an output from the counter section to control the analog measuring functions.
2. Control logic. This section contains the gating and pulse generating circuits necessary to route the inputs to the counter and tb control the counter functions.
3. Master clock. This consists of a crystal oscillator and a frequency divider chain to provide the main timing for all the measurements.

4. Input wave shaper. This circuit is used during frequency and period measurement to convert the input signal into a form suitable for connecting to logic circuitry.

5. Timer control. This unit is used to start and stop the counter for time measurement.

6. Voltmeter. This unit contains the high input impedance stage, the rectifier and the dual-ramp voltage-to-time converter used for voltage and resistance measurement.

7. Resistance and capacitance unit. This consists of the current source used for resistance and capacitance measurement and the circuitry necessary to obtain a capacitance-to-time conversion.

The following specification is regarded as a target for the performance. For dcV measurement (automatic polarity indication), input resistance is 11.1 MΩ, range is from 199.9 mV to 199.9 V in four steps, accuracy is ±0.1% full-scale deviation (fsd) with ±0.2% of reading. For acV measurement, input impedance is 10 MΩ in parallel with 25 pF, range is the same as for dcV, measurement value is of average, calibrated rms, accuracy is ±0.1% fsd with ±0.5% of reading, and frequency ranges are 50 Hz–10 kHz ±1% and 10 Hz–50 kHz ±5%. For resistance measurement, range is from 1.999 K to 1.999 MΩ in four steps, and accuracy is ±0.1% fsd with ±0.5% of reading. For capacitance measurement, range is 1999 pF to 1.999 μF in four steps, and accuracy is ±0.1% fsd with ±0.5% of reading. The counter/timer unit has a frequency range from 0 to 5 MHz, and a period/time interval of 20 μs minimum; accuracies for frequency, period, and time interval are ±1 part in $10^5 \pm 1$ count, ±1% ±1 count, and ±1 part in $10^5 \pm 1$ count, respectively; input level and impedance for frequency/period measurements are 10 mV to 10 V and 100K in parallel with 10 pF, respectively; input level for time interval measurement is DTL input; gate times are from 10 μs to 1 s.

14.2 DIGITAL MEASURING SYSTEMS AND MULTIMETER BLOCK DIAGRAMS

Frequency Measurement

The signal whose frequency is to be measured is applied to the input of the wave shaping module in Fig. 14.2-1. This either amplifies or limits the input signal, depending on its amplitude, and then converts it into a rectangular wave form Ⓐ having a peak-to-peak amplitude of 5 V.

The master clock frequency Ⓑ has a period equal to the desired count duration. For example, if the count duration is to be 10 ms, a frequency of 100 Hz will be selected. In order to open the count gate for the correct time, this clock frequency is divided by two Ⓒ before it is applied to the count

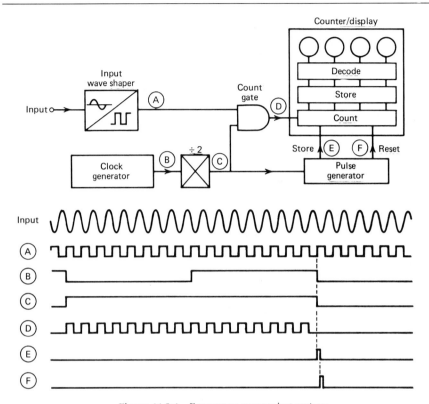

Figure 14.2-1   Frequency-measuring system.

gate and also to the control pulse circuit which generates the "store" Ⓔ and "reset" Ⓕ commands.

Assuming that the counter has been set to zero, the sequence of operation is as follows. The count gate is disabled for one clock period by the output of the divide-by-two. This connects the shaped input waveform to the input of the counter so that it counts the number of cycles during one clock period. At the end of this period, the negative-going edge of the timing signal Ⓒ causes the pulse generator to generate two successive pulses. The first of these Ⓔ commands the counter unit to "store" and display the state of the count section. The second Ⓕ "reset" the count section to zero ready for the next cycle of operation. This process will then restart when the timing signal Ⓒ goes positive once more. Hence, the unit counts and updates the display on alternate clock periods and, with a constant input frequency, produces a steady reading.

Period and Time Interval Measurements

The major difference between period and frequency measurements is that the roles of the clock generator and input wave shaper are reversed, as in

Fig. 14.2-2. Instead of counting the number of input cycles during one clock period, the number of clock pulses during one input cycle is connected. As with frequency measurement, the input waveform is "squared up" Ⓐ by the input wave shaper. It is then divided by two Ⓑ and fed to the count gate and to the control pulse generator. The output from the clock generator is also fed to this gate so that, when it is enabled by the input, clock pulses Ⓒ are fed to the counter. The "store," "display," and "reset" functions are the same as for frequency measurement. This period measurement facility has its chief application at low frequencies where the normal counter would be very inaccurate. For example, a frequency of 5 Hz measured using a 1-s count period could only be measured to an accuracy of ± 1 cycle or ±20%. By measuring period (200 ms), however, the accuracy could be very much improved. In practice, the accuracy could be better than 0.1% provided that there was no noise present on the waveform to be measured. The chief disadvantage of period measurement is that the result is the reciprocal of the required answer.

The only difference between the period- and time-measuring functions is that, whereas period is measured continuously on a cyclic basis, time is measured as the interval between two separately applied impulses, as shown in Fig. 14.2-3. To prevent the time information from being upset by contact

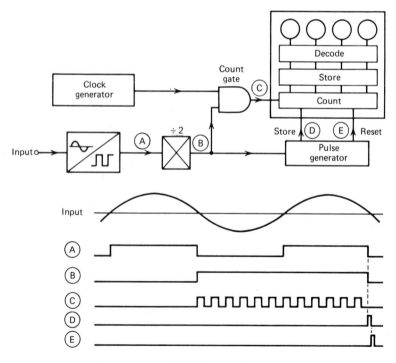

Figure 14.2-2   Period-measuring system.

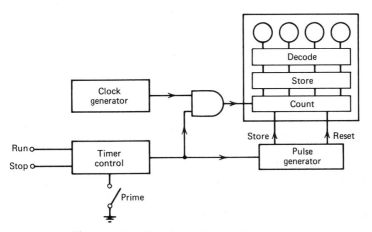

Figure 14.2-3 Time-interval-measuring system.

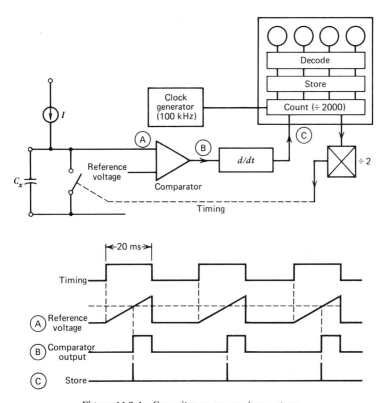

Figure 14.2-4 Capacitance-measuring system.

bounce or other spurious inputs, the timer control circuit is arranged to work on a "one-shot" basis so that it needs priming before each measurement.

Voltage and Resistance Measurements

The principles of the voltage and resistance measuring systems have been discussed in Section 9.8.

Capacitance Measurement

If current I and voltage V are constant in the relationship $C = It/V$, then capacitance $C = kt$, where k is constant and t is time. This simple relationship suggests that it is possible to measure capacitance in terms of the time required for the voltage drop across the capacitor, charged from a constant-current source, to reach a predetermined level. The method of implementing this technique is illustrated in Fig. 14.2-4.

At the beginning of the measurement cycle, the capacitor under test is completely discharged. The shorting switch across the capacitor is then opened allowing the current from the constant-current source to flow into the capacitor. This, in turn, causes the voltage across the capacitor to rise linearly with time. The comparator detects when the voltage across the capacitor equals the reference voltage and causes the control logic to send a "store" command to the counter thus displaying the time taken and, hence, the capacitance value. At the halfway point during the cycle, the switch is closed once more so that the capacitor is discharged, ready for the next measurement cycle.

This method of measurement cannot resolve the effects of leakage resistance. When it is used to measure low values of capacitance, very low

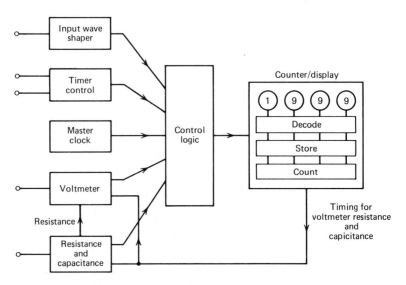

Figure 14.2-5   (Diag. 1) Block diagram of the multimeter.

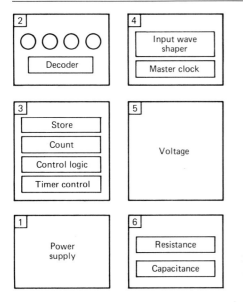

Figure 14.2-6 (Diag. 2) Block diagram rearranged to show the sections of the instrument circuit.

currents or very short periods are needed. Therefore, the top range has been limited to 1.999 μF, and the lowest range was chosen to be 1999 pF. Since leakage generally occurs in electrolytic capacitors, this method is not suitable for measuring them.

Block Diagram of the Multimeter

The instrument falls naturally into seven blocks, as shown in Fig. 14.2-5. In practice, it is convenient to rearrange some of these blocks so that they fit onto six circuit boards, as shown in Fig. 14.2-6. Boards 1, 2, and 3 form the basic digital display section.

14.3   BASIC DIGITAL DISPLAY SECTION

Decoder/Display

The decoder/display diagram is shown in Fig. 14.3-1. Either the seven-segment LED displays or the seven-segment incandescent filament displays (e.g., Minition 3015F) may be employed. They are suitable for use with the TTL, BCD-to-seven-segment decoder SN7447A. The display consists of four of these indicators. Three of them are driven by the decoders (7447A) but the fourth is driven by a transistor, as it only needs to indicate "1" or "0."

Counter/Store

The timing waveforms generated by the control circuit are shown in Fig. 14.3-2. The diagram of the counter/store and control logic is shown in Fig.

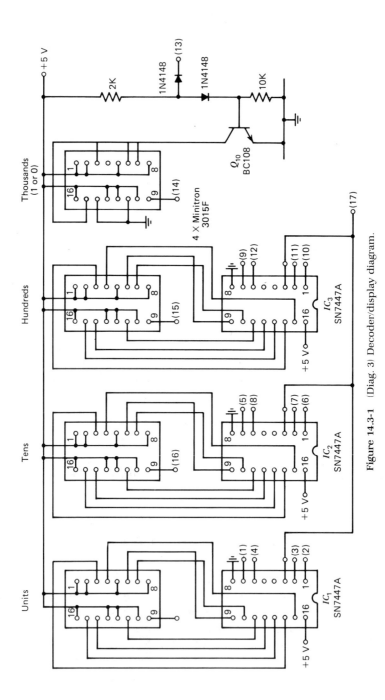

Figure 14.3-1 (Diag. 3) Decoder/display diagram.

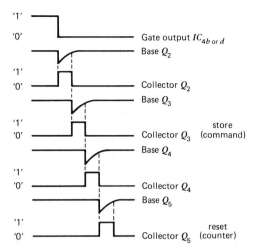

Figure 14.3-2 (Diag. 4) Timing waveforms generated by the control circuit.

14.3-3. In order to give the 1999 display capability needed for the instrument, the counter must divide by 2000. However, a further divide by two is necessary to drive the voltmeter and capacitance meter so that the total division required is 4000. Figure 14.3-3 shows that this division is achieved by using three decade dividers, type SN7490(A), followed by a dual JK flip-flop, type SN7473. The BCD outputs from the decade dividers are fed to three four-bit latches, type SN7475, which act as the memory for the three least significant digits of the counter. The memory for the most significant digit is a one-bit latch made from a quad two-input gate SN7400. This latch consists of two switches IC_{14a} and IC_{14b} feeding IC_{14c} and IC_{14d}, which are cross-coupled to form an RS (set–reset) flip–flop. When the command input is high, gates IC_{14a} and IC_{14b} are enabled so that the outputs from the RS flip-flops follow the inputs. When the command input goes low, the switches are disabled and the RS flip-flop retains the state that was present at the instant when the transition occurred. The command inputs to IC_{11}, IC_{12}, IC_{13}, IC_{14} are driven by the gates in IC_5, which, in turn, receive their inputs either from the voltmeter/capacitance meter or from the pulse generating chain Q_2–Q_5.

Control Logic

This circuit performs the switching functions illustrated in Figs. 14.2-1–14.2-3. Logic gates have been used here as they allow the switching to be achieved simply by switching the control inputs to "1," gate enabled, or "0," gate disabled. In this way the switching can be done at dc, so that there is no necessity to take signals back and forth between the front panel and the circuit board. The input to the count gate IC_{4a} comes from IC_{1a} (input from wave shaper for frequency counting) or IC_{1b} (input from master clock for

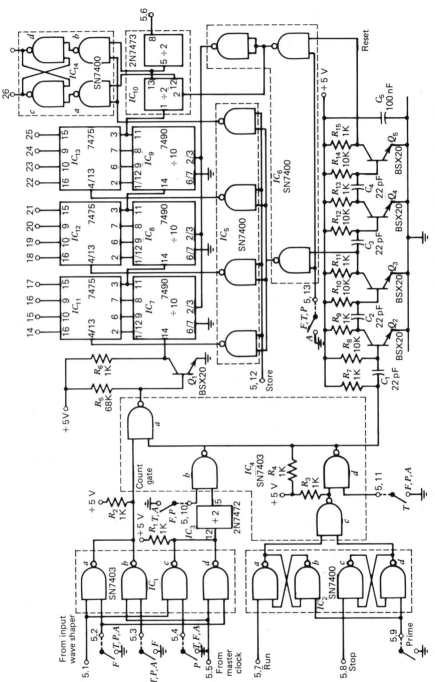

Figure 14.3-3 (Diag. 5) Counter/store and control logic. Switch S_1, shown dotted, indicates the switching necessary to achieve the various functions: F—frequency counter; T—timer; P—period meter; A—voltage, resistance, and capacitance meter. Connections: 5.1 to 5.13. [From Waddington, D. E. O'N., *Wireless World*, 161 (April, 1973).]

time, period or voltage/resistance/capacitance measurements). Similarly the input to the count-gate controller IC_{4b} comes from the divide-by-two IC_3, from either IC_{1c} or IC_{1d}. The output from IC_{4b} also feeds the control-pulse generating chain (Q_2, Q_3, Q_4, and Q_5), which works as follows.

The output from IC_{4b} in Fig. 14.3-3 is normally high so that the 22 pF capacitor C_1 is fully charged. When this output goes low, the base of Q_2 is taken negative by the charge on C_1, Q_2 thereby being switched off. This capacitor now discharges through the 10K resistor R_8 and at a time approximately equal to $0.7\,RC$, Q_2 will switch on again causing a similar switching action to take place at Q_3. This, in turn, will be followed by Q_4 and finally Q_5. The reason for using four of these differentiators instead of the two obviously necessary is that there are delays in each gate. Separating the pulses by a controlled interval ensures that the circuit has "settled" before each instruction (i.e., "store" or "reset") is given

Timer Control

This circuit has been arranged so that once the "run" and "stop" inputs have been connected to ground neither will operate again until the circuit has been reset. By doing this the effects of any multiple contacts, such as bounce in switches, can be eliminated. The circuit works as follows: When the "Prime" switch is momentarily closed, the RS flip-flops $IC_{2a,b}$ and $IC_{2c,d}$ are set so that the output of IC_{2a} is at 0 and IC_{2d} at 1. The 0 causes the output of IC_{4c} to be at 1, IC_{4d} at 0, and thus no pulses from the master clock can get through the count gate IC_{4a}. When the "Run" input is grounded, the RS flip-flop $IC_{2a,b}$ changes state so that the output at IC_{2a} goes to 1, enables the count gate IC_{4a}, and the counter starts counting clock pulses. When the "Stop" input is grounded, the output at IC_2 goes to 0, disabling the count gate and thus stopping the counter. The output of IC_{4d} also goes to 0, triggering the timing chain described above.

In order to permit the "wired-OR" function to be implemented, IC_1 and IC_4 use "free collector" gates type SN7403. However, DTL gates such as the $\mu A946$ can be used here in which case the 1K pull-up resistor can be omitted. However, in order to allow C_1 to charge up quickly, the pull-up resistor on the output of IC_{4d} should be retained.

14.4 INPUT WAVE SHAPER AND MASTER CLOCK

Input Wave Shaper

The circuit of the input wave shaper is shown in Fig. 14.4-1. This has been designed to accept a wide range of input voltages, 20 mV to 100 V, so that it was necessary to include some form of limiting as near to the input as possible. This obviates the necessity for gain control, but it has the disadvantage that the input resistance varies with input level. The variation, how-

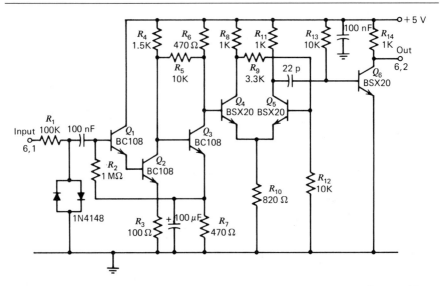

Figure 14.4-1 (Diag. 6) Input wave shaper. [From Waddington, D. E. O'N., *Wireless World*, 162, (April, 1973).]

ever, should not pose any real problems since, for inputs below 600 mV peaks, the resistance is of the order of 1 MΩ and, above this level, it never falls below 100K. The high-frequency response of this arrangement could be improved, if required, by connecting a small capacitor in parallel with the 100K resistor (R_1), but it is difficult to assess the exact value needed. The input amplifier consists of three transistors Q_1–Q_3. The ac gain is set at 100 by the emitter resistor R_3 and the shunt feedback resistor R_5, and the dc conditions are set by the overall feedback via the 1 MΩ resistor R_2. The output from the collector of Q_3 is fed directly to the Schmitt trigger Q_4 and Q_5. In order to control the shape of the waveform to be counted, the output from the trigger is differentiated and amplified by Q_6. The pulse width at the output will be approximately 170 ns, so that the highest frequency that can be counted will be limited to about 5 MHz.

Master Clock

In the circuit of the master clock, shown in Fig. 14.4-2, it will be noted that both the master clock and the input wave shaper run off the +5-V rail so that, if only the counter/timer functions are required, the plus and minus 12-V sections of the power supply can be emitted.

Because the master clock provides the standard for all the time and frequency measurements, the accuracy of these is critically dependent on this section. Accordingly, a crystal-controlled oscillator is used to provide the basic standard. The crystal is operated in the series resonant mode, coupling the output at the collector of Q_1 to the base of Q_2. The amplitude of the positive feedback to the base of Q_1 is held substantially constant by the

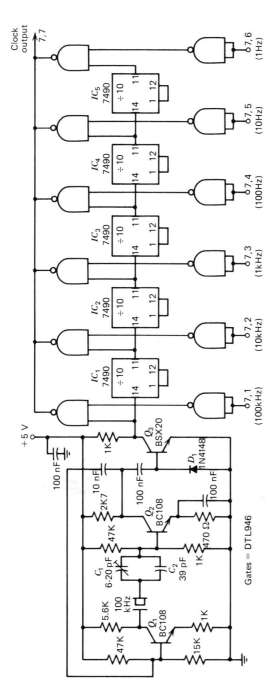

Figure 14.4-2 (Diag. 7) Master clock and divider chain. [From Waddington, D. E. O'N., *Wireless World*, 162 (April, 1973).]

limiting effect of the diode D_1 and the base–emitter diode of Q_3. Selecting the value of the parallel combination of C_1 and C_2 to be 50 pF will give a frequency of 100 kHz $\pm$ 3 Hz or an accuracy of 3 parts in 10^5. The output, which is taken from the collector of Q_3, has a rectangular waveform and is suitable for feeding directly to the logic gates and frequency-divider chain. If the counter functions of the instrument are not needed, the frequency-divider chain can be omitted, and this output is then fed directly to the input of the counter chain (base of Q_1, Fig. 14.3-3). Normally, however, this output is fed to the frequency-divider chain that consists of five decade dividers connected in cascade thus providing output frequencies of 100 kHz, 10 kHz, 1 kHz 100 Hz, 10 Hz, and 1 Hz. Each frequency output is fed to one input of a two-input gate, the outputs of the gates being connected together to give a "wired-OR" function. The other input to each gate is fed from an individual inverter. Grounding the input to an inverter causes its output to go high, thus enabling the gate to which it is connected and selecting the appropriate frequency to be fed to the output. As a result, the six frequencies can be selected by a single-pole six-position switch that connects the appropriate inverter inputs to ground. It should not be necessary to hold the ungrounded inverter inputs high as there should not be enough electrical noise within the instrument to cause errors. For this frequency selection three-quad two-input gates type μA946 were employed since they can be used to give the "wired-OR" function without additional resistors. However, if pull-up resistors are added to the circuit, open collector TTL gates type SN7403 can be used.

14.5   DUAL-RAMP DIGITAL VOLTMETER

Input Amplifier, Rectifier, and Polarity Indicator Circuits

This circuit is shown in Fig. 14.5-1. The input attenuation is divided into two sections: a 10:1 switch of amplifier gain by negative feedback, and a 100:1 attenuator at the input to the amplifier. The ACV/DCV switch at the input has been included to compensate for the fact that average-reading voltmeters are normally calibrated in rms so that a factor of 1.11:1 has to be allowed for. This also makes it possible to switch an isolating capacitor in series with the input when measuring ac voltage. In order to achieve a high input resistance, a differential pair of FETs is used to feed an op-amp type μA741, the gain being stablized by negative feedback. In order to reduce lead lengths, this sections was built on the back of the voltage-range-selector switch.

The output from the input amplifier is fed to a circuit, which is sometimes called an "absolute value rectifier" since it gives an output that is effectively equal to the modulus of the input signal. In more familiar terms, it is a full-wave rectifier that works down to dcV. The action is as follows:

If a sine wave (A) is applied to the input, as in Fig. 14.5-2, the signal at

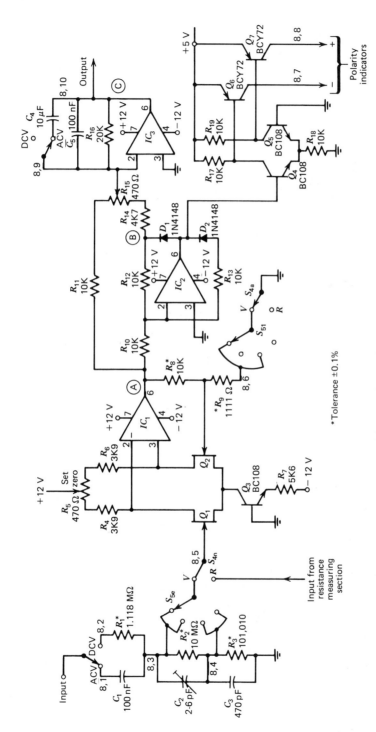

Figure 14.5-1 (Diag. 8) Input amplifier, rectifier, and polarity indicator circuits. [Waddington, D. E. O'N., *Wireless World*, 163 (April, 1973).]

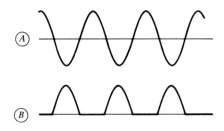

Figure 14.5-2   Rectifier waveforms.

B will consist of a half-wave rectified waveform having the same peak amplitude as A but inverted in phase. These two signals are added at the input of the amplifier IC_3 in the ratio 2:1 so that the output (C) is a full-wave rectified version of the input. Thus the output at C will be negative regardless of the input polarity. The 470-Ω variable resistor R_{15} is used to ensure that A and B are added in the correct proportions. A Signetics type N5556V integrated circuit is used for IC_2 since it has a better slew rate than μA741C and thus gives a better frequency response. The output from pin 6 of IC_2 is used to drive a comparator Q_4/Q_5, which switches the polarity indicators. With a dcV input, only one polarity indicator will light at a time, but with acV both will light giving an indication that the input is alternating. A capacitor is connected between the input and output of IC_3 to give a low-pass filter effect. When the instrument is switched to acV, the value of this capacitor is increased to 10 μF to provide smoothing. Of necessity, this increases the time necessary for the reading to settle.

Dual-Ramp Analog-to-Digital Converter (ADC)

The ADC section is shown in Fig. 14.5-3a, where the three-position switch shown in Fig. 9.8-2 is replaced by two FET switches Q_8 and Q_9, the third position of the switch being provided by the 33K ground-return resistor, R_9, when both FETs are turned off. The reference voltage is provided by a low-temperature-coefficient Zener diode D_3. The absolute voltage of this Zener diode is not important, as its effective value can be set by adjusting the 6.8K variable resistor R_2. In order to prevent oscillations occurring at the instant when the output from the integrator IC_4 reaches zero, the comparator IC_5 has been made to act as a Schmitt trigger by positive feedback from the gate IC_{6a}. The timing input to this gate had to be included to enable the zero of the instrument to be set and overrides the backlash (hysteresis) that is essential to the operation of the Schmitt trigger, so ensuring that the comparator works as a zero-crossing detector (see the final paragraph). The "store" command is derived by differentiating the output of IC_{6b} and amplifying it using Q_{12}.

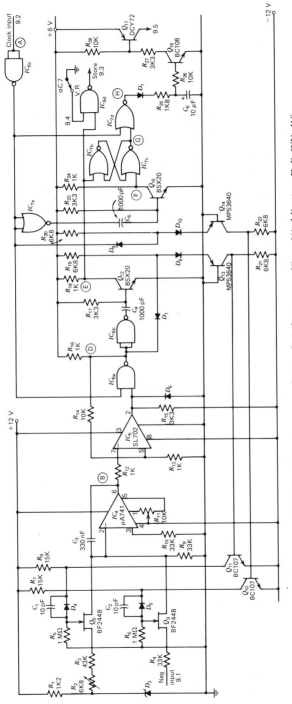

Figure 14.5-3 (a) (Diag. 9) Dual-ramp analog-to-digital converter. [From Waddington, D. E. O'N., *Wireless World*, 163 (April, 1973).]

The gates of IC_7 are used to give the out-of-range indication as follows: The RS flip-flop in Fig. 14.5-3a is set by pulse F in Fig. 14.5-3b, which is generated by differentiating the falling edge of the clock input. It is then reset by the "store" pulse E so that IC_{7d} is again disabled, and when the clock goes positive once more, there is no output at H. However, if E does not arrive in time (i.e., the input voltage is too high), both inputs to IC_{7d} go low, so that its output, H, goes high. This turns on Q_{16} and thus causes the overload lamp to light. The capacitor C_6 acts as pulse stretcher to keep the lamp alight even though the output at H does not remain high continuously during the overload period but switches on and off at the clock rate. "NOR" gates have been used in this section to reduce the package count. Had "NAND" gates been used, it would have been necessary to put inverter in series with the inputs to the RS flip-flop.

The inverting zero-crossing detector is additionally explained as follows. A basic zero-crossing detector determines if an input voltage is greater than zero or less than zero. In response to this determination, the output voltage can assume only two possible states. The output assumes the positive state if $v_i < 0$ and the negative state if $v_i > 0$. A basic inverting zero-crossing detector with hysteresis is shown in Fig. 14.5-4. Since R_f is connected from the op-amp output to the noninverting input, it provides a small amount of positive feedback. Hysteresis is provided by the feedback path. Note that v_0 vs v_i always travels clockwise around the box. If v_i is less than zero and is becoming more positive, it has to cross zero and rise to $\Delta v_1 = R_p V_{Z1}/(R_p + R_f)$ before the output switches states. The arrows in Fig. 14.5-4b show that this is the only path by which this change if state can occur. After

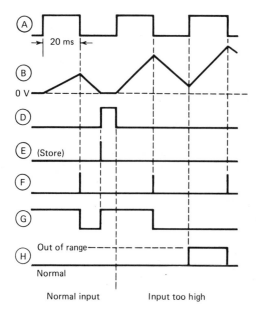

Figure 14.5-3 (b) Operation of the "out-of-range" indicator.

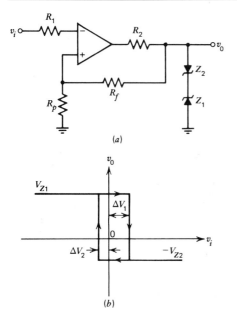

(a)

(b)

Figure 14.5-4 (a) Inverting zero-crossing detector with hysteresis and (b) its transfer function.

the transition has taken place, the output cannot return to the positive state unless v_i has a negative noise spike of at least

$$v_i(\text{noise, peak}) = \Delta v_1 + \Delta v_2 = \frac{R_p(V_{Z1} + V_{Z2})}{R_p + R_f}. \qquad (14.5\text{-}1)$$

14.6 RESISTANCE- AND CAPACITANCE-MEASURING CIRCUITS

Resistance-Measuring Circuit

The resistance-measuring circuit, shown in Fig. 14.6-1, uses the main voltmeter circuit with the input attenuator disconnected and the gain of the input amplifier set to $\times 1$, together with a constant-current source that can be switched to give four different values of current, 1 μA (1 MΩ), 10 μA (100K), 100 μA (10K) and 1 mA (1K). The simplest possible circuit was used here: a transistor with its base tied to a reference voltage and switched resistors in series with its emitter. The reference voltage is provided by a 4.7-V Zener diode, chosen because its temperature coefficient tracks that of base–emitter voltage of a transistor fairly closely. The transistor chosen is a BC252, a *pnp* transistor with a very high β even at low values of collector current. For the resistance measurement, the input of the FET voltmeter and amplifier is connected directly to the collector of the current source transistor. The resistance to be measured is then connected because this junction and ground, and the voltmeter reads the voltage drop across the resistor. The voltage range and the current through the resistor have been chosen so that

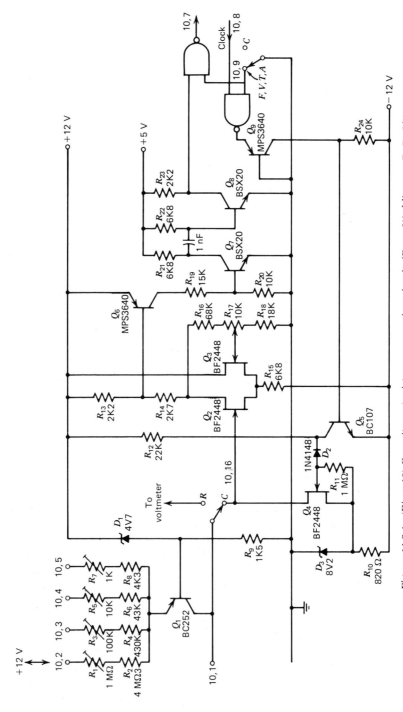

Figure 14.6-1 (Diag. 10) Capacitance/resistance-measuring circuit. [From Waddington, D. E. O'N., *Wireless World*, 164 (April, 1973).]

508

the instrument reads resistance directly. As the input resistance of the FET amplifier is greater than 100 MΩ, it does not affect the accuracy of any great extent.

Capacitance Meter

The capacitance meter uses the same current source as the resistance-measuring circuit. With a 1 μA charging current, the voltage drop across a 2000 pF capacitor will reach 10 V in 20 ms. As the time required for the counter to reach full scale is 20 ms (= 2 kc/100 kHz), the comparator is arranged so that it detects when the voltage across the capacitor reaches 10 V. With only plus and minus 12-V lines this 10-V requirement could have been awkward. However, by making the initial charge on the capacitor -8.2 V, a voltage developed across the Zener diode D_3, the positive excursion of the voltage across the capacitor need only be 1.8 V. This is suitable to feed to the input of Q_2, which is cross coupled with Q_3 to form a Schmitt-trigger circuit. FETs are used here since their high input impedance does not load the capacitor. The backlash in the trigger circuit does not affect the accuracy of the measurement since the comparator only detects the transition as the voltage across the capacitor goes positive. The discharge of the capacitor is effected by switching Q_4 on. The output from the trigger circuit is amplified by Q_6 before it is differentiated and limited by Q_7. The gates are used to switch the clock input and the output of this section on or off as required.

FET Switches

In both the voltmeter and capacitance meter, use is made of FET switches. In order for these switches to work adequately, two conditions must be satisfied: when the switch is "on," the input signal must never reach sufficient amplitude to start to turn the switch off and, conversely, when the switch is off, the signal amplitude must not be sufficient to turn it on! Hence, to ensure that the switch operates correctly, the input signal to the switch should be as high as practical and the switching signal to the gate of the FET should be as large as possible, in this case 24-V peak-to-peak. In the voltmeter section the signal swing is minimized by placing the FET at the "virtual ground" point of the integrator. In the capacitance meter, the FET is connected so that the voltage developed across it cannot affect the switching. In both cases, the switching control voltage at the gate is made to switch between the positive and negative rails. In order to do this it is necessary to translate the voltage from the logic circuits (0 and $+5$ V) to a suitable level. This is done as shown in Fig. 14.6-2. When the output from the gate IC_1 is low, D_1 is turned on so that the base–emitter junction of Q_1 is reverse biased and Q_1 is turned off, turning Q_2 off so that its collector "goes" to the positive rail. This turns the FET Q_3 on. When the output from IC_1 goes high, D_1 is reverse biased and a current determined by the value of R_1 flows into the emitter of Q_1, turning it on. This, in turn, bottoms Q_2 so that its collector voltage goes to the negative, turning the FET Q_3 off. The diode

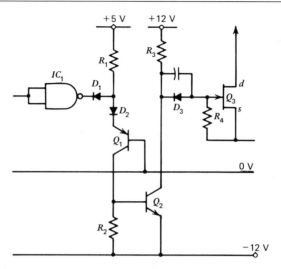

Figure 14.6-2   Method of driving FET switches from 5-V lines.

D_3 is included in the circuit so that, when the FET is in the "on" condition, no gate current that could introduce errors can be drawn. If this circuit is used with DTL gates, provided that no other inputs are to be driven, R_1, D_1, and D_2 can be omitted.

14.7   INTERCONNECTION OF THE MULTIMETER SECTIONS

The interconnection of the multimeter sections is shown in Fig. 14.7-1. The instrument has two function switches, one for the Frequency/Time/Period (FIP) section and the other for the Voltage/Resistance/Capacitance (VRC) section, each having a position that transfers control to the other. Similarly there are two range switches whose functions are shown in Table 14.7-1. Range indication is accomplished almost entirely by decimal point switching.

In general, the interconnection of the boards within the instrument is not critical but it is as well to observe the following precautions:

1. Wire the supply lines to each board with separate leads from the power supply so that there are no common supply paths.

2. Wire the grounds separately to a common point at the junction of the chassis, input smoothing capacitors, and center tap of mains transformer (if this kind is used). In addition to $+12$ V and -12 V supplies, a 5-V dc source is required. The 5-V earth should, of course, go to its own stabilizer earth point.

3. Do not screen the FP input lead as this will impair the sensitivity at high frequencies.

4. Keep the lead from the capacitance-measuring input terminal as short as possible to reduce strays.

Table 14.7-1 Range Switches

FTP

Function	Position 1	Position 2	Position 3	Position 4	Position 5	Position 6	Wafer
Frequency (F)	Test	MHz	MHz	kHz	kHz		
Time (T)	ms	ms	ms		S	S	S_{1a}
Period (P)	ms	ms	ms		S	S	S_{1b}
Select T		10 kHz	1 kHz	100 Hz	10 Hz	1 Hz	S_{1c}
Base	100 kHz						
Decimal point frequency	—	4.15	4.14	4.16	4.15	4.14	
Decimal point T/P	4.15	4.16	—	4.15	4.16	—	

VRC

Function	Position 1	Position 2	Position 3	Position 4	Wafer
Voltage	mV	V	V	V	
Resistance	K	K	K	K	
Capacitance	nF	nF	nF	nF	
Decimal point volts	4.16	4.14	4.15	4.16	S_{5a}
Decimal point R/C	—	4.16	4.15	4.14	S_{5b}
I (source) (R)	10.2 (1 µA)	10.3 (10 µA)	10.4 (100 µA)	10.5 (1 mA)	S_{5c}
I (source) (C)	10.5 (1 mA)	10.4 (100 µA)	10.3 (10 µA)	10.2 (1 µA)	S_{5d}
Voltmeter input attenuator	8.3	8.3	8.4	8.4	S_{5e}
Voltmeter gain	8.6	—	8.6	—	S_{5f}

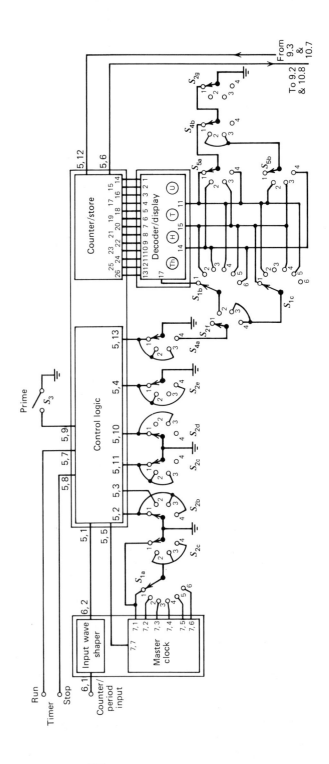

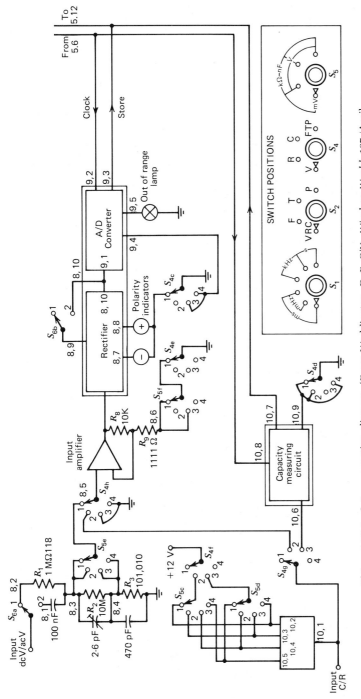

Figure 14.7-1 Interconnection diagram. [From Waddington, D. E. O'N., *Wireless World*, 167 (April, 1973).]

14.8   7106/7107 SINGLE-CHIP DIGITAL METERS

Introduction

The 7106 and 7107 ICs (manufactured by Teledyne, Intersil, etc.) contain all the active circuitry for a $3\frac{1}{2}$ digit panel meter on a single CMOS chip. The 7106 is designed to interface with a liquid-crystal display (LCD), while the 7107 is intended for the light-emitting-diode (LED) display. Both circuits contain BCD to seven-segment decoders, display drivers, a clock, and a reference. To build a high-performance panel meter (with auto-zero and auto polarity features), it is only necessary to add a display, four resistors, four capacitors, and an input filter if required.

The 7106/7107 ADCs are of integration (dual-slope) type. The 7106 with liquid-crystal display and 7107 with LED display are shown in Figs. 14.8-1 and 14.8-2, respectively. The circuit board layouts and assembly drawings are shown in Fig. 14.8-3 (7106) and Fig. 14.8-4 (7107). The 7106 EV/KIT and 7107 EV/KIT are offered by Teledyne Semiconductor; each kit contains the appropriate IC, a circuit board, a display (LCD/1706, LED/1707), passive components, and miscellaneous hardware.

Liquid-Crystal Display (7106)

The 7106 generates the symmetrical square wave to the back plane (BP) internally. The user should generate the decimal point from the plane drive by inverting the BP (pin 21) output.

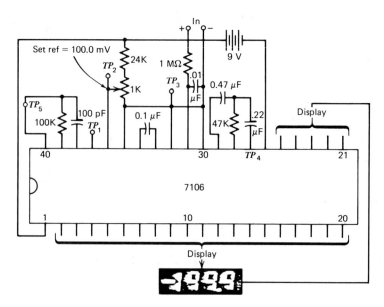

Figure 14.8-1   7106 with liquid crystal display. (Courtesy of Teledyne Semiconductor.)

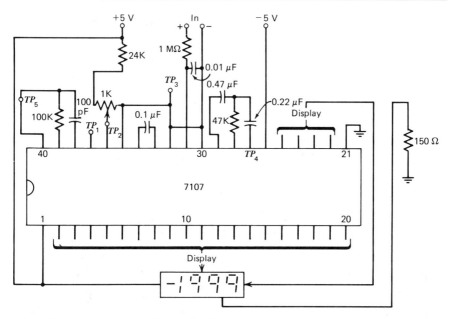

Figure 14.8-2   7107 with LED display. (Courtesy of Teledyne Semiconductor.)

In some displays, a satisfactory decimal point can be achieved by tying the decimal plane to COMMON (pin 32). This pin is internally regulated at about 2.8 V below V^+. Prolonged use of this technique, however, may permanently burn-in the decimal, because COMMON is not exactly midway between BP high and BP low. In applications where the decimal point remains fixed, a simple MOS inverter shown in Fig. 14.8-5 can be used. For instruments where the decimal point must be shifted, a quad exclusive OR gate shown in Fig. 14.8-6 is recommended. Note that in both instances, TEST (pin 37, TP1) is used as V^- for the inverters. This pin is capable of sinking about 1 mA and is approximately 5 V below V^+. The BP output (pin 21) oscillates between V^+ and TEST.

Light-Emitting-Diode Display (7107)

The 7107 will sink 8 mA per segment. This drive produces a bright display suitable for most applications. A fixed decimal point can be turned on by tying the appropriate cathode to ground through a 150-Ω resistor. The circuit boards supplies with the kits will accommodate either HP 0.3 displays or the MAN 3700 types. Note that the HP has the decimal point cathode on pin 6, whereas the MAN 3700 has the decimal point cathode on pin 9. Not all the decimal points are brought out to jumper pads. It may be necessary to wire directly from the 150-Ω resistor to the display. For multiple range instruments, a 7400 series CMOS quad gate should be used.

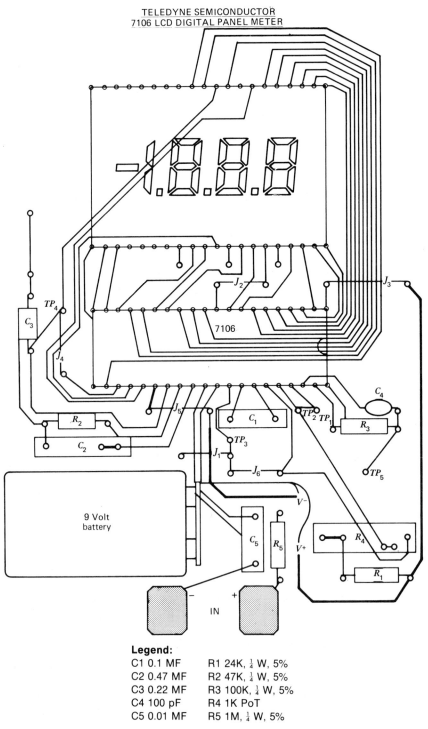

Legend:

C1 0.1 MF	R1 24K, $\frac{1}{4}$ W, 5%
C2 0.47 MF	R2 47K, $\frac{1}{4}$ W, 5%
C3 0.22 MF	R3 100K, $\frac{1}{4}$ W, 5%
C4 100 pF	R4 1K PoT
C5 0.01 MF	R5 1M, $\frac{1}{4}$ W, 5%

Figure 14.8-3 7106—Circuit board layout and component placement. (Courtesy of Teledyne Semiconductor.)

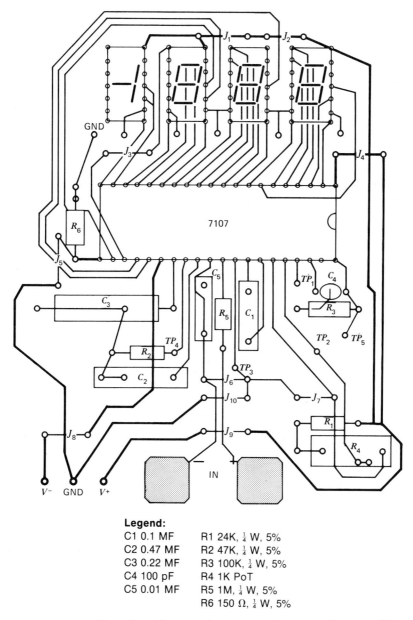

Legend:

C1 0.1 MF	R1 24K, $\frac{1}{4}$ W, 5%
C2 0.47 MF	R2 47K, $\frac{1}{4}$ W, 5%
C3 0.22 MF	R3 100K, $\frac{1}{4}$ W, 5%
C4 100 pF	R4 1K PoT
C5 0.01 MF	R5 1M, $\frac{1}{4}$ W, 5%
	R6 150 Ω, $\frac{1}{4}$ W, 5%

Figure 14.8-4 7107—Circuit board layout and component placement. (Courtesy of Teledyne Semiconductor.)

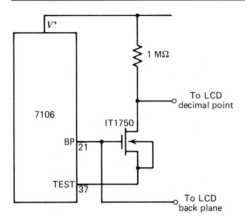

Figure 14.8-5 Simple MOS inverter for fixed decimal point.

Full Scale Readings

200 mV Full Scale. The kits have been optimized for 200 mV Full Scale. The component values supplied are those specified in Figs. 14.8-3 and 14.8-4. The Teledyne Semiconductor input noise of 10 μV is the best of any competitive IC as is shown by the complete absence of the last digit jitter.

2.000 V Full Scale. The component values in Table 14.8-1 change the integrator time constant and reference and the auto zero capacitor time constant. In addition, the decimal point jumper should be changed so that the display reads 1.999.

Clock

Setting the clock oscillator at precisely 48 kHz will result in the optimum line frequency 60-Hz noise rejection (Fig. 14.8-7). Since the integration period is an integral number of the line frequency period, the *RC* oscillator

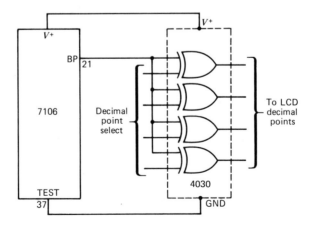

Figure 14.8-6 Exclusive-OR gate for decimal point drive.

Table 14.8-1　Component Values for Full-Scale Options

Component (Type)	200.0 mV full scale	2.000 V full scale
C_2 (Mylar)	0.47 μF	0.047 μF
R_1	24K	1.5K[a]
R_2	47K	470K

[a] Changing R_1 to 1.5K will reduce the battery life of the 7106 kit. As an alternative, the potentiometer can be changed to 25 K.

supplied in the kit runs at approximately 48 kHz, giving a measurement frequency of three readings per second. Countries with 50-Hz line frequencies should set the clock to 40 kHz by increasing the value of the 100K resistor across pins 39 and 40 to 120K.

An external clock can also be used. In the 7106, the internal logic is referenced to TEST. External clock waveforms should therefore swing between TEST and V^- (Fig. 14.8-8a). In the 7107, the internal logic is referenced to GND so any generator whose output swings from ground to $+5$ V will work well (Fig. 14.8-8b).

Capacitors

The dual-slope technique cancels the effects of long-term stability and temperature coefficient. The integration capacitor should be a low-dielectric-loss type. Inexpensive polypropelene capacitors have the low-dielectric-loss characteristics and are recommended. Mylar capacitors may be used for C_1 (reference) and C_2 (auto zero).

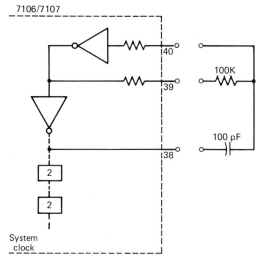

Figure 14.8-7　7106/7107 internal oscillator/clock.

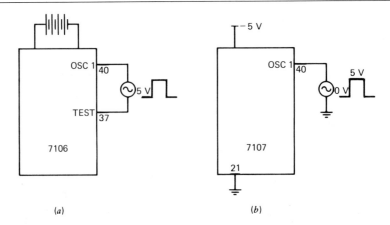

Figure 14.8-8   External clock options: (a) 7106; (b) 7107.

Reference

The voltage between V^+ and COMMON is internally regulated at about 2.8 V. This reference is adequate for many applications.

For 200 mV full scale, the voltage applied between REF HI and REF LO should be set at about 100 mV. For 2.000 V full scale, set the reference voltage at 1.0 V. The reference inputs are floating and the only restriction on the applied voltage is that it should lie in the range V^- to V^+. For calibration, place 190.0 mV on input and adjust REF pot (R_4) for 1900 readout.

For greater temperature stability, an external reference can be added as shown in Figs. 14.8-9a and 14.8-9b.

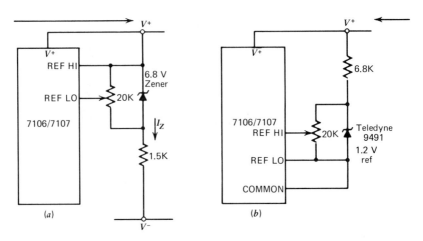

Figure 14.8-9   Using an external reference.

Power Supplies

The 7106 kit is intended to be operated from a 9-V battery. INPUT LO is shorted to COMMON, causing V^+ to sit at 2.8 V positive with respect to INPUT LO, and V^- at 6.2 V negative with respect to INPUT LO.

The 7107 kit should operated from ± 5 V. Noisy supplies should be by-passed with 6.8-μF tantalum capacitors to ground at the point where the supplies enter the board.

If a -5 V source is unavailable, a suitable negative supply can be gen-erated locally using the circuit shown in Fig. 14.8-10.

Preliminary Tests

1. Rubbing alcohol or another appropriate cleaning agent should be used to remove impurities on PC boards.

2. In order to ensure that unused segments on the LCD displays do not turn on, tie them to the back plane (pin 22).

3. Auto Zero—With the inputs shorted, the display should read zero. The negative sign will be on about one-half of the time, showing the input to be exactly zero volts.

4. Polarity—A negative sign indicates a negative reading. No sign in-dicates a positive reading.

5. Overrange—For inputs greater than full scale, only 1 or -1 will be displayed. The three least significant digits will be suppressed.

6. Calibration—The instrument should be calibrated at 1900 counts by using a high-quality $4\frac{1}{2}$ DVM.

Applications

Input Attenuator. To measure voltages greater than 2 V, an input atten-uator is needed as shown in Fig. 14.8-11. The full scale sensitivity is given

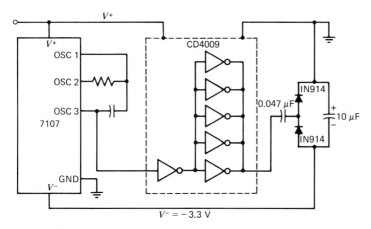

Figure 14.8-10   Generating negative supply from $+5$ V.

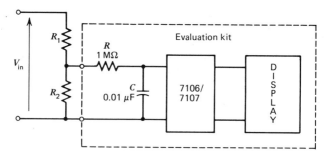

Figure 14.8-11   Input attenuator for $V_{in} \geq 2.0$ V. The RC filter introduces a negligible 1 μV error.

by

$$V_{in} \text{ (full scale)} = 1.999 \, V_{ref} \times \frac{R_1 + R_2}{R_2} . \tag{14.8-1}$$

It is important that R_1 and R_2 remain fixed for the calibration period of the instrument. Metal film resistors with good long-term drift characteristics and low temperature coefficients are recommended.

The input attenuator reduces the input resistance of the circuit from greater than 10^{12} Ω to $(R_1 + R_2)$. This place an upper limit of about 10 MΩ on the input resistance that can readily be achieved when using an attenuator before the A/D input current causes offset errors.

To measure full scale voltage less than 199.9 mV, an op-amp is used prior to the 7106/7 inputs. Note that the auto zero circuitry within the IC cannot take care of the op-amp offset or voltage drift. For example, the use of the 308A will add 1 μV/°C voltage drift typical.

Figure 14.8-12 shows a circuit with ±20 mV full scale and an input re-

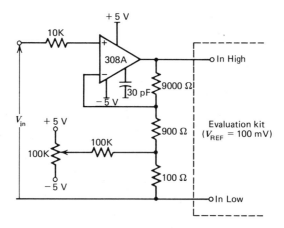

Figure 14.8-12   An op-amp used prior to the 7106/7 inputs for measuring full scale voltage < 199.9 mV.

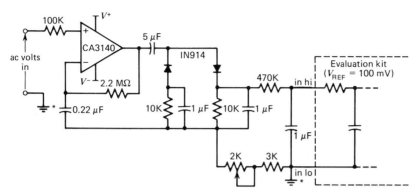

Figure 14.8-13 Ac-to-dc converter. GND*—in the absence of split supply operation, TEST (pin 37) can be used as ground.

sistance greater than 10 MΩ. For scale factors between 100 mV and 1 mV per least significant digit (LSD), the reference voltage required is given by

$$V_{ref} = (\text{Voltage change represented by one LSD}) \times 10^3. \quad (14.8\text{-}2)$$

For scale factors greater than 1 mV/LSD, the most straightforward approach is to use an input attenuator in conjunction with a 1-V reference.

AC Voltage Measurements.   It is necessary to build an AC-to-DC converter as shown in Fig. 14.8-13 in order to measure AC voltages with the 7106/7.

Multirange DVMs.   One of the commonly used multirange DVM circuits is shown in Fig. 14.8-14. In this circuit, any switch contact resistance appears

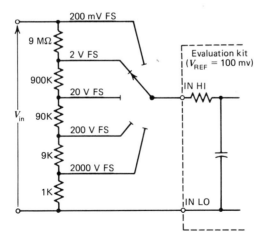

Figure 14.8-14 Multirange digital voltmeter. *Caution:* High voltages can be lethal. Proper operating precautions must be observed by the user. Teledyne assumes no liability for unsafe operation.

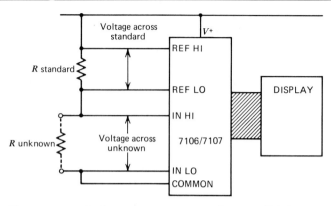

Figure 14.8-15 Resistance measurement with 7106/7 digital meter.

in series with the 7106/7 input resistance. Since the input resistance is greater than $10^{12}\ \Omega$, errors due to the switch are negligible.

Resistance Measurements. A typical resistance-measurement circuit is shown in Fig. 14.8-15. The unknown resistance is put in series with a known standard and a current passes through the pair. The voltage developed across the unknown is applied to the input (between IN HI and IN LO), and the voltage across the known resistor applied to the reference input (REF HI and REF LO). If the unknown equals the standard, the display will read 1000. The displayed reading can be determined from the following expression:

$$\text{Displayed reading} = \frac{R\ \text{unknown}}{R\ \text{standard}} \times 1000. \qquad (14.8\text{-}3)$$

The display will overrange for R unknown $\geq 2 \times R$ standard.

Current Measurements. The use of a shunt resistor converts the current to a voltage. The relationship between the current and the displayed reading

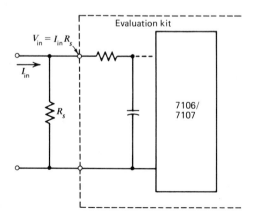

Figure 14.8-16 Current-measurement circuit.

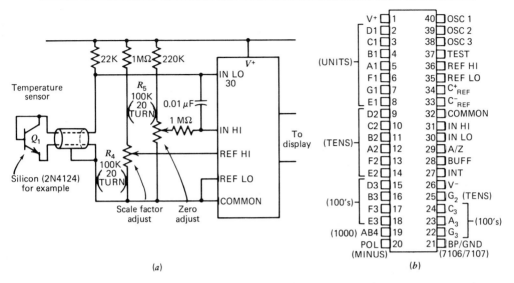

Figure 14.8-17 (a) Digital thermometer circuit and (b) 7106/7 pin assignment. (Courtesy of Teledyne Semiconductor.)

for the circuit of Fig. 14.8-16 is found by

$$\text{Displayed reading} = \frac{I_{\text{in}} \times R_s}{V_{\text{ref}}} \times 1000. \qquad (14.8\text{-}4)$$

When measuring current the 199.9 mV scale is used. This limits the voltage drop to 100 μV per count.

Temperature Measurements. Figure 14.8-17a shows a digital-thermometer circuit with a diode-connected transistor as the temperature transducer. V_{BE} has a temperature coefficient of -2.1 mV/°C. A scale factor of 0.1°C/count may be obtained by setting the reference at 210 mV. At 0°C and 100 μA bias current, the diode-connected transistor will have a forward voltage drop of approximately 550 mV. A fixed 500 mV source is set up to offset the diode drop. In the circuit of Fig. 14.8-17, adjust R_5 to give 000.0 output reading with Q_1 at 0°C. Then adjust R_4 for a 100.0 reading with Q_1 at 100°C.

BIBLIOGRAPHY

Comer, D. J., *Electronics with Integrated Circuits*. Addison-Wesley, Reading, Massachusetts, 1981, Chap. 7.

Intersil ICL7106/7107 3-1/2 Digit Single Chip A/D Convertor Data Acquisition. Intersil, Inc., 10710N, Tantau Ave., Cupertino, California 95014, 1980.

Stout, D. F. and M. Kaufman, *Handbook of Operational Amplifier Circuit Design*. McGraw-Hill, New York, 1976, Chap. 6.

Teledyne 7106/7107 Digital Meter Applications Including KIT Assembly Instructions, AN-11. Teledyne Semiconductor, 1300 Terra Bella Ave., Mountain View, California 94043, Dec., 1981.

Waddington, D. E. O'N., "Digital Multimeter," *Wireless World*, 155–170 (April, 1973).

Questions

14-1 Describe the principle of dual-ramp analog-to-digital converter.

14-2 What is the difference between period and frequency measurements?

14-3 Discuss the timer control operation.

14-4 How does the master clock work?

14-5 How does the rectifier in Fig. 14.5-1 work?

14-6 Explain the inverting zero-crossing detector with hysteresis.

14-7 Draw a simple current measurement circuit and explain its operation.

14-8 Discuss the operation of the capacitance meter shown in Fig. 14.6-1.

14-9 Describe the method of driving FET switches from 5-V lines.

14-10 What precautions should be observed when the boards within the instrument are interconnected?

14-11 What are the preliminary tests in building the 7106/7 digital meter?

14-12 How does the AC-to-DC converter in Fig. 14.8-13 work?

14-13 Describe the operation principle of the thermometer shown in Fig. 14.8-17a.

Problems

14.2-1 A constant voltage of 2 V is applied to the input of the integrator in Fig. 9.8-2a for 50 ms. Find the output voltage if $R_1 = 50\text{K}$ and $C_1 = 0.5\ \mu\text{F}$.

Answer.

$$V_o = \frac{-1}{R_1 C_1} \int_0^t V_i\, dt = \frac{-V_i}{R_1 C_1} \int_0^{0.05} dt = \frac{(-2\text{V})t}{(50\text{K})(0.5\ \mu\text{F})} \Big|_0^{0.05}$$

$$= \frac{(-2\ \text{V})(0.05\ \text{s})}{25 \times 10^{-3}} - 0 = \frac{-0.1 \times 10^3}{25} = -4\ \text{V}.$$

14.5-1 In Fig. 14.5-4, the ideal positive and negative trip points are represented by Δv_1 and Δv_2, respectively. Find Δv_1 and Δv_2.

Answer. In the ideal inverting zero-crossing detector, the positive trip point for v_i is

$$\Delta v_1 = \frac{R_p V_{Z1}}{R_p + R_f},$$

and the negative trip point for v_i is

$$\Delta v_2 = \frac{-R_p V_{Z2}}{R_p + R_f}.$$

14.8-1 The multirange digital voltmeter of Fig. 14.8-14 has the ranges of (a) 200 mV FS, (b) 2 V FS, (c) 20 V FS, (d) 200 V FS, and (e) 2000 V FS. Find the values of R_1 and R_2 for each range, using Eq. (14.8-1).

Answer. (a) $R_1 = 0$, $R_2 = 10$ MΩ; (c) $R_1 = 9900$K; $R_2 = 100$K.

14.8-2 Assume that $R_s = 1000$ Ω, $I_{in} = 200$ μA, and $V_{ref} = 100$ mV in the current-measurement circuit of Fig. 14.8-16. Find the displayed reading.

Answer. 2000.

15

Introduction
to the TV Terminal
Using a Microprocessor

15.1 BASIC TV TERMINALS

The TV terminal (TVT) is primarily used as the terminal of a computer or microprocessor (MPU) for displaying the output information on the TV screen. There are two methods for TV terminal interfacing: one is that using an' rf modulator, and the other is of direct video technique. A typical TVT block diagram is shown in Fig. 15.1-1.

The memory stores the external input information, which, after a suitable conversion, will be displayed when applied to the TV set. One of the most commonly used memories is 2102 MOS RAM. In order that our information can be received by the usual TV set, the TVT must have a system timing circuit for converting and transmitting the signals. Since the timing frequency generally exceeds the extent capably handled by the microprocessor, it is necessary to add an auxiliary circuit containing seven or eight TTL or CMOS devices for the horizontal synchronization between line and line.

The information stored in the memory is primarily of characters. The character generator and the video output stage are used to divide these characters into a group of dots, which appear on the screen as a view of complete characters. The character generator output, passing through a shift register, becomes a number of serial data. These data mix with the sync signal, resulting in the output video signal. This video signal can be applied to the video amplifier stage of a TV set, or modulated with an rf carrier and then sent to the antenna terminal.

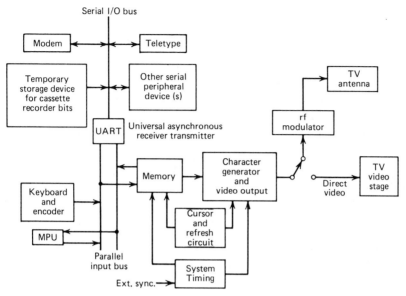

Figure 15.1-1 Block diagram of TVT using either an rf modulator or the direct video method.

The cursor and refresh circuits are used to send the new information into the memory and the corresponding display section. In order that the user knows the location of the next character appearing on the TV screen, the cursor circuit should display a flashing square at a suitable position on the screen. Both the cursor and refresh circuits have two different constructions: one is the frame-rate system in which only one character per sweep ($\frac{1}{60}$ s) is inserted, the other is DMA (direct memory access) system in which the information of a frame is transferred within $\frac{1}{60}$ s. The DMA system is much faster and, therefore, more complicated than the frame-rate type.

In designing TVT we usually prefer parallel operation, with which 8 bits are conveyed each time. The parallel operation is not only faster and cheaper than the serial operation, but also directly compatible with the MPU data bus, keyboard, etc. When using a serial input, the input bits must pass through a serial interface so that the operation is converted to parallel type. An IC designed for this purpose is called the universal asynchronous receiver transmitter (UART).

ICs Used in TVTs

Most ICs used in TVTs are made by the manufacturers listed in Table 15.1-1. The following ICs are used in the basic TV terminal.

1. Baud-Rate Generator. The number of bits per second being transmitted is commonly referred to as the baud rate. Typical rates are 100, 150, 300, 600, and 1200 bits/s or baud. The baud-rate generator is usually crystal

Table 15.1-1   IC Manufacturers and Their Addresses

Advanced Micro Devices 901 Thompson Place Sunnyvale, CA 94086	MOS Technology, Inc. 950 Rittenhouse Road Norristown, PA 19401
American Microsystems, Inc. 3800 Homestead Road Santa Clara, CA 95051	Mostek Corporation 1215 West Crosby Road Carrolton, TX 75006
Cal-Tex Semiconductor 3090 Alfred Street Santa Clara, CA 95950	Motorola Semiconductor Products, Inc. Box 20912 Phoenix, AZ 85036
Cermetek, Inc. 660 National Avenue Mountain View, CA 94040	National Semiconductor 2900 Semiconductor Drive Santa Clara, CA 95051
Electronic Arrays, Inc. 550 East Middlefield Road Mountain View, CA 94043	Nitron Corporation 10420 Bubb Road Cupertino, CA 95014
Exar Integrated Systems, Inc. 750 Palomar Avenue Sunnyvale, CA 94086	RCA Solid State Division Somerville, NJ 08876
Fairchild Semiconductor 464 Ellis Street Mountain View, CA 94040	Rockwell International Microelectronic Device Div. 3310 Miraloma Avenue Anaheim, CA 92803
General Instrument Corporation 600 West John Street Hicksville, NY 11802	Signetics 811 East Arques Avenue Sunnyvale, CA 94086
Harris Semiconductor Box 883 Melbourne, FL 32901	SMC Microsystems 35 Marcus Boulevard Hauppauge, NY 11787
Intel Corporation 3065 Bowers Avenue Santa Clara, CA 95051	Synertek 3050 Coronado Drive Santa Clara, CA 95051
Intersil, Inc. 10900 North Tantau Avenue Cupertino, CA 95014	Texas Instruments Box 5012 Dallas, TX 75222
Monolithic Memories, Inc. 1165 East Arques Avenue Sunnyvale, CA 94086	Western Digital Corporation 3128 Red Hill Avenue Newport Beach, CA 92663

controlled to maintain an exact reference frequency. This frequency is used to control the serial-parallel interface of the UART, telephone modem, teletype (TTY), cassette recorder, etc. Standard Microsystem 5061: 18 pins; two outputs; two reference frequencies. Motorola 14411: 24 pins; 16 different sync signals; frequency selection: $\times 1$, $\times 8$, $\times 16$, $\times 64$.

2. Character Generator. The memory stores the characters of ASCII (American Standard Code for Information Interchange) and a few of timing instructions. They are all converted to the dot-character shapes by the character generator for the information transmitted to the TV. Signetics or GI 2513 converts 64 characters to 5×7 dot-characters; $V_{CC} = +5$ V. Standard Microsystem 5004 contains a section for temporary output storage and uses 7×9 dot-matrix; $V_{CC} = +5$ V. Monolothic Memory 6072 converts 128 characters to 7×9 dot-characters.

3. Keyboard Encoder. It is used to convert the open and close keying signal into the ASCII code. GI and Standard Microsystem 2376; 3600; encoder for 88 keys.

4. Output Drive Stage and Receiver. AMD 26S10 (quad), equivalent to Motorola 3443, TI T5138. Motorola (TI, AMD) 1488, 1489 used in RS232-C standard.

5. PROM. Intersil 5600; NS5330; TI74188: BJT; permanent memory; capacity $= 32 \times 8$ bits; access delay $= 50$ ns. EPROM (Erasable PROM) Intersil (AMD) 1702; NS5202: memory capacity $= 2048$ bits $(= 256 \times 8)$; data are erased when irradiated with ultraviolet light.

6. RAM with 1024 bits, the main memory. Intel (AMD) 2101; TI4039; 22 pins; 256 words; 4 bits/word. Intel (AMD, Fairchild, Synertek) 2102; Intersil 7552; TI4033; 16 pins, 1 bit/word. Intel 5101, CMOS, for long term storage with battery source.

7. UART
      American Microsystem S1883
      General Instruments AY-5-1012
      Signetics 2536
      Standard Microsystems COM 2502
      Texas Instruments TMS 6012
      Western Digital TR 1602

15.2   BASIC TELEVISION PRINCIPLES

Interlaced Scanning for TV Raster

The objective of a TVT is to convert the characters or bits into a video signal that can be received by a TV set. In a TV receiver a television picture is formed by scanning an electron beam across the face of the picture tube. A scanning line is one sweep of the electron beam from the left of the picture tube to the right and is initiated by the horizontal synchronization, as shown in Fig. 15.2-1. The horizontal sync pulse causes termination of a line, horizontal retrace (flyback) of the electron beam back to the left side of the screen, and the start of the new line. During the retrace time the beam is blanked so that the retrace will not be seen. The time allotted for each complete line (including retrace) is 63.5 µs. Of this about 16% is taken by retrace, leaving 53.5 µs of usable line. Video information in the form of a

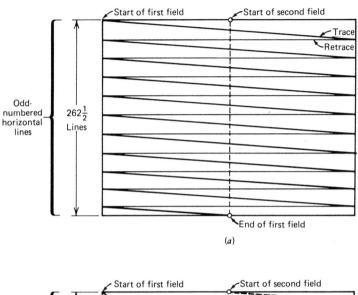

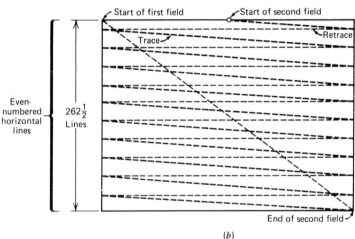

Figure 15.2-1 Interlaced scanning process for the basic raster: (a) structure of the first field; (b) structure of the second field. A complete frame contains two consecutive fields.

voltage fed to the picture tube controls the brightness of the beam as it is swept across the screen.

To minimize flicker, a frame (complete picture) is broken down into two fields. To trace out a field, the electron beam is slowly deflected from the top of the screen to the bottom as it rapidly sweeps horizontal lines. This vertical sweep is allotted 16.67 ms (60 Hz), thus there are $262\frac{1}{2}$ lines in one field. In a manner similar to the horizontal sync, the vertical sync causes the beam to be returned from the bottom of the screen to the top to start a new field. The beam is blanked during vertical retrace which takes about

1250 μs. This leaves 242 usable lines in each field. The two consecutive fields are interlaced with each other, so that the horizontal lines of one field fit in between the horizontal lines of the other field. This is called the interlaced scanning method. The result is 30 frames every second of about 484 usable (525 total) lines each. The horizontal-line frequency is 15,750 Hz for B/W TV and 15,735 Hz for color TV.

The video composite signal applied to the input of the video amplifier contains the information needed to generate the vertical and horizontal sync, blanking, and video. The TVT may simulate this signal by supplying a composite waveform containing the same information normally present except sound.

Interlace Modes

When the same information is stored in both fields of a scan frame, the display is in the interlace mode. This mode is effective in improving the quality of the picture. If the even lines of a character are displayed in the even fields and the odd lines displayed in the odd fields, the mode is an interlace sync and video mode. This effectively doubles the character time density on the screen since no information is duplicated. There are some restrictions in interlace mode.

Noninterlace Mode

Noninterlace mode commonly uses 262 or 264 scanning lines per frame. Since 60 frames per second are maintained precisely, the horizontal line frequency is 15,720 or 15,840 Hz. The horizontal and vertical sync signals must be locked each other.

Dot-Matrix Pattern

The scanning nature of the TV raster requires that the video (or brightness) information be sent in serial form to control the electron beam as it sweeps lines across the screen. Suppose, for example, that the character "K" is to be displayed as shown in Fig. 15.2-2. The first line can be represented as 10001—ones signifying light spots (dots) and zeros signifying dark spots. The remaining six lines are similarly to be represented as a series of dots and dark spaces. When the seven lines are displayed one above the other, the 7×5 character "K" is seen. This method is called the dot-matrix. The tedious job of deciding where to put the dots (ones versus zeros) to generate a given character is done by an IC called the character generator (containing ROM and some control circuit).

Color Technique

A horizontal line frequency of nearly 15,735 Hz may result in a good color display in a TVT system. Basically, it needs the addition of a color subcarrier (about 3.58 MHz) to the video output. In order to produce various colors the subcarrier should alter its phase with respect to itself after the last sync pulse has elapsed. Figure 15.2-3 shows the difference between the frequency responses of the color TV and the B/W TV. The B/W TV has a 4-MHz bandwidth of video baseband and a 4.5-MHz narrow sound subcarrier. The video signal is amplitude modulated, while the sound is frequency-modulated. The total rf channel is 6 MHz, as shown in Fig. 15.2-3*b*. A color subcarrier is added at 3.57945 MHz, as shown in Fig. 15.2-3*c*. The video signal to which no color subcarrier is added, is referred to as the luminance (brightness). The chroma information is modulated on the subcarrier. The color subcarrier with its modulation is called the chrominance, which determines the hue and saturation of a color.

Hue corresponds to the resultant subcarrier phase with respect to a reference point. Saturation corresponds to the resultant subcarrier amplitude. Brightness corresponds to the amplitude of the monochrome (black and white) video signal (Y).

After each horizontal sync pulse, there is a color burst of eight cycles continuous at least, as shown in Fig. 15.2-4. This color burst is a sync signal, utilized as timing reference in the chroma circuits of a color TV set. The magnitude of the color burst should be 25% of the maximum amplitude, corresponding to the peak-to-peak value of the horizontal sync pulse.

When a color TV set is tuned to a color-broadcast signal, the waveform applied to the video detector consists of a composite color signal modulated on the picture-IF carrier. This composite color signal is demodulated through the video detector to develop the envelope information. By means of a bandpass amplifier (whose response peaked at the high-frequency end of the video

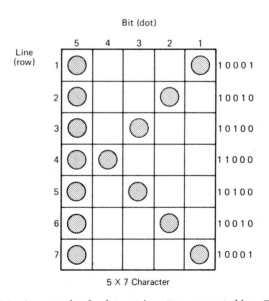

Figure 15.2-2   An example of a dot-matrix pattern generated by a TV display.

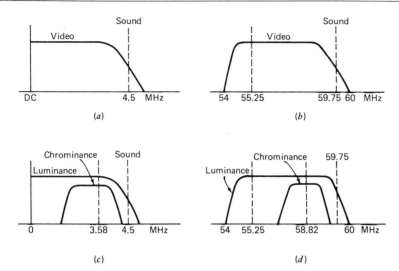

Figure 15.2-3 Difference between the frequency responses of the black-white (B/W) TV and color TV: (*a*) B/W TV—base band. (*b*) B/W TV—second channel. (*c*) Color TV—base band. (*d*) Color TV—second channel.

channel), the chroma signal is taken from the demodulated video signal. Then the chroma signal is synchronously demodulated by the chroma demodulators, which respond to a band of video frequencies from 3.0 to 4.2 MHz, with the 3.58 MHz as reference. The ratio of the currents through the red, blue, and green electron guns in the picture tube is determined by the phase and amplitude of the chroma signal, which correspond to the hue and saturation of the color, respectively.

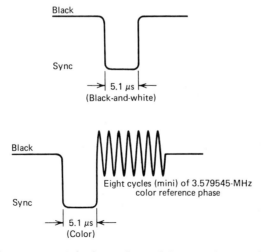

Figure 15.2-4 Color burst after each horizontal sync pulse.

Table 15.2-1   Colors Corresponding to
Phases and Delays of Color Sync Signals

Color	Approximate Phase	Approximate Delay
Burst	0°	0
Yellow	15°	12 ns
Red	75°	58 ns
Magenta	135°	105 ns
Blue	195°	151 ns
Cyan	255°	198 ns
Green	315°	244 ns

Because the chroma section has a considerably narrower bandwidth than the B/W section, the chroma signal is comparatively delayed in its arrival at the color picture tube. Hence, a delay line is inserted in the B/W channel. The delay time provided is typically 1 μs. This delay provides optimum "color fit" in the black and white image. In addition to the delay line, a sound-subcarrier trap and a color-subcarrier trap are used to attenuate the responses at 4.5 MHz and 3.58 MHz, respectively. The output luminance signals simultaneously modulate the cathodes of three electron guns to determine the brightness of the picture tube.

Basically, a color demodulator is a phase detector and an amplitude detector. After each horizontal sync pulse, the 3.58-MHz reference signal in the TV set immediately finds the color burst and compares with the latter at the phase detector, so that the reference signal and the color burst interlock. Colors corresponding to the phases and delays of color sync signals are listed in Table 15.2-1. Since there are not many required colors, the phase-shift of the reference signal can be accomplished by using appropriate digital gates rather than analog circuits.

15.3   CATHODE-RAY CONTROLLER (CRTC) MC6845

CRTC Block Diagram

The block diagram of CRTC chip MC6845 is shown in Fig. 15.3-1. The chip is designed to control a variety of CRT display devices. It provides programmable formatting of three different types of raster-scan modes. The 6845 can provide addressing of 16K of memory and will refresh dynamic memory. The CRTC has additional addressing for a 2515 character-generator ROM, and is designed to be operated most efficiently in a 6909 MPU environment.

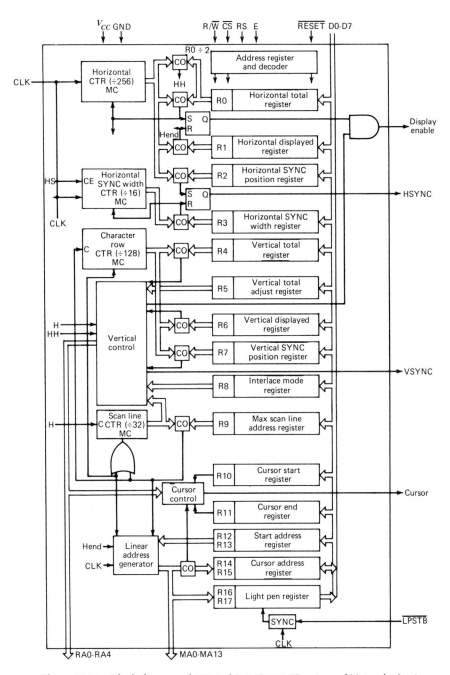

Figure 15.3-1 Block diagram of CRTC chip MC6845. (Courtesy of Motorola, Inc.)

Pinout Lines and Signals

R/$\overline{\text{W}}$ Line.

The read/write line determines whether the addressed register within the CRTC is written to or read by the MPU.

Chip Select ($\overline{\text{CS}}$).

The chip select line enables communication to the CRTC internal register file when a valid device decode address is available from the processor.

Register Select (RS).

The register select line selects either an address register or a data register within the internal register file of the CRTC.

Enable (E) Lines.

The enable line enables the data bus I/O drivers and clocks data to and from the CRTC.

Reset Line.

The reset line clears all counters in the CRTC, puts all outputs to the low level, stops device operation, and leaves the control registers unchanged.

Data Bus D_0–D_7.

The data bus lines are tristate lines which pass parallel data to and from the CRTC control registers. Note: In the tristate logic, the output of a device is either in an active state (1-state or 0-state), or in the high-impedance (OFF) state. Hence, if several devices are connected to a bus and only one device is operating, the remaining devices should be held in the high-impedance state. The purpose of tristate logic is to isolate components from the buses of a microprocessor system, thus creating a shared bus.

Display Enable.

The display enable signal indicates that the CRTC is providing addressing to information in the active display area.

Horizontal Sync (HSYNC).

The horizontal sync pulse is generated for the video display monitor or video processing unit to indicate the time when a scan line has begun.

Vertical Sync (VSYNC).

The vertical sync signal is necessary in the video processing unit for composite video generations. It indicates the start of a vertical scan page.

Cursor Line.

The cursor line indicates to external video processing logic that the cursor is to be displayed.

Light Pen Strobe ($\overline{\text{LPSTB}}$).

The light pen strobe line will latch the current video display refresh memory address to indicate where the light pen was positioned on the screen.

Refresh Memory Addresses.

These 14 lines (MA_0–MA_{13}) will refresh display within a 16K area of refresh memory. This area is the video display memory area.

Raster Addresses.

These five lines (RA_0–RA_4) address the character ROM for the row of the character that is to be displayed.

Clock Line (CLK).

The clock line is the sync signal which relates all CRTC signals.

Internal Register File

The register initialization for a common 80×24 line display is listed in Table 15.3-1. The register file is described as follows.

The address register is an indirect register and indicates the address of some other register in the bank of 18 registers.

The horizontal total register determines the horizontal scan frequency. It is expressed as the period of one scan line in character time units.

The horizontal displayed register indicates the number of displayed characters per line. The scan line does not include horizontal retrace time.

The horizontal sync position register indicates where the horizontal sync pulse commences on the horizontal scan line.

The horizontal sync width register controls the width of the horizontal sync pulse for customizing to a variety of CRT monitors.

The vertical total register is a coarse control for the vertical scan frequency. It is normally set close to 50 or 60 Hz and expressed as a time period related to character line times.

The vertical total adjust register is a fine control for the adjustment of the overall vertical period and is expressed as a number of scan lines.

The vertical displayed register indicates the number of actual displayed character rows on the screen.

The vertical sync position register indicates to the CRTC when the vertical sync pulse is generated in terms of the total vertical scan time. It is expressed in character row times.

The interlace mode register selects one of three display modes, nonin-

Table 15.3-1 Typical 80 × 24 Screen Format Initialization of CRTC (Courtesy of Motorola, Inc.)[a]

Register	Register File	Program Unit[a]	Calculation[b]	Programmed Value	
				Decimal	Hex
R0	H Total	T_c	102 × 0.527 = 53.76 μs	102 − 1 = 101	N_{ht} = $65
R1	H Displayed	T_c	80 × 0.527 = 42.16 μs	80	N_{hd} = $50
R2	H Sync Position	T_c	86 × 0.527 = 45.32 μs	86	N_{hsp} = $56
R3	H Sync Width	T_c	9 × 0.527 = 4.74 μs	9	N_{hsw} = $09
R4	V Total	T_{cr}	25 × 645.12 = 16.13 ms	25 − 1 = 24	N_{vt} = $18
R5	V Total Adjust	T_{sl}	10 × 53.76 = 0.54 ms	10	N_{adj} = $0A
R6	V Displayed	T_{cr}	24 × 645.12 = 15.48 ms	24	N_{vd} = $18
R7	V Sync Position	T_{cr}	24 × 645.12 = 15.48 ms	24	N_{vsp} = $18
R8	Interface Mode	—		—	$00
R9	Max Scan Line Address	T_{sl}		11	N_{sl} = $08
R10	Cursor Start	T_{sl}		0	$00
R11	Cursor End	T_{sl}		11	$08
R12	Start Address (H)	—			$00
R13	Start Address (L)	—		128	$80
R14	Cursor (H)	—			$00
R15	Cursor (L)	—		128	$80

[a] Clock Period = T_c = 0.527 μs.
Scan Line Period = T_{sl} = (N_{ht} + 1) × T_c = 102 × 527 μs = 53.76 μs.
Character Row Period = T_{cr} = N_{sl} × T_{sl} = 12 × 53.76 μs = 645.12 μs.
[b] These are typical values for the Motorola M3000 Monitors. Values may vary for other monitors.

terlace or one of two types of interlace modes. Information on how to control the modes is given in Table 15.3-2.

The maximum scan line address register determines the number of scan lines per character row. This includes spacing between lines of the field.

The cursor start register controls the cursor format and operates as indicated in Table 15.3-2. Bit 6 is the blink enable control. Bit 5 is the blink rate and display control. The other five bits control which scan line starts the cursor display.

The cursor end register indicates the scan line that marks the end of the cursor display.

The start address registers determine in video memory where the display starts after vertical blanking time.

The cursor address registers (high byte and low byte) include the location of the cursor in the video memory.

The light pen registers (high byte and low byte) indicate the content of the video memory address register when the light pen option is implemented.

The CRTC Used Inside an Interactive Terminal

When used inside an interactive terminal, the CRTC chip is primarily connected to reflect interfacing to a 6809 MPU.

The code listed below is from Motorola MC6845 Data Specification Publication AD1-465 (1977). The code in conjunction with Table 15.3-1 performs a typical register initialization for a common 80 × 24 line frame display.

```
           NAM     CRTINT
           ORG     $0
           CLRB                        /CLEAR COUNTER
           LDX     #$20                /POINT X TO CONSTANTS
    CRTI1  STB     $9000               /CRTC ADDR REG
           LDA     0,X                 /GET NEXT CONSTANT
           STA     $9001               /CRTC REG
           INX                         /MOVE POINTER
           INCB                        /ALL DONE?
           CMPB    #$10
           BNE     CRTI1
           SWI
           ORG     $20
    CRTTAB FCB     $65,$50,$56,$9      /CONSTANTS
           FCB     $18,$0A,$18,$18
           FCB     0,$0B,0,$0B
           FDB     $80,$80
           END
```

The video block information is located in an access-protected area for storage of video information. During power-up of the MPU–CRTC circuit,

Table 15.3-2   Cursor Start Register
(Courtesy of Motorola, Inc.)

Bit 6	Bit 5	Cursor Display Mode
0	0	Nonblink
0	1	Cursor Non Display
1	0	Blink, $\frac{1}{16}$ Field Rate
1	1	Blink, $\frac{1}{32}$ Field Rate

the registers should be initialized by execution of a ROM-based routine that loads the registers correctly to the size of the display. Register initialization for a common 80 × 24 line display is given in Table 15.3-1.

15.4   A TV TERMINAL USING MOTOROLA MC6801

Major Devices Used for the TVT

A low-cost TV terminal can be formed using an MC6801 microprocessor (MPU), an MC6847 video display generator (VDG), and MC1372 rf modulator, a TV set, etc.

The MC6801 MPU is well suited for application as a terminal controller. Its four ports, on-chip programmable timer, ROM, RAM, and SCI (serial communications interface) comprise all the essentials for controlling a terminal. The four ports allow for easy keyboard and mode-select switch interface while still retaining full 64K byte (8-bit word) address capability. The programmable timer is well suited for cursor and bell timing, and the SCI needs only RS-232 interface buffers for use by the terminal. The on-board ROM is 2K bytes, allowing plenty of "software room" beyond the 500 bytes necessary for basic terminal operation. The on-chip RAM is 128 bytes, suitable for program stack and scratch-pad memory.

The MC6847 VDG and the MC1372 rf modulator make possible use of a conventional, unmodified TV set as a monitor. The MC1372 interfaces directly with the TV set 75-Ω antenna terminals. The TV terminal described here uses a total of only 15 devices. This number can be reduced to 13 by using transistors for the RS-232 interface.

Interface Specificiation RS-232-C

The most commonly used serial interface specification is RS-232-C, proposed by the Electronic Industries Association (EIA). The RS-232-C specification describes the electrical, mechanical, and functional characteristics of the data interchange. The RS-232-C standard specifies a number of restrictions, among which there are three main ones: (1) The driver circuits on the interchange should be able to withstand open circuits or shorts without damage; (2) the receiver should be able to withstand an input signal of up to ±25 V without damage; and (3) the input impedance of the receiver should not be less than 3K or more than 7K. RS-232-C provides good-quality communications over single-wire transmission lines up to 50 ft at data rates up to 20K bits per second. To be able to communicate well in a noisy atmosphere, RS-232-C recommends using large-signal voltage swings. Actually, the signal should swing positive and negative with respect to ground. Typically ±12- or ±15-V signal levels are used. The negative voltage represents the logic 1 (mark) and the positive voltage represents 0 (space).

The Terminal

The TVT provides the user with the choice of two data formats: Industry Standard Mark/Space (NRZ, Non-Return to Zero) and Biphase. As shown in Fig. 15.4-1, each format consists of a start bit (low), eight data bits, and one stop bit (high). Most users prefer the NRZ format, which is a universal standard. The biphase format, however, has advantages that will be useful in many applications; it is more immune to noise and can tolerate a bit-rate drift of up to 25%.

Four baud rates per MPU operating frequency are software selectable, as shown in Fig. 15.4-2. Baud rates not listed can be generated by selecting a crystal to give the desired baud rate or by using an external clock to drive the SCI, an MC6801 option. Baud rate, format, and SCI clock source options are controlled by the rate and mode control register in the MC6801, as shown in Fig. 15.4-2. For simplicity, user control of this register is determined by a set of four DIP (dual-in-line package) switches (S0–S3) which are set by the user in the same format as called for by the register. Other options, paging or scrolling and full or half duplex, are also selected by DIP switches (S4–S5). These switches are continually polled by software and can be changed "on the fly."

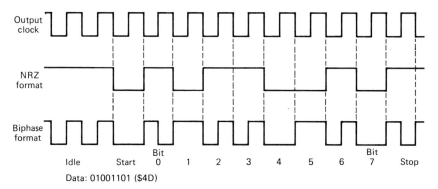

Figure 15.4-1   SCI data formats. (Courtesy of Motorola, Inc.)

7	6	5	4	3	2	1	0	
X	X	X	X	CC1	CC0	SS1	SS0	ADDR: $0010

Rate and Mode Control Register (RMCR)

	$4f_0$—	2.4576 MHz	4.0 MHz	4.9152 MHz
SS1:SS0	E	614.4 kHz	1.0 MHz	1.2288 MHz
0 0	+16	26 μs/38,400 baud	16 μs/62,500 baud	13.0 μs/76,800 baud
0 1	+128	208 μs/4800 baud	128 μs/7812.5 baud	104.2 μs/9600 baud
1 0	+1024	1.67 ms/600 baud	1.024 ms/976.6 baud	833.3 μs/1200 baud
1 1	+4096	6.67 ms/150 baud	4.096 ms/244.1 baud	3.33 ms/300 baud

SCI Bit Times and Rates

CC1:CC0	Format	Clock source	Port 2 Bit 2
00	Biphase	Internal	Not used
01	NRZ	Internal	Not used
10	NRZ	Internal	Output
11	NRZ	External	Input

SCI Format and Clock Source Control

Figure 15.4-2 SCI baud rate and format options. (Courtesy of Motorola, Inc.)

The terminal recognizes most widely used control characters, including

Line Feed (LF)
Carriage Return (CR)
Backspace (BS)
Bell (BEL)
Clear Screen (SUB)
Cursor Forward (FF)
Cursor Up (VT)

The cursor blinks and is nondestructive; that is, it can be moved over any displayed character without changing the character. The cursor will simply blink over the character, with the character appearing each time the cursor is "off."

The display format is controlled by the MC6847 VDG which displays a complement of 64-bit ASCII characters in a 32 (across) by 16 (down) format.

15.5   ASSEMBLY OF THE TVT USING MC6801

The MC6801 MPU can be configured to operate in three basic modes:
(1) single-chip mode (single-chip microcomputer), (2) expanded nonmultiplexed mode (256-byte external address capability), and (3) expanded
multiplexed mode (64K-byte external address capability). Several expanded
multiplexed modes give the user choices of combinations of on-chip or
external RAM, ROM, and interrupt vectors. A summary of these operating
modes is shown in Table 15.5-1. A vector is a location finder (address)
usually indicating the starting point of a particular operation or routine. An
automatic prestored address is used in the interrupt process.

The TVT presented here uses an MC6801 operated in Mode 2. This mode
configures the MPU for 128 bytes of on-chip RAM, external ROM, external
interrupt vectors, and uses an MCM2708 EPROM for easy software devel-
opment. Fully developed software can be "masked" into the MC6801 on-
chip ROM. The MPU can then be operated in expanded multiplexed Mode
6, which uses the on-chip ROM and RAM, with only an MC6801 mode select
design change. An MC6803 can be used instead of an MC6801 in terminals
using an external EPROM or ROM. Masking is a technique for sensing
specific binary conditions and ignoring others, typically accomplished by
placing 0's in bit positions of no interest and 1's bit positions to be sensed.

The MC6801 has four ports. In the expanded multiplexed modes, Port 1
is used for general I/O, Port 2 for mode selection and SCI/Timer I/O or for
general I/O, Port 3 for multiplexed data and lower-order addresses (A0/D0–
A7/D7), and Port 4 for higher-order addresses (A8–A15).

Port 2 has five pins (#8–#12) associated with it, as shown in the schematic

Table 15.5-1   Operating Mode Summary (Courtesy of Motorola, Inc.)

Mode	P22	P21	P20	ROM	RAM	Interrupt vectors	Bus Mode	Operating Mode
7	H	H	H	I	I	I	I	Single-chip
6	H	H	L	I	I	I	Mux	Multiplexed
5	H	L	H	I	I	I	NMux	Nonmultiplexed
4	H	L	L	I	I	I	I	Single-chip test
3	L	H	H	E	E	E	Mux	Multiplexed/no RAM nor ROM
2	L	H	L	E	I	E	Mux	Multiplexed/RAM
1	L	L	H	I	I	E	Mux	Multiplexed/RAM and ROM
0	L	L	L	I	I	I	Mux	Multiplexed test

Legend: I—Internal; E—External; Mux—Multiplexed; NMux—Nonmultiplexed; L—Logic
"0"; H—Logic "1".

of Fig. 15.5-1. Pins 8 (*P20*), 9 (*P21*), and 10 (*P22*) of Port 2 are dual-purpose pins. On the rising edge of $\overline{\text{RESET}}$, the MC6801 latches the logical state of these pins as the upper three bits of the Port 2 register. These are read-only bits, and have no pins directly associated with them. They are the "mode control" bits and their states determine the operating mode of the MPU. The mode select voltages are applied to pins 8, 9, and 10 through pull-up resistors so that immediately after $\overline{\text{RESET}}$, pin 8 can be used for the Timer input, pin 9 for the Timer output, and pin 10 for the SCI clock input (if used).

Pins 11 and 12 of Port 2 are used by the SCI. MC1488 and MC1489 line drivers are used as the RS-232 interface. The on-board SCI controls all serial communications, serving the function of a UART.

Seven pins of I/O Port 1 are used for keyboard ASCII input. The Timer is used to latch the keyboard strobe. The keyboard should generate a strobe at least 2 MPU cycles in duration with each keyboard entry. This strobe is tied to pin 10, the Timer input. An input edge detector tied to this pin sets a flag (Input Capture Flag) in the MPU Timer Control/Status Register each time a key is depressed. The software polls this flag for keyboard servicing. A flag is a bit or bits stored for later use, indicating the existence of a previous or a present condition. Polling is an important multiprocessing method used to identify the source of interrupt requests.

A simple bell circuitry using two NAND gates is used to generate a 4-kHz tone. Software gates the bell using the MPU Timer output, pin 9.

The terminal operating mode switches are read through an MC74LS244 octal three-state driver at address $3800. A simple switch and pull-up arrangement is used for mode selection.

The display circuitry consists of the MC6847 Video Display Generator, the MC1372 modulator, two MCM2114 RAMs, and associated circuitry. The VDG is operated in the Alphanumeric Internal mode to display characters in a 32 (across) by 16 (down) format and in the Semigraphics 4 mode to display the cursor. Both interlaced (MC6847Y) and noninterlaced (MC6847) versions of the VDG are available.

The VDG is clocked by the MC1372 at a 3.58 MHz color-burst frequency. It reads data sequentially from display RAM and, in this terminal, uses an on-chip character generator to produce the alphanumeric displays. Two MCM2114s are used as the display RAM. Only 512 bytes of display RAM are needed for the Alphanumeric and Semigraphic 4 modes, so only nine VDG address lines are used. A tenth address line (A9), held high by the VDG, is connected to the MCM2114s to ensure proper address decoding.

The MPU also has access to the display RAM in order to change displayed characters. To avoid contention for the display RAM the $\overline{\text{MS}}$ (Memory Select) pin of the VDG is pulled low whenever the MPU access the RAM, forcing the VDG address port to a high impedance. To avoid a noisy display during MPU writes to the RAM, the MPU reads the state of $\overline{\text{HS}}$ (Horizontal Sync), which goes low during horizontal retrace, through Port 1 bit 7 (pin

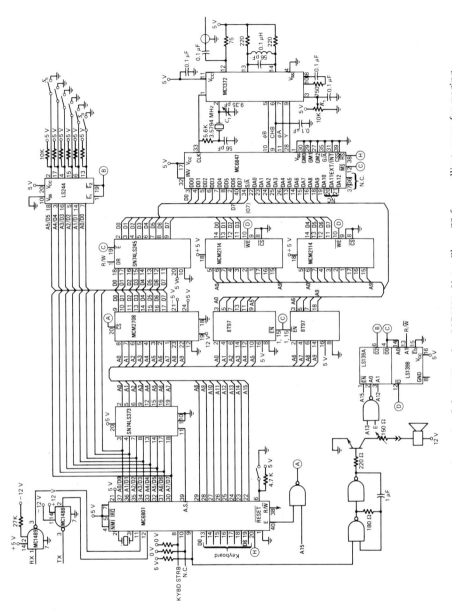

Figure 15.5-1 Schematic of the TVT using MC6801. Note: Close S5 for scrolling, open for paging. Close S4 for half duplex, open for full duplex. See Fig. 15.4-2 for the positions of S0–S4. (Courtesy of Motorola, Inc.)

20). The MPU will write to the display RAM only during horizontal retrace when $\overline{HS}$ is low.

The VDG outputs chrominance (ϕA, ϕB, Y) and chroma bias (CHB) to the MC1372 Modulator which generates composite video for the television. The modulator is clocked by a 3.58 MHz crystal and its TTL compatible clock output (pin 1) is used to drive the VDG.

C1 is used to adjust the clock frequency; R1 is used to adjust the clock duty cycle.

The software of the TVT using MC6801 is described in Appendix B. To understand this, one must have a sound background in programming concepts.

15.6   REVIEW OF PROGRAMMING CONCEPTS

Computer Programs and Instructions

A computer program can be defined as a series of instructions or statements prepared in a form acceptable to the computer, the purpose of which is to achieve a certain result. A computer instruction can be defined as a set of characters that define an operation. The operation is a specific action which a computer performs whenever an instruction calls for it (e.g., division, addition, subtraction, 0Ring, etc.). The number of different operations that a computer can perform and the speed with which it can perform such operations provide a measure of how powerful the computer is. A character is one symbol of a set of elementary symbols. Usually, symbols include the decimal digits 0 through 9, the letters A through Z, punctuation marks, dollar signs, commas, operation symbols, and any other single symbols that a computer may read, store, or write. In computer programming, it is commonly for one to use the symbols corresponding to the entire typewriter keys.

Computer programs are often referred to as software (although software actually contains the user programs and the routines, as mentioned previously). The ROM that contains the program may be referred to as firmware. Once the program has been "burned" into a ROM, it is very difficult to change.

Computer instructions can be expressed in a variety of forms. They can be expressed as binary numbers (e.g., 00111110), hexadecimal numbers (e.g., 3EH), mnemonic code (e.g., OUT 3EH), full words [e.g., OUTPUT ACCUMULATOR DATA TO DEVICE #3E (HEX)], or full mathematical expressions.

Most of the microprocessor systems available today utilize up to 16 address lines. In the hexadecimal (hex) system (base 16), each group of four digits is represented by one hex symbol. Thus, 16 binary digits can be represented with four hex symbols. The relationship between base 10, base 16, and base 2 is shown in Fig. 15.6-1.

Base 10 (decimal)	Base 16 (hexadecimal)	Base 2 (binary)
0	0	0000
1	1	0001
2	2	0010
3	3	0011
4	4	0100
5	5	0101
6	6	0110
7	7	0111
8	8	1000
9	9	1001
10	A	1010
11	B	1011
12	C	1100
13	D	1101
14	E	1110
15	F	1111

Figure 15.6-1 Relationship between base 10, base 16, and base 2.

Mnemonics

Mnemonic is a term that describes something used to assist the human memory. Mnemonic code is defined as the computer instructions written in a form the programmer can easily remember, but which must be converted into machine language later by a computer or by the user. Mnemonic language is a programming language that is based on easily remembered symbols and that can be assembled into machine language by a computer. Mnemonic operation is defined as computer instructions that are written in meaningful notation. Most manufacturers of microprocessors define a two-, three-, or four-letter code (known as a mnemonic) that describes the function of each instruction. In Table 6-3, the mnemonic code ABA is assigned to describe an instruction, that is "to add the accumulator A to the accumulator B, and place the result to the accumulator A." The ABA operation may be represented by the hex code 1B, which appears in the ROM memory 0001 1011. Hence, when the microprocessor decodes the instruction, it says, "I must add the accumulators A and B together, place the result in accumulator A, and then go to the next sequential location in the memory for my next instruction."

Machine Language

Each computer instruction is written as a series of 1's and 0's that specifically characterize that instruction and no other. Such a binary representation of a computer instruction is called machine language or machine code.

Programs written with mnemonic codes are known as source programs. It is from these source programs that binary code is generated. The source code converting to machine language can be accomplished in two ways. One is to refer to the manufacturer's programming manual and manually look up the machine language equivalent for each mnemonic code in the source listing. This manual technique, called "hand assembling," is very often used by the hobbyist today. The second method of converting from mnemonic code to machine language is with an "assembler." An assembler is an independent program designed to convert a source program into a machine language program. Each type of MPU available on the market today must have an assembler specifically tailored for its MPU.

Bytes

In general, a word is the number of bits that a computer can manipulate simultaneously. If the number of bits in a word is eight, we usually employ the term byte rather than word. Most microcomputers, since they have 8-bit words, naturally deal in 8-bit bytes. A 16-bit memory address is treated as two 8-bit memory address bytes, an 8-bit HI byte and an 8-bit LO byte.

ASCII

The term ASCII stands for American Standard Code for information interchange. Note in Fig. 15.6-2 that the ASCII code for an "A" is "1000001." This means that if you wish the typewriter that is tied to your system to type an "A," you must transmit the signal "1000001" on the data lines to the typewriter.

Flowcharts

Before we actually start writing a program, we must learn how to organize our problems for computer solution. Perhaps the best way to do this is to use flowcharts. A flowchart is a layout of our problem-solving method, or a graphical presentation illustrating the logical steps, calculations, and decisions, in sequence, that must be performed to accomplish a specific task. A program can be written after the flowchart has been generated.

Figure 15.6-3 is a flowchart illustrating the steps and decisions by a computer. The symbols for this flowchart are commonly used. The oval symbol is used at the beginning or end of the program. The circle symbol is used as connectors between pages of flowchart. The rectangular box is used for an operation. The diamond-shaped symbol is a decision symbol. Arrows indicate flow lines.

Memory

Two kinds of memory commonly used are the random-access memory (RAM) and the read-only memory (ROM). The RAM is also called the

Character	ASCII Code	Character	ASCII Code
@	1000000	FORM FEED	0001100
A	1000001	CARRIAGE RETURN	0001101
B	1000010	RUBOUT	1111111
C	1000011	SPACE	0100000
D	1000100	!	0100001
E	1000101	''	0100010
F	1000110	#	0100011
G	1000111	$	0100100
H	1001000	%	0100101
I	1001001	&	0100110
J	1001010	'	0100111
K	1001011	(	0101000
L	1001100	)	0101001
M	1001101	*	0101010
N	1001110	+	0101011
O	1001111	'	0101100
P	1010000	−	0101101
Q	1010001	.	0101110
R	1010010	/	0101111
S	1010011	0	0110000
T	1010100	1	0110001
U	1010101	2	0110010
V	1010110	3	0110011
W	1010111	4	0110100
X	1011000	5	0110101
Y	1011001	6	0110110
Z	1011010	7	0110111
[	1011011	8	0111000
\	1011100	9	0111001
]	1011101	:	0111010
↑	1011110	;	0111011
NULL	0000000	<	0111100
HORIZ TAB	0001001	=	0111101
LINE FEED	0001010	>	0111110
VERT TAB	0001011	?	0111111

Figure 15.6-2   ASCII codes (7-bit).

read/write memory; it is a semiconductor memory into which logic 0 and logic 1 states can be written (stored) and read out again (retrieved). The ROM is a semiconductor memory from which digital data can be repeatedly read out, but cannot be written into as in the case for read/write memory. Actually, the ROM in microprocessors may be a special kind of memory called an erasable programmable read-only memory, or EPROM. The RAMs

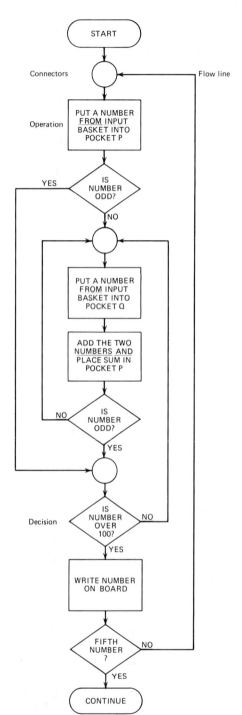

Figure 15.6-3 Flowchart illustrating the steps and decisions by a computer.

are used to store program variable data and instructions. The ROMs are almost always used to store the microprocessor-system's initialization and operating routines as well as codes for special control. Memory address is defined as the storage location of a memory word (not byte).

BIBLIOGRAPHY

Andrews, M., *Programming Microprocessor Interfaces for Control and Instrumentation*. Prentice-Hall, Englewood Cliffs, New Jersey, 1982.

Bishop, R., *Basic Microprocessors and the 6800*. Hayden Book Company, Rochelle Park, New Jersey, 1979.

Gantt, C. W., Jr., Build a Television Display,'' *BYTE,* **16** (Jan., 1976).

Lancaster, D., "Television Interface," *BYTE,* **20** (Oct., 1975).

Morales, A. J., *A Low-Cost Terminal Using the MC6801, AN-798 Application Note*. Motorola Semiconductor Products Inc., 1980.

Streitmatter, G. A., *Microprocessor Software: Programming Concepts and Techniques*. Reston Publishing, Reston, Virginia, 1981.

Questions

15-1 What are the two methods for TV terminal interfacing?

15-2 What is the purpose of an IC UART?

15-3 What is a baud-rate generator?

15-4 What is the function of a character generator?

15-5 What is the function of a keyboard encoder?

15-6 Describe the interlaced scanning method.

15-7 What is the color-burst signal?

15-8 Simply describe the CRTC MC6845.

15-9 What does the RS-232-C standard specify?

15-10 What is a byte?

15-11 Which of the following instructions are machine language? (a) HALT. (b) INC. (c) 76H. (d) 11010011.

15-12 What is a flowchart?

Problems

15.2-1 Draw the 7 × 5 character "S" as a dot-matrix pattern. The first and seventh lines are represented by 01110. What are the similar representations for the remaining five lines?

Answer.

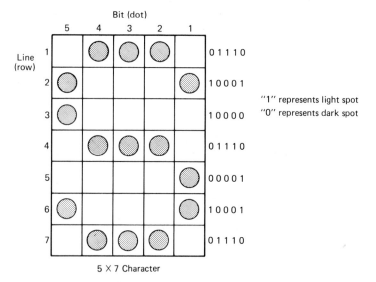

"1" represents light spot
"0" represents dark spot

5 X 7 Character

15.2-2   Draw the 7 × 5 character "R" as a dot-matrix pattern. Give a suitable representation for each line referring to Problem 15.2-1.

Appendix A

Motorola MC6800
Instructions

Motorola MC6800 instructions (Courtesy of Motorola, Inc.): accumulator and memory instructions listed in Table 6-3; other instructions listed below.

Jump and Branch Instructions

Jump and Branch Operations	Mnemonic	Relative OP	~	#	Index OP	~	#	Extnd OP	~	#	Inher OP	~	#	Branch Test	H (5)	I (4)	N (3)	Z (2)	V (1)	C (0)
Branch Always	BRA	20	4	2										None	•	•	•	•	•	•
Branch If Carry Clear	BCC	24	4	2										$C = 0$	•	•	•	•	•	•
Branch If Carry Set	BCS	25	4	2										$C = 1$	•	•	•	•	•	•
Branch If = Zero	BEQ	27	4	2										$Z = 1$	•	•	•	•	•	•
Branch If ≥ Zero	BGE	2C	4	2										$N \oplus V = 0$	•	•	•	•	•	•
Branch If > Zero	BGT	2E	4	2										$Z + (N \oplus V) = 0$	•	•	•	•	•	•
Branch If Higher	BHI	22	4	2										$C + Z = 0$	•	•	•	•	•	•
Branch If ≤ Zero	BLE	2F	4	2										$Z + (N \oplus V) = 1$	•	•	•	•	•	•
Branch If Lower Or Same	BLS	23	4	2										$C + Z = 1$	•	•	•	•	•	•
Branch If < Zero	BLT	2D	4	2										$N \oplus V = 1$	•	•	•	•	•	•
Branch If Minus	BMI	2B	4	2										$N = 1$	•	•	•	•	•	•
Branch If Not Equal Zero	BNE	26	4	2										$Z = 0$	•	•	•	•	•	•
Branch If Overflow Clear	BVC	28	4	2										$V = 0$	•	•	•	•	•	•
Branch If Overflow Set	BVS	29	4	2										$V = 1$	•	•	•	•	•	•
Branch If Plus	BPL	2A	4	2										$N = 0$	•	•	•	•	•	•
Branch To Subroutine	BSR	8D	8	2										See Special Operations	•	•	•	•	•	•
Jump	JMP				6E	4	2	7E	3	3				See Special Operations	•	•	•	•	•	•
Jump To Subroutine	JSR				AD	8	2	BD	9	3					•	•	•	•	•	•
No Operation	NOP										01	2	1	Advances Prog. Cntr. Only	•	•	•	•	•	•
Return From Interrupt	RTI										3B	10	1		•	•	⑩	•	•	•
Return From Subroutine	RTS										39	5	1	See special Operations	•	•	•	•	•	•
Software Interrupt	SWI										3F	12	1		•	S	•	•	•	•
Wait for Interrupt	WAI										3E	9	1		•	⑪	•	•	•	•

Index Register and Stack Manipulation Instructions

Index Register and Stack Pointer Operations	Mnemonic	Immed OP	Immed ~	Immed #	Direct OP	Direct ~	Direct #	Index OP	Index ~	Index #	Extnd OP	Extnd ~	Extnd #	Inher OP	Inher ~	Inher #	Boolean/Arithmetic Operation	H (5)	I (4)	N (3)	Z (2)	V (1)	C (0)
Compare Index Reg	CPX	BC	3	3	9C	4	2	AC	6	2	BC	5	3				$(X_H/X_L) - (M/M+1)$	•	•	⑦	↕	⑧	•
Decrement Index Reg	DEX													09	4	1	$X - 1 \rightarrow X$	•	•	•	↕	•	•
Decrement Stack Pntr	DES													34	4	1	$SP - 1 \rightarrow SP$	•	•	•	•	•	•
Increment Index Reg	INX													08	4	1	$X + 1 \rightarrow X$	•	•	•	↕	•	•
Increment Stack Pntr	INS													31	4	1	$SP + 1 \rightarrow SP$	•	•	•	•	•	•
Load Index Reg	LDX	CE	3	3	DE	4	2	EE	8	2	FE	6	3				$M \rightarrow X_H, (M + 1) \rightarrow X_L$	•	•	⑨	↕	R	•
Load Stack Pntr	LDS	8E	3	3	9E	4	2	AE	8	2	BE	6	3				$M \rightarrow SP_H, (M + 1) \rightarrow SP_L$	•	•	⑨	↕	R	•
Store Index Reg	STX				DF	5	2	EF	7	2	FF	6	3				$X_H \rightarrow M, X_L \rightarrow (M + 1)$	•	•	⑨	↕	R	•
Store Stack Pntr	STS				9F	5	2	AF	7	2	BF	6	3				$SP_H \rightarrow M, SP_L \rightarrow (M + 1)$	•	•	⑨	↕	R	•
Indx Reg → Stack Pntr	TXS													35	4	1	$X - 1 \rightarrow SP$	•	•	•	•	•	•
Stack Pntr → Indx Reg	TSX													30	4	1	$SP + 1 \rightarrow X$	•	•	•	•	•	•

Condition Code Register Manipulation Instructions

Conditions Code Register		Inher			Boolean	5	4	3	2	1	0
Operations	Mnemonic	OP	~	#	Operation	H	I	N	Z	V	C
Clear Carry	CLC	0C	2	1	0 → C	•	•	•	•	•	R
Clear Interrupt Mask	CLI	0E	2	1	0 → I	•	R	•	•	•	•
Clear Overflow	CLV	0A	2	1	0 → V	•	•	•	•	R	•
Set Carry	SEC	0D	2	1	1 → C	•	•	•	•	•	S
Set Interrupt Mask	SEI	0F	2	1	1 → I	•	S	•	•	•	•
Set Overflow	SEV	0B	2	1	1 → V	•	•	•	•	•	S	•
Acmltr A → CCR	TAP	06	2	1	A → CCR			⑫			
CCR → Acmltr A	TPA	07	2	1	CCR → A	•	•	•	•	•	•

Condition Code Register Notes:
(Bit set if test is true and cleared otherwise)
① (Bit V) Test: Result = 10000000?
② (Bit C) Test: Results ≠ 00000000?
③ (Bit C) Test: Decimal value of most significant BCD Character greater than nine? (Not cleared if previously set.)
④ (Bit V) Test: Operand = 10000000 prior to execution?
⑤ (Bit V) Test: Operand = 01111111 prior to execution?
⑥ (Bit V) Test: Set equal to result of N ⊕ C after shift has occurred.
⑦ (Bit N) Test: Sign bit of most significant (MS) byte of result = 1?
⑧ (Bit V) Test: 2's complement overflow from subtraction of MS bytes?
⑨ (Bit N) Test: Result less than zero? (Bit 15 = 1)
⑩ (All) Load Condition Code Register from Stack. (See Special Operations)
⑪ (Bit I) Set when interrupt occurs. If previously set, a Non-Maskable Interrupt is required to exit the wait state.
⑫ (ALL) Set according to the contents of Accumulator A.

Index Register and Stack Manipulation Instructions

Pointer Operations	Mnemonic	Immed OP	~	#	Direct OP	~	#	Index OP	~	#	Extnd OP	~	#	Inher OP	~	#	Boolean/Arithmetic Operation	H (5)	I (4)	N (3)	Z (2)	V (1)	C (0)
Compare Index Reg	CPX	BC	3	3	9C	4	2	AC	6	2	BC	5	3				$(X_H/X_L) - (M/M + 1)$	•	•	⑦	↔	⑧	•
Decrement Index Reg	DEX													09	4	1	$X - 1 \to X$	•	•	•	↔	•	•
Decrement Stack Pntr	DES													34	4	1	$SP - 1 \to SP$	•	•	•	•	•	•
Increment Index Reg	INX													08	4	1	$X + 1 \to X$	•	•	•	↔	•	•
Increment Stack Pntr	INS													31	4	1	$SP + 1 \to SP$	•	•	•	•	•	•
Load Index Reg	LDX	CE	3	3	DE	4	2	EE	8	2	FE	6	3				$M \to X_H, (M + 1) \to X_L$	•	•	⑨	↔	R	•
Load Stack Pntr	LDS	8E	3	3	9E	4	2	AE	8	2	BE	6	3				$M \to SP_H, (M + 1) \to SP_L$	•	•	⑨	↔	R	•
Store Index Reg	STX				DF	5	2	EF	7	2	FF	6	3				$X_H \to M, X_L \to (M + 1)$	•	•	⑨	↔	R	•
Store Stack Pntr	STS				9F	5	2	AF	7	2	BF	6	3				$SP_H \to M, SP_L \to (M + 1)$	•	•	⑨	↔	R	•
Indx Reg → Stack Pntr	TXS													35	4	1	$X - 1 \to SP$	•	•	•	•	•	•
Stack Pntr → Indx Reg	TSX													30	4	1	$SP + 1 \to X$	•	•	•	•	•	•

Condition Code Register Manipulation Instructions

Conditions Code Register		Inher			Boolean	5	4	3	2	1	0
Operations	Mnemonic	OP	~	#	Operation	H	I	N	Z	V	C
Clear Carry	CLC	0C	2	1	0 → C	•	•	•	•	•	R
Clear Interrupt Mask	CLI	0E	2	1	0 → I	•	R	•	•	•	•
Clear Overflow	CLV	0A	2	1	0 → V	•	•	•	•	R	•
Set Carry	SEC	0D	2	1	1 → C	•	•	•	•	•	S
Set Interrupt Mask	SEI	0F	2	1	1 → I	•	S	•	•	•	•
Set Overflow	SEV	0B	2	1	1 → V	•	•	•	•	S	•
Acmltr A → CCR	TAP	06	2	1	A → CCR			⑫			
CCR → Acmltr A	TPA	07	2	1	CCR → A	•	•	•	•	•	•

Condition Code Register Notes:
(Bit set if test is true and cleared otherwise)
① (Bit V) Test: Result = 10000000?
② (Bit C) Test: Results ≠ 00000000?
③ (Bit C) Test: Decimal value of most significant BCD Character greater than nine? (Not cleared if previously set.)
④ (Bit V) Test: Operand = 10000000 prior to execution?
⑤ (Bit V) Test: Operand = 01111111 prior to execution?
⑥ (Bit V) Test: Set equal to result of N ⊕ C after shift has occurred.
⑦ (Bit N) Test: Sign bit of most significant (MS) byte of result = 1?
⑧ (Bit V) Test: 2's complement overflow from subtraction of MS bytes?
⑨ (Bit N) Test: Result less than zero? (Bit 15 = 1)
⑩ (All)   Load Condition Code Register from Stack. (See Special Operations)
⑪ (Bit I) Set when interrupt occurs. If previously set, a Non-Maskable Interrupt is required to exit the wait state.
⑫ (ALL)   Set according to the contents of Accumulator A.

Appendix B

Software of the TVT Using MC6801

The software initializes the MC6801 (see Fig. 15.5-1), services characters from the RS-232 communication link, and, from the keyboard, controls character display and controls the bell.

INITIALIZATION

The software first configures Port 1, used to read the keyboard, as a data input port by clearing the Port 1 Data Direction Register. It then clears the Timer Control/Status Register. This disables all timer interrupts and programs the Time Input Capture Flag to become Set whenever a high-to-low transition is applied to the timer input pin. This pin is tied to the keyboard strobe.

The software then enables the SCI transmitter and receiver and programs Port 2 for Timer and SCI I/O. It then reads the terminal mode switches and jumps to subroutine CHGWO, which loads the mode word into STATWO, a scratch-pad register, for later comparison.

The display RAM consists of 512 bytes located in addresses $2000 through $21FF. For scrolling purposes the software loads display RAM locations $2200 through $2220 with ASCII "blanks." It then jumps to subroutine BLANK, which clears the screen, then loads the index register with $2000. The index register is used as a screen pointer and $2000 is the first screen location.

MAIN PROGRAM

The main program is essentially a loop that writes the cursor; branches to subroutine TIMER, which controls the terminal; erases the cursor; then branches back to TIMER.

Since the terminal interprets only 6-bit ASCII, data bit seven in the display RAM is left free for use as a VDG control bit and is used by this terminal to control S/Ā for displaying the cursor. When S/Ā is low the VDG is in the Alphanumeric Internal mode and will display ASCII characters. When S/Ā is high the VDG is in the Semigraphics 4 mode and will display a color block in one of the eight colors. The software used in this terminal selects a green cursor by writing $80 into the display RAM location pointed to by the index register.

Subroutine TIMER first checks for changes in the terminal operating mode by jumping to subroutine CHKSTA. CHKSTA reads the switches and compares their settings to the last setting stored in register STATWO. If switch selections have changed, CHKSTA will load a new value into STATWO and reprogram the SCI. Otherwise, TIMER continues by loading register TEMPX with a value ($7FF) that controls cursor duration. This value will be decremented to zero at which time the cursor will be removed. With each decrement the keyboard and SCI are serviced by subroutines CHKC and SERRX.

The cursor is removed by replacing it with the contents of SAVCHR, a register that stores the character in the location pointed to by the index register (screen pointer). TIMER is used once again to provide delay.

Subroutine CHKC services the keyboard by first checking the Timer Input Capture Flag for the presence of a keyboard strobe. If the flag is not set, the program returns to subroutine TIMER. If the flag is set, CHKC clears the flag and transmits the keyboard character out of the serial port. CHKC then reads STATWO and tests for full duplex selection. If the mode is half duplex, the character is displayed by subroutine DISPL. If the mode is full duplex, no character is displayed at this time but will be displayed by subroutine SERRX, which services the SCI.

CHKC then jumps to subroutine ENDSCN to test for end of screen and pages or scrolls if necessary according to the user mode selection.

Subroutine SERRX services SCI input characters by testing the Receiver Data Register Full flag in the Transmit/Receive Control and Status Register. If no character is present, the program returns to TIMER. If a character is present, it is displayed by DISPL. The program then Jumps to ENDSCN to test for end of screen, then returns to TIMER.

The program flowchart shown in Fig. A2-1 offers a detailed outline of the program. Table A2-1 gives the program listing.

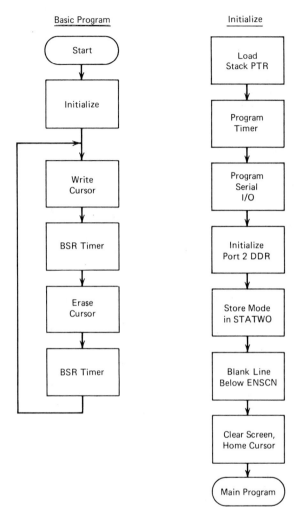

Basic Program

Start

Initialize

Write
Cursor

BSR Timer

Erase
Cursor

BSR Timer

Initialize

Load
Stack PTR

Program
Timer

Program
Serial
I/O

Initialize
Port 2 DDR

Store Mode
in STATWO

Blank Line
Below ENSCN

Clear Screen,
Home Cursor

Main Program

Figure A2-1 Program flowchart. (Courtesy of Motorola Semiconductor Products, Inc.)

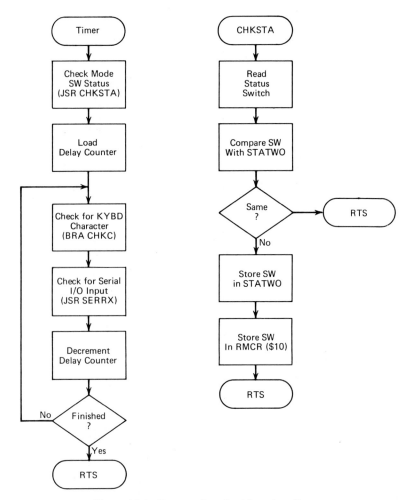

Figure A2-1 Program flowchart (*continued*).

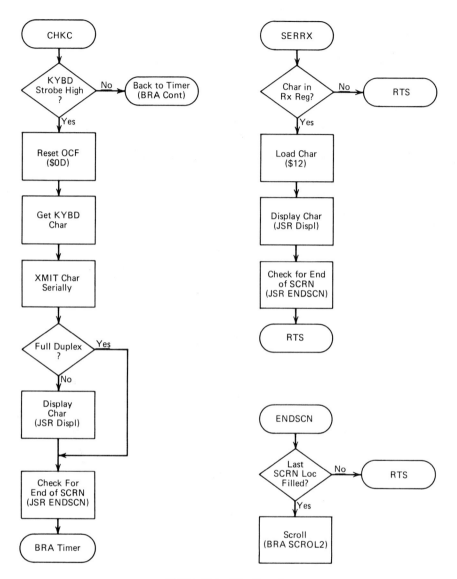

Figure A2-1 (*Continued*)

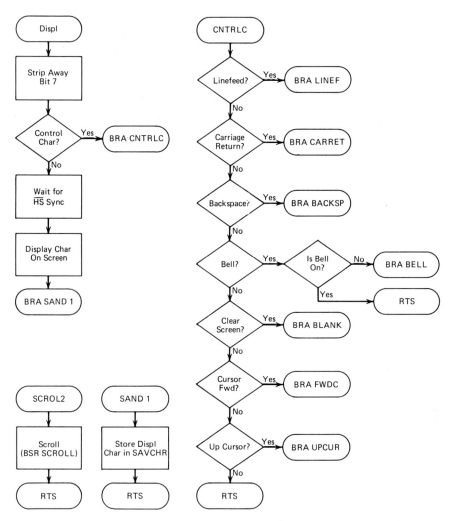

Figure A2-1 (*Continued*)

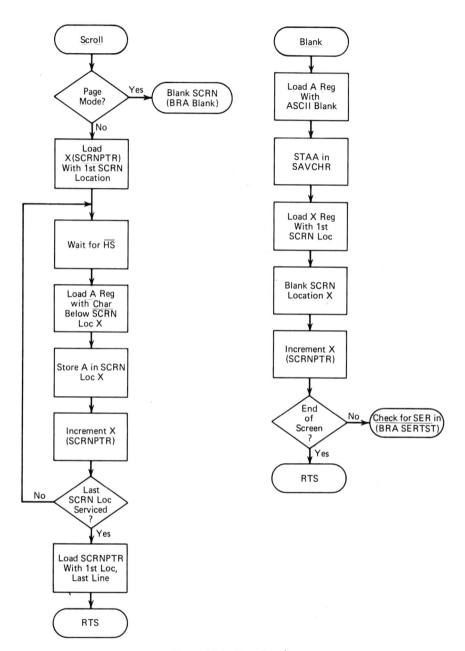

Figure A2-1 (*Continued*)

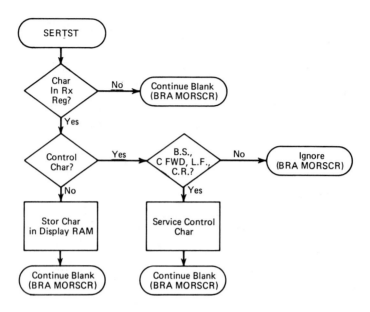

Figure A2-1 *(Continued)*

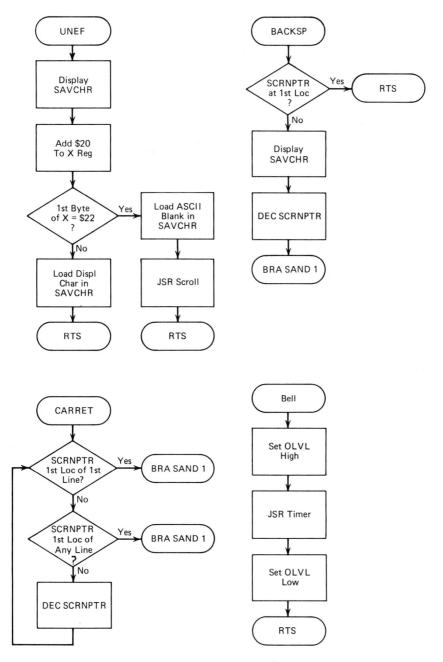

Figure A2-1   (Continued)

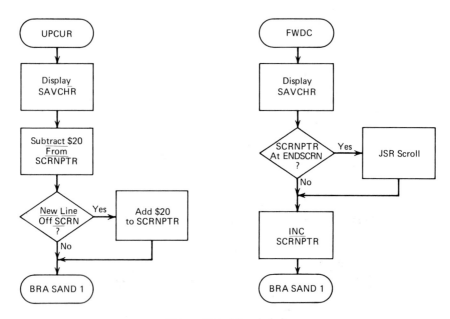

Figure A2-1 (*Concluded*)·

```
PAGE  001    DMBTRM

00001                        *THIS PROG DISPLAYS AND XMITS KYBD CHAR,
00002                        *RECEIVES AND DISPLAYS SER INPUT CHAR,
00003                        *SCROLLS, BLINKS CURSOR, LINEFEEDS, MOVES
00004                        *CURSOR FWD, BACKSPACES CURSOR, MOVES CURSOR UP,
00005                        *CARRIAGE RETURNS, GIVES USER CHOICE OF FOUR
00006                        *SERIAL I/O FORMATS AND A CHOICE OF HALF OR
00007                        *FULL DUPLEX,RINGS BELL,GIVES
00008                        *USER CHOICE OF SCROLLING OR PAGING
00009                        *A.J.MORALES   MAR2,79

00011                             OPT      Z01
00012                             NAM      DMBTRM
00013A FC00                       ORG      $FC00
00014          00FF  A SAVCHR EQU      $00FF        STORE CHAR UNDER CURSOR
00015          00FE  A STATWO EQU      $00FE        STORES TERMINAL OPERATION MODE
00016          00FC  A TEMPX  EQU      $00FC        X-REGISTER SCRATCH PAD
00017          00FA  A TEMPX2 EQU      $00FA        X-REGISTER SCRATCH PAD
00018          1800  A SWITCH EQU      $1800        ADDR FOR SWITCH READ
00019          00F7  A STACK  EQU      $00F7

00021                        ******INITIALIZE******

00023A FC00 8E 00F7  A            LDS      #STACK
00024A FC03 86 00    A            LDAA     #$00
00025A FC05 97 00    A            STAA     $00          PORT 1 READ
00026A FC07 97 08    A            STAA     $08          PROGRAM PTM
00027A FC09 86 0A    A            LDAA     #$0A
00028A FC0B 97 11    A            STAA     $11          ENABLE SER TX,RX
00029A FC0D 86 12    A            LDAA     #$12
00030A FC0F 97 01    A            STAA     $01          PORT 2 DDR
00031A FC11 B6 1800  A            LDAA     SWITCH       CHECK SWITCH STATUS
00032A FC14 BD FD18  A            JSR      CHGWO        LOAD TERM. MODE IN STATWO
00033                        *LINE AFTER EDNDSCRN IS BLANKED FOR SCROLL PURPOSES
00034A FC17 CE 2200  A            LDX      #$2200
00035A FC1A 86 20    A            LDAA     #$20         ASCII BLANK
00036A FC1C 97 FF    A            STAA     SAVCHR
00037A FC1E A7 00    A FILL  STAA     0,X
00038A FC20 08                    INX
00039A FC21 8C 2220  A            CPX      #$2220
00040A FC24 26 F8 FC1E            BNE      FILL
00041A FC26 8D 48 FC70            BSR      BLANK

00043                        ******MAIN PROGRAM*******

00045                        *WRITE CURSOR
00046A FC28 86 80    A WRITEC LDAA     #$80
00047A FC2A A7 00    A            STAA     0,X
00048A FC2C 8D 08 FC36            BSR      TIMER
00049                        *ERASE CURSOR
00050A FC2E 96 FF    A ERASEC LDAA     SAVCHR
00051A FC30 A7 00    A            STAA     0,X
00052A FC32 8D 02 FC36            BSR      TIMER
00053A FC34 20 F2 FC28            BRA      WRITEC
00054                        *TIMER ROUTINE PROVIDES TIMING FOR CURSOR BLINK,
00055                        *CHECKS FOR INPUT FROM KYBD AND SERIAL I/O, AND
00056                        *CHECKS FOR OPERATOR CHANGES OF STATUS.
00057A FC36 BD FD10  A TIMER  JSR      CHKSTA      CHECK FOR TRM MODE CHANGES
00058A FC39 3C                    PSHX
```

569

```
PAGE   002   DMBTRM

00059A FC3A CE 07FF  A           LDX     #$07FF     TIMER DELAY PARAMETER
00060A FC3D 20 0E FC4D MORE      BRA     CHKC       CHECK FOR INPUT FROM KYBD
00061A FC3F DF FC    A CONT      STX     TEMPX
00062A FC41 38                   PULX
00063A FC42 BD FD03  A           JSR     SERRX      CHECK FOR SERIAL INPUT
00064A FC45 3C                   PSHX
00065A FC46 DE FC    A           LDX     TEMPX
00066A FC48 09                   DEX
00067A FC49 26 F2 FC3D           BNE     MORE       FINISHED TIMING?
00068A FC4B 38                   PULX
00069A FC4C 39                   RTS
00070                    *CHECK FOR INPUT FROM KYBD
00071A FC4D 96 08    A CHKC      LDAA    $08        TEST FOR STROBE VIA ICF ON PTM
00072A FC4F 2B 02 FC53           BMI     YESC
00073A FC51 20 EC FC3F           BRA     CONT       CONTINUE TIMER ROUTINE
00074                    *THIS ROUTINE SERVICES CHAR FROM KYBD
00075A FC53 38        YESC       PULX
00076A FC54 96 0D    A           LDAA    $0D        RESET ICF
00077A FC56 96 02    A           LDAA    $02        KYBD ASCII
00078A FC58 97 13    A           STAA    $13
00079A FC5A D6 FE    A           LDAB    STATWO
00080A FC5C C4 10    A           ANDB    #$10
00081A FC5E 27 03 FC63           BEQ     FULLD      TEST FOR FULL DUPLEX
00082A FC60 BD FD1D  A           JSR     DISPL
00083A FC63 8D 02 FC67 FULLD     BSR     ENDSCN
00084A FC65 20 CF FC36           BRA     TIMER
00085                    *TEST FOR LAST SCREEN LOC AND MAKE SCROLL DECISION
00086A FC67 8C 2200  A ENDSCN CPX        #$2200     LAST SCRN LOC FILLED?
00087A FC6A 27 01 FC6D           BEQ     SCROL2
00088A FC6C 39                   RTS
00089A FC6D 8D 71 FCE0 SCROL2 BSR        SCROLL
00090A FC6F 39                   RTS
00091                    *THIS ROUTINE BLANKS ENTIRE SCREEN
00092                    *AND SERVICES INCOMING SERIAL CHAR
00093                    *DURING THE ROUTINE
00094A FC70 86 20    A BLANK      LDAA    #$20
00095A FC72 97 FF    A           STAA    SAVCHR
00096A FC74 CE 2000  A           LDX     #$2000
00097A FC77 DF FC    A           STX     TEMPX
00098A FC79 A7 00    A MORSCR STAA       0,X
00099A FC7B 08                   INX
00100A FC7C 8C 2200  A           CPX     #$2200     END OF SCREEN?
00101A FC7F 26 07 FC88           BNE     SERTST     CHECK FOR INPUT SER CHAR
00102A FC81 DE FC    A           LDX     TEMPX
00103A FC83 A6 00    A           LDAA    0,X
00104A FC85 97 FF    A           STAA    SAVCHR
00105A FC87 39                   RTS
00106A FC88 D6 11    A SERTST LDAB      $11
00107A FC8A 2A ED FC79           BPL     MORSCR     MORE BLNKING IF NO SER CHAR
00108A FC8C 3C                   PSHX
00109A FC8D DE FC    A           LDX     TEMPX
00110A FC8F D6 12    A           LDAB    $12
00111A FC91 C4 7F    A           ANDB    #$7F
00112A FC93 37                   PSHB
00113A FC94 C4 60    A           ANDB    #$60
00114A FC96 27 09 FCA1           BEQ     CONTRL     BRA IF CONTROL CHAR
00115A FC98 33                   PULB
00116A FC99 E7 00    A           STAB    0,X
```

```
PAGE  003   DMBTRM

00117A FC9B 08                INX
00118A FC9C DF FC    A         STX    TEMPX
00119A FC9E 38                PULX
00120A FC9F 20 D8 FC79         BRA    MORSCR
00121A FCA1 33          CONTRL PULB
00122A FCA2 C1 08    A         CMPB   #$08      TEST FOR BACKSPACE
00123A FCA4 27 0F FCB5         BEQ    BACKS2
00124A FCA6 C1 0C    A         CMPB   #$0C      TEST FOR CURSOR FWD
00125A FCA8 27 15 FCBF         BEQ    FWDC2
00126A FCAA C1 0A    A         CMPB   #$0A      TEST FOR LINE FEED
00127A FCAC 27 16 FCC4         BEQ    LINEF2
00128A FCAE C1 0D    A         CMPB   #$0D      TEST FOR CARRIAGE RETURN
00129A FCB0 27 19 FCCB         BEQ    CARET2
00130A FCB2 38          GOBACK PULX
00131A FCB3 20 C4 FC79         BRA    MORSCR
00132A FCB5 8C 2000  A  BACKS2 CPX    #$2000
00133A FCB8 27 F8 FCB2         BEQ    GOBACK
00134A FCBA 09                DEX
00135A FCBB DF FC    A         STX    TEMPX
00136A FCBD 20 F3 FCB2         BRA    GOBACK
00137A FCBF 08          FWDC2  INX
00138A FCC0 DF FC    A         STX    TEMPX
00139A FCC2 20 EE FCB2         BRA    GOBACK
00140A FCC4 C6 20    A  LINEF2 LDAB   #$20
00141A FCC6 3A                ABX
00142A FCC7 DF FC    A         STX    TEMPX
00143A FCC9 20 E7 FCB2         BRA    GOBACK
00144A FCCB 8C 2000  A  CARET2 CPX    #$2000
00145A FCCE 27 E2 FCB2         BEQ    GOBACK
00146A FCD0 3C                PSHX
00147A FCD1 DF FA    A         STX    TEMPX2
00148A FCD3 DC FA    A         LDD    TEMPX2
00149A FCD5 C4 1F    A         ANDB   #$1F
00150A FCD7 26 03 FCDC         BNE    MAS8
00151A FCD9 38                PULX
00152A FCDA 20 D6 FCB2         BRA    GOBACK
00153A FCDC 38          MAS8   PULX
00154A FCDD 09                DEX
00155A FCDE 20 EB FCCB         BRA    CARET2
00156A FCE0 D6 FE    A  SCROLL LDAB   STATWO
00157A FCE2 C4 20    A         ANDB   #$20
00158A FCE4 27 8A FC70         BEQ    BLANK
00159A FCE6 CE 2000  A         LDX    #$2000
00160                  *THE FOLLOWING LOOP SYNCHS SCROLL TO HORIZ
00161                  *SYNCH FROM VDG TO MINIMIZE NOISE ON SCREEN
00162                  *DURING WRITE TO DISPLAY RAM
00163A FCE9 D6 02    A  NOTYET LDAB   $02       PORT1 I/O
00164A FCEB C4 80    A         ANDB   #$80      ISOLATE H SYNCH
00165A FCED 27 FA FCE9         BEQ    NOTYET    WAIT FOR HIGH
00166A FCEF D6 02    A  SAND4  LDAB   $02
00167A FCF1 C4 80    A         ANDB   #$80
00168A FCF3 26 FA FCEF         BNE    SAND4     WAIT FOR LOW
00169A FCF5 A6 20    A         LDAA   32,X
00170A FCF7 A7 00    A         STAA   0,X
00171A FCF9 08                INX
00172A FCFA 8C 2200  A         CPX    #$2200
00173A FCFD 26 EA FCE9         BNE    NOTYET
00174A FCFF CE 21E0  A         LDX    #$21E0    SCRNPTR TO 1ST LOC,LAST LINE
```

```
PAGE   004   DMBTRM

00175A FD02 39              RTS
00176                       *CHECK FOR SERIAL INPUT
00177A FD03 96 11    A SERRX LDAA   $11       TX/RX CONTR STATUS REG
00178A FD05 2B 01 FD08      BMI     SERDIS    TEST FOR BIT 7 (ICF)
00179A FD07 39              RTS
00180A FD08 96 12    A SERDIS LDAA  $12       RECEIVER REGISTER
00181A FD0A 8D 11 FD1D      BSR     DISPL
00182A FD0C BD FC67  A      JSR     ENDSCN
00183A FD0F 39              RTS
00184                       *HERE STATUS SWITCHES ARE CHECKED.  CALLING ADDR
00185                       *$3800 SELECTS NO MEMORY. STATUS SWITCHES
00186                       *ARE READ VIA LS244 LATCH
00187A FD10 B6 1800  A CHKSTA LDAA  SWITCH
00188A FD13 91 FE    A      CMPA    STATWO    HAVE SWITCHES CHANGED?
00189A FD15 26 01 FD18      BNE     CHGWO
00190A FD17 39              RTS
00191                       *CHANGE CONTROL WORD IN STATWO
00192A FD18 97 FE    A CHGWO STAA   STATWO
00193A FD1A 97 10    A      STAA    $10       SERIAL MODE CONTROL REG
00194A FD1C 39              RTS
00195                       *THE DISPLAY ROUTINE FIRST CHECKS FOR CONTRL
00196                       *CHAR.  IF CONTRL CHAR, CHAR IS TESTED FOR
00197                       *ACTION; OTHERWISE CHAR IS DISPLAYED.
00198A FD1D 84 7F    A DISPL ANDA   #$7F
00199A FD1F 36              PSHA
00200A FD20 84 60    A      ANDA    #$60
00201A FD22 27 12 FD36      BEQ     CNTRLC    TEST FOR CONTROL CHAR
00202A FD24 32              PULA
00203                       *WAIT FOR HORIZONTAL SYNCH
00204A FD25 D6 02    A SAND2 LDAB   $02
00205A FD27 C4 80    A      ANDB    #$80
00206A FD29 27 FA FD25      BEQ     SAND2
00207A FD2B D6 02    A SAND3 LDAB   $02
00208A FD2D C4 80    A      ANDB    #$80
00209A FD2F 26 FA FD2B      BNE     SAND3
00210A FD31 A7 00    A      STAA    0,X
00211A FD33 08              INX
00212A FD34 20 4C FD82      BRA     SAND1
00213                       *THE FOLLOWING ROUTINE DECODES CONTROL CHAR
00214A FD36 32         CNTRLC PULA
00215A FD37 81 0A    A      CMPA    #$0A      TEST FOR LINE FEED
00216A FD39 27 27 FD62      BEQ     LINEF
00217A FD3B 81 0D    A      CMPA    #$0D      TEST FOR CARRIAGE RETURN
00218A FD3D 27 3A FD79      BEQ     CARRET
00219A FD3F 81 08    A      CMPA    #$08      TEST FOR BACKSPACE
00220A FD41 27 5C FD9F      BEQ     BACKSP
00221A FD43 81 07    A      CMPA    #$07      TEST FOR BELL
00222A FD45 26 07 FD4E      BNE     CLRSCR
00223A FD47 96 08    A      LDAA    $08
00224A FD49 C4 01    A      ANDB    #$01
00225A FD4B 27 5F FDAC      BEQ     BELL      IS BELL ALREADY ON?
00226A FD4D 39              RTS
00227A FD4E 81 1A    A CLRSCR CMPA  #$1A      TEST FOR CLEAR SCREEN
00228A FD50 26 07 FD59      BNE     MORECH
00229A FD52 C6 20    A      LDAB    #$20
00230A FD54 D7 FF    A      STAB    SAVCHR
00231A FD56 7E FC70  A      JMP     BLANK
00232A FD59 81 0C    A MORECH CMPA  #$0C      TEST FOR CURSOR FWD
```

PAGE 005 DMBTRM

```
00233A FD5B 27 5B FDB8          BEQ    FWDC
00234A FD5D 81 0B      A        CMPA   #$0B       TEST FOR UP CURSOR
00235A FD5F 27 67 FDC8          BEQ    UPCUR
00236A FD61 39                  RTS               DEFAULT BACK
00237A FD62 D6 FF    A LINEF    LDAB   SAVCHR
00238A FD64 E7 00      A        STAB   0,X
00239A FD66 C6 20      A        LDAB   #$20
00240A FD68 3A                  ABX               INCREMENT SCRNPTR 1 LINE
00241A FD69 DF FC      A        STX    TEMPX
00242A FD6B DC FC      A        LDD    TEMPX
00243A FD6D 81 22      A        CMPA   #$22       SCRNPTR OFF SCREEN?
00244A FD6F 26 11 FD82          BNE    SAND1
00245A FD71 8D 24 FD97          BSR    SCROL1     IF SCRNPTR OFF SCRN,SCROLL
00246A FD73 DC FC      A        LDD    TEMPX
00247A FD75 C4 1F      A        ANDB   #$1F       GET HORIZ POS OF SCRNPTR
00248A FD77 3A                  ABX
00249A FD78 39                  RTS
00250A FD79 D6 FF    A CARRET   LDAB   SAVCHR
00251A FD7B E7 00      A        STAB   0,X
00252A FD7D 8C 2000  A MAS5     CPX    #$2000     SCRNPTR ALREADY AT LIMIT?
00253A FD80 26 05 FD87          BNE    MAS3
00254                 *SAND1    STORES CHAR UNDER CURSOR
00255A FD82 E6 00    A SAND1    LDAB   0,X
00256A FD84 D7 FF      A        STAB   SAVCHR
00257A FD86 39                  RTS
00258A FD87 3C         MAS3     PSHX
00259A FD88 DF FC      A        STX    TEMPX
00260A FD8A DC FC      A        LDD    TEMPX
00261A FD8C C4 1F      A        ANDB   #$1F       SCRNPTR AT 1ST LINE LOC?
00262A FD8E 26 03 FD93          BNE    MAS4
00263A FD90 38                  PULX
00264A FD91 20 EF FD82          BRA    SAND1
00265A FD93 38         MAS4     PULX
00266A FD94 09                  DEX               DECREMENT SCRNPTR
00267A FD95 20 E6 FD7D          BRA    MAS5       TEST SCRNPTR LOC AGAIN
00268A FD97 C6 20    A SCROL1   LDAB   #$20
00269A FD99 D7 FF      A        STAB   SAVCHR
00270A FD9B BD FCE0    A        JSR    SCROLL
00271A FD9E 39                  RTS
00272A FD9F 8C 2000  A BACKSP   CPX    #$2000
00273A FDA2 26 01 FDA5          BNE    MAS2
00274A FDA4 39                  RTS
00275A FDA5 D6 FF    A MAS2     LDAB   SAVCHR
00276A FDA7 E7 00      A        STAB   0,X
00277A FDA9 09                  DEX
00278A FDAA 20 D6 FD82          BRA    SAND1
00279A FDAC 86 01    A BELL     LDAA   #$01
00280A FDAE 97 08      A        STAA   $08        SET OLVL HIGH NEXT COMPARE
00281A FDB0 BD FC36    A        JSR    TIMER      PROVIDES BELL DURATION
00282A FDB3 86 00      A        LDAA   #$00
00283A FDB5 97 08      A        STAA   $08        SET OLVL LOW NEXT COMPARE
00284A FDB7 39                  RTS
00285A FDB8 D6 FF    A FWDC     LDAB   SAVCHR
00286A FDBA E7 00      A        STAB   0,X
00287A FDBC 8C 21FF    A        CPX    #$21FF     END OF SCREEN?
00288A FDBF 26 04 FDC5          BNE    MAS6
00289A FDC1 BD FCE0    A        JSR    SCROLL
00290A FDC4 09                  DEX
```

```
PAGE   006   DMBTRM

00291A FDC5 08          MAS6    INX
00292A FDC6 20 BA FD82          BRA     SAND1
00293A FDC8 D6 FF    A UPCUR    LDAB    SAVCHR
00294A FDCA E7 00    A          STAB    0,X
00295A FDCC DF FC    A          STX     TEMPX
00296A FDCE DC FC    A          LDD     TEMPX
00297A FDD0 83 0020  A          SUBD    #$20      MOVE SCRNPTR UP 1 LINE
00298A FDD3 DD FC    A          STD     TEMPX
00299A FDD5 DE FC    A          LDX     TEMPX
00300A FDD7 47                  ASRA
00301A FDD8 81 10    A          CMPA    #$10      SCRNPTR OFF SCREEN?
00302A FDDA 26 02 FDDE          BNE     LIMIT
00303A FDDC 20 A4 FD82          BRA     SAND1
00304              *LIMIT RESTORES SCRNPTR TO TOP LINE WHEN
00305              *ATTEMPT IS MADE TO MOVE CURSOR OFF
00306              *SCRN VIA UPCUR
00307A FDDE DF FC    A LIMIT    STX     TEMPX
00308A FDE0 DC FC    A          LDD     TEMPX
00309A FDE2 C3 0020  A          ADDD    #$20
00310A FDE5 DD FC    A          STD     TEMPX
00311A FDE7 DE FC    A          LDX     TEMPX
00312A FDE9 20 97 FD82          BRA     SAND1
00313                           END
TOTAL ERRORS 00000


    FCB5 BACKS2 00123 00132*
    FD9F BACKSP 00220 00272*
    FDAC BELL   00225 00279*
    FC70 BLANK  00041 00094*00158 00231
    FCCB CARET2 00129 00144*00155
    FD79 CARRET 00218 00250*
    FD18 CHGWO  00032 00189 00192*
    FC4D CHKC   00060 00071*
    FD10 CHKSTA 00057 00187*
    FD4E CLRSCR 00222 00227*
    FD36 CNTRLC 00201 00214*
    FC3F CONT   00061*00073
    FCA1 CONTRL 00114 00121*
    FD1D DISPL  00082 00181 00198*
    FC67 ENDSCN 00083 00086*00182
    FC2E ERASEC 00050*
    FC1E FILL   00037*00040
    FC63 FULLD  00081 00083*
    FDB8 FWDC   00233 00285*
    FCBF FWDC2  00125 00137*
    FCB2 GOBACK 00130*00133 00136 00139 00143 00145 00152
    FDDE LIMIT  00302 00307*
    FD62 LINEF  00216 00237*
    FCC4 LINEF2 00127 00140*
    FDA5 MAS2   00273 00275*
    FD87 MAS3   00253 00258*
    FD93 MAS4   00262 00265*
    FD7D MAS5   00252*00267
    FDC5 MAS6   00288 00291*
    FCDC MAS8   00150 00153*
    FC3D MORE   00060*00067
```

```
PAGE   007    DMBTRM

FD59 MORECH 00228 00232*
FC79 MORSCR 00098*00107 00120 00131
FCE9 NOTYET 00163*00165 00173
FD82 SAND1  00212 00244 00255*00264 00278 00292 00303 00312
FD25 SAND2  00204*00206
FD2B SAND3  00207*00209
FCEF SAND4  00166*00168
00FF SAVCHR 00014*00036 00050 00095 00104 00230 00237 00250 00256 00269 00275
            00285 00293
FD97 SCROL1 00245 00268*
FC6D SCROL2 00087 00089*
FCE0 SCROLL 00089 00156*00270 00289
FD08 SERDIS 00178 00180*
FD03 SERRX  00063 00177*
FC88 SERTST 00101 00106*
00F7 STACK  00019*00023
00FE STATWO 00015*00079 00156 00188 00192
1800 SWITCH 00018*00031 00187
00FC TEMPX  00016*00061 00065 00097 00102 00109 00118 00135 00138 00142 00241
            00242 00246 00259 00260 00295 00296 00298 00299 00307 00308 0031(
            00311
00FA TEMPX2 00017*00147 00148
FC36 TIMER  00048 00052 00057*00084 00281
FDC8 UPCUR  00235 00293*
FC28 WRITEC 00046*00053
FC53 YESC   00072 00075*
```

FURTHER DEVELOPMENT OF THE TERMINAL

The terminal can be further developed to meet many user requirements with few hardware and software changes. Two improvements are particularly worth considering: interrupt drive and graphics capability. A detailed description of the VDG operating modes is given in Table A2-2.

Table A2-2   Detailed Description of

MS	G/A	S/A	EXT/INT	GM2	GM1	GM0	CSS	INV	Character Color	Background	Border	Display Mode
1	0	0	0	x	x	x	0 1	0 1 0 1	Green Black Orange Black	Black Green Black Orange	Black Black	32 Characters per row 16 Character rows
1	0	0	1	x	x	x	0 1	0 1 0 1	Green Black Orange Black	Black Green Black Orange	Black Black	32 Characters per row 16 Character rows
1	0	1	0	x	x	x	x	x	Lx C2 C1 C0 Color 0 x x x Black 1 0 0 0 Green 1 0 0 1 Yellow 1 0 1 0 Blue 1 0 1 1 Red 1 1 0 0 Buff 1 1 0 1 Cyan 1 1 1 0 Magenta 1 1 1 1 Orange		Black	64 Display elements per row 32 rows of Display elements
1	0	1	1	x	x	x	0 1	x	Lx C1 C0 Color 0 x x Black 1 0 0 Green 1 0 1 Yellow 1 1 0 Blue 1 1 1 Red 0 x x Black 1 0 0 Buff 1 0 1 Cyan 1 1 0 Magenta 1 1 1 Orange		Black	64 Displays elements per row 48 rows of Display elements
1	1	x	x	0	0	0	0 1	x	C1 C0 Color 0 0 Green 0 1 Yellow 1 0 Blue 1 1 Red 0 0 Buff 0 1 Cyan 1 0 Magenta 1 1 Orange		Green Buff	64 Display elements per row 64 rows of Display elements
1	1	x	x	0	0	1	0 1	x	Lx Color 0 Black 1 Green 0 Black 1 Buff		Green Buff	128 Display elements per row 64 rows of Display elements
1	1	x	x	0	1	0	0 1	x	Same Color as Graphics One		Green Buff	128 Display elements per row 64 rows of Display elements

VDG Modes (Courtesy of Motorola, Inc.)

Screen Detail	VDG Data Bus	Comments
Internal Alphanumerics	extra · ASCII Code	The ALPHANUMERIC INTERNAL mode uses an internal character generator (which contains the following five dot by seven dot characters: @ABCDEFGHIJKLMNOPQRST UVWXYZ[\]↑↑→SP!"#$%&'()° + . − , 0123456789:;< = >?. The six bit ASCII code leaves two bits free and these may be externaly connected to the mode pins (G/$\overline{A}$, S/$\overline{A}$, EXT/$\overline{INT}$, GM2, GM1, GM0, CSS or INV).
(8×12)	One Row of Custom Characters	The ALPHANUMERIC EXTERNAL mode uses an external character generator as well as a row counter. Thus, custom character fonts or graphic symbol sets with up to 256 different 8 × 12 dot "characters" may be displayed.
One Element (L_3 L_2 / L_1 L_0)	C_2 C_1 C_0 L_3 L_2 L_1 L_0 (extra)	The SEMIGRAPHICS FOUR mode uses an internal "coarse graphics" generator in which a rectangle (eight dots by twelve dots) is divided into four equal parts. The luminance of each part is determined by a corresponding bit on the VDG data bus. The color of illuminated parts is determined by three bits.
One Element (L_5 L_4 / L_3 L_2 / L_1 L_0)	C_1 C_0 L_5 L_4 L_3 L_2 L_1 L_0	The SEMIGRAPHIC SIX mode is similar to the SEMIGRAPHIC FOUR mode with the following differences. The eight dot by twelve dot rectangle is divided into six equal parts. Color is determined by the two remaining bits.
E_3 E_2 E_1 E_0	C_1 C_0 C_1 C_0 C_1 C_0 C_1 C_0	The COLOR GRAPHICS ONE mode uses a maximum of 1024 bytes of display RAM in which one pair of bits specifies one picture element.
L_7 L_6 L_5 L_4 L_3 L_2 L_1 L_0	L_7 L_6 L_5 L_4 L_3 L_2 L_1 L_0	The RESOLUTION GRAPHICS ONE mode uses a maximum of 1024 bytes of display RAM in which one pair of bits specifies one picture element
E_3 E_2 E_1 E_0	C_1 C_0 C_1 C_0 C_1 C_0 C_1 C_0	The COLOR GRAPHICS TWO mode uses a maximum of 2048 bytes of display RAM in which one pair of bits specifies one picture element.

577

$\overline{\text{MS}}$	G/$\overline{\text{A}}$	S/$\overline{\text{A}}$	EXT/$\overline{\text{INT}}$	GM2	GM1	GM0	CSS	INV	Character Color	Background	Border	Display Mode
											VDG Pins → Color → TV	

$\overline{\text{MS}}$	G/$\overline{\text{A}}$	S/$\overline{\text{A}}$	EXT/$\overline{\text{INT}}$	GM2	GM1	GM0	CSS	INV	Character Color	Background	Border	Display Mode
1	1	x	x	0	1	1	0	x	Same color as Resolution Graphics One		Green	128 Display elements per row
							1				Buff	96 rows of Display elements
1	1	x	x	1	0	0	0	x	Same color as Color Graphics One		Green	128 Display elements per row
							1				Buff	96 rows of Display elements
1	1	x	x	1	0	1	0	x	Same color as Resolution Graphics One		Green	128 Display elements per row
							1				Buff	192 rows of Display elements
1	1	x	x	1	1	0	0	x	Same color as Color Graphics One		Green	128 Display elements per row
							1				Buff	192 rows of Display elements
1	1	x	x	1	1	1	0	x	Same color as Resolution Graphics One		Green	256 Display elements per row
							1				Buff	192 rows of Display elements

(*Continued*)

Screen Detail	VDG Data Bus	Comments
→|2|← L_7 ... L_0	L_7 L_6 L_5 L_4 L_3 L_2 L_1 L_0	The RESOLUTION GRAPHICS TWO mode uses a maximum of 1536 bytes of display RAM in which one bit specifies one picture element.
→|2|← E_3 ... E_0	C_1 C_0 C_1 C_0 C_1 C_0 C_1 C_0	The COLOR GRAPHICS THREE mode uses a maximum of 3072 bytes of display RAM in which one pair of bytes specifies one picture element.
→|2|← L_7 ... L_0	L_7 L_6 L_5 L_4 L_3 L_2 L_1 L_0	The RESOLUTION GRAPHICS THREE mode uses a maximum of 3072 bytes of display RAM in which one bit specifies one picture element.
→|2|← E_3 ... E_0	C_1 C_0 C_1 C_0 C_1 C_0 C_1 C_0	The COLOR GRAPHICS SIX mode uses a maximum of 6144 bytes of display RAM in which one pair of bits specifies one picture element.
→|1|← L_7 ... L_0	L_7 L_6 L_5 L_4 L_3 L_2 L_1 L_0	The RESOLUTION GRAPHICS SIX mode uses a maximum of 6144 bytes of display RAM in which one bit specifies one picture element.

Appendix C

Bibliography: Suggested References for Further Reading

Bell, D. A., *Solid-State Pulse Circuits,* 2nd ed. Reston Publishing Company, Reston, Virginia, 1981.

Bibbero, R. J., *Micropressors in Instruments and Control.* Wiley, New York, 1977.

Chute, G. M. and R. D. Chute, *Electronics in Industry.* McGraw-Hill, New York, 1979.

Doebelin, E. O., *Measurement Systems: Application and Design,* 2nd ed. McGraw-Hill, New York, 1975.

Hall, E. L., *Computer Picture Processing and Recognition.* Academic Press, New York, 1978.

Hanna, N. N., "Application of FETs in Temperature Compensation of DC Amplifiers," *IEEE Transactions on Instrumentation and Measurement,* **IM-28**(1), 32–36 (1979).

Hicks, R., J. R. Schenken, and M. A. Steinrauf, *Laboratory Instrumentation.* Harper and Row, New York, 1974.

Kuo, B., *Automatic Control Systems,* 3rd ed. Prentice-Hall, Englewood Cliffs, New Jersey, 1975.

Lue, J. T., A Simple Circuit of a Crystal Oscillator and Its Application to a High-Resolution NMR Spectrometer, *IEEE Transactions on Instrumentation and Measurement,* **IM-25**(1), 76–78 (1976).

McGovern, P. A., "Nanosecond Passive Voltage Probes," *IEEE Transactions on Instrumentation and Measurement,* **IM-26**(1), 46–52 (1977).

Masuda, Y., M. Nishikawa, and B. Ichijo, "New Methods of Measuring Capacitance and Resistance of Very High Loss Materials at High Frequencies, *IEEE Transactions on Instrumentation and Measurement,* **IM-29**(1), 28–36 (1980).

Morgenthaler, L. P. and T. J. Poulos, "A Microcomputer System for Control of an Atomic Absorption Spectrometer, *Am. Lab.,* **8**, 37–45 (1975). See also A. L. Robinson, *Science* **195**, 1314–1318, 1367 (1977).

Nakane, H., S. Omori, and I. Yokoshima, "Improving the Frequency Characteristics of RF Standard Magnetic-Field Generator Employing Loop Antenna, *IEEE Transactions on Instrumentation and Measurement,* **IM-26**(1), 25–28 (1977).

Ney, M. and F. E. Gardial, "Automatic Monitor for Microwave Resonators," *IEEE Transactions on Instrumentation and Measurement,* **IM-26**(1), 10–13 (1977).

Peatman, J. B., *Microcomputer-Based Design.* McGraw-Hill, New York, 1977.

Rosenthal, L. A., "Improved Frequency Meter Circuit," *IEEE Transactions on Instrumentation and Measurement,* **IM-26**(4), 421 (1977).

Sachse, H. B., *Semiconducting Temperature Sensors and Their Application.* Wiley, New York, 1975.

Shrader, R. L., *Electronic Communication.* McGraw-Hill, New York, 1980.

Watanabe, K., M. Madihian, and T. Yamamoto, "A Cascade Phase Sensitive Detector—Phase Response, *IEEE Transactions on Instrumentation and Measurement,* **IM-29**(1), 3–6 (1980).

Wobschall, D., *Circuit Design for Electronic Instrumentation.* McGraw-Hill, New York, 1979.

Index